Ecological Studies, Vol. 189

Analysis and Synthesis

Ecological Studies

Volumes published since 2001 are listed at the end of this book.

E. Granéli J.T. Turner (Eds.)

Ecology of Harmful Algae

With 45 Figures, 13 in Color, and 15 Tables

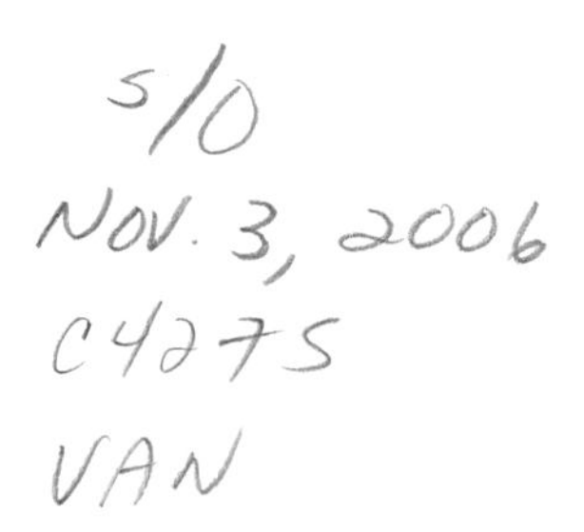

Prof. Dr. Edna Granéli
University of Kalmar
Department of Marine Sciences
391 82 Kalmar
Sweden

Prof. Dr. Jefferson T. Turner
University of Massachusetts Dartmouth
Biology Department and School for
Marine Science and Technology
285 Old Westport Road
North Dartmouth MA 02747
USA

Cover illustration: Factors affecting harmful algae (circle in the middle) gains and losses. On the upper half the HA gains include their intrinsic ability to utilize inorganic and organic compounds (mixotrophy), nutrients from anthropogenic origin, and under adverse conditions release allelochemical compounds that kill other algae (allelopathy) or their grazers. On the lower half the losses the HA might suffer, in this case no blooms will be formed or damage to the environment will occur.

ISSN 0070-8356
ISBN-10 3-540-32209-4 Springer Berlin Heidelberg New York
ISBN-13 978-3-540-32209-2 Springer Berlin Heidelberg New York

Springer is a part of Springer Science+Business Media
springer.com

Printed in The Netherlands

Editor: Dr. Dieter Czeschlik, Heidelberg, Germany
Desk editor: Dr. Andrea Schlitzberger, Heidelberg, Germany
Cover design: *design & production* GmbH, Heidelberg, Germany
Typesetting and production: Friedmut Kröner, Heidelberg, Germany

31/3152 YK – 5 4 3 2 1 0 – Printed on acid free paper

Preface

In the open sea, primary production is almost totally based on photosynthesis by pelagic unicellular or colonial microalgae, collectively known as phytoplankton. Benthic algae are important primary producers only in extremely shallow water where sunlight sufficient for photosynthesis penetrates to the bottom. Thus, phytoplankton are the basis of aquatic food chains.

Tens to hundreds of species of phytoplankton belonging to different taxonomic units usually coexist in natural assemblages. Phytoplankton are microscopic, ranging in size from less than 1 μm to more than 100 μm, with generation times of no more than a few days. Thus, phytoplankton populations exhibit large temporal variations in response to abiotic factors such as light, temperature, nutrients, and water movement, and biotic factors such as grazing, competition, parasitism, and microbial attack. Normally, the standing crop of phytoplankton remains low because these loss factors generally balance rapid intrinsic rates of growth through cell division.

Occasionally, increases in one or a few species can overcome losses, such that a given species can dominate phytoplankton assemblages and cause blooms lasting for several weeks or more. Such blooms are due to combinations of favorable phytoplankton growth, increased physical concentration by hydrographic or meteorological processes, and reduced losses due to factors such as viruses, sedimentation, and grazing.

Some phytoplankton blooms can cause adverse effects. These include oxygen depletion, reduced water quality aesthetics, clogging of fish gills, or toxicity. Blooms of such Harmful Algae (HA) cause Harmful Algal Blooms (HABs). Of the approximately 5,000 known species of phytoplankton, only some 300 species form HABs that are deleterious to aquatic ecosystems in one way or another, and only about 80 of these species are known to be toxin producers. Some phytoplankton toxins can be accumulated and/or transported in food chains to higher trophic levels where they contaminate shellfish, making them unsuitable for human consumption, or poison upper-level consumers, including fish, seabirds, marine mammals, and humans. The economic effects of such blooms, including losses to fisheries, tourism, monitoring, and health care can be substantial. In Europe, such losses annually approach 862 million Euros, and in the USA 82 million dollars (see Chap. 30).

Harmful algae have been the subjects of scientific and societal interest for centuries. Because blooms of toxic dinoflagellates were known to occasionally discolor water red or brownish-red, they were, and still are, known as "red tides." Water discoloration was noted for the lower Nile in the Bible, and Darwin made microscopic observations of discolored water during the voyage of the HMS *Beagle.* However, the frequency of HABs, and the locations affected may be increasing worldwide. In recent years, increases in the numbers of HA species able to produce toxins have been detected, and new toxins continue to be chemically characterized.

It is often assumed that phytoplankton toxins evolved to deter their zooplankton grazers. However, most phytoplankton species, including many toxin producers, appear to be routinely grazed by many zooplankters in natural mixed phytoplankton assemblages. Other HA toxins appear to be involved primarily in allelopathy, being released in the dissolved state into sea water and causing deleterious effects on other competitor phytoplankton species. Some HA toxins may be secondary metabolites that are only coincidentally toxic. Thus, the role of phytoplankton toxins in the ecology of the algae that produce them remains unclear.

HA toxin levels can vary depending on concentrations of nutrients in the water such as nitrogen and phosphorus. In some cases, HA intracellular toxin levels increase in cells grown under unbalanced nutrient conditions. This may be because toxins are the molecules that algal cells use to store or retain sparse nutrients, or because cells under nutrient stress transfer nitrogen from chlorophyll molecules to toxin molecules, causing reductions in rates of cell division but building up toxin levels in the remaining non-dividing cells. Alternatively, some HA species may produce higher amounts of toxins under nutrient-stressed conditions, thereby more effectively reducing losses to grazers and/or by releasing greater amounts of allelochemical substances to neutralize co-occurring phytoplankton species that are competitors for sparse nutrients.

Unbalanced nitrogen and phosphorus conditions are often recurrent in coastal waters due to increased anthropogenic discharges of a given nutrient, relative to others. Thus, it is possible that even if HABs have not increased in occurrence, the deleterious effects of these blooms may have increased their impacts due to increased toxicity due to unbalanced or anthropogenically altered nutrient ratios.

Despite increased research activity, the last major organized published synthesis of HA ecology was the volume originating from the NATO workshop on harmful algal blooms held in Bermuda in 1996 (Physiological Ecology of Harmful Algal Blooms; Springer-Verlag, 1998). Although the reviews in the volume from the Bermuda meeting were excellent and comprehensive for the time, they are now almost a decade old and somewhat dated by recent developments. Accordingly, we were approached by Springer-Verlag with a request to compile an updated synthesis of HA ecology, organized primarily around processes and questions, rather than organisms. Thus, we invited a global assemblage of active HA researchers to contribute to the chapters in this volume, and many of these same specialists had also contributed to the previous Bermuda meeting volume. All chapters in this volume were peer-reviewed. We hope that this volume will complement other recent reviews and syntheses in *Harmful Algae* and other journals, and in international HA meeting volumes to identify gaps in our present understanding of HA ecology and suggest areas for additional research.

Kalmar, Sweden *Edna Granéli*
Dartmouth, USA *Jefferson T. Turner*

May, 2006

Contents

Part A Harmful Algae and Their Global Distribution

1 An Introduction to Harmful Algae 3
E. Granéli and J.T. Turner

References 7

2 Molecular Taxonomy of Harmful Algae 9
S. Janson and P.K. Hayes

2.1 Introduction 9
2.2 Dinophyta (Dinoflagellates) 10
2.2.1 General Morphology 10
2.2.2 *Dinophysis* 11
2.2.3 *Alexandrium* 11
2.2.4 *Protoperidinium, Prorocentrum* 12
2.2.5 *Karenia, Karlodinium, Takayama* 13
2.2.6 *Amphidinium, Cochlodinium, Gyrodinium* 14
2.3 Cyanobacteria (Blue-Green Algae) 14
2.3.1 *Anabaena, Aphanizomenon, Nodularia* 14
2.3.2 *Microcystis* 15
2.3.3 *Trichodesmium* 16
2.4 Bacillariophyta (Diatoms) 17
2.4.1 *Amphora, Pseudo-nitzschia, Nitzschia* 17
2.5 Concluding Remarks 17
References 18

3 **The Biogeography of Harmful Algae** 23
N. LUNDHOLM and Ø. MOESTRUP

3.1 Biogeography and Species Concepts 23
3.1.1 Genetic Variation 24
3.2 Biogeographical Distribution 25
3.3 Distribution of Harmful Species 26
3.3.1 Dinoflagellates 26
3.3.2 Diatoms 27
3.3.3 Haptophytes 29
3.3.4 Raphidophyceans 29
3.3.5 Cyanobacteria 31
References 32

4 **Importance of Life Cycles in the Ecology of Harmful Microalgae** 37
K.A. STEIDINGER and E. GARCÉS

4.1 Introduction 37
4.2 Phases of Phytoplankton Bloom Development and Life Cycles 39
4.2.1 Initiation 39
4.2.2 Growth and Maintenance 41
4.2.3 Dispersal/Dissipation/Termination 44
4.3 Environmental Factors versus Biological Factors Affecting Transition 44
4.4 Status of Knowledge and Direction Needed 45
References 47

Part B The Ecology of Major Harmful Algae Groups

5 **The Ecology of Harmful Dinoflagellates** 53
J.M. BURKHOLDER, R.V. AZANZA, and Y. SAKO

5.1 Introduction 53
5.2 General Ecology 54
5.2.1 Motility 54
5.2.2 Temperature, Light, Salinity and Turbulence 55
5.2.3 Nutrition: the Continuum from Auxotrophy to Parasitism . 56

5.3 Blooms, Including Toxic Outbreaks 59
5.4 Human Influences 60
5.5 Conceptual Frameworks to Advance Understanding 61
References 64

6 The Ecology of Harmful Flagellates Within Prymnesiophyceae and Raphidophyceae 67
B. Edvardsen and I. Imai

6.1 Introduction 67
6.2 Class Prymnesiophyceae (Division Haptophyta) 67
6.2.1 Taxonomy, Morphology and Life History 67
6.2.2 Distribution and Abundance 68
6.2.3 Autecology and Ecophysiology 69
6.2.4 Toxicity and Toxins 70
6.2.5 Ecological Strategies 71
6.3 Class Raphidophyceae (Division Heterokontophyta) 72
6.3.1 Taxonomy, Morphology and Life History 72
6.3.2 Distribution and Abundance 73
6.3.3 Autecology and Ecophysiology 74
6.3.4 Toxicity 75
6.3.5 Ecological Strategies 75
References 77

7 The Ecology of Harmful Diatoms 81
S.S. Bates and V.L. Trainer

7.1 Introduction 81
7.2 Toxin-Producing Diatoms, Genus *Pseudo-nitzschia* 82
7.3 Domoic Acid in the Marine Food Web 83
7.4 Physiological Ecology of *Pseudo-nitzschia* spp. 84
7.5 Molecular Tools for Studying *Pseudo-nitzschia* 86
7.6 Conclusions and Directions for Future Research 87
References 88

8 Ecology of Harmful Cyanobacteria 95
H.W. Paerl and R.S. Fulton III

8.1 Introduction 95
8.2 Environmental Factors Controlling CyanoHABs 97

8.2.1 Nutrients 97
8.2.2 Physical-Chemical Factors: Salinity and Turbulence 102
8.2.3 Salinity and Turbulence 102
8.3 CyanoHAB Interactions with Micro/Macroorganisms 104
8.4 CyanoHAB Management 106
References 107

9 Brown Tides 111
C. J. Gobler and W. G. Sunda

9.1 Background 111
9.2 Nutrients and Physical Factors 113
9.3 Sources of Cell Mortality 117
References 120

Part C The Ecology and Physiology of Harmful Algae

10 Harmful Algal Bloom Dynamics in Relation to Physical Processes 127
F.G. Figueiras, G.C. Pitcher, and M. Estrada

10.1 Introduction 127
10.2 Physical Constraints: From Diffusion to Advection 128
10.3 Life-Forms 129
10.4 Algal Communities 130
10.5 Retention and Transport 131
10.5.1 Retention-Reduced Exchange 131
10.5.2 Transport 133
References 136

11 Ecological Aspects of Harmful Algal In Situ Population Growth Rates 139
W. Stolte and E. Garcés

11.1 Introduction 139
11.2 Ecological Interpretation of In Situ Growth Rate Measurements 140

11.3 In Situ Growth Rates; Variation Among Taxonomic Groups . 143
11.4 Are Harmful Algal Species *r*- or *K*-Strategists? 147
11.5 Conclusions 149
References 149

12 Harmful Algae and Cell Death 153
M.J.W. Veldhuis and C.P.D. Brussaard

12.1 Introduction 153
12.2 Mortality of HABs 156
12.3 Death Due to HABs 157
12.4 Mechanisms to Avoid Cell Mortality 158
12.5 Ecological Implications 159
References 160

13 The Diverse Nutrient Strategies of Harmful Algae: Focus on Osmotrophy 163
P. M. Glibert and C. Legrand

13.1 Introduction and Terminology 163
13.2 Osmotrophy Pathways and Methods to Explore Them 164
13.3 Cellular Costs and Benefits of Osmotrophy 167
13.4 Ecological Significance of Osmotrophy 168
13.5 A Comment on Evolutionary Aspects of Osmotrophy 170
13.6 Conclusions 171
References 171

14 Phagotrophy in Harmful Algae 177
D. Stoecker, U. Tillmann, and E. Granéli

14.1 Introduction 177
14.2 Phagotrophy and its Advantages 180
14.3 Relationship of Phagotrophy to Toxicity 182
14.4 Significance of Phagotrophy 184
References 185

15 Allelopathy in Harmful Algae: A Mechanism to Compete for Resources? . . . 189
E. GRANÉLI and P.J. HANSEN

15.1 Harmful Algal Species Known of Allelopathy . . . 189
15.2 Approaches to Demonstrate/Study Allelopathy – Pitfalls and Strength/Weaknesses of Experimental Approaches . . . 189
15.3 Which Toxins are Involved in the Allelopathic Effects? . . . 192
15.4 Influence of Abiotic and Biotic Factors on Allelopathy . . . 194
15.4.1 Abiotic Factors . . . 194
15.4.2 Biotic Factors . . . 196
15.5 Ecological Significance of Allelopathy in Marine Ecosystems 198
References . . . 199

16 Trace Metals and Harmful Algal Blooms . . . 203
W.G. SUNDA

16.1 Introduction . . . 203
16.2 Chemistry and Availability of Metals . . . 204
16.3 Trace Metals as Limiting Nutrients . . . 205
16.4 Trace Metal Toxicity . . . 207
16.5 Trace Metal Effects on HABs: Domoic Acid Production in *Pseudo-nitzschia* . . . 208
16.6 Trace Metal Effects on Other HAB Species . . . 210
References . . . 211

17 Molecular Physiology of Toxin Production and Growth Regulation in Harmful Algae . . . 215
A. CEMBELLA and U. JOHN

17.1 Introduction . . . 215
17.2 Phycotoxin Biosynthesis . . . 216
17.3 Growth and Regulation of Toxin Production . . . 217
17.4 Toxin Production Through the Cell Cycle . . . 219
17.5 Molecular Approaches to Growth and Toxin Expression . . 220
17.6 Current and Future Perspectives . . . 223
References . . . 226

18 **Chemical and Physical Factors Influencing Toxin Content** . 229
E. Granéli and K. Flynn

18.1 Introduction 229
18.2 Growth Stage and Toxin Production 229
18.3 Physical Factors Influencing Toxin Content 230
18.4 Inorganic Nutrients and Toxin Content 231
18.5 Organic Matter and Toxin Content 237
18.6 Conclusions 238
References 239

19 **Relationships Between Bacteria and Harmful Algae** 243
M. Kodama, G.J. Doucette, and D.H. Green

19.1 Introduction 243
19.2 Diversity of Algal-Associated Bacteria 244
19.2.1 Bacteria Associated with Harmful Algal Species 244
19.2.2 Spatio-Temporal Relationships Between Bacteria and Algae 246
19.3 Bacterial Influences on Algal Growth, Metabolism, and Toxins 247
19.3.1 Bacterial Effects on Algal Growth 247
19.3.2 The Role of Bacteria in Toxin Production 248
19.3.3 Bacterially-Mediated Release and Metabolism of Algal Toxins 249
19.4 Potential Implications of Interactions Among Bacteria . . . 250
19.5 Future Directions/Research Needs/Critical Questions 251
References 252

Part D Harmful Algae and the Food Web

20 **Harmful Algae Interactions with Marine Planktonic Grazers** 259
J.T. Turner

20.1 Introduction 259
20.2 Planktonic Grazers 260
20.2.1 Heterotrophic Dinoflagellates and other Flagellates 260
20.2.2 Tintinnids and Aloricate Ciliates 261
20.2.3 Rotifers 261

20.2.4 Copepods and other Mesozooplankton 262
20.3 HAB Toxin Accumulation in Zooplankton 263
20.4 Selective Grazing and Feeding Deterrence by Harmful Algae 263
20.5 Impact of Zooplankton Grazing on Formation and Termination of HA Blooms 264
20.6 Conclusions 265
References 266

21 Pathogens of Harmful Microalgae 271
P.S. Salomon and I. Imai

21.1 Introduction 271
21.2 Viruses 271
21.2.1 Host Specificity 273
21.3 Algicidal Bacteria 273
21.3.1 Modes of Algicidal Activity and Specificity 273
21.3.2 Ecology of Algicidal Bacteria and Harmful Microalgae . . . 274
21.3.3 Seaweed Beds as Prevention of HABs 275
21.4 Parasitic Fungi 275
21.4.1 Host Specificity 276
21.5 Parasitic Protists 276
21.5.1 Host Specificity 278
21.5.2 Host Avoidance of Parasitic Infection 278
21.6 Conclusions and Future Perspectives 279
References 280

22 Phycotoxin Pathways in Aquatic Food Webs: Transfer, Accumulation, and Degradation 283
G. J. Doucette, I. Maneiro, I. Riveiro, and C. Svensen

22.1 Introduction 283
22.2 Bacteria 283
22.3 Zooplankton 285
22.4 Bivalves 286
22.5 Benthic Invertebrates (Non-Bivalves) 287
22.6 Fishes 288
22.7 Seabirds and Marine Mammals 289
22.8 Summary and Conclusions 290
References 293

Part E Studying and Mitigating Harmful Algae: New Approaches

23 Molecular Approaches to the Study of Phytoplankton Life Cycles: Implications for Harmful Algal Bloom Ecology 299
R. W. Litaker and P. A. Tester

23.1 Introduction . . . 299
23.2 Identifying Life Cycle Stages Using Fluorescence In Situ Hybridization (FISH) . . . 299
23.3 Nuclear Staining to Determine Ploidy and Growth Rates . . 301
23.4 Genomic Approaches to Identifying Mitotic and Meiotic Life Cycle Stages . . . 302
23.5 Measuring Genetic Recombination During Sexual Reproduction . . . 305
23.6 Future Application of Reverse Transcriptase Assays and DNA Microarrays in Life Cycle Studies . . . 305
23.7 Conclusions . . . 307
References . . . 307

24 Laboratory and Field Applications of Ribosomal RNA Probes to Aid the Detection and Monitoring of Harmful Algae . . . 311
K. Metfies, K. Töbe, C. Scholin, and L.K. Medlin

24.1 Introduction . . . 311
24.2 Ribosomal RNA Sequences as Markers for Phylogenetic Studies and Species Identification . . . 312
24.3 Fluorescent in Situ Hybridization (FISH) for Identifying Intact Cells . . . 312
24.3.1 TSA-FISH for Flow Cytometry . . . 314
24.3.2 TSA-FISH for Solid Phase Cytometry . . . 315
24.4 Detecting Many Species Simultaneously Using DNA Probe Arrays . . . 316
24.4.1 Microarrays on Glass Slides and Fluorescence Detection . . 316
24.4.2 Handheld Array Device That Uses Electro-Chemical Detection . . . 318
24.4.3 DNA Probe Arrays for Autonomous Detection of Species Using the Environmental Sample Processor (ESP) 319
24.5 Conclusions . . . 320
References . . . 321

25 **Mitigation and Controls of HABs** . . . 327
H.G. Kim

25.1 Introduction . . . 327
25.2 Mitigation Strategies and Control of HABs . . . 328
25.2.1 Precautionary Impact Preventions . . . 328
25.2.2 Direct and Indirect Bloom Controls . . . 329
25.2.3 Contingency Plans for Fish Culture . . . 334
25.3 Conclusions . . . 335
References . . . 335

Part F Human Impact on Harmful Algae and Harmful Algae Impact on Human Activity

26 **The Complex Relationships Between Increases in Fertilization of the Earth, Coastal Eutrophication and Proliferation of Harmful Algal Blooms** . . . 341
P.M. Glibert and J.M. Burkholder

26.1 Introduction . . . 341
26.2 Global Trends in Population, Agricultural Fertilizer Usage and Implications for Export to Coastal Waters . . . 341
26.3 Nutrient Limitation versus Eutrophication: Basic Conceptual Framework . . . 343
26.4 Nutrient Loading, Nutrient Composition, and HABs . . . 344
26.5 Factors Complicating the Relationship Between Eutrophication and HABs . . . 347
26.6 Conclusions . . . 350
References . . . 351

27 **"Top-Down" Predation Control on Marine Harmful Algae** . . . 355
J.T. Turner and E. Granéli

27.1 Introduction . . . 355
27.2 "Top-down" Predators . . . 357
27.2.1 Medusae . . . 357
27.2.2 Ctenophores . . . 358
27.2.3 Fishes . . . 358
27.3 Case Studies . . . 359

27.3.1 Black Sea 359
27.3.2 Mesocosm Studies 360
27.4 Conclusions 362
References 363

28 Climate Change and Harmful Algal Blooms 367
B. Dale, M. Edwards, and P. C. Reid

28.1 Introduction 367
28.2 Evidence from the Past 369
28.3 Results from Plankton Records 370
28.4 Results from the Sedimentary Record of Dinoflagellate Cysts 372
28.5 Conclusions 375
References 376

29 Anthropogenic Introductions of Microalgae 379
G. Hallegraeff and S. Gollasch

29.1 Potential Transport Vectors for Microalgae 379
29.2 Vector Surveys for Microalgae 380
29.3 Evidence for Successful Establishment of Non-Indigenous Microalgae 381
29.3.1 Absence in Historic Samples 381
29.3.2 Sediment Cyst Cores 381
29.3.3 Increasing Molecular Evidence 382
29.4 Management Options to Reduce Risk of Introductions . . . 383
29.4.1 Warning System for HABs in Ballast-Water-Uptake Zones . 383
29.4.2 Ballast Water Exchange Studies on Phytoplankton 384
29.4.3 Treatment Options 386
29.5 Conclusions 388
References 388

30 The Economic Effects of Harmful Algal Blooms 391
P. Hoagland and S. Scatasta

30.1 Introduction 391
30.2 Scientific Concerns 392
30.3 Economic Concerns 392
30.4 Why Measure Economic Losses? 393
30.5 Economic Losses 394

30.6 Economic Impacts 397
30.7 Estimates of National Economic Effects 398
30.8 Conclusions 401
References 402

Subject Index 403

Contributors

AZANZA, R.V.
Marine Science Institute, University of the Philippines, Diliman 1101, Quezon City, Philippines, e-mail: rhod@upmsi.ph

BATES, S.S.
Fisheries and Oceans Canada, Gulf Fisheries Centre, PO Box 5030, Moncton, New Brunswick, E1C 9B6, Canada, e-mail: batess@dfo-mpo.gc.ca

BRUSSAARD, C.P.D.
Royal Netherlands Institute for Sea Research, PO Box 59, 1790 AB Den Burg, The Netherlands, e-mail: corina.brussaard@nioz.nl

BURKHOLDER, J.M.
Center for Applied Aquatic Ecology, 620 Hutton Street, Suite 104, North Carolina State University, Raleigh, North Carolina 27606, USA, e-mail: joann_burkholder@ncsu.edu

CEMBELLA, A.
Alfred Wegener Institute for Polar and Marine Research, Am Handelshafen 12, Bremerhaven, Germany, e-mail: acembella@awi-bremerhaven.de

DALE, B.
Geoscience Department, University of Oslo, PB 1047 Blindern, 0316 Oslo, Norway, e-mail: barri.dale@geo.uio.no

DOUCETTE, G.J.
NOAA/National Ocean Service, 219 Fort Johnson Rd., Charleston, South Carolina 29412, USA, e-mail: greg.doucette@noaa.gov

Edvardsen, B.

Department of Biology, Plankton Biology, University of Oslo, PO Box 1066 Blindern, 0316 Oslo, Norway, e-mail: bente.edvardsen@bio.uio.no

Edwards, M.

Sir Alister Hardy Foundation for Ocean Science, Citadel Hill, Plymouth, UK, e-mail: maed@sahfos.ac.uk

Estrada, M.

Institut de Ciències del Mar, CMIMA (CSIC), Pg Marítim de la Barceloneta 37-49, 08003 Barcelona, Spain, e-mail: marta@icm.csic.es

Figueiras, F.G.

Instituto de Investigacións Mariñas, CSIC, Eduardo Cabello 6, 36208 Vigo, Spain, e-mail: paco@iim.csic.es

Flynn, K.

Institute of Environmental Sustainability, University of Wales Swansea, Singleton Park, Swansea SA2 8PP, Wales, UK, e-mail: k.j.flynn@swansea.ac.uk

Fulton iii, R.S.

Division of Environmental Sciences, St. Johns River Water Management District, Palatka, Florida 32178-1429, USA, e-mail: rfulton@sjrwmd.com

Garcés, E.

Departament de Biologia Marina i Oceanografia, Institut de Ciències del Mar, CMIMA, Passeig Marítim de la Barceloneta, 37-49 08003 Barcelona, Spain, e-mail: esther@icm.csic.es

Glibert, P.M.

University of Maryland Center for Environmental Science, Horn Point Laboratory, PO Box 775, Cambridge, Maryland 21613, USA, e-mail: glibert@hpl.umces.edu

Gobler, C.J.

Marine Sciences Research Center, Stony Brook University, Stony Brook, New York 11790-5000, USA, e-mail: cgobler@notes.cc.sunysb.edu

Gollasch, S.

Bahrenfelder Str. 73a, 22765 Hamburg, Germany, e-mail: sgollasch@aol.com

Granéli, E.

Department of Marine Sciences, University of Kalmar, 391 82, Kalmar, Sweden, e-mail: edna.graneli@hik.se

Green, D.H.

Scottish Assoc. for Marine Science, Dunstaffnage Marine Laboratory Oban, Argyll Scotland PA37 1QA, UK, e-mail: david.green@sams.ac.uk

Hallegraeff, G.

University of Tasmania, Hobart Tas 7001, Australia, e-mail: hallegraeff@utas.edu.au

Hansen, P.J.

Marine Biological Laboratory, University of Copenhagen, 3000 Helsingør, Denmark, e-mail: pjhansen@bi.ku.dk

Hayes, P.K.

School of Biological Sciences, University of Bristol, Woodland Road, Bristol, BS8 1UG, UK, e-mail: Paul.Hayes@bristol.ac.uk

Hoagland, P.

Marine Policy Center, Woods Hole Oceanographic Institution, Woods Hole, Massachusetts, USA, e-mail: phoagland@whoi.edu

Imai, I.

Division of Applied Biosciences, Graduate School of Agriculture, Kyoto University, Kyoto 606-8502, Japan, e-mail: imai1ro@kais.kyoto-u.ac.jp

Janson, S.

Marine Science Division, Department of Biology and Environmental Science, University of Kalmar, 39182 Kalmar, Sweden, e-mail: sven.janson@hik.se

John, U.

Alfred Wegener Institute for Polar and Marine Research, Am Handelshafen 12, Bremerhaven, Germany, e-mail: ujohn@awi-bremerhaven.de

Kim, H.G.

Department of Oceanography, Pukyong National University, Busan, Korea, e-mail: hgkim7592@yahoo.co.kr

Kodama, M.

Kitasato University, School of Fisheries Sciences, Sanriku, Iwate 022-0101, Japan, e-mail: kodama@kitasato-u.ac.jp

Legrand, C.

Marine Science Division, Department of Biology and Environmental Science, University of Kalmar, 39182 Kalmar, Sweden, e-mail: catherine.legrand@hik.se

Litaker, R.W.

National Ocean Service, NOAA, 101 Pivers Island Road, Beaufort, North Carolina 28516, USA, e-mail: wayne.litaker@noaa.gov

Lundholm, N.

Department of Phycology, Biological Institute, University of Copenhagen, Denmark, e-mail: nlundholm@bi.ku.dk

Maneiro, I.

Edificio de Ciencias Experimentais, Universidad de Vigo, 36310 Vigo, Spain, e-mail: imaneiro@uvigo.es

Medlin, L.K.

Alfred Wegener Institute, Am Handelshafen 12, 27570 Bremerhaven, Germany, e-mail: lkmedlin@awi-bremerhaven.de

Metfies, K.

Alfred Wegener Institute, Am Handelshafen 12, 27570 Bremerhaven, Germany, e-mail: kmetfies@awi-bremerhaven.de

Moestrup, Ø.

Department of Phycology, Biological Institute, University of Copenhagen, Denmark, e-mail: moestrup@bi.ku.dk

Paerl, H.W.

Institute of Marine Sciences, University of North Carolina at Chapel Hill, Morehead City, North Carolina 28557, USA, e-mail: hpaerl@email.unc.edu

Pitcher, G.C.

Marine and Coastal Management, Private Bag X2, Rogge Bay 8012, Cape Town, South Africa, e-mail: gpitcher@deat.gov.za

Reid, P.C.

Sir Alister Hardy Foundation for Ocean Science, Citadel Hill, Plymouth, UK, e-mail: chris.reid@port.ac.uk

Riveiro, I.

Edificio de Ciencias Experimentais, Universidad de Vigo, 36310 Vigo, Spain, e-mail: iriveiro@uvigo.es

Sako, Y.

Faculty of Agriculture, Kyoto University, Kitashirakawa, Oiwake-Cho, Sakyo-Ku, Kyoto 606-8502, Japan, e-mail: sako@kais.kyoto-u.ac.jp

Salomon, P.S.

Marine Science Department, University of Kalmar, 391 82 Kalmar, Sweden, e-mail: paulo.salomon@ens.ufsc.br

Scatasta, S.

Environmental Economics and Natural Resources Group, Wageningen University, Wageningen, The Netherlands, e-mail: sara.scatasta@wur.nl

Scholin, C.

Monterey Bay Aquarium Research Institute, (MBARI), 7700 Sandholdt Rd, Moss Landing, California 95039-0628, USA, e-mail: scholin@mbari.org

Steidinger, K.A.

Florida Institute of Oceanography/Florida Fish and Wildlife Conservation Commission, St. Petersburg, Florida, 33701 USA, e-mail: karen.steidinger@myfwc.com

Stoecker, D.

UMCES, Horn Point Laboratory, PO Box 775, Cambridge, Maryland 21664, USA, e-mail: stoecker@hpl.umces.edu

Stolte, W.

Dept of Biology and Environmental Sciences, University of Kalmar, 39182 Kalmar, Sweden, e-mail: willem.stolte@hik.se

Sunda, W.G.

National Ocean Service, NOAA, 101 Pivers Island Road, Beaufort, North Carolina 28516, USA, e-mail: bill.sunda@noaa.gov

Svensen, C.

Department of Aquatic Bioscience, Norwegian College of Fishery Science, University of Tromsø, 9037 Tromsø, Norway, e-mail: camillas@nfh.uit.no

Tester, P.A.

National Ocean Service, NOAA, 101 Pivers Island Road, Beaufort, North Carolina 28516, USA, e-mail: pat.tester@noaa.gov

Tillmann, U.

Alfred Wegener Institute, Am Handelshafen 12, 27570 Bremerhaven, Germany, e-mail: utillmann@awi-bremerhaven.de

Töbe, K.

Alfred Wegener Institute, Am Handelshafen 12, 27570 Bremerhaven, Germany, e-mail: ktoebe@awi-bremerhaven.de

Trainer, V.L.

NOAA Fisheries, Northwest Fisheries Science Center, 2725 Montlake Boulevard East, Seattle, Washington 98112, USA, e-mail: vera.l.trainer@noaa.gov

Turner, J.T.

Biology Department and School for Marine Science and Technology, University of Massachusetts Dartmouth, North Dartmouth, Massachusetts 02747, USA, e-mail: jturner@umassd.edu

Veldhuis, M.J.W.

Royal Netherlands Institute for Sea Research, PO Box 59, 1790 AB, Den Burg, The Netherlands, e-mail: veldhuis@nioz.nl

Part A
Harmful Algae and Their Global Distribution

1 An Introduction to Harmful Algae

E. GRANÉLI and J.T. TURNER

Harmful algae have been the subjects of scientific and societal interest for centuries. Because blooms of toxic dinoflagellates were known to occasionally discolor water red or brownish red, they were, and still are known as "red tides." Water discoloration was noted for the lower Nile in the Bible (Exodus 7:20–21), and Darwin made microscopic observations of discolored water from an apparent dinoflagellate bloom off Chile during the voyage of the HMS *Beagle* ("*Some of the water placed in a glass was of a pale reddish tint and, examined under a microscope, was seen to swarm with minute animalculae darting about and often exploding. Their shape is oval and contracted in the middle by a ring of vibrating curved ciliae.*") (Galtsoff 1949, 1954).

In this book, the term "harmful algae" (HA) is used in a broad sense, referring to algae that can cause a variety of deleterious effects on aquatic ecosystems, including negative aesthetic effects such as beach fouling, oxygen deficiency, clogging of fish gills, or poisoning of various organisms. A direct effect of some HA blooms can be oxygen deficiency in deep waters, which in turn, causes mass mortality of benthic animals and fish kills (Granéli et al. 1989).

Some red-tide dinoflagellates and other harmful algae produce powerful toxins that can cause fish kills or shellfish poisoning. Included are PSP (paralytic shellfish poisoning), DSP (diarrhetic shellfish poisoning), ASP (amnesic shellfish poisoning), and NSP (neurotoxic shellfish poisoning), as well as other yet-uncharacterized toxins (see Turner and Tester 1997; Wright and Cembella 1998; Cembella 2003). Such toxicity can cause shellfish intoxication, leading to human fatalities, as well as vectorial intoxication whereby toxins are accumulated and transported through pelagic food webs by feeding interactions, leading to mortality of fish, seabirds, or marine mammals. In some cases, toxic blooms of flagellates of the genera *Chrysochromulina* or *Prymnesium* can disrupt entire ecosystems (Edvardsen and Paasche 1998).

In addition to toxicity, there are other adverse effects prompting the recent use of the more inclusive term "harmful algae." Such additional effects include organic loading leading to anoxia, such as in the 1976 bloom of *Ceratium tripos* off New York or the 1987–88 *Ceratium* spp. blooms in the Kattegat, beach

Ecological Studies, Vol. 189
Edna Granéli and Jefferson T. Turner (Eds.)
Ecology of Harmful Algae

fouling associated with massive blooms of *Phaeocystis* spp. off northern Europe, irritation of fish gills leading to suffocation by spines of *Chaetoceros* spp., or disruption of ecosystems by brown tides in Narragansett Bay, Long Island embayments, or the Laguna Madre of Texas. The economic impact of such blooms can be substantial.

Are harmful algae unique, compared to other phytoplankton? They certainly are in terms of the above-mentioned adverse effects that are of interest to humans, but other than producing toxins or other noxious chemicals, are they substantially different from other phytoplankters? Possibly they are not. Harmful algae (HA) as well as other species of phytoplankton and other organisms are all following their own autecological agendas, which together comprise community synecological dramas. Further, HA phytoplankton comprise only a small proportion of all phytoplankton species. Of the known 5,000 named living phytoplankton species (Sournia et al. 1991), known HAB species comprise some 300 species that can cause water discoloration, and only some 80 species that produce toxins that can cause human shellfish poisoning (Hallegraeff 2003).

Why are some phytoplankton toxic? It is often assumed that production of these toxins evolved to deter grazers. However, if such toxins poison primarily consumers of shellfish, or other upper-level consumers such as seabirds, marine mammals, and humans, rather than primary grazers of phytoplankton, such as bivalves and zooplankton (Turner et al. 1998), can these toxins be considered effective grazing deterrents? Other HA toxins appear to be involved primarily in allelopathy, being released in the dissolved state into seawater and causing deleterious effects on other competitor phytoplankton species (Fistarol et al. 2003, 2004; Legrand et al. 2003). HA toxins may be secondary metabolites that are only coincidentally toxic, being primarily associated with other processes such as nitrogen storage, nucleic acid biosynthesis, bioluminescence, chromosomal structural organization, ion channel transport across membranes, bacterial endosymbiosis, or pheromones inducing sexuality during bloom decline, rather than serving as grazer deterrents (Cembella 1998).

HA taxa seem to possess various attributes that enable them to form massive blooms that can dominate their ecosystems for extended periods of time. What are these attributes? Do toxins serve as deterrents that poison zooplankton grazers that might otherwise control HA blooms? Do toxins act as allelopathogens to wage chemical warfare upon other phytoplankton species that are competitors of HA species for light and nutrients? Why are toxins produced by scores of species from various microalgal groups, including dinoflagellates, diatoms, cyanobacteria, raphidophytes, pelagophytes, haptophytes, chrysophytes, and prymnesiophytes? Are HA blooms primarily due to meteorological or physical oceanographic anomalies that sporadically cause unusually high concentrations of HA species that are otherwise comparatively rare? Are HA blooms in response to anthropogenic nutrient loading from agriculture or sewage? Questions such as these, together with suggestions that HA blooms are possibly increasing in frequency and geographic extent in response to anthropogenic activities, have prompted increased interest in HA bloom ecology over the last decade. This increased interest in HA is not just confined to the scientific community, but also extends to the general public and governments that support scientists.

The recent flowering of interest in HA blooms has prompted a renaissance in the study of phytoplankton ecology. Because it is important to know whether an algal bloom is caused by an environmentally benign species or one that can poison or kill humans or other organisms of interest to them, such as fish, seabirds, or marine mammals, the importance of "species" has returned to the study of phytoplankton. This is after several decades of banishment of taxonomy from a field that largely viewed phytoplankton cells as no more than chlorophyll containers, uptakers of radioisotopes, or as "particles" that served as food for zooplankters. Indeed, many phytoplankton ecologists appear to have recovered from the seduction in the 1960s and 70s by fluorometers, liquid scintillators, and electronic particle counters, and returned to microscopes and flow cytometers in attempts to better quantify and identify the taxa we study. Because (as students often complain), under the microscope, many different things "all look alike," HA phytoplankton ecology has been at the forefront of prompting all phytoplankton ecology to join the recent revolution in molecular biology in order to better identify and study phytoplankton species of interest. We now know much more about not only HA phytoplankton, but all phytoplankton then we did a few short years ago.

The recent growth in HA phytoplankton ecology as a scientific discipline is evidenced by the increasing frequency and size of its international meetings. The first international conference on harmful algae was held in Boston, Massachusetts, USA in 1974, with approximately 100 coauthors of less than 50 papers listed in the table of contents. Since then, at subsequent conferences (2nd, Miami, Florida, USA, 1978), (3rd, St. Andrews, New Brunswick, Canada, 1985), (4th, Lund, Sweden, 1989), (5th, Newport, Rhode Island, USA, 1991), (6th, Nantes, France, 1993), (7th, Sendai, Japan, 1995), (8th, Vigo, Spain, 1997), (9th, Hobart, Tasmania, Australia, 2000), (10th, St. Petersburg, Florida, USA, 2002), (11th, Cape Town, South Africa, 2005), participation has steadily grown to a maximum (in St. Petersburg) of 629 published abstracts of talks or posters in the conference program with participants from 48 countries. In addition, other notable international meetings addressing HA blooms included (to name a few) the International Symposium on Red Tides in Takamatsu, Japan (1987), the "Novel Phytoplankton Blooms" meeting on Long Island, New York, USA (1988), the "Physiological Ecology of Harmful Algal Blooms" meeting in Bermuda (1996), two symposia on harmful marine algae in the United States in Woods Hole, Massachusetts, USA (2000 and 2003), nine Canadian workshops on harmful marine algae (up through 2005), the Harmful Algae Management and Mitigation Conference in Subic Bay, Philippines (1999), several Gordon Conferences, at least six conferences on toxic cyanobacteria (up through 2004), and special sessions at meetings of ASLO (American Society of Limnology & Oceanography) and/or AGU (American Geophysical Union) or TOS (The Oceanography Society), EUROHAB workshops (1998 Kalmar, Sweden, 2003, Amsterdam, Netherlands), the GEOHAB Open Science Meeting on HABs and Eutrophication (Baltimore, Maryland, USA, 2005), and others. There is also a new journal, *Harmful Algae*, which began publishing in 2002. Further evidence for the growth of this discipline is the increased research funding that is being invested by governments throughout the world for the study of harmful algae.

Despite this increased activity, the last major organized published synthesis of HA ecology was the volume from the Bermuda meeting in 1996 (Anderson et al. 1998). This volume addressed most major areas of HA science, through a combination of organism-based reviews (for example, *Alexandrium* complex and related species, fish-killing taxa such as *Chattonella* spp., *Heterosigma akashiwo*, *Gymnodinium breve*, *Pfiesteria piscicida*, *Prymnesium* spp. and *Chrysochromulina* spp., and other taxa, including species of *Phaeocystis*, *Dinophysis*, *Pseudo-nitzschia*, and *Noctiluca*). Other reviews focused on subjects that applied to various HAB species such as genetic variation, chemistry and physiology of various toxins, phagotrophy, and interactions of various HAB taxa with grazers, parasites, nutrients, trace elements, turbulence, and bacteria.

Although the reviews in the volume from the Bermuda meeting were excellent and comprehensive for the time, they are now almost a decade old and somewhat dated by recent developments. Accordingly, we were approached by Springer-Verlag with a request to compile an updated synthesis of HA ecology, organized primarily around processes and questions, rather than organisms. Thus, we invited a global assemblage of active HA researchers to contribute to the chapters in this volume, and many of these same specialists had also contributed to the previous Bermuda meeting volume. All chapters in this volume were peer-reviewed, by 1–3 reviewers in addition to the editors. We hope that this volume will complement other recent reviews and syntheses in *Harmful Algae* and other journals and in international HA meeting volumes to identify gaps in our present understanding of HA ecology and to suggest areas for additional research.

Acknowledgements. We are most grateful to Christina Esplund and Roseni de Carvalho for the invaluable help they gave during the entire time this book was compiled. From the re-drawing of figures, to improving photographic resolution, text layout, etc., they have indefatigably worked with all chapters. With smiles on their faces, they took up the challenge posed by the idiosyncrasies of almost all of the scientists involved in this book. Thank you!

References

Anderson DM, Cembella AD, Hallegraeff GM (1998) Physiological ecology of harmful algal blooms. NATO ASI Series 41. Springer, Berlin Heidelberg New York, 662 pp

Cembella AD (1998) Ecophysiology and metabolism of paralytic shellfish toxins in marine microalgae. In: Anderson DM, Cembella AD, Hallegraeff GM (eds) Physiological ecology of harmful algal blooms. NATO ASI Series 41. Springer, Berlin Heidelberg New York, pp 381–403

Cembella AD (2003) Chemical ecology of eukaryotic microalgae in marine ecosystems. Phycologia 42:420–44

Edvardsen B, Paasche E (1998) Bloom dynamics and physiology of *Prymnesium* and *Chrysochromulina*. In: Anderson DM, Cembella AD, Hallegraeff GM (eds) Physiological ecology of harmful algal blooms. NATO ASI Series 41. Springer, Berlin Heidelberg New York, pp 193–208

Fistarol GO, Legrand C, Granéli E (2003) Allelopathic effect of *Prymnesium parvum* on a natural plankton community. Mar Ecol Prog Ser 255:115–125

Fistarol GO, Legrand C, Selander E, Hummert C, Stolte W, Granéli E (2004) Allelopathy in *Alexandrium* spp.: effect on a natural plankton community and on algal monocultures. Aquat Microb Ecol 35:45–56

Galtsoff PS (1949) The mystery of the red tide. Sci Monthly 68:109–117

Galtsoff PS (1954) Red tide. US Dept Interior, Fish Wildl Serv, Spec Sci Rept 46:1–52

Granéli E, Carlsson P, Olsson P, Sundström B, Granéli W, Lindahl O (1989) From anoxia to fish poisoning: the last ten years of phytoplankton blooms in Swedish marine waters. In: Cosper EM, Bricelj VM, Carpenter EJ (eds) Novel phytoplankton blooms: causes and impacts of recurrent brown tides and other unusual blooms. Springer, Berlin Heidelberg New York, pp 407–427

Hallegraeff GM (2003) Harmful algal blooms: a global overview. In: Hallegraeff GM, Anderson DM, Cembella AD (eds) Manual on harmful marine microalgae, vol 11, 2nd edn. IOC-UNESCO. Paris, pp 25–49

Legrand C, Rengefors K, Fistarol GO, Granéli E (2003) Allelopathy in phytoplankton – biochemical, ecological and evolutionary aspects. Phycologia 42:406–419

Sournia A, Chretiennot-Dinet MJ, Ricard M (1991) Marine phytoplankton: how many species in the world ocean? J Plankton Res 13:1093–1099

Turner JT, Tester PA (1997) Toxic marine phytoplankton, zooplankton grazers, and pelagic food webs. Limnol Oceanogr 42:1203–1214

Turner JT, Tester PA, Hansen PJ (1998) Interactions between toxic marine phytoplankton and metazoan and protistan grazers. In: Anderson DM, Cembella AD, Hallegraeff GM (eds) Physiological ecology of harmful algal blooms. NATO ASI Series 41. Springer, Berlin Heidelberg New York, pp 453–474

Wright JLC, Cembella AD (1998) Ecophysiology and biosynthesis of polyether marine biotoxins. In: Anderson DM, Cembella AD, Hallegraeff GM (eds) Physiological ecology of harmful algal blooms. NATO ASI Series 41. Springer, Berlin Heidelberg New York, pp 427–451

2 Molecular Taxonomy of Harmful Algae

S. Janson and P.K. Hayes

2.1 Introduction

Harmful algae are not a homogenous group of organisms that can be classified as a taxonomic unit. In fact, algae are not even a natural taxonomic grouping. We define eukaryotic algae as organisms that have a permanent plastid, chlorophyll-*a* as their primary photosynthetic pigment and that lack a sterile covering of cells around the reproductive cells (Lee 1999) and "prokaryotic algae" as cyanobacteria (blue-green algae). The taxa considered in this chapter are primarily the species listed by the Intergovernmental Oceanographic Commission (IOC) found at www.bi.ku.dk/IOC. To be consistent with the list, only planktonic species are included here. It is sometimes hard to delineate species that are typically harmful from those that are not. Therefore, we use genera as the primary taxonomic unit and combine closely related species where their harmful status is uncertain. The taxonomy of harmful algae has recently been reviewed in several chapters in the volume edited by Hallegraeff et al. (2003). The present chapter focuses on molecular taxonomy, especially the relationships between observed morphological features and genetic characters. The molecular taxonomy of all harmful algae cannot be covered in this chapter, so we highlight examples from three main groups, the dinoflagellates, cyanobacteria, and diatoms. Molecular data for these organisms and other species can be retrieved from www.ncbi.nlm. nih.gov.

Analysis of molecular sequences can be used to resolve the evolutionary relationships and taxonomic position for species that have few distinct morphological characteristics. Since the pioneering work on bacteria by Woese (1987), the analysis of ribosomal RNA genes has revolutionized our understanding of the phylogeny and taxonomy of morphologically depauperate organisms. The small subunit ribosomal RNA gene (ssu rDNA) is still the most commonly used sequence for molecular taxonomy: here we will use the term 18S rDNA for the eukaryotic and 16S rDNA for the prokaryotic ssu rDNA, respectively. The sequence for the gene encoding of the large subunit

Ecological Studies, Vol. 189
Edna Granéli and Jefferson T. Turner (Eds.)
Ecology of Harmful Algae

ribosomal RNAs from eukaryotes is collectively called 28S rDNA here. The 28S rDNA has higher information content than the 18S rDNA and is therefore preferred for many taxonomic studies. The 5.8S rDNA is located in the region between the 18S and 28S rDNAs. There are two apparently non-coding regions that are transcribed together with the three rRNA encoding genes; they are called intergenic transcribed spacers (ITS). The ITS regions are generally less conserved and thus allow better resolution between closely related species, or different strains in one species, than do either the 18S or the 28S rDNA, and they are often used at the population level.

There are other genes or loci that can be useful in taxonomy. Comparison of protein coding genes can provide valuable information, but generally only if they are homologous genes, i.e., they share the same evolutionary history. Comparisons of different genes, and the use of different methods of analysis, can give conflicting results (see for example Taylor 2004) and some loci may be inappropriate for use as taxonomic markers, for example, genes encoding proteins under strong directional selection. When distinguishing taxa among algae, plastid gene sequences have been used less than has been the case for higher plants. The pigmentation of the plastid, however, has been used extensively to classify algae and is still considered a valid character (Daugbjerg et al. 2000; de Salas et al. 2003). One complication with using plastid sequences in one major group of harmful algae, the dinoflagellates, is the high evolutionary rate in peridinin-containing plastids (Zhang et al. 2000), which makes it difficult to align sequences and thus to perform valid phylogenetic analyses. Another complication is that plastid sequences, e.g., both 16S rDNA and the protein-coding *psbA* gene sequences, from different species can be identical (Takishita et al. 2002).

2.2 Dinophyta (Dinoflagellates)

2.2.1 General Morphology

The dinoflagellates are characterized by having two flagella, often organized as one transverse and one longitudinal. The transverse flagellum resides in a groove in the cell called the girdle and the corresponding longitudinal groove is known as the sulcus. The girdle divides the cell into an upper "epicone" and a lower "hypocone". The relationship between these two sections in terms of shape, size, ornamentation, and surface structure is an important feature in morphological taxonomy. Intracellular characteristics are the presence of sac-like vacuoles at the periphery, the amphiesmal vesicles, or alveoles. The presence of alveoles unites the super-group alveolates, comprising mainly ciliates, apicomplexa, and dinoflagellates. The dinoflagellate cell wall may be either armored by cellulose plates (in the alveoles) or naked without such plates. The amor is termed the theca, and consequently the plates are called thecal plates.

2.2.2 *Dinophysis*

The genus *Dinophysis* is characterized by having a polarized cell morphology where the hypocone is the major part of the cell and the epicone is less than one-tenth of the whole cell. Morphology within *Dinophysis* is highly variable and the delineation between species is based on the cell shape and size. Many of the species described so far need to be confirmed using molecular methods. For the 18S rDNA sequences, differences of 0.3–0.9 % between the species *Dinophysis norvegica, D. acuminata, D. acuta* and *D. fortii* have been reported (Edvardsen et al. 2003). Based on molecular data, Edvardsen et al. (2003) suggested that morphologically intermediate forms between *D. acuminata* and *D. norvegica* might also be genetic intermediates, produced by interbreeding between species (Edvardsen et al. 2003). The same study revealed that the 28S rDNA sequence was identical in both *D. fortii* and *D. acuta*, bringing into question the validity of their separation at the species level. The low variation in 18S rDNA sequences within the entire species group suggests that they have separated recently, or that they have a slow rate of evolution, perhaps mediated by the possession of a large number of 18S rRNA gene copies. Sequencing data from different *Dinophysis* spp. from Sweden and North America revealed a similar pattern. Here, the more variable 28S rDNA sequences showed low variation between *D. norvegica* and *D. acuminata* (<1 %), and an intra-species variation within *D. acuminata* of between 0 and 0.7 % (Rehnstam-Holm et al. 2002). Similarly, sequences of the 28S rDNA from *D. acuminata, D. caudata, D.* cf. *dens, D. fortii, D. sacculus,* and *D. tripos* isolated from different geographic regions, revealed distinct clades for nearly all species, but with very short distances between them (Guillou et al. 2002).

2.2.3 *Alexandrium*

The genus *Alexandrium* comprises species with a conspicuous girdle and sulcus, and many small thecal plates: species of *Alexandrium* are primarily defined on the basis of the morphology of certain thecal plates. The three *Alexandrium* species that are commonly reported in HAB events, *A. catenella, A. fundyense* and *A. tamarense,* form what is known as the "*A. tamarense* species complex". These three toxic species were found to be more closely related to each other than to three non-toxic species (*A. affine, A. insuetum* and *A. pseudogonyaulax*) based on their rDNA-ITS and 5.8S rDNA sequences (Adachi et al. 1996). In that study, *A. catenella* and the majority of *A.*

tamarense sequences formed discrete clusters, with *A. fundyense* within the *A. tamarense* cluster, and with some strains of *A. tamarense* outside of this cluster. This study, and results from 28S rDNA sequencing, reveals a geographical correlation between strains within the species complex, rather than a correlation with morphology (Scholin et al. 1995; Medlin et al. 1998). The main distinctive morphological character separating *A. fundyense* and *A. tamarense* is a "ventral pore" (a small cavity right above the junction where the two flagella are anchored in the cell). The validity of this character has been questioned by Hansen et al. (2003) who examined a number of *A. minutum* strains that were indistinguishable based on 28S rDNA sequences, but where the ventral pore was either present or absent in different strains. Hansen et al. (2003) also found no correlation between the type of toxin produced and genetic affiliation, rather the toxin profile varied with growth conditions.

It appears that the rDNA sequences within *Alexandrium* show variability, which could reflect a true genetic divergence in species with similar morphology. In other words, morphological descriptions underestimate the true diversity in this case. It should be kept in mind, however, that hybridization between sequences from closely related species could occur in vitro while processing samples and during PCR. True hybridization could occur in nature, and bottleneck effects and genetic drift or positive selection could create distinct genotypes occupying different areas in space and/or time. Some of the anomalies might also be due to misidentified species in the literature and databases.

2.2.4 *Protoperidinium, Prorocentrum*

The genus *Protoperidinium* appear to be polyphyletic, based on the sequence of the 28S rDNA, forming sister groups with gymnodinoid and prorocentroid dinoflagellates (Daugbjerg et al. 2000). The only documented toxic species of *Protoperidinium* is *P. crassipes*. Limited studies of two isolates of this one toxic species have shown that it is a sister group to *P. divergens* and that it encompasses very little divergence at the level of its 18S rDNA (Yamaguchi and Horiguchi 2005).

The conventional taxonomy of *Prorocentrum* species is complex and relies on differences between taxa that can only be resolved by electron microscopy. In one study of nine species of *Prorocentrum* and their respective 18S rDNA sequences (Grzebyk and Sako 1998), the strains were divided between two distinct groups, the first comprising benthic species (*P. lima, P. arenarium, P. maculosum,* and *P. concavum*) and the second group comprising planktonic and bentho-planktonic species (*P. micans, P. minimum, P. mexicanum* and *P. panamensis*). The sequence of bentho-planktonic *P. emarginatum* appeared as a sister group to the two main groups and may have diverged early in the radiation of the genus. *P. donghaiense* was recently compared genetically with other *Prorocentrum* species and seemed to be identical to *P. dentatum* at the level of the 18S rDNA, but with slight differences between their rDNA-ITS sequences (Lu et al. 2005). Similarly, *P. minimum* was closely related to *P. donghaiense* and *P. dentatum* based on its 18S rDNA, but was divergent at the rDNA-ITS. This suggests, as would be expected, that the rDNA-ITS is better at resolving closely related *Prorocentrum* species, than other, more conserved regions of the rDNA.

2.2.5 *Karenia, Karlodinium, Takayama*

The members of the genus *Karenia* are naked and characterized by a straight apical groove (Daugbjerg et al. 2000). Both morphological and genetic (28S rDNA) characters have been used to describe three new species of *Karenia* (Haywood et al. 2004). The shape of the upper and lower parts of the cells (epi- and hypotheca) were useful characters to differentiate *K. selliformis, K. bidigitata,* and *K. papilionacea* as new species, whereas six strains of *K. mikimotoi* were clustered together, as were two strains of *K. brevis* (Haywood et al. 2004). Another species, *K. umbella*, has been described recently (de Salas et al. 2003), again based on both morphological data and the sequence of the 28S rDNA. This species is related to *K. longicanalis*, but it is larger and it has a distinctive thecal morphology, for example a longer epicone with finger-like sulcal intrusions.

In samples from natural populations, the 28S rDNA data for *K. mikimotoi* points to a homogenous group with no or small variations in the observed sequences (Guillou et al. 2002): isolates identified as *K. mikimotoi* from French coastal waters were very similar in gene sequence to those identified previously from other geographic locations, such as Denmark, England and New Zealand.

Sequence data for the 28S rDNA of *Karlodinium*, characterized by a short apical groove, includes the type species *K. micrum* (Daugbjerg et al. 2000). Analysis of 28S rDNA from samples from natural populations off the French coast showed a clustering of four *Karlodinium* sequences with small variations between them (Guillou et al. 2002). It is possible that all four sequences were from representatives of *K. micrum*, because they were clustered with high support and very short branches separated them.

A fifth gymnodinoid genus based on the apical groove shape and molecular data has been described recently. The new genus *Takayama* includes two species with a sigmoid apical groove: these are distinct from *Akashiwo* (spiroid apical groove), *Gymnodinium* (horseshoe-shaped apical groove), *Karenia*, and *Karlodinium*. One species, *T. tasmanica*, is characterized by an S-shaped apical groove and a central pyrenoid surrounded by starch, while the other species, *T. helix*, has a shallow sigmoid apical groove and plastids with individual pyrenoids (de Salas et al. 2003). The 28S rDNA sequences differed by 3.8 % between the two species, and the sequences formed a well-supported cluster with *Karlodinium* spp. as a sister group.

2.2.6 *Amphidinium, Cochlodinium, Gyrodinium*

Members of *Gyrodinium* are heterotrophic with a large displacement of the girdle. The rDNA-ITS regions from species of this genus form a monophyletic cluster (Shao et al. 2004), and 28S rDNA sequences confirm this (Hansen and Daugbjerg 2004). Similarly, the single sequence available from *Cochlodinium,* for the toxic species *C. polykrikoides,* suggests that it forms a distinct genus, with high bootstrap support (Shao et al. 2004).

The genus *Amphidinium* comprises naked benthic species. In a recent study, 12 *Amphidinium* species were subject to phylogenetic analyses based on 28S rDNA sequences. This analysis revealed that *Amphidinium* species with minute left-deflected epicones were monophyletic, including the type species *A. operculatum* (Jørgensen et al. 2004). The epicone shape and size was the only morphological character that united the *Amphidinium sensu stricto* species. The morphology of the plastids and pyrenoids were not useful characters. In this analysis, *A. operculatum* appears on a long branch, suggesting that it is distantly related to the other species. The distinct nature of *A. operculatum* was confirmed by Murray et al. (2004) who further showed insignificant variation between strains: *A. carterae* and *A. massartii,* on the other hand, showed a high variation between strains.

2.3 Cyanobacteria (Blue-Green Algae)

2.3.1 *Anabaena, Aphanizomenon, Nodularia*

Large populations of species of *Anabaena* and *Aphanizomenon* are found in both fresh and brackish waters, including the Baltic Sea. Several species of *Anabaena* seem to be present in the Baltic Sea, with *A. lemmermannii* typically found in the open sea, and *A. inequalis* in near-shore areas where the influence of land-derived nutrients is more marked. The sequences from the regulatory *hetR* gene from these two species have been compared (Janson and Granéli 2002): one isolate of *A. lemmermannii* was distinct from two isolates of *A. inequalis.* Even though the two isolates of *A. inequalis* were morphologically indistinguishable, they diverged at the *hetR* locus. As part of the same study *Aphanizomenon* spp. were analyzed: on the basis of morphological characters, they were divided into the variants *Aph. flos-aquae* var *flos-aquae* and var *klebahnii.* These two variants were separated by 2.5 % difference in sequence at *hetR*, and the Baltic Sea *Aphanizomenon* sp. was 0.2% different from *Aph. flos-aquae* var *klebahnii*. The morphological characters used to separate the variants were cell sizes and shape of the colony. The Baltic Sea *Aphanizomenon* sp. would fall under *Aph. flos-aquae* var *klebahnii* using these criteria. These data demonstrate that diversity within the genera *Anabaena* and *Aphanizomenon* could be underestimated if only morphological characters are used. The hidden diversity suggested by *hetR* sequences is supported by a study of *Aphanizomenon flos-aquae* var *klebahnii* isolates from a Finnish lake, where several different rDNA-ITS types were identified (Laamanen et al. 2002). In this study, the rDNA-ITS diversity was somewhat lower in the Baltic Sea *Aphanizomenon* samples than in the lake, but there was no detectable difference in morphology: this is consistent with the study of Barker et al. (2000a) who found no sequence variation at the *cpcBA*-IGS locus in samples of *Aphanizomenon* collected directly from the water column in the Baltic Sea.

Nodularia is a genus of heterocyst-forming cyanobacteria, some members of which can form extensive blooms, particularly in brackish waters. The Baltic Sea is one area where such blooms occur. Three morphologically defined species have been described: *N. baltica, N. litorea*, and *N. spumigena*. Several studies suggest that all toxic strains belong to the same cluster or species, the bloom-forming planktonic *N. spumigena*, as shown by 16S rDNA data and sequence variation in several protein-coding genes and non-coding spacer regions (Barker et al. 1999; Bolch et al. 1999; Lehtimäki et al. 2000; Moffitt et al. 2001; Janson and Granéli 2002). Most studies point to zero, or a very low, variation within the 16S rRNA gene, with the exception of Moffitt et al. (2001) where significant variation was detected within the 16S rRNA gene, even though strains found to be polymorphic at this locus were indistinct at the more variable *cpcBA*-IGS (see Bolch et al. 1999). The *cpcBA*-IGS, rDNA-ITS and *gvpA*-IGS were analyzed in 13 strains of *N. litorea/N. spumigena* from the Baltic Sea (Barker et al. 1999) where it was reported that phenotypic groupings based on cell sizes were incongruent with the grouping of genotypes, calling into question the validity of *N. litorea*. A subsequent population genetic analysis of a large number of Baltic Sea *Nodularia* filaments supports the occurrence of a single planktonic species, *N. spumigena* (Barker et al. 2000b; Hayes et al. 2002).

2.3.2 *Microcystis*

Microcystis is a genus of unicellular colony-forming cyanobacteria where the taxonomy is still in a state of considerable flux and confusion. The *Microcystis aeruginosa* species complex dominates in many freshwater systems in both temperate and tropical regions. The life cycle of this species includes both pelagic and benthic stages in temperate regions, with the pelagic stage predominating in the summer. As would be expected for such a globally important toxic organism, *Microcystis* has been subject to several studies seeking to reconcile its molecular and morphological taxonomy. The shape of the colony has been used extensively to classify *Microcystis* species in many taxonomic treatments. However, it now appears that this may be a poor character to define species, and, as a consequence of this, five species of *Microcystis* previously distinguished by their colony shape have been merged into *M. aeruginosa* (Otsuka et al. 2001). Another factor is the spatial distribution of cells in the water, in that a region or depth can be specifically occupied by genetically distinct strains as seen for marine unicellular cyanobacteria (Rocap et al. 2002). In a study of a French lake using rDNA-ITS sequences, it was shown that there was no distinct spatial clustering of populations (Humbert et al. 2005). The production of the toxin microcystin is an example of where the use of genetic analysis has not always been helpful in accurately predicting the distribution of a specific phenotypic character. Some studies show that strains that produce microcystin differ in rDNA-ITS sequence from non-toxic strains (Janse et al. 2004). This is in contrast to earlier findings using the *cpcBA*-IGS as a high-resolution marker (Tillett et al. 2001). In practice, it is only the presence of the toxin-associated genes, e.g., *mcyAB* that is likely to provide an unambiguous indication of toxic potential.

2.3.3 *Trichodesmium*

Trichodesmium is the most abundant and conspicuous nitrogen-fixing cyanobacterial genus in tropical oceans. Some interesting features of the taxonomy and phylogeny of these filamentous non-heterocystous species remained hidden until molecular methods were adopted. The first surprise was that the closest neighbor, based on 16S rDNA, was the freshwater species *Oscillatoria* sp. PCC 7515 (Wilmotte et al. 1994), however, *Oscillatoria* sp. PCC 7515 could be a tropical species, similar to *Trichodesmium*, but since it was isolated from a water tank in a greenhouse, its real habitat remains to be confirmed. Another surprise in the *Trichodesmium* phylogeny was the finding that the largest and smallest of the five described species were closest together in both 16S rDNA and *hetR* sequences (Janson et al. 1999). The five species were grouped into three clusters: the first included *T. contortum* and *T. tenue*, *T. thiebautii* and *T. hildebrandtii* fell into a second, with *T. erythraeum* as a third. The whole genus showed a variation of around 3% in 16S rDNA, and 10% in *hetR* sequences (Janson et al. 1999). A further surprise was the finding that the species *Katagnymene pelagica* and *K. spiralis* seemed to be consistently clustered with *Trichodesmium* with both 16S rDNA and *hetR* sequences supporting this relationship (Lundgren et al. 2005). This means that among free-living filamentous cyanobacteria, there is only one group of closely related species in the tropical areas of the oceans.

2.4 Bacillariophyta (Diatoms)

2.4.1 *Amphora, Pseudo-nitzschia, Nitzschia*

In one study, the 28S rDNA sequences from all the toxic species of *Amphora, Pseudo-nitzschia,* and *Nitzschia* were included (Lundholm et al. 2002). The *P. australis* and *P. seriata* sequences were very similar and clustered together, while *P. multiseries* formed a distinct cluster as a sister group to *P. pungens* (non-toxic). The sequences from *N. navis-varingica* and *A. coffeaeformis* were, as expected, distantly related to the *Pseudo-nitzschia* spp. sequences. This suggests that the ability to produce domoic acid has either evolved independently several times, that the necessary genes have been laterally transferred, or that multiple losses have occurred. In addition, strains of *P. multistriata* from the Gulf of Naples have been found to produce domoic acid (Orsini et al. 2002). In both studies the *P. multistriata* sequences formed a sister group to *P. australis.*

2.5 Concluding Remarks

The use of molecular taxonomy is growing rapidly, both because gene sequences provide excellent tools for solving difficult taxonomic problems, and because technological advances are making it ever easier to obtain gene sequences. However, molecular data have not solved all taxonomic problems associated with algae. Closely related species are hard to separate even with the highly variable rDNA-ITS sequences, as in *Dinophysis norvegica* – *D. acuminata* and *Prorocentrum* spp. among the dinoflagellates. The locus of choice, at least for dinoflagellates, seems to be the 28S rRNA gene. Perhaps it would be a good idea to also include rDNA-ITS and 5.8S rRNA sequences for higher resolution. The rDNA locus shows a good correlation with certain morphological characters, such as the apical groove shape. Perhaps some structures are governed by protein structures that are directly linked to their coding sequences, and other structures depend on environmental conditions. The ventral pore in some dinoflagellates seems to be an important example of the latter, which would explain some of the difficulties in resolving the *Alexandrium tamarense* species complex.

Among cyanobacteria, there are many more loci used in phylogenies, population genetics, and taxonomy. This is beneficial because single-locus studies can be prone to misinterpretation because of horizontal gene transfer events. In this chapter, we have given several examples where the *cpcBA*-IGS locus is incongruent with other loci (see also Gugger et al. 2005). Thus, this locus should not be used to study phylogeny and evolution. There are several possible reasons for this: (1) the *cpcBA*-IGS locus shows anomalies in itself; there are regions of coding and non-coding regions and they perform differently in phylogenies, (2) the *cpcBA*-IGS locus is possibly transferred frequently between similar strains, or (3) there is a strong selective pressure to evolve new pigments.

All molecular analyses of taxonomy and diversity could be improved with larger sample sizes. A few genera that particularly need further attention are the dinoflagellates *Akashiwo* and *Heterocapsa,* which contain bloom-forming species with unknown diversity. Another example is the cyanobacterial species *Aphanizomenon* sp. in the Baltic Sea. This species is a major component of the phytoplankton biomass, but we know little about its population biology. One thing that is clear is that the taxonomy of algae should not be based only on morphological data (see e.g., Wilcox 1998), although some morphological characters do seem to make phylogenetic sense, such as the apical groove of dinoflagellates. Molecular taxonomy also provides a framework to study organisms in nature, using molecular probes as outlined in Chap. 24 of this volume.

References

Adachi M, Sako Y, Ishida Y (1996) Identification of the toxic dinoflagellates *Alexandrium catenella* and *A. tamarense* (Dinophyceae) using DNA probes and whole-cell hybridization. J Phycol 32:1049–1052

Barker GLA, Hayes PK, O'Mahony SL, Vacharapiyasophon P, Walsby AE (1999) A molecular and phenotypic analysis of *Nodularia* (cyanobacteria) from the Baltic Sea. J Phycol 35:931–937

Barker GLA, Konopka A, Handley BA, Hayes PK (2000a) Genetic variation in *Aphanizomenon* (cyanobacteria) colonies from the Baltic Sea and North America. J Phycol 36:947–950

Barker GLA, Handley BA, Vacharapiyasophon P, Stephens JR, Hayes PK (2000b) Allele specific PCR shows that genetic exchange occurs among genetically diverse *Nodularia* (cyanobacteria) filaments in the Baltic Sea. Microbiology 146:2865–2875

Bolch CJS, Orr PT, Jones GJ, Blackburn SI (1999) Genetic, morphological, and toxicological variation among globally distributed strains of *Nodularia* (cyanobacteria). J Phycol 35:339–355

Daugbjerg N, Hansen G, Larsen J, Moestrup Ø (2000) Phylogeny of some of the major genera of dinoflagellates based on ultrastructure and partial LSU rDNA sequence data, including the erection of three new genera of unarmoured dinoflagellates. Phycologia 39:302–317

de Salas MF, Bolch CJS, Botes L, Nash G, Wright SW, Hallegraeff GM (2003) *Takayama* gen. nov (Gymnodiniales, Dinophyceae), a new genus of unarmored dinoflagellates with sigmoid apical grooves, including the description of two new species. J Phycol 39:1233–1246

Edvardsen B, Shalchian-Tabrizi K, Jakobsen KS, Medlin LK, Dahl E, Brubak S, Paasche E (2003) Genetic variability and molecular phylogeny of *Dinophysis* species (Dinophyceae) from Norwegian waters inferred from single-cell analyses of rDNA. J Phycol 39:395–408

Grzebyk D, Sako Y (1998) Phylogenetic analysis of nine species of *Prorocentrum* (Dinophyceae) inferred from 18S ribosomal DNA sequences, morphological comparisons, and description of *Prorocentrum panamensis*, sp. nov. J Phycol 34:1055–1068

Gugger M, Molica R, Le Berre B, Dufour P, Bernard C, Humbert JF (2005) Genetic diversity of *Cylindrospermopsis* strains (Canobacteria) isolated from four continents. Appl Environ Microbiol 71:1097–1100

Guillou L, Nezan E, Cueff V, Denn EEL, Cambon-Bonavita MA, Gentien P, Barbier G (2002) Genetic diversity and molecular detection of three toxic dinoflagellate genera (*Alexandrium*, *Dinophysis*, and *Karenia*) from French coasts. Protist 153:223–238

Hallegraeff GM, Anderson DM, Cembella AD (eds) (2003) Manual on harmful marine microalgae. UNESCO Publishing, Paris, 793 pp

Hansen G, Daugbjerg N (2004) Ultrastructure of *Gyrodinium spirale*, the type species of *Gyrodinium* (Dinophyceae), including a phylogeny of *G. dominans*, *G. rubrum* and *G. spirale* deduced from partial LSU rDNA sequences. Protist 155:271–294

Hansen G, Daugbjerg N, Franco JM (2003) Morphology, toxin composition and LSU rDNA phylogeny of *Alexandrium minutum* (Dinophyceae) from Denmark, with some morphological observations on other European strains. Harmful Algae 2:317–335

Hayes PK, Barker GLA, Batley J, Beard SJ, Handley BA, Vacharapiyasophon P, Walsby AE (2002) Genetic diversity within populations of cyanobacteria assessed by analysis of single filaments. Antonie Van Leeuwenhoek 81:197–202

Haywood AJ, Steidinger KA, Truby EW, Bergquist PR, Bergquist PL, Adamson J, MacKenzie L (2004) Comparative morphology and molecular phylogenetic analysis of three new species of the genus *Karenia* (Dinophyceae) from New Zealand. J Phycol 40:165–179

Humbert JF, Duris-Latour D, Le Berre B, Giraudet H, Salencon MJ (2005) Genetic diversity in *Microcystis* populations of a French storage reservoir assessed by sequencing of the 16S-23S rRNA intergenic spacer. Microb Ecol 49:308–314

Janse I, Kardinaal WEA, Meima M, Fastner J, Visser PM, Zwart G (2004) Toxic and non-toxic *Microcystis* colonies in natural populations can be differentiated on the basis of rRNA gene internal transcribed spacer diversity. Appl Environ Microbiol 70:3979–3987

Janson S, Granéli E (2002) Phylogenetic analyses of nitrogen-fixing cyanobacteria from the Baltic Sea reveal sequence anomalies in the phycocyanin operon. Int J Sys Evol Microbiol 52:1397–1404

Janson S, Bergman B, Carpenter EJ, Giovannoni SJ, Vergin K (1999) Genetic analysis of natural populations of the marine diazotrophic cyanobacterium *Trichodesmium*. FEMS Microbiol Ecol 30:57–65

Jørgensen MF, Murray S, Daugbjerg N (2004) *Amphidinium* revisited. I. Redefinition of *Amphidinium* (Dinophyceae) based on cladistic and molecular phylogenetic analyses. J Phycol 40:351–365

Laamanen MJ, Forsstrøm L, Sivonen K (2002) Diversity of *Aphanizomenon flos-aquae* (cyanobacterium) populations along a Baltic Sea salinity gradient. Appl Environ Microbiol 68:5296–5303

Lee RE (1999) Phycology. Cambridge University Press, Cambridge, 614 pp

Lehtimäki J, Lyra C, Suomalainen S, Sundman P, Rouhiainen L, Paulin L, Salkinoja-Salonen M, Sivonen K (2000) Characterization of *Nodularia* strains, cyanobacteria from brackish waters, by genotypic and phenotypic methods. Int J Sys Evol Microbiol 50:1043–1053

Lu DD, Goebel J, Qi YZ, Zou JZ, Han XT, Gao YH, Li YG (2005) Morphological and genetic study of *Prorocentrum donghaiense* Lu from the East China Sea, and comparison with some related *Prorocentrum* species. Harmful Algae 4:493–505

Lundgren P, Janson S, Jonasson S, Singer A, Bergman B (2005) Unveiling of novel radiations within *Trichodesmium* cluster by *hetR* gene sequence analysis. Appl Environ Microbiol 71:190–196

Lundholm N, Daugbjerg N, Moestrup Ø (2002) Phylogeny of the Bacillariaceae with emphasis on the genus *Pseudo-nitzschia* (Bacillariophyceae) based on partial LSU rDNA. Eur J Phycol 37:115–134

Medlin LK, Lange M, Wellbrock U, Donner G, Elbrächter M, Hummert C, Luckas B (1998) Sequence comparisons link toxic European isolates of *Alexandrium tamarense* from the Orkney Islands to toxic North American stocks. Eur J Protistol 34:329–335

Moffitt MC, Blackburn SI, Neilan BA (2001) rRNA sequences reflect the ecophysiology and define the toxic cyanobacteria of the genus *Nodularia*. Int J Sys Evol Microbiol 51:505–512

Murray S, Jørgensen MF, Daugbjerg N, Rhodes L (2004) *Amphidinium* revisited. II. Resolving species boundaries in the *Amphidinium operculatum* species complex (Dinophyceae), including the descriptions of *Amphidinium trulla* sp nov and *Amphidinium gibbosum*. comb. nov. J Phycol 40:366–382

Orsini L, Sarno D, Procaccini G, Poletti R, Dahlmann J, Montresor M (2002) Toxic *Pseudo-nitzschia multistriata* (Bacillariophyceae) from the Gulf of Naples: morphology, toxin analysis and phylogenetic relationships with other *Pseudo-nitzschia* species. Eur J Phycol 37:247–257

Otsuka S, Suda S, Shibata S, Oyaizu H, Matsumoto S, Watanabe MM (2001) A proposal for the unification of five species of the cyanobacterial genus *Microcystis* Kutzing ex Lemmermann 1907 under the Rules of the Bacteriological Code. Int J Syst Evol Microbiol 51:873–879

Rehnstam-Holm AS, Godhe A, Anderson DM (2002) Molecular studies of *Dinophysis* (Dinophyceae) species from Sweden and North America. Phycologia 41:348–357

Rocap G, Distel DL, Waterbury JB, Chisholm SW (2002) Resolution of *Prochlorococcus* and *Synechococcus* ecotypes by using 16S-23S rDNA internal transcribed spacer (ITS) sequences. Appl Environ Microbiol 68:1180–1191

Scholin CA, Hallegraeff GM, Anderson DM (1995) Molecular evolution of the *Alexandrium tamarense* 'species complex' (Dinophyceae): Dispersal in the North American and West Pacific regions. Phycologia 34:472–485

Shao P, Chen YQ, Zhou H, Yuan J, Qu LH, Zhao D, Lin YS (2004) Genetic variability in Gymnodiniaceae ITS regions: implications for species identification and phylogenetic analysis. Mar Biol 144:215–224

Takishita K, Koike K, Maruyama T, Ogata T (2002) Molecular evidence for plastid robbery (Kleptoplastidy) in *Dinophysis*, a dinoflagellate causing diarrhetic shellfish poisoning. Protist 153:293–302

Taylor FJR (2004) Illumination or confusion? Dinoflagellate molecular phylogenetic data viewed from a primarily morphological standpoint. Phycol Res 52:308–324

Tillett D, Parker DL, Neilan BA (2001) Detection of toxigenicity by a probe for the microcystin synthetase A gene (*mcyA*) of the cyanobacterial genus *Microcystis*, comparison of toxicities with 16S rRNA and phycocyanin operon (phycocyanin intergenic spacer) phylogenies. Appl Environ Microbiol 67:2810–2818

Wilcox TP (1998) Large-subunit ribosomal RNA systematics of symbiotic dinoflagellates: morphology does not recapitulate phylogeny. Mol Phylogen Evol 10:436–448

Wilmotte A, Neefs J-M, DeWachter R (1994) Evolutionary affiliation of the marine nitrogen-fixing cyanobacterium *Trichodesmium* sp. strain NIBB 1067, derived by 16S ribosomal RNA sequence analysis. Microbiology 140:2159–2164

Woese CR (1987) Bacterial evolution. Microbial Rev 51:221–271
Yamaguchi A, Horiguchi T (2005) Molecular phylogenetic study of the heterotrophic dinoflagellate genus *Protoperidinium* (Dinophyceae) inferred from small subunit rRNA gene sequences. Phycol Res 53:30–42
Zhang ZD, Green BR, Cavalier-Smith T (2000) Phylogeny of ultra-rapidly evolving dinoflagellate chloroplast genes: A possible common origin for sporozoan and dinoflagellate plastids. J Mol Evol 51:26–40

3 The Biogeography of Harmful Algae

N. Lundholm and Ø. Moestrup

3.1 Biogeography and Species Concepts

The fundamental goal of biogeography is to describe how species are distributed and to consider the processes that may explain the distribution patterns. This obviously requires a well-defined species concept, a requirement that cannot always be met since more than one species concept is in use, and the different concepts are not always congruent. We will briefly address this problem.

Although the classical species concept is based on morphology, the variation allowed within a species is not fixed; rather it depends on the interpretation of the specialists working on the group. It may therefore vary among taxonomic groups (Mann 1999).

In the biological species concept, a species is defined as a population that is reproductively isolated from other populations. However, reproductive isolation cannot be studied in extinct species, asexual species, and in species that do not mate in the laboratory. In most cases, so little information on breeding capabilities is known that the default species concept has been the morphological species concept (Gosling 1994). In addition, hybridization takes place between populations that still remain distinct due to low viability of hybrids, or hybridization found in the laboratory apparently does not occur in the field due to geographical, seasonal, or other barriers. Hence, discontinuous variation can be found between populations that potentially or actually interbreed (Mann 1999).

In the phylogenetic species concept (Davis 1997), species are defined as the smallest groupings of organisms that are hierarchically related. In other words, variation is hierarchical between species and non-hierarchical (= reticulate) within species.

There is often agreement between the different concepts (e.g., Davis 1997; Mann 1999; Coleman 2000), but conflicts appear for example when species defined by the phylogenetic species concept are not reproductively isolated. The phylogenetic species concept emphasizes differentiation (genetic, mor-

Ecological Studies, Vol. 189
Edna Granéli and Jefferson T. Turner (Eds.)
Ecology of Harmful Algae

phological, and ecological) rather than reproductive isolation, and the latter may be preceded by differentiation (Gosling 1994). Other disagreements occur when molecular and morphological speciation do not occur at the same rate. Morphologically identical subclades may be phylogenetically distinct but reproductively isolated (e.g., Coleman 2000; Gonzalez et al. 2001). Such discrepancies must be considered before distribution patterns can be understood.

3.1.1 Genetic Variation

A way to address the species problem is to study genetic variation within an apparent species/species complex. That phytoplankton may show large intraspecific genetic variation was first demonstrated by Brand (1981) and others, and later confirmed by different techniques (e.g., Rynearson and Armbrust 2004). Large genetic variation has been found in harmful taxa such as species of *Pseudo-nitzschia* (Evans et al. 2004) and *Alexandrium* (Cembella et al. 1988), and in *Gambierdiscus toxicus* (Chinain et al. 1997).

One of the few population-genetic studies performed has shown that locally distinct populations occur in *Ditylum brightwellii* (Rynearson and Armbrust 2004). Populations from two interconnected estuaries differed physiologically and genetically, indicating that genetic interchange between the populations was limited. Differential selection was suggested as a factor allowing local genetic differentiation and subsequently speciation (Rynearson and Armbrust 2004). In other words, different environmental regimes or niches may lead to speciation. The speciation process may result in morphologically different species or in cryptic species, as the selection pressure may or may not be accompanied by morphological changes (Rynearson and Armbrust 2004).

Whether most planktonic microorganisms are cosmopolitan as sometimes claimed, must await an improved understanding of the genetic variation at species level. However, the problem of delineating species will persist since evolution is an ongoing process.

A relation between genetic variation and geographical distance has sometimes been found, in which the genetic variation among strains from the same locality is smaller than between strains from distant localities. This applies to *Cylindrospermopsis raciborskii* (Neilan et al. 2003), the *Alexandrium tamarense* species complex (John et al. 2003) (see further below), and *Ostreopsis ovata* (Pin et al. 2001). In other cases, no relation exists between genetic diversity and biogeography (Casamatta et al. 2003). Hence, high dispersal rates do not necessarily lead to high gene flow and low genetic divergence, because geology, distance, selection, and local genetic drift may counteract the effect of dispersal (Palumbi 1992).

3.2 Biogeographical Distribution

Assessment of species biogeography is hampered not only by disagreement on species concepts but also by insufficient knowledge about variation patterns and lack of data from geographically distant areas. Also, the biogeography of species is not fixed since species are being continuously spread, e.g., by ships' ballast water.

A cosmopolitan distribution of microbial morphospecies has been suggested as a general feature of organisms below 1 mm due to their enormous population sizes and hence increased chance of spreading (Finlay and Fenchel 2004, and references therein). This idea of 'everything is everywhere' regards morphospecies. However, reproductive isolation may be molecularly based and there is evidence that rapid evolution may occur in genes involved in sexual reproduction (Armbrust and Galindo 2001). Strong local selection may, among other evolutionary forces, induce speciation in spite of a high dispersal rate, and speciation may occur without morphological change (Mann 2001). Many species are undoubtedly cosmopolitan. However, present taxonomic knowledge is very limited and biodiversity almost certainly heavily underestimated.

Despite the lack of a solid taxonomy and other drawbacks, there is sufficient evidence for the existence of several distribution patterns in marine phytoplankton (e.g., Semina 1997) and other unicellular organisms (Foissner 1999). Some species, like the dinoflagellate *Alexandrium ostenfeldii* appear to be cosmopolitan. Its wide geographic occurrence has been confirmed by growth experiments that found it both eurythermal and euryhaline (Jensen and Moestrup 1997), and by molecular data (John et al. 2003). Other taxa are restricted to warm waters, e.g., the toxic species *Alexandrium tamiyavanichii* from tropical South East Asia. Others again, like several cold-water species of the genus *Fragilariopsis* show a uni-polar distribution, being restricted to either the northern or the southern hemisphere. The warm tropical waters serve as a barrier to dispersal of cold-water species, which may explain the existence of unipolar species.

Finally, some species show a bipolar distribution that may have arisen after dispersion via deep cold water at the equator, via atmospheric circulation, or via biological vectors like birds, etc. Resting cysts and other resistant stages in the life cycle may have been involved. Another hypothesis explains bipolar distribution as going back to former periods when the ice margins of the two hemispheres were closer together. Molecular evidence for bipolar distributions has been provided for some foraminiferans and the dinoflagellate *Polarella glacialis* (Darling et al. 2000; Montresor et al. 2003).

3.3 Distribution of Harmful Species

3.3.1 Dinoflagellates

Lack of molecular data prevents assessment of the distribution of the many species of *Dinophysis*. More data are available on the genus *Alexandrium*, particularly on the so-called "*tamarense* complex", as well as certain species of *Gymnodinium*, on which we will therefore focus.

Considerable uncertainty regards the taxonomy of the *tamarense* complex, comprising the morphotypes *A. catenella*, *A. tamarense*, and *A. fundyense*. Molecular studies have not supported separation into three species. Lilly (2003), in a study on the D1-D2 LSU rDNA from 110 isolates, found the isolates to fall into five groups. Two groups were restricted to a single locality (Tasmania and Italy, respectively), and both belonged to the *tamarense* morphotype. The third group also comprised *tamarense* morphotypes only, and the strains were restricted to Europe (i.e., Sweden and Ireland in the north; Spain in the south). A single strain of this group occurred in Japan, suggesting human-induced spreading. Remarkably, all examined strains of these three groups were found to be non-toxic. Both of the remaining two groups comprised *tamarense*, *fundyense*, and *catenella* morphotypes, and all examined strains were toxic. In other words, toxicity was confined to these groups. Both groups are widely distributed. The American group is the only genetic group in the Americas, and extends into the North Atlantic as far east as Scotland, and to the North Pacific as far west as Japan and Korea. Its distribution recalls that of *A. ostenfeldii* mentioned above, occurring from the Arctic to warm water, although strains from truly tropical waters have not yet been discovered. The American clade also occurs in South Africa, indicating spreading with the circumpolar current. The second group/clade comprising all three morphotypes is mainly confined to East Asia, from Thailand to Korea. This East Asian genotype has been found sporadically elsewhere, in Port Phillip Bay, Australia, on the North Island of New Zealand and in the French Mediterranean. It appears to be temperate-tropical. The two toxic genotypes overlap in distribution in Korea and Japan.

Whether the differences in distribution are real must await data from some of the many areas that remain undersampled. Thus no genetic information exists about the *tamarense* complex in Africa outside South Africa, and in a large area extending from the eastern Mediterranean over the Middle East to Indonesia, and in most of tropical America. Future recordings of material from the *tamarense* complex should include both information about morphotype and genotype. Studies on other genes are needed to critically evaluate the idea of five genotypes.

The large genus *Gymnodinium* includes a group characterized by microreticulate cysts. One species, *G. catenatum*, is a serious PSP toxin-producer, while *G. nolleri* and *G. microreticulatum*, appear to be non-toxic.

The three species are closely related and have been repeatedly confused, resulting in erroneous conclusions about their geographic distribution. Identifications based solely on measurements of cyst diameter are not entirely reliable, as the cyst diameter overlaps and considerable variation has been detected. The three species show a somewhat erratic geographic distribution, which may reflect that cysts have been recently introduced into new areas. Thus, the presence of *G. catenatum* in Australia is believed to be due to spreading with ships' ballast water to Tasmania, and from there to the southeastern mainland of Australia and the North Island of New Zealand. Cysts are absent from sediments in Tasmania except near Hobart, where ballast water is usually discharged.

All three species are confined to temperate and tropical waters. *Gymnodinium catenatum* is very widely distributed, particularly in warmer water, Tasmania being its coldest area of occurrence. In Europe, *G. catenatum* does not occur north of the Iberian Peninsula, but it is a serious cause of PSP in western North Africa.

In contrast, *G. microreticulatum* is an East Asian/Australian species. It has recently been reported in Portugal, and cysts collected in Uruguay germinated into this species (Bolch and Reynolds 2002). Do these isolated reports indicate recent introductions or has the species been overlooked?

Gymnodinium nolleri is a species of Western Europe, the Mediterranean, and the Middle East, with a single report from the Benguela current, Namibia. Thus, present indications are that *G. nolleri* and *G. microreticulatum* are/have been species of limited geographic distribution, while *G. catenatum* is (at least now) distributed in warmer waters of all continents.

3.3.2 Diatoms

The genus *Pseudo-nitzschia* is one of the most studied genera and comprises most harmful diatoms known. Due to the abundance of data, we will discuss this genus in some detail. Until recently, ten species were known to produce the neurotoxin domoic acid (*P. australis, P. calliantha, P. delicatissima, P. fraudulenta, P. galaxiae, P. multiseries, P. multistriata, P. pungens, P. seriata* and *P. turgidula* (Lundholm and Moestrup 2004 in Moestrup 2004), but *P. cuspidata* has recently been added (Bill et al. 2005).

Pseudo-nitzschia seriata is a potentially toxic species that has caused toxic blooms and has only been found in cold waters of the northern hemisphere (see Hasle and Lundholm 2005). Despite reports to the contrary, it has never been confirmed outside of the Atlantic Ocean, which is a very unusual distribution pattern. Phylogenetically and morphologically, *P. seriata* appears to be well defined (Hasle and Lundholm 2005).

Species from warmer waters include *P. brasiliana*, *P. subpacifica*, and *P. subfraudulenta*, while *P. pungens* is more or less cosmopolitan (e.g., Hasle 2002). Species such as *P. australis*, *P. delicatissima*, *P. fraudulenta*, *P. multiseries*, *P. americana*, and *P. calliantha* are probably also cosmopolitan, despite the paucity of observations from tropical and subtropical waters.

High levels of genetic diversity were found among isolates of the toxic *P. calliantha*, *P. delicatissima*, and *P. multiseries* (Evans et al. 2004; Orsini et al. 2004). Similarly, a high level of genetic diversity among 464 isolates of *P. pungens* from the North Sea was found by microsatellite markers. The variation did not include cryptic speciation, as low levels of genetic differentiation were found both spatially and temporally. Rather, the strains seemed to constitute one large unstructured population (Evans et al. 2005). There was no major difference between isolates from the North Sea and Atlantic Canada, suggesting that few barriers to gene flow exist between these two areas (Evans et al. 2005). Conversely, Pacific and Canadian Atlantic isolates of the very similar species *P. multiseries* showed barriers to gene flow (Evans et al. 2004). *Pseudo-nitzschia multiseries* may comprise several populations, but phylogenetic and morphological studies have confirmed species delineation (Lundholm et al. 2002; Orsini et al. 2002). *Pseudo-nitzschia multiseries* is distributed similarly to *P. pungens* (Hasle 2002).

The high level of genetic diversity within *Pseudo-nitzschia* species partly explains the problems with specificity of molecular probes applied in regions other than those in which they were developed (Parsons et al. 1999; Orsini et al. 2002; Lundholm et al. 2003). Other problems are due to the existence of several morphologically similar species. Thus, when combining morphological and molecular data, *P. pseudodelicatissima* was found to be a species complex, including *P. cuspidata* and two new species, *P. calliantha* and *P. caciantha* (Lundholm et al. 2003). The species diversity within the complex is still not fully described (Lundholm et al. 2003; Kaczmarska et al. 2005). *Pseudo-nitzschia calliantha* is most probably cosmopolitan, being recorded from subpolar to tropical waters, but data on the other species are insufficient to evaluate their distributions.

Based on ITS rDNA analyses, cryptic diversity was found in *P. delicatissima* from the Bay of Naples. Some of this diversity was accompanied by morphological variation (Orsini et al. 2004) and comprises new species (Lundholm et a. 2006). The remaining variation represents cryptic variation. Genetic diversity may perhaps also explain the observed variation in size and shape of *P. galaxiae*, where temporal variation in appearance of morphotypes corresponded to different environmental conditions (Cerino et al. 2005).

In conclusion, the toxic *Pseudo-nitzschia* species seem to show different distribution patterns. What complicates the picture further are the long-term shifts observed in the occurrence of certain species: *P. multiseries* in Canadian and Scandinavian waters and *P. delicatissima* in Norwegian waters (Hasle 2002). The recent appearance of *P. australis* in Northern Atlantic waters could also be part of a long-term shift, or it could perhaps be the result of new introductions to the area (see Hasle 2002).

3.3.3 Haptophytes

Prymnesium is, on a global scale, a seriously harmful genus. It comprises six confirmed species, of which four (*P. calathiferum, P. faveolatum, P. parvum* (including *P. patelliferum*) and *P. zebrinum*) are potential fish killers (Fresnel et al. 2001; Moestrup 2004). *Prymnesium parvum* and *P. patelliferum* are separated by small morphological differences, and studies based on growth characteristics, toxicity, and molecular data indicate that the two taxa belong to a single life cycle (Larsen 1999). Hence, *P. patelliferum* has been given the rank of form (*P. parvum* f. *patelliferum*). *Prymnesium saltans*, the type species of *Prymnesium*, has been suggested to be conspecific with *P. parvum*, but lack of material has prevented further studies (Moestrup and Thomsen 2003). *Prymnesium parvum* is widely distributed, extremely euryhaline, and very temperature tolerant. It blooms mainly in brackish or coastal waters, especially in nutrient-rich and relatively warmer waters (Edvardsen and Paasche 1998), and has been reported from both tropical and temperate localities (Edvardsen and Paasche 1998). *Prymnesium calathiferum* is known from New Zealand, Vietnam, eastern Africa and Martinique (Moestrup and Thomsen 2003), and it is probably a warm-water species of subtropical and tropical waters. *Prymnesium faveolatum* is presently known only from Malta, Greece, France, and Spain and *P. zebrinum* from three places in France (Fresnel et al. 2001). Studies on the biogeography of *Prymnesium* and other haptophytes are generally few and population studies are much needed.

The geographic distribution of *Chrysochromulina* is poorly known, since identification generally requires careful examination of the species-characteristic scales. Many species are very widely distributed (many probably cosmopolitan) but warm-water and cold-water species also exist. The harmful species *C. polylepis* does not thrive in warm water.

3.3.4 Raphidophyceans

The Raphidophyceae comprises some eight toxic species (*Chattonella antiqua, C. globosa, C. marina, C. ovata, C. subsalsa, C. verruculosa, Fibrocapsa japonica* and *Heterosigma akashiwo* (Moestrup 2004; Hiroishi et al. 2005). They are often difficult to recognize in preserved samples, which results in misidentifications. Caution is therefore needed before biogeographical conclusions are made. Raphidophyceans often co-occur and share ecophysiological features to the extent that Smayda (1998) discussed a "raphidophyte niche."

Heterosigma akashiwo is distributed in temperate and subtropical to tropical waters in both the northern and southern hemispheres (Smayda 1998; Imai and Itakura 1999, and references herein). The few records from tropical areas may be explained by cysts requiring low temperatures for survival (in temperate areas below 10 °C; Smayda 1998). The lack of records from cold waters agrees with the motile stages occurring mainly during the warmer seasons of temperate regions (Imai and Itakura 1999). Studies have revealed physiological and ecological differences between strains of *H. akashiwo* (Smayda 1998). However, data based on ITS rDNA showed that 18 of 19 strains collected from localities worldwide were identical, indicating that the species may have been spread in recent times (Connell 2000). Viable cysts of *H. akashiwo* are some of the most common cells in ships' ballast tanks (Connell 2000).

Toxic *Chattonella* species have had an enormous economic impact, particularly in Japan (Okaichi 1989). *Chattonella marina* has been reported from tropical, subtropical, and temperate waters worldwide and in winter, it occurs as cysts (Marshall and Hallegraeff 1999, and references herein). Physiological differences among strains from geographically diverse locations suggest the presence of ecophenotypes (Marshall and Hallegraeff 1999; Marshall and Newman 2002). *Chattonella marina* and *C. antiqua* are practically identical in part of the LSU and the ITS regions (99.99 % in ITS), to the extent that, combined with the morphological similarities, they may be conspecific (Connell 2000; Hirashita et al. 2000). Studies on sterols and fatty acids support conspecificity (Marshall et al. 2002). The genetic difference between *C. subsalsa* and *C. marina/C. antiqua* was found to be 9.8 %.

Species delineation of the remaining *Chattonella* species is uncertain and distribution records are sparse (Hallegraeff and Hara 2003). Studies combining morphological and molecular data are badly needed.

Fibrocapsa japonica has been observed in Pacific and Atlantic waters of both hemispheres, and it is probably distributed in temperate and subtropical waters worldwide (Kooistra et al. 2001). It has not been reported from tropical or polar waters, but may have been overlooked in the tropics. A high degree of similarity (only a few polymorphic sites) in ITS rDNA was found among 16 strains of *F. japonica* from geographically diverse locations (Kooistra et al. 2001). Intra-individual genetic polymorphism was found to increase with more recent isolation date, indicating recent dispersal of the species.

3.3.5 Cyanobacteria

Cyanobacteria are often considered to be distributed worldwide. Komàrek, probably the foremost expert on cyanobacteria, has reformulated this to 'cosmopolitan species exist only in those cases where the corresponding biotopes are widely distributed over the whole globe' (Komàrek and Anagnostidis 2005). The species concept is not always clear, and within proven genotypes, many eco- and morphotypes exist. Tropical regions in particular comprise many different biotopes, which have resulted in numerous specialized eco- and morphotypes, some of which are endemic.

The toxicity of the different morphotypes is diverse, and the same morphospecies may produce different toxins in different parts of the world, or they may be non-toxic. The cases mentioned below comprise two freshwater species. For other examples of cyanobacteria, see Chap. 2.

Cylindrospermopsis raciborskii is a freshwater species known to produce the alkaloid cylindrospermopsin, PSP toxins, and some unknown compounds. This species was long thought to be restricted to the tropics but it has now been found in many temperate regions. Material from the Americas, Europe, and Australia has been examined with modern techniques, confirming that all belong to the same species, although the morphology, at least in culture, is somewhat variable (Neilan et al. 2003). In 16S rDNA, the examined material was 99.1 % identical. However, when using a technique with higher resolution (HIP1 PCR), the material fell into three distinct groups, an Australian, a European, and an American (Neilan et al. 2003), supporting results from nifH sequencing (Dyble et al. 2002). Australian and European groups were closely related, and anthropogenic spreading of the species from Australia to Europe was hypothesized, explaining the apparently recent invasion of *C. raciborskii* to Europe (Neilan et al. 2003). The toxin profiles differed between the three populations: in Australia *C. raciborskii* produced cylindrospermopsin. The Brazilian strain produced only PSP toxins. The European strains formed none of these compounds, but exerted a toxic effect on mice. Neither the European nor the American strains possessed the genes involved in biosynthesis of cylindrospermopsin (Neilan et al. 2003). Subsequently, strains from Thailand and Japan were found to cluster with the Australian strains and also produced cylindrospermopsin (Chonudomkul et al. 2004). No geographical subdivision between the Japanese, Thai, or Australian strains was found; however, incorporation of more strains is needed to support the division into biogeographical populations.

Another toxic freshwater blue-green alga, *Anabaena circinalis*, is distributed worldwide in temperate and warmer waters. Morphological identification of *Anabaena* species is difficult, but molecular data support the morphological definition of *A. circinalis* (Fergusson and Saint 2000), which has been examined from widely different locations (Australia, Europe, Japan, and the USA). In 16S rDNA, the material agreed 98 %, indicating that it may belong to the same species. The clones were not phylogenetically separated into geographical areas but by their ability to form PSP toxins. The PSP toxic and the non-toxic clones formed two distinct gene clusters (Beltran and Neilan 2000), although both clusters contained exceptions.

In conclusion, although the adverse effects of harmful algae have resulted in much more research than in other groups of unicellular algae, our present knowledge on the geographical distribution leaves much to be desired. Information on genetic and morphological diversity and variation is needed for most harmful taxa, in particular cyanobacteria, raphidophyceans, and unarmored dinoflagellates.

References

Armbrust EV, Galindo HM (2001) Rapid evolution of a sexual reproduction gene in centric diatoms of the genus *Thalassiosira*. Appl Environ Microbiol 67:3501–3513

Beltran EC, Neilan BA (2000) Geographical segregation of the neurotoxin-producing cyanobacterium *Anabaena circinalis*. Appl Environ Microbiol 66:4468–4474

Bill B, Lundholm N, Connell L, Baugh KA, Trainer VL (2005) Domoic acid in *Pseudo-nitzschia cuspidata* from Washington State coastal waters. Abstract from the 3rd Symposium on Harmful Algae in the US. Monterey, CA. Oct. 2–7, 2005, p 77

Bolch CJS, Reynolds MJ (2002) Species resolution and global distribution of microreticulate dinoflagellate cysts. J Plankton Res 24:565–578

Brand LE (1981) Genetic variability in reproduction rates in marine phytoplankton populations. Evolution 35:1117–1127

Casamatta DA, Vis ML, Sheath RG (2003) Cryptic species in cyanobacterial systematics: a case study of *Phormidium retzii* (Oscillatoriales) using RAPD molecular markers and 16S rDNA sequence data. Aquat Bot 77:295–309

Cembella AD, Taylor FJR, Therriault J-C (1988) Cladistic analysis of electrophoretic variants within the toxic dinoflagellate genus *Protogonyaulax*. Bot Mar 31:39–51

Cerino F, Orsini L, Sarno D, Dell'Aversano C, Tartaglione L, Zingone A (2005) The alternation of different morphotypes in the seasonal cycle of the toxic diatom *Pseudo-nitzschia galaxiae*. Harmful Algae 4:33–48

Chinain M, Germain M, Sako Y, Pauillac S, Legrand AM (1997) Intraspecific variation in the dinoflagellate *Gambierdiscus toxicus* (Dinophyceae) I. Isozyme analysis. J Phycol 38:1106–1112

Chonudomkul D, Yongmanitchai W, Theeragool G, Kawachi M, Kasai F, Kaya K, Watanabe MM (2004) Morphology, genetic diversity, temperature tolerance and toxicity of *Cylindrospermopsis raciborskii* (Nostocales, Cyanobacteria) strains from Thailand and Japan. FEMS Microbiol Ecol 48:345–355

Coleman AW (2000) The significance of a coincidence between evolutionary landmarks found in mating affinity and a DNA sequence. Protist 151:1–9

Connell L (2000) Nuclear ITS region of the alga *Heterosigma akashiwo* (Chromophyta: Raphidophyceae) is identical in isolates from the Atlantic and Pacific basins. Mar Biol 136:953–960

Darling KF, Wade CM, Stewart IA, Kroon D, Dingle R, Brown AJL (2000) Molecular evidence for genetic mixing of arctic and antarctic subpolar populations of planktonic foraminifers. Nature 405:43–47

Davis JI (1997) Evolution, evidence and the role of species concepts in phylogenetics. Syst Bot 22:373–403

Dyble J, Paerl HW, Neilan BA (2002) Genetic characterization of *Cylindrospermopsis raciborskii* (Cyanobacteria) isolates from diverse geographic origins based on nifH and cpcBA-IGS nucleotide sequence analysis. Appl Environ Microbiol 68:2567–2571

Edvardsen B, Paasche E (1998) Bloom dynamics and physiology of *Prymnesium* and *Chrysochromulina*. In: Anderson DM, Cembella AD, Hallegraeff GM (eds) Physiological ecology of harmful algal blooms. NATO ASI Series 41. Springer, Berlin Heidelberg New York, pp 193–208

Evans KM, Bates SS, Medlin LK, Hayes PK (2004) Microsatellite marker development and genetic variation in the toxic marine diatom *Pseudo-nitzschia multiseries* (Bacillariophyceae). J Phycol 40:911–920

Evans KM, Kühn SF, Hayes PK (2005) High levels of genetic diversity and low levels of genetic differentiation in North Sea *Pseudo-nitzschia pungens* (Bacillariophyceae) populations. J Phycol 41:506–514

Fergusson KM, Saint CP (2000) Molecular phylogeny of *Anabaena circinalis* and its identification in environmental samples by PCR. Appl Environ Microbiol 66:4145–4148

Finlay BJ, Fenchel T (2004) Cosmopolitan metapopulations of free-living microbial eukaryotes. Protist 155:237–244

Foissner W (1999) Protist diversity: estimates of the near-imponderable. Protist 150:363–368

Fresnel J, Probert I, Billard C (2001) *Prymnesium faveolatum* sp. nov. (Prymnesiophyceae), a new toxic species from the Mediterranean Sea. Vie Millieu 51:89–97

Gonzalez MA, Coleman AW, Gómez PI, Montoya R (2001) Phylogenetic relationship among various strains of *Dunaliella* (Chlorophyceae) based on nuclear ITS rDNA sequences. J Phycol 37:604–611

Gosling, EM (1994) Speciation and wide-scale genetic differentiation. In: Beaumont AR (ed) Genetics and evolution of aquatic organisms. Chapman and Hall, London, pp 1–15

Hallegraeff GM, Hara Y (2003) Taxonomy of harmful marine raphidophytes. In: Hallegraeff GM, Anderson DM, Cembella AD (eds) Manual on harmful marine microalgae, vol 11, 2nd edn. IOC-UNESCO, Paris, pp 511–522

Hasle GR (2002) Are most of the domoic acid-producing species of the diatom genus *Pseudo-nitzschia* cosmopolites. Harmful Algae 1:137–146

Hasle GR, Lundholm N (2005) *Pseudo-nitzschia seriata* f. *obtusa* (Bacillariophyceae) raised in rank based on morphological, phylogenetic and distributional data. Phycologia 44:96–107

Hirashita T, Ichimi K, Montani S, Nomura M, Tajima S (2000) Molecular analysis of ribosomal RNA gene of red tide algae obtained from the Seto Inland Sea. Mar Biotechnol 2:267–273

Hiroishi S, Okada H, Imai I, Yoshida T (2005) High toxicity of the novel bloom-forming species *Chattonella ovata* (Raphidophyceae) to cultured fish. Harmful Algae 4:783–787

Imai I, Itakura S (1999) Importance of cysts in the population dynamics of the red tide flagellate *Heterosigma akashiwo* (Raphidophyceae). Mar Biol 133:755–762

Jensen MØ, Moestrup Ø (1997) Autecology of the toxic dinoflagellate *Alexandrium ostenfeldii*: life history and growth at different temperatures and salinities. Eur J Phycol 32:9–18

John U, Fensome RA, Medlin LK (2003) The application of a molecular clock based on molecular sequences and the fossil record to explain biogeographic distributions within the *Alexandrium tamarense* "species complex" (Dinophyceae). Mol Biol Evol 20:1015–1027

Kaczmarska I, LeGresley MM, Martin JL, Ehrman J (2005) Diversity of the diatom genus *Pseudo-nitzschia* Peragallo in the Quoddy region of the Bay of Fundy. Harmful Algae 4:1–19

Komàrek J, Anagnostidis K (2005) Cyanoprokaryota. 2. Teil: Oscillatoriales. Süsswasserflora von Mitteleuropa 19/2. Elsevier, München, 759 pp

Kooistra WHCF, Boer MK, Vrieling EG, Connell LB, Gieskes WWC (2001) Variation along ITS markers across strains of *Fibrocapsa japonica* (Raphidophyceae) suggests hybridisation events and recent range expansion. J Sea Res 46:213–222

Larsen A (1999) *Prymnesium parvum* and *P. patelliferum* (Haptophyta) – one species Phycologia 38:541–543

Lilly EL (2003) Phylogeny and biogeography of the toxic dinoflagellate *Alexandrium*. PhD Thesis, MIT and WHOI Joint Program in Oceanography/Applied Ocean Science and Engineering. 226 pp

Lundholm N, Moestrup Ø, Kotaki Y, Hoef-Emden K, Scholin C, Miller C (2006) Inter- and intraspecific variation of the *Pseudo-Nitzschia delicatissima* complex (Bacillariophyceae) illustrated by rRNA probes, morphological data and phylogenetic analyses. J Phycol 42:464–481

Lundholm N, Daugbjerg N, Moestrup Ø (2002) Phylogeny of the Bacillariaceae with emphasis on the genus *Pseudo-nitzschia* (Bacillariophyta) based on partial LSU rDNA. Eur J Phycol 37:115–134

Lundholm N, Moestrup Ø, Hasle GR, Hoef-Emden K (2003) A study of the *Pseudo-nitzschia pseudodelicatissima/cuspidata* complex (Bacillariophyceae): what is *P. pseudodelicatissima*? J Phycol 39:797–813

Mann DG (1999) The species concept in diatoms. Phycologia 38:437–495

Mann DG (2001) Freshwater algae: Taxonomy, biogeography and conservation. Phycologia 40:387–389

Marshall JA, Hallegraeff GM (1999) Comparative ecology of the harmful alga *Chattonella marina* (Raphidophyceae) from South Australia and Japanese waters. J Plankton Res 21:1809–1822

Marshall JA, Newman S (2002) Differences in photoprotective pigment production between Japanese and Australian strains of *Chattonella marina* (Raphidophyceae) J Exp Mar Biol Ecol 272:13–27

Marshall JA, Nichols PD, Hallegraeff GM (2002) Chemotaxonomic survey of sterols and fatty acids in six marine raphidophyte algae. J Appl Phyc 14:255–265

Moestrup Ø, Thomsen HA (2003) Taxonomy of toxic haptophytes (prymnesiophytes) In: Hallegraeff GM, Anderson DM, Cembella AD (eds) Manual on harmful marine microalgae, vol 11, 2nd edn. IOC-UNESCO, Paris, pp 433–464

Moestrup Ø (ed) (2004) IOC Taxonomic Reference List of Toxic Plankton Algae, Intergovernmental Oceanographic Commission of UNESCO; www.bi.ku.dk/ioc/default.asp

Montresor M, Lovejoy C, Orsini L, Procaccini G, Roy S (2003) Bipolar distribution of the cyst-forming dinoflagellate *Polarella glacialis*. Polar Biol 26:186–194

Neilan BA, Saker ML, Fastner J, Töröknes A, Burns BP (2003) Phylogeography of the invasive cyanobacterium *Cylindrospermopsis raciborskii*. Mol Ecol 12:133–140

Okaichi T (1989) Red tide problems in the Seto Inland Sea. In: Okaichi T, Anderson DM, Nemoto T (eds) Red tides: biology, environmental science and toxicology. Elsevier, New York, pp 137–142

Orsini L, Sarno D, Procaccini G, Poletti R, Dahlsmann J, Montresor M (2002) Toxic *Pseudo-nitzschia multistriata* (Bacillariophyceae) from the Gulf of Naples: morphology, toxin analysis and phylogenetic relationships with other *Pseudo-nitzschia* species. Eur J Phycol 37:247–257

Orsini L, Procaccini G, Sarno D, Montresor M (2004) Multiple rDNA ITS-types within the diatom *Pseudo-nitzschia delicatissima* (Bacillariophyceae) and their relative abundances across a spring bloom in the Gulf of Naples. Mar Ecol Prog Ser 271:87–98

Palumbi SR (1992) Marine speciation on a small planet. Trends Ecol Evol 7:114–118

Parsons ML, Scholin CA, Miller PE, Doucette GJ, Powell CL, Fryxell GA, Dortch Q, Soniat TM (1999) *Pseudo-nitzschia* species (Bacillariophyceae) in Louisiana coastal waters: molecular probes field trials, genetic variability, and domoic acid analyses. J Phycol 35:1368–1378

Pin LC, Teen LP, Ahmad A, Usup G (2001) Genetic diversity of *Ostreopsis ovata* (Dinophyceae) from Malaysia. Mar Biotechnol 3:246–255

Rynearson TA, Armbrust EV (2004) Genetic differentiation among population of the planktonic marine diatom *Ditylum brightwelli* (Bacillariophyceae). J Phycol 40:34–43

Semina HJ (1997) An outline of the geographical distribution of oceanic phytoplankton. Adv Mar Biol 32:528–563

Smayda TJ (1998) Ecophysiology and bloom dynamics of *Heterosigma akashiwo* (Raphidophyceae). In: Anderson DM, Cembella AD, Hallegraeff GM (eds) Physiological ecology of harmful algal blooms. NATO ASI Series 41. Springer, Berlin Heidelberg New York, pp 113–131

4 Importance of Life Cycles in the Ecology of Harmful Microalgae

K.A. Steidinger and E. Garccés

4.1 Introduction

Species survive today because they have adapted to their physical-geological-chemical-biological environment through space-time. A species continually adapts as it evolves, or it becomes extinct. The basis of this process in biological units is the ability to produce like forms, principally through sexual reproduction and inheritance of genetic traits that help anchor populations through time. The reason sexual reproduction is significant in plants, protists, or microorganisms, is that asexual reproduction can sustain lineages for only several hundred thousand years, but sexual reproduction and outcrossing can sustain lineages for millions of years (Grant 1981). Asexuals have reduced genetic diversity and are less successful in adapting to a changing environment; they are short-lived, evolutionarily (Holsinger 2000). Many harmful microalgae lines are evolutionarily old and therefore successful. Herring fishes also survive, even though populations can wax and wane and survival can involve an adaptive life history. Hjort (1914) theorized that herring in temperate waters are programmed to spawn and produce larvae at a time and place of abundant planktonic food. If the two pelagic links are not in synchrony in space-time, there will be a poor year-class. Cushing (1990) called this the match/mismatch hypothesis. The same concept may be applicable to life-stage transitions in microalgae, e.g., the transition from benthic resting cell to planktonic vegetative cell. Life cycles are the most rudimentary expression of a species and its genetic component is often the chain of evidence for its survival. The preceding is mentioned to emphasize the significance of life cycles, particularly sexual life cycles, in all biota. In eukaryotic organisms, sexual reproduction involves gametes (isogamous, anisogamous, oogamous), syngamy, and meiosis. In certain prokaryotic cyanobacteria, there is exchange of genetic material through conjugation. Alternations of life stages are a general feature of the algae with different ploidy stages occupying different

Ecological Studies, Vol. 189
Edna Granéli and Jefferson T. Turner (Eds.)
Ecology of Harmful Algae

niches in space-time. Some of these cycles are complex and polymorphic; many of them are poorly characterized and understood.

Reynolds and Smayda (1998) put forth models of planktonic dinoflagellate succession and assemblages based on adaptations evolved over time to survive environmental changes. The adaptive strategies were expressed as a C-S-R concept where the C adaptation was being a good competitor, the S adaptation was being able to tolerate stress and the R adaptation was the ability to tolerate disturbance. These adaptive strategies involve both *r* and K strategists in each type of adaptation. Smayda and Reynolds (2001, 2003) took this further and characterized nine life-form types and introduced "5 Rules of Assembly", whereby specific habitat conditions select for specific dinoflagellate life-form types based on abiotic factors such as nutrient availability, irradiance, and turbulence. Rule 5 of Smayda and Reynolds (2003) states, "...the selection of species within a given life form is stochastic, i.e., being at the right place at the right time and at suitable inoculum levels". This concept is at the base of this presentation when dinoflagellate life cycles are discussed; dinoflagellates dominate the number of harmful algae species. Diatoms on the other hand do not follow these rules presumably because to date it is thought that they have limited habitat selection. Pitcher (1990), however, presented evidence for diatom resting spores seeding upwelling systems in the southern Benguela Current. Some flagellates also select for habitat types, for example the prymnesiophyte *Phaeocystis*.

Harmful microalgae represent a small percentage of known microalgae, but each year more species are discovered that produce toxins. Some of these are new to science and may have been cryptic species, while others are already described but their toxic properties were not known. There are some subtle differences between HA and non-HA species, but there are more common denominators particularly regarding habitat exploitation and adaptations. Little is known of the adaptative life cycles. For example, of about 1,500 to 2,000 extant dinoflagellates (Taylor 1990; Gomez 2004) less than 100 species are known to produce toxins, and not all strains are toxic. Of the known toxic species, life cycles of less than half have been studied (Garcés et al. 2002, in Garcés et al. 2002a). The first described complete life cycle for a harmful microalgal species was Walker and Steidinger (1979, in Garcés et al. 2004) who described all the stages of *Alexandrium monilatum* from the vegetative cell to the hypnozygote, including the formation of gametes. They verified the culture observations by taking cysts from natural sediments and inducing excystment.

A species' ability to produce toxins does not directly translate to its ability to bloom. Although most currently known toxic species are bloom species, there are many non-toxic species that form blooms. This raises the issue of what a bloom is. Smayda (1997) raised the question of "what is a bloom?" and concluded it was not just a species biomass issue but that it should be framed in an ecological context and its harmful effects quantified. A bloom in the context of this paper is a higher-than-normal abundance above background levels that has some harmful effect. In the case of *Karenia brevis* in the eastern Gulf of Mexico, the concept of a bloom has to be framed by "where" the bloom is occurring, e.g., inshore or offshore because the blooms start offshore (>18 km) and gradually increase in biomass until there are fish kills offshore, either on the bottom or in the water column. Fish kills generally occur at about 100,000 cells L^{-1} (Quick and Henderson 1975; Roberts 1979) and this should be the bloom descriptor for offshore waters. For inshore waters at the mouths of bays, where recreational and commercial shellfish beds are regulated, a *K. brevis* bloom should be defined as cell counts above 5,000 L^{-1} because this is the level at which shellfish beds are closed to harvesting. Satellite detection of *K. brevis* surface blooms is at 100,000 cells L^{-1} (Tester et al. 1998), and this is a level that is useful in forecasting bloom movement. The species can be found at non-bloom periods at background levels of <1,000 cells L^{-1} (Steidinger 1975; Geesey and Tester 1993) down to 50 m. A bloom is a relative term where a species population increases over time and this increase leads to an observable or recordable effect. An effective concentration may be several hundred cells L^{-1} (*Dinophysis* DSP) or several thousand cells per liter (*Alexandrium* PSP). A rhetorical question at this point is "What is the probability of an encounter of two gametes of a HA species at a specific cell concentration" or are life cycle transitions density-dependent? The significant point here is that the "bloom," regardless of its biomass, goes through sequential developmental phases such as initiation, growth, maintenance, and dispersal/dissipation/termination. These different phases can act as "triggers" for the induction of transitions in life stages of a microalgal species and there may be feedback mechanisms such as cell density and "infochemicals".

4.2 Phases of Phytoplankton Bloom Development and Life Cycles

4.2.1 Initiation

Initiation is the first phase of a planktonic phytoplankton bloom development and it requires an inoculum of cells to seed a bloom. Without the inoculum there would be no increased population. The inoculum can be from several sources and may involve different life stages (Anderson and Morel 1979; Montresor 1992; Garcés et al. 1998). Smayda and Reynolds (2003) summarized blooms by presenting three categories: holoplanktonic, meroplanktonic and advected sources. Holoplanktonic infers that the source population exists all year long at low concentrations and has a wide temperature tolerance; it does not mean that there is not a sexual cycle or that the maintenance life stage is a vegetative cell. Meroplanktonic infers an alternation between planktonic and benthic forms. This alternation could be between dimorphic haploid planktonic stages and diploid benthic resting stages for dinoflagellates, or benthic temporary resting cells such as in some diatoms (Sicko-Goad et al. 1989; Mann 2002, in Garcés et al. 2002a). Many neritic and estuarine microalgae have meroplanktonic coupling of stages that involves asexual and sexual processes. The timing of other events such as dormancy and maturation can affect timing of blooms as modulated by abiotic factors such as light and temperature (Montresor 1992; Zingone et al. 2002, in Garcés et al. 2002a). Dormancy itself can last days (12 days for *Gymnodinium catenatum*, Moita and Amorin 2002, in Garcés et al. 2002a) to months or years and depend on the species and abiotic factors such as temperature. The third category, advected populations, actually may be a later stage of a bloom rather than an initial stage. It may be a higher-than-normal abundance of a species that has been passively concentrated at a boundary layer or discontinuity and advected to another area, such as an embayment. Zingone et al. (2002, in Garcés et al. 2002a) pointed out that there are three possible scenarios for cysts, i.e., they can germinate continually, sporadically, or seasonally. At the beginning of cyst formation, there can be many with little viability or there can be few with high viability. Although dormant cells can be in a resting state for years, viability of any cyst population can vary as can hatchability (Anderson 1998).

In the study of photosynthetic *Lingulodinium polyedrum* cultures, Figueroa and Bravo (2005a) found that this heterothallic dinoflagellate could produce two types of resting cysts, one an ecdysal sexual 2n stage and the other a spiny resting cyst or hypnozygote (2n). If the medium was phosphate depleted, spiny cysts resulted from syngamy, but if nutrients were replete, ecdysal cysts were produced. The ecdysal product took up to 72 h to germinate, and 24 to 48 h later, the germling produced two cells presumably by mitotic division. Hypnozygotes were dormant for 2–4 months and excystment depended on exogenous factors, as well as endogenous factors. Figueroa and Bravo (2005a) concluded that the ecdysal process took a sexual 2n product rapidly to a 1n product. This represents Type V in the terminology of Smayda and Reynolds (2001). In other studies, not all planozygotes in culture transition into hypnozygotes; some die, and some divide.

Most species are either homothallic or heterothallic, e.g., an advantage for homothallism relates to increased gamete encounters and reduced energy expenditure. Occasionally, a species has been documented to be both homothallic and heterothallic. Homothallism and heterothallism both involve sexual reproduction and could provide DNA repair. Probert (2002, in Garcés et al. 2002a) thought that HA species could select for homothallism for maintenance of populations if the habitat did not promote environmental concentrations of cells for sustaining blooms. In one study on heterothallic *Gymnodinium nolleri* (Figueroa and Bravo 2005b), an alternation of life stages proceeded from vegetative to gamete to planozygote, and from there the next product depended on nutritional status of the external medium. If nutrients were high, the 2n planozygote divided producing presumably two 1n cells; if however, nutrients were low, a hypnozygote resulted. Nutritional status of culture experiments can influence what path a species will follow.

Karenia mikimotoi is known to aggregate at pycnoclines in thin layers and constitutes a year-round population off the Bay of Biscay and at the perimeter of the Celtic Sea (Raine 2002, in Garcés et al. 2002a). It is thought that because small cells of *K. mikimotoi* have higher growth rates (1.0 division day^{-1} vs. 0.6 divisions day^{-1}, Gentien 2002, in Garcés et al. 2002a) that they contribute to blooms. This represents Type IV of Smayda and Reynolds (2001). Dinoflagellate "small cells" can represent gametes or asexual disproportionate daughter cells. In some cases, the small cells act as anisogametes and fuse with larger cells, or they can grow and mitotically divide, e.g., in *Dinophysis* (Reguera and Gonzalez-Gil 2001). In the case of the *Dinophysis* small cells, they have been identified as other species and are now known to be part of the life cycle (Reguera et al. 2004). This represents Type VII of Smayda and Reynolds (2001).

In diatoms, population size decreases until a threshold level for sexuality, e.g., gametogenesis, fertilization, and production of auxospores. Auxospores can re-seed the water column and lead to asexual division and high concentrations. Centric diatoms are oogamous while pennates are isogamous and can have non-flagellated gametes. Transition is controlled by cell size. The window of opportunity for sexual induction reported in heterothallic *Pseudo-nitzschia multiseries* in culture is 63 % (120 μm long for sexual induction and the largest vegetative cells are about 190 μm long (Davidovich and Bates 1998). Davidovich and Bates (2002) stated that sexuality in *Pseudo-nitzschia* in culture occurred in the exponential phase and that increased gamete production was associated with increased photoperiod length and speculated that cells needed additional energy to undergo sexuality. Another diatom strategy for re-establishing population biomass is forming thick-walled resting cells that settle to the bottom or accumulate at pycnoclines. These cells are non-zygotic. Induction of this resting stage appears to be related to nitrogen levels and light and perhaps predator avoidance (Mann 2002, in Garcés et al. 2002a). McQuoid and Hobson (1996), Montresor (1992), and Pitcher (1990) described other diatom life cycles. As with dinoflagellates and other microalgae, the information on whether bacteria influence sexual transitions is very limited.

4.2.2 Growth and Maintenance

Life cycles involve growth strategies to exploit different environments at different times. Different microalgal groups and different species have different adaptations. Typically during growth, the vegetative asexual haploid or diploid form of the microalga is the stage responding to environmental conditions while growing and increasing population biomass. In some cases, sexual stages can revert to an asexual stage and continue to reproduce asexually and therefore grow. Growth can be division of single cells or increases in colony size through cell division and polysaccharide production of a matrix. For single cells, growth can be measured over a range of <1 to >3 divisions per day, depending on the microalgal group and species. Typically, the smaller the cell volume the higher the cell division rate. However, within microalgal groups, in situ growth rates can vary by an order-of-magnitude, and even within species, differences are substantial (see Stolte and Garcés, Chap. 11). Cyanobacteria have two interesting alternatives for growth besides binary fission and division cysts; some genera produce heterocysts and others produce akinetes. Heterocysts fix atmospheric nitrogen and therefore are an adaptation to nutrient-poor waters whereas akinetes represent resting cells. Not much is known about akinete formation or germination. The point to be made is that experimental growth studies are not usually multifactorial for nutrients, availability of prey, photoperiod, light quality and quantity, temperature, salinity, turbulence and other environmental factors whereas in situ growth studies are. Unfortunately there are few in situ measurements available. In the case of HA heterotrophs that fed on prey, e.g., *Pfiesteria* and its close relatives, stage transitions can depend on food availability (Parrow and Burkholder 2003a, 2003b). Most taxa are characterized by an alternation between dormant/quiescent phases and growth phases. Growth stages are responsible for biomass increases while resting stages are capable of withstanding hostile environmental conditions. A heteromorphic life stage can represent an advantage, since it allows the allocation of the species biomass into stages of different size ranges, morphology, and survival-defense capabilities. These stages have markedly different physiological characteristics, i.e., single cells and colonies, flagellate and coccoid stages, planktonic and benthic.

Traditionally, temporary cysts of dinoflagellates have been considered nonmotile stages, formed by exposure of motile, vegetative cells to unfavorable conditions such as mechanical shock, or sudden changes in temperature. Studies on ecdysal cysts of *Pfiesteria piscicida* (Litaker et al. 2002), *Pseudopfiesteria shumwayae* (Parrow and Burkholder 2003b) and *Alexandrium taylori* (Garcés et al. 1998) have considered this stage as part of the asexual reproduction. However, there are studies on *Pfiesteria,* cryptoperidiniopsoids and *Lingulodinium polyedrum* that show temporary cysts as a product of the sexual cycle (Parrow and Burkholder 2004; Figueroa and Bravo 2005a), In the case of *A. taylori*, it exhibited a daily shift from a motile stage at the water surface to a non-motile stage, the ecdysal cyst, in the sediments. The production of ecdysal cysts could be advantageous by allowing a stock of the population to be stored in sediments (Garcés et al. 1998, 2002b). Moreover, these authors suggest that temporary cysts in the *Alexandrium* genus could be a means to avoid predation or attack by viruses, bacteria and parasites (Garcés 2002b). It has even been suggested that survival of temporary pellicle cysts of dinoflagellates in mussel and oyster feces have indicated that such processes may serve as a potential dispersion of the species (Laabir and Gentien 1999).

Phaeocystis globosa is a haptophyte that produces large (millimeters), hollow, buoyant gelatinous colonies of thousands of cells that produce bioactive compounds capable of deterring predators. Ungrazed monospecific *Phaeocystis* blooms can lead to marine mortalities due to low oxygen and increased viscosity conditions created from the blooms. *Phaeocystis* reaches high concentrations under nutrient-enriched conditions such as upwelling or eutrophication, but only after a light threshold has been reached. Nutrients also regulate the duration of blooms. Individual motile *Phaeocystis* need a substrate to attach to, wherein they become non-motile, divide and produce polysaccharides that constitute the gelatinous mass. Colonies can disintegrate (presumably under stress) and sink. *Phaeocystis* has four cell types but is not thought to produce bottom-resting stages. Triggers for transition between 2n and 1n cells are thought to be nutrient levels, irradiance, and turbulence. As in many other HA species, there is an alternating haploid-diploid life cycle. Raphidophytes, have motile and non-motile stages, and the non-motile stages can be associated with mucous and the benthos (see Lancelot et al. 2002 and Peperzak 2002 in Garcés et al. 2002a for a discussion on *Phaeocystis)*.

The growth phase may also be a sexual reproductive cue if there are infochemicals released that trigger gamete formation in a developing population, particularly if the growth phase is confined to a boundary, stratum, or water mass that provides a restricted medium for growth and accumulation. In coastal Argentina, *Alexandrium* cysts were found on the bottom below a current boundary offshore (Carreto et al. 1985) suggesting that at initial population growth from excystment there was a mechanism for gamete formation and encystment. In another area, Chesapeake Bay, USA, encystment of *Gymnodinium uncatenum* was speculated to be at the beginning of the bloom (Tyler et al. 1981). Vertical migration or distribution has also been shown to be a successful strategy in accessing potentially limiting nutrients. Maintenance and concentration are phases in bloom development that can be passive in that populations are physically aggregated and kept at boundaries even though they may be transported along boundaries within a circulation feature (Smayda and Reynolds 2001). In order for such aggregated increased biomass to be maintained, increased nutrient sources must be available and usable. All growth, whether it be divisions per day (μ) in HA and non-HA populations, needs to be balanced against losses as in other biotic yields. Loss factors in phytoplankton include cell lysis, grazing, sedimentation, and advection.

4.2.3 Dispersal/Dissipation/Termination

Different life stages and life-cycle adaptations allow a species to extend its tolerance of environmental conditions, its distribution, and its survival. Dispersal can be key to survival, but maintenance of populations within a key area, realm, or strata may help ensure survival for sexually reproducing populations. The integrity of a water mass, such as a gyre, a discontinuity layer, or waters downstream of upwelling event, for certain periods of time may be crucial to success. When that integrity is disrupted, dissipation and dilution of populations and their favorable environmental conditions can occur. Species–species interactions in a water mass can condition that water mass to favor one species or one group of microalgae over another, based on physiological efficiencies and tolerances (Walsh and Steidinger 2001; Mulholland et al. 2004).

Dilution of conditioning factor(s) can make the environment no longer conducive to growth and vegetative reproduction. Termination of blooms is often attributed to nutrient-depleted water, zooplankton predation, advection, and life-stage transitions, such as vegetative planktonic form to benthic resting form. When water masses are advected, whole blooms can be translocated. In some studies, bloom termination resulted from zooplankton grazing, prompting a transition from vegetative to resting stages, while other studies showed that exudates from zooplankton inhibited dinoflagellate excystment (Rengefors et al. 1998). Transition from planktonic to benthic stages can be regulated by endogenous or exogenous mechanisms. Cyst formation in *A. minutum* started in a period with high vegetative cell densities in the water column. Once production was initiated, encystment fluxes remained constant for 2 weeks, over the periods of maintenance and decline of the bloom (Garcés et al. 2004).

4.3 Environmental Factors versus Biological Factors Affecting Transition

Major transition points in the life cycle of most microalgae are the vegetative cell to resting cell, and the latter can be a temporary resting stage or a more-resistant dormant state. The dormant state can be a zygote with a variable dormancy period. Regulating environmental factors include temperature, salinity, light, nutrient availability, turbulence, and other stress variables. The first transition is a vegetative cell to a gamete, followed by gamete fusion and the production of a zygote. This involves syngamy and in haplont life cycles the 1n gametes fuse to form a 2n motile or non-motile zygote. There can be one stage or two-stage meiosis with resultant haplonts to begin the cycle again. Although nutrient stress, such as nitrogen or phosphorus or both, has been used to induce gamete formation in various microalgae, gamete formation occurs in nutrient-replete waters (Garcés et al. 2002a). Transmission electron microscopy of dinoflagellate cysts has revealed the ultrastructure of resting stages, including reduced membranes and organelle structure, reduced number of organelles, reduced nuclear organization and presumably reduced metabolism (Bibby and Dodge 1972; Kennaway and Lewis 2004). Ultrastructure studies have also revealed internal bacteria in either the motile cells or cysts of *Alexandrium*, and that association raises the question of the source of toxin production. Are toxins from the HA cell or bacteria? However, the toxicity of *A. tamarense* has been attributed to an inheritable character (LIFEHAB) and is associated with the light-induced G1 phase of growth (Garcés et al. 2002a).

More importantly in bloom chronology is the excystment or benthic to planktonic transition, whether it involves dinoflagellate hypnozygotes, diatom auxospores, chrysophyte spores, raphidophyte resting cells or some other benthic stage. Here there are clear endo- and exogenous modulators and regulators like temperature, light, oxygen, day length, available nutrients, and endogenous rhythms.

Biological modulators and regulators can be species–species interactions, release of infochemicals, and internal clocks or what are called endogenous rhythms. Matrai et al. (2005) have shown that excystment of *Alexandrium* populations from the eastern Gulf of Maine exhibited a circannual endogenous rhythm with an average period of 11 months. Dormancy of resting stages often cannot be broken, although it can be entrained and accelerated. This alone indicates self-regulation and internal-feedback mechanisms whether they are levels of reserves, or timed lapses in relation to specific-sensed variables. Another self-regulating mechanism may be planktonic cell density thresholds for gamete production.

4.4 Status of Knowledge and Direction Needed

As is apparent from the contents of this chapter, research on dinoflagellates dominates information on HA life cycles, and within the dinoflagellates much of the information relates to encystment and excystment. Little information is available on the life cycles of HA diatoms, prymnesiophytes, raphidophytes, pelagophytes and cyanophytes. In relation to sequential phases of bloom development, most information is on the initiation phase. In many cases, it is not known why blooms terminate, or how! The executive summary of Garcés et al. (2002a) outlines the major needs in HA life cycle research and monitoring. For example, one of the most fundamental goals is to have a clearer picture of species, mating systems, population dynamics, cause–effect relationships, and distributions. Descriptive models exist, but mathematical models for species population dynamics are few. In conclusion, Garcés et al. (2002a) stated that "Multiscale physical-chemical-biological interactions in life-stage transitions require that the scale of the relevant processes dictate the scale of sampling for environmental parameters".

One of the most productive paths to take would be to create meaningful monitoring programs of vegetative cells in the water column and resting stages in the benthos, whether they are asexual or sexual. These databases can be merged on a regional basis and made widely available. This is already underway for some regions. Some efforts to plot distributions in a georeferenced format are being recommended for global space-time patterns.

External and internal cues can trigger sexual transitions, or even asexual reproduction. But what are the genetics, the feedback loops, the infochemicals, the thresholds, etc., that provide for a gamete to revert from programmed fusion with another gamete, or a planozygote to revert from programmed meiosis since both these cells can, under certain conditions, proceed to divide asexually? Is it an on-off gene with regulator genes or another genetic switch? The answers to these questions can help answer variation in seasonality and succession, among others. It will also be important to know the different genotypes in resting pools, whether in the benthos or in identifiable strata or circulation features. If a species blooms in the spring and fall, are they both the same genotype? This is one area where the complimentary sciences of morphology and genetics can help fine-tune identification of morphotypes and genotypes of specific species. New techniques with molecular probes may help advance accurate identification of different strains within a species and different life cycle stages (see Litaker and Tester, Chap. 23).

Because of the potential influence of eutrophication on the occurrence and frequency of HABs, particularly in estuaries and nearshore areas, looking at resistant stages in marine sediments over vertical timelines may prove productive.

In the literature, there are various numbers given for an effective inoculum size to seed blooms, but how accurate are these estimates? Do they vary with different groups and different species? Remember Smayda and Reynolds' (2003) Rule of Assembly 5: "The selection of a species within a given life form is stochastic, i.e., being at the right place at the right time and at suitable inoculum levels". The fact that there are resting stages in sediments that can re-seed the water column does not guarantee successful inoculation or the fact that there are so many "overwintering cells" in the plankton does not guarantee that this condition alone will support a bloom. Inoculum size is a critical part of the equation.

Acknowledgements. Thanks to the European Commission FP 6, SEED projects (GOCE-CT-2005-003875) for financial support to Esther Garcés.

References

Anderson DM (1998) Physiology and bloom dynamics of toxic *Alexandrium* species, with emphasis on life cycle transitions. In: Anderson DM, Cembella AD, Hallegraeff GM (eds) Physiological ecology of harmful algal blooms. Springer, Berlin Heidelberg New York, pp 29–48

Anderson DM, Morel D (1979) The seeding of two red tide blooms by the germination of benthic *Gonyaulax tamarensis* hypnocysts. Est Coast Shelf Sci 8:279–293

Bibby BT, Dodge JD (1972) The encystment of a freshwater dinoflagellate: a light and electron microscopical study. Brit Phycol J 7:85–100

Carreto JI, Negri RM, Benavides HR, Akselman R (1985) Toxic dinoflagellate blooms in the Argentine Sea. In: Anderson DM, White AW, Baden DG (eds) Toxic dinoflagellates. Elsevier, New York, pp 147–152

Cushing DH (1990) Plankton production and year-class strength in fish populations: an update of the match/mismatch hypothesis. Adv Mar Biol 26:249–292

Davidovich NA, Bates SS (1998) Sexual reproduction in the pennate diatom *Pseudo-nitzschia multiseries* and *P. pseudodelicatissima* (Bacillariophyceae). J Phycol 34:126–137

Davidovich NA, Bates SS (2002) *Pseudo-nitzschia* life cycle and the sexual diversity of clones in diatom populations In: Garcés E , Zingone A, Dale B, Montresor M, Reguera B (eds) Proc of the LIFEHAB workshop: Life history of microalgal species causing harmful algal blooms. Office Publ Eur Comm, Luxembourg, pp 31–36

Figueroa RI, Bravo I (2005a) Sexual reproduction and two different encystment strategies of *Lingulodinium polyedrum* (Dinophyceae) in culture. J Phycol 41:370–379

Figueroa RI, Bravo I (2005b) A study of sexual reproduction and determination of mating type of *Gymnodinium nolleri* (Dinophyceae) in culture. J Phycol 41:74–83

Garcés E, Delgado M, Masó M, Camp J (1998) Life history and in situ growth rates of *Alexandrium taylori* (Dinophyceae, Pyrrophyta). J Phycol 34:880–887

Garcés E, Zingone A, Montresor M, Reguera B, Dale B (eds) (2002a) LIFEHAB: Life histories of microalgal species causing harmful blooms. Office Publ Eur Comm, Luxembourg, 208 pp

Garcés E, Masó M, Camp J (2002b) Role of temporary cysts in the population dynamics of *Alexandrium taylori* (Dinophyceae). J Plankton Res 24:681–686

Garcés E, Bravo I, Vila M, Figueroa RI, Masó M, Sampedro N (2004) Relationship between vegetative cells and cyst production during *Alexandrium minutum* bloom in Arenys de Mar harbour (NW Mediterranean). J Plankton Res 26:637–645

Geesey M, Tester PA (1993) *Gymnodinium breve*: ubiquitous in Gulf of Mexico waters? In: Smayda TJ, Shimizu Y (eds) Toxic phytoplankton blooms in the sea. Elsevier, Amsterdam, pp 251–255

Gomez F (2004) A list of free-living dinoflagellate species in the world's oceans. Acta Bot Croat 64:129–212

Grant V (1981) Plant speciation, 2nd edn. Columbia University Press, New York, 563 pp

Hjort J (1914) Fluctuations in the great fisheries of northern Europe viewed in the light of biological research. Rapp Reun Cons Int Explor Mer 20:1–13

Holsinger KE (2000) Reproductive systems and evolution in vascular plants. PNAS 97:7037–7042

Kennaway GL, Lewis JM (2004) An ultrastructural study of hypnozygotes of *Alexandrium* species (Dinophyceae). Phycologia 43:353–363

Laabir M, Gentien P (1999) Survival of toxic dinoflagellates after gut passage in the Pacific oyster *Crassostrea gigas* Thunburg. J Shellfish Res 18:217–222

Litaker RW, Vandersea MW, Kibler SR, Madden VJ, Noga EJ, Tester PA (2002) Life cycle of the heterotrophic dinoflagellate *Pfiesteria piscicida* (Dinophyceae). J Phycol 38:442–463

Matrai P, Thompson B, Keller M (2005) *Alexandrium* spp. from eastern Gulf of Maine populations: circannual excystment of resting cysts. Deep-Sea Res (in press)

McQuoid MR, Hobson LA (1996) Diatom resting stages. J Phycol 32:889–902

Montresor M (1992) Life histories in diatoms and dinoflagellates and their relevance in phytoplankton ecology. OEBALIA suppl XVII:241–257

Mulholland MR, Heil CA, Bronk DA, O'Neil JH, Bernhardt P (2004) Does nitrogen regeneration from the N2 fixing cyanobacteria *Trichodesmium* spp. fuel *Karenia brevis* blooms in the Gulf of Mexico? In: Steidinger KA, Landsberg JH, Tomas CR, Vargo GA (eds) Harmful Algae 2002. Florida Fish and Wildlife Conservation Comm, Florida Inst Oceanogr, IOC-UNESCO, St. Petersburg, USA, pp 47–49

Parrow MW, Burkholder JM (2003a) Estuarine heterotrophic cryptoperidiniopsoids (Dinophyceae): life cycle and culture studies. J Phycol 39:678–696

Parrow MW, Burkholder JM (2003b) Reproduction and sexuality in *Pfiesteria shumwayae* (Dinophyceae). J Phycol 39:697–711

Parrow MW, Burkholder JM (2004) The sexual life cycles of *Pfiesteria piscicida* and Cryptoperidiniopsoids (Dinophyceae). J Phycol 40:664–673

Pitcher G (1990) Phytoplankton seed populations off the Cape Peninsula upwelling plume, with particular reference to resting spores of *Chaetoceros* (Bacillariophyceae) and their role in seeding upwelling waters. Est Coast Shelf Sci 31:283–301

Quick JA Jr, Henderson GE (1975) Evidences of new "ichthyointoxicative" phenomena in *Gymnodinium breve* red tides. In: LoCicero VR (ed) Proc 1st Int Conf on Toxic Dinoflagellate Blooms, Mass Sci Tech Foun, Wakefield, Massachusetts, pp 413–422

Reguera B, González-Gil S (2001) Small cell and intermediate cell formation in species of *Dinophysis* (Dinophyceae, Dinophysiales). J Phycol 37:318–333

Reguera B, González-Gil S, Delgado M (2004) Formation of *Dinophysis dens* Pavillard and *D. diegensis* Kofoid from laboratory incubations of *Dinophysis acuta* Ehrenberg and *D. caudata* Saville-Kent. In: Steidinger KA, Landsberg JA, Tomas CR, Vargo GA (eds) Harmful algae 2002. Florida Fish and Wildlife Cons Comm, Florida Inst Oceanogr, IOC-UNESCO, St. Petersburg, USA, pp 440–442

Rengefors K, Karlsson I, Hansson LA (1998) Algal cyst dormancy: a temporary escape from herbivory. Proc Royal Lond Soc B 265:1353–1358

Reynolds CS, Smayda TJ (1998) Principles of species selection and community assembly in the phytoplankton: further explorations of the Mandala. In: Reguera B, Blanco J, Fernández ML, Wyatt T (eds) Harmful algae. Xunta de Galicia and IOC-UNESCO, Santiago de Compostela, pp 8–10

Roberts BS (1979) Occurrence of *Gymnodinium breve* red tides along the west and east coasts of Florida during 1976 and 1977. In: Taylor DL, Seliger HH (eds) Toxic dinoflagellate blooms. Elsevier-North Holland, pp 199–202

Sicko-Goad L, Stoermer EF, Kociolek JP (1989) Diatom resting cell rejuvenation and formation: time course, species records and distribution. J Plankton Res 11:375–389

Smayda TJ (1997) What is a bloom? A commentary. Limnol Oceanogr 42:1132–1136

Smayda TJ, Reynolds CS (2001) Community assembly in marine phytoplankton: application of recent models to harmful dinoflagellate blooms. J Plankton Res 23:447–461

Smayda TJ, Reynolds CS (2003) Strategies of marine dinoflagellate survival and some rules of assembly. J Sea Res 49:95–106

Steidinger K (1975) Basic factors influencing red tides. In: LoCicero VR (ed) Proceedings of the 1st international conference on toxic dinoflagellate blooms. Mass Sci Tech Found Wakefield, Massachusetts, USA, pp 153–162

Taylor FJR (1990) Phylum Dinoflagellata. In: Margulis L, Corliss JO, Melkonian M, Chapman DJ, McKhann HI (eds) Handbook of Protoctista. Jones and Bartlett, Boston, pp 419–437
Tester PA, Stumpf RP, Steidinger KA (1998) Ocean color imagery: what is the minimum detection level for *Gymnodinium breve* blooms? In: Reguera B, Blanco J, Fernandez ML, Wyatt T (eds) Harmful algae. Xunta de Galicia and IOC- UNESCO, Santiago de Compostela, pp 149–151
Tyler MA, Seliger HH (1981) Selection for a red tide organism: physiological responses to the physical environment. Limnol Oceanogr 26:310–324
Walsh JJ, Steidinger KA (2001) Saharan dust and Florida red tides: the cyanophyte connection. J Geophys Res 106:11597–11612

Part B
The Ecology of Major Harmful Algae Groups

5 The Ecology of Harmful Dinoflagellates

J.M. BURKHOLDER, R.V. AZANZA, and Y. SAKO

5.1 Introduction

Dinoflagellates are mostly estuarine and marine, with only ~250–300 of the ~2,000 known species inhabiting freshwaters (Graham and Wilcox 2000; Carty 2003). Most species are considered beneficial; dinoflagellates are dominant primary producers in the tropical and subtropical oceans, and are also abundant in late spring/summer plankton of temperate and subarctic seas, and in ice communities from the Antarctic to northern temperate lakes. About half of the known free-living species are exclusively heterotrophic (Gaines and Elbrächter 1987) and, thus, poorly fit the classical definition of algae. In fact, Graham and Wilcox (2000) referred to dinoflagellates as "fundamentally heterotrophic protists." Relatively few are harmful (~185 species; Smayda and Reynolds 2003), defined here as producing potent toxins (~60 estuarine and marine species; Burkholder 1998) and/or other bioactive substances that adversely affect beneficial organisms (Smayda 1997; Rengefors and Legrand 2001); causing disease or death of beneficial aquatic life by predation or parasitism; and/or causing undesirable changes in habitats. Although most harmful dinoflagellates are planktonic, free-living benthic species are important ciguatoxin producers, and various parasitic taxa also have a benthic habit, growing attached to or inside prey.

In this chapter, general ecological features of harmful dinoflagellates are reviewed, followed by consideration of frameworks to advance understanding. A caveat is merited: generalizations about a given species are often based upon the characteristics of one strain as representative of all strains of that species in nature. The complex reality, instead, is that major intraspecific variability is common in harmful dinoflagellates, as in other harmful algae (reviewed in Burkholder et al. 2001; Burkholder and Glibert 2005) – so much so that opposite interpretations presented as conclusions have been gained from considering only individual strains, to the detriment of ecological understanding. High intraspecific variability in life-history traits, behavior, growth, nutrition, blooms, toxicity, and genetics has been documented for

Ecological Studies, Vol. 189
Edna Granéli and Jefferson T. Turner (Eds.)
Ecology of Harmful Algae

many harmful dinoflagellates. Strain differences are a fundamental characteristic in the ecology of these species, and should be a critical consideration in forming interpretations.

5.2 General Ecology

The greatest diversity and abundance of harmful dinoflagellates occurs in estuaries and coastal marine waters, coinciding with higher nutrient supplies from land sources and/or upwelling (Taylor and Pollingher 1987; Graham and Wilcox 2000). These organisms tend to be large in comparison to other phytoplankton, averaging about 45 μm on the major cell axis (estimated from Hallegraeff 2002), although some can be as small as ~5–7 μm. Their size and higher cellular surface-to-volume ratios generally result in lower affinities for dissolved nutrients than smaller taxa, and may have selected for mixophagotrophy in photosynthetic species (Smayda 1997). Nevertheless, many species are capable of rapid growth. Review of the experimental literature revealed that contrary to previous generalizations, only 15% of larger free-living harmful species had growth rates >1.0 day^{-1}; the fastest reported, *Prorocentrum minimum*, was 3.54 day^{-1} (Smayda 1997). As a survival mechanism, some species rapidly form temporary cysts in response to sudden adverse conditions, as well as other types of cysts with resistant coverings as part of their sexual and asexual life histories (see synopsis in Parrow and Burkholder 2003). The larger size of many species (and in some cases, production of toxins and other bioactive substances) may deter various predators. Nevertheless, grazers are a major influence on the ecology of harmful dinoflagellates (Chaps. 20, 22, and 27, this book).

5.2.1 Motility

Harmful dinoflagellates exhibit directed motion in response to chemical stimuli, gravity, and light (the latter in photosynthetic species, which generally are attracted to low light and repelled by high light; Cullen and MacIntyre 1998; Carty 2003). Vertical migration, which involves geotaxis, a circadian rhythm, and chemosensory behavior, is exhibited by some photosynthetic species that move to shallower depths during the day and to deeper waters at night for nutrient acquisition and predator avoidance. This complex behavior depends on the species and environmental conditions. Remarkable distances (relative to the cell size) of up to 16 m day^{-1} can be traversed, with swimming velocities up to 1–2 m h^{-1} or more (280–560 μm s^{-1}; Eppley et al. 1968; Kamykowski et al. 1998). Certain heterotrophic species have shown the opposite directional behavior, moving to deeper waters during light periods for predator avoidance, and to shallow waters at night for prey acquisition. They also have demonstrated strong chemosensory responses to prey (Gaines and Elbrächter 1987; Burkholder et al. 2001).

5.2.2 Temperature, Light, Salinity and Turbulence

Harmful dinoflagellates are basically indistinguishable from other dinoflagellates from the same habitats in their responses to temperature, salinity, and light (Taylor and Pollingher 1987; Smayda 1997). Freshwater species have poor salinity tolerance, whereas estuarine taxa are euryhaline; some coastal marine toxigenic species can also tolerate a wide salinity range, at least in culture (Taylor and Pollingher 1987; Carty 2003). Photosynthetic taxa can rapidly adjust to fluctuating light; for example, they can photoadapt to low irradiance by increasing the size and/or the number of photosynthetic units (Smayda 1997).

The previous generalization that dinoflagellates are especially sensitive to (negatively affected by) turbulence (see Lewis and Hallett 1997; Estrada and Berdalet 1998) was based on tests with much higher turbulence than is found in most natural conditions. In recent research, three-dimensional turbulence fields were experimentally established that incorporated both spatial and temporal intermittence. The turbulence intensities used were within the range of natural conditions, and the study included a broad taxonomic range of harmful dinoflagellate species (different sizes, morphs). Cell numbers and net growth rates of the various dinoflagellates tested, including harmful taxa, were unaffected or stimulated by high turbulence ($\varepsilon \sim 10^{-4}$ m^2 s^{-3}, simulating moderate gale conditions), in comparison to cell numbers and net growth rates in low turbulence ($\varepsilon \sim 10^{-8}$ m^2 s^{-3}; Sullivan and Swift 2003). Kamykowski et al. (1998) suggested that small-scale turbulence most likely is not a major factor controlling bloom formation, except perhaps in surf zones and in shallow estuaries subjected to strong winds.

Turbulence can also affect dinoflagellate predator–prey interactions. Using natural levels over 24 to 48-h periods, Havskum et al. (2005) found that photosynthesis-supported growth of a strain of the mixotrophic dinoflagellate predator, *Fragilidium subglobosum*, was unaffected by any turbulence level tested. In contrast, at high turbulence ($\varepsilon \geq 0.05$ cm^2 s^{-3}), net population growth of the prey species *Ceratium tripos* decreased and its vertical distribution changed. At very high turbulence ($\varepsilon = 1$ cm^2 s^{-3}), *C. tripos* sank to the bottom of the water column, where it provided rich food patches for *F. subglobosum*. Thus, at low prey densities (5–8 *C. tripos* cells mL^{-1}), growth and ingestion rates of *F. subglobosum* were elevated at the highest turbulence level, and were comparable to growth and ingestion rates at high water-column prey densities ($\gg$10 *C. tripos* cells mL^{-1}).

5.2.3 Nutrition: the Continuum from Auxotrophy to Parasitism

Auxotrophy. Photosynthetic dinoflagellates are auxotrophic, requiring vitamins that are produced by other organisms (Graham and Wilcox 2000). They vary in dominant xanthophylls, mostly containing peridinin and also dinoxanthin and diadinoxanthin. Peridinin is three-dimensionally packaged with chlorophyll and protein to form an efficient "PCP" light-harvesting" complex (Hofmann et al. 1996). In contrast, fucoxanthin, an accessory pigment that absorbs light in the "green window" and enables its use by chlorophyll-*a* in photosynthesis, predominates in some toxigenic species that do not contain peridinin (e.g., *Karenia brevis, K. mikimotoi*, and *Karlodinium micrum*). Photosynthetic dinoflagellates, including harmful taxa, have form II of Rubisco (ribulose 1,5-bisphosphate carboxylase/oxygenase), found in anaerobic proteobacteria. Form II of Rubisco has much lower specificity for CO_2 over O_2 than the Rubisco I found in most photosynthetic protists (Palmer 1996). Thus, Rubisco II may enable survival across a wider range of habitats and stressful conditions, while a carbon-concentrating mechanism(s) may help to compensate for the lower efficiency of carbon fixation (Palmer 1996; Leggat et al. 1999).

Harmful auxotrophic dinoflagellates, like other photosynthetic organisms, tend to be limited primarily by the macronutrient phosphorus in freshwaters and by nitrogen in estuarine and marine waters, and they are capable of storing P and N (Graham and Wilcox 2000; Lomas and Glibert 2000). Organic as well as inorganic forms are significant sources of these nutrients (see Chapter 13 and 14, this book). Micronutrients, including forms of selenium and iron complexed with humic substances in particular, have also been shown to influence blooms in some harmful phototrophic dinoflagellates (e.g., Boyer and Brand 1998; Doblin et al. 2000).

Mixotrophy. It has been suggested that many, if not most, photosynthetic dinoflagellates are actually mixotrophic, relying upon a combination of phototrophy and phagotrophy (Granéli and Carlsson 1998; Stoecker 1999; see Chap. 14). Mixotrophy in harmful species began to be rigorously examined in the 1990s (Jacobson 1999). Apparently it occurs among certain free-living and parasitic photosynthetic dinoflagellates (reviewed in Cachon and Cachon 1987; Jacobson 1999; Stoecker 1999). From the opposite perspective, the heterotroph *Pfiesteria piscicida* can sometimes retain kleptochloroplasts from cryptomonad prey, which function for short periods (Lewitus et al. 1999a). Kleptoplastidic *P. piscicida* was also found to be capable of uptake of ^{15}N-nitrate (Lewitus et al. 1999b). Mixotrophy is often difficult to assess because some species have low feeding rates or feed only under certain conditions that are poorly simulated in culture; some feed rapidly (within fractions of a second) on certain prey; and other organelles can obscure food vacuoles (Stoecker 1999). Although mixotrophy occurs in most harmful dinoflagellates examined thus far, for most species the relative importance of photosynthesis, dissolved organic nutrients, and feeding are unknown. The evidence suggests that at least in some species, mixotrophy is important: for example, in culture *Karlodinium micrum* grew as a phototroph without prey, and its maximum growth rate was twice as high when given prey, although it was unable to grow in darkness with prey (Stoecker 1999).

Heterotrophy. Obligate heterotrophs represent the greatest void in present understanding about dinoflagellate ecology, in part because many of them have not been cultivable even in undefined media (Maestrini 1998). Some species have complex life histories as well, with benthic and planktonic forms that are sporadically manifested among strains that are difficult to relate as the same species (e.g., Popovsky and Pfiester 1990).

The classic work of Lessard and Swift (1985) demonstrated the potential ecological importance of heterotrophic dinoflagellates as predators in coastal food webs, and opened a new area of research, as most studies on dinoflagellate heterotrophy had focused on the autecological rather than ecosystem level (Jeong 1999). Even so, except for excellent early descriptions, the sparse autecological research that existed prior to the 1980s focused on osmotrophy (Jacobson 1999). In addition, until the mid-1990s it was thought that the theca in armored dinoflagellates was a barrier that minimized phagotrophy (Jacobson 1999). It is now known that an array of fascinating prey capture mechanisms enable heterotrophic dinoflagellates (including mixotrophs) to engulf prey that are similar in size or much larger (reviewed in Hansen and Calado 1999). Harmful free-living heterotrophs include certain toxigenic *Protoperidinium* and *Pfiesteria* spp., and parasitic taxa (below). Of the free-living taxa, "trapping-type" predators, *Protoperidinium* spp., use a pallium (feeding veil) to capture prey, whereas "searching-type" *Pfiesteria* spp. use an extended organelle (peduncle) to bore into prey cells and suction the contents, a process referred to as myzocytosis (Gaines and Elbrächter1987) (Fig. 5.1).

Heterotrophic dinoflagellates fall along a predator-parasite continuum (Coats 1999). About 150 species are parasitic, and many others have been encountered but not yet formally described (Coats 1999). Like some of the heterotrophic free-living dinoflagellates, the cryptic behavior and complex life cycles of parasitic dinoflagellates often render them difficult to characterize and, often, even difficult to detect or diagnose (Cachon and Cachon 1987). Most parasitic dinoflagellates are estuarine and marine, and believed to be cosmopolitan in distribution. Their prey range from various algae (including, in potentially beneficial actions, harmful dinoflagellates; e.g., Nishitani et al. 1985) to an array of other protists, invertebrates, and vertebrates; some are prey specialists, others prey generalists. While most are obligate heterotrophs, at least eight genera have one or more photosynthetic life-history stages, and some are believed to retain kleptochloroplasts. Photosynthesis in some parasitic species can supply up to half of the energy needed for growth (reviewed in Coats 1999).

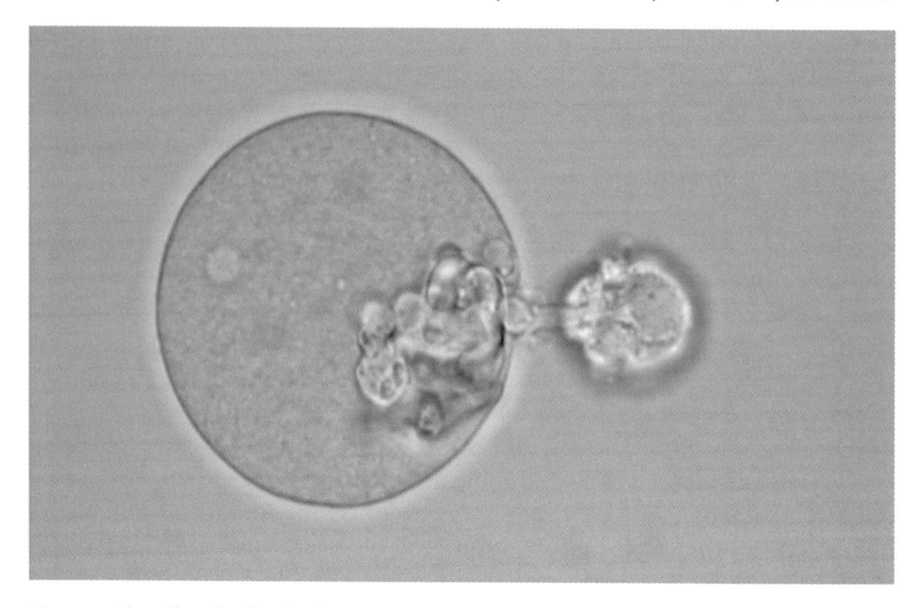

Fig. 5.1. Flagellated cell of *Pfiesteria shumwayae* (*right*) beginning to suction the contents from a spherical fish cell (*left*) using its peduncle, creating turbulence in the fish cell cytoplasm. *Scale bar* 10 μm. From Parrow and Burkholder (2003)

Most information on parasitic dinoflagellates consists of descriptions of their impacts; for most species, little is known about natural controls or favorable environmental factors that enable persistence during periods when they are not associated with their hosts. Thus far, few species have been shown to produce cysts (Coats 1999). Infections spread mostly by flagellated stages (dinospores, a term used for the zoospores of parasitic species). Parasitic species that infect crustaceans can have population-level effects, causing sexual castration and mortality of the hosts and destruction of the eggs. For example, *Ichthyodinium chabelardi* has caused mass mortality of sardines in the Mediterranean (Cachon and Cachon 1987). This species was described as the only parasitic dinoflagellate that significantly affects natural fish populations (Coats 1999), a generalization that may change as knowledge expands. Virulent strains of *Amyloodinium ocellatum* parasitize warm-water estuarine and marine fish, and are especially lethal to cultured fish in closed systems (reviewed in Coats 1999). *Piscinoodinium pillulare* similarly affects cultured freshwater fish. In coastal Alaska, *Hematodinium* sp. has caused increasing concern in "bitter crab" disease: It infects tanner crabs (*Chionoecetes bairdi*) during molt; once established it is 100 % lethal, and the meat becomes unmarketable before the crabs die. Similar impacts are caused along coastal Europe by *Hematodinium* infestations of velvet swimming crabs (*Necora puber*).

5.3 Blooms, Including Toxic Outbreaks

The term "bloom" is confusing when applied to harmful dinoflagellates. High-biomass nontoxic blooms in coastal marine waters have caused anoxia / hypoxia and mass death of finfish and shellfish. Perhaps the most famous was a bloom of *Ceratium tripos* in New York Bight in 1976, which was associated with prolonged bottom-water anoxia and fish/shellfish kills over 13,000 km^2 (Mahoney and Steimle 1979). Toxic photosynthetic dinoflagellates can form high-biomass blooms of up to 10^9 cells L^{-1} (up to 400–500 μg chlorophyll-*a* L^{-1} (Taylor and Pollingher 1987). More commonly, however, outbreaks of toxic species occur in low abundance and cannot be detected by water-column discoloration. The toxins can be concentrated when cells are filtered by shellfish (many dinoflagellate toxins), or when benthic populations are consumed and the toxins bioaccumulate through the food web to apex finfish predators (ciguatoxins) (reviewed by Hallegraeff 1993; Burkholder 1998). Dinoflagellate toxins are among the most potent biotoxins known; they can cause disease and death in shellfish and finfish as well as people who consume contaminated seafood (see Chap. 22, this book).

All toxigenic species thus far are estuarine or marine; most are mixotrophs that occur in tropical to warm-temperate waters, but some become abundant in cold temperate regions after the decline of spring diatom blooms. Photosynthetic species produce an array of saxitoxins and derivatives, brevetoxins, ciguatoxins, diarrhetic toxins, and other toxins. In addition, a small number of toxigenic species are heterotrophs (*Protoperidinium crassipes* – azaspiracid toxins, James et al 2003; *Pfiesteria* spp. – *Pfiesteria* toxin(s), Burkholder et al. 2005). Toxin production by freshwater taxa has not been verified. Bacteria may be important in toxin production, but few toxigenic dinoflagellate species have been tested for toxin in bacteria-free cultures. Based on limited information, in some cases bacteria can produce toxin (e.g., saxitoxins; Doucette et al. 1998); in others, toxin production can occur in the absence of bacteria, but is enhanced by their presence (Bates et al. 1995; Burkholder et al. 2005; and see Chap 19, this book).

Production of other bioactive substances (allelochemicals that influence intraspecific competition, and anti-predator allelopathic substances; Smayda 1997, and see Chap. 15, this book) appears to be much more common. This research area has focused on saltwater taxa, but a recent study by Rengefors and Legrand (2001) also showed allelochemical production by freshwater *Peridinium aciculiferum*. Toxins often can be vectored through the food web and result in indiscriminate die-offs of many fauna that are not predators. In contrast, allelochemicals are usually directly targeted (Smayda 1997). Thus, advantages can be discerned for production of allelochemicals, but toxins may or may not directly target predators (see Chap. 20, this book), and their potential metabolic/ecological benefit to the dinoflagellates or dinoflagellate/bacteria associations that produce them is often unclear.

The sets of factors that trigger bloom initiation are poorly known for most harmful dinoflagellates, in part because most blooms are sporadic and sufficient data on pre-bloom and initial conditions are usually not available (Walsh et al. 2001). Given the lack of such critical information, Cembella (1998a, p. 650) described the challenge of modeling efforts to predict blooms: "Attempts at dynamic modeling using conventional input parameters (nutrient uptake kinetics, grazing rates, specific growth rates, etc.)... [thus far] have been of little utility." Bloom maintenance, senescence, and dissipation are better understood, including influences of factors ranging from self-shading and intraspecific competition for nutrients, to aggregation/dispersal by currents and frontal patterns, to grazing and disease (Smayda 1997; Kamykowski et al. 1998). Factor interactions often are not well quantified for field conditions, however, and mixotrophy and intraspecific variability generally have not been considered in modeling. As another impediment, environmental and genetic controls on toxin production, by all toxigenic dinoflagellate species, are generally poorly known, inconsistent, and the subject of considerable debate (Plumley 1997; Cembella 1998b; Burkholder et al. 2005). Overall, progress continues to be made toward reliably predicting blooms, but the status of modeling harmful dinoflagellates remains as Franks (1997) earlier described it: The most general form of a coupled physical-biological model (with detailed boundary conditions, a turbulence-closure submodel, and a detailed biological model) has not yet been successfully created for any harmful dinoflagellate species.

5.4 Human Influences

Human activities may potentially affect harmful dinoflagellate populations in several major ways (Hallegraeff 1993; Burkholder 1998; Glibert et al. 2005; see Chaps. 26–30, this book): Increased turbidity from watershed development has depressed populations of benthic harmful dinoflagellates such as certain ciguatoxin producers, while in some planktonic systems flagellates have been favored over beneficial diatoms. Development has also increased supplies of macronutrients N and P in receiving waters, including organic forms such as urea (see Chap. 26), and has altered trace metal inputs and the character of humic acid chelators. These shifts have favored some harmful phototropic dinoflagellate species. In addition, certain free-living harmful heterotrophic dinoflagellates can be indirectly stimulated by nutrient pollution, through increased abundance of algal prey or other mechanism(s) (Yih and Coats 2000; Burkholder et al. 2001). Industrialization has accelerated warming trends that have begun to affect all geographic regions. Increased shipping and ballast water exchange has enhanced transport of species. Overfishing and degraded water quality have dramatically altered food webs and led, in some cases, to reduced pressure on harmful dinoflagellates via elimination/disease of predators. At the same time, increased finfish and shellfish aquaculture has enhanced opportunities for parasitic and toxigenic dinoflagellates, evidenced, for example, in toxic *Heterocapsa circularisquama* attacks on cultured shellfish in Japan. The lack of baseline information about ecological interactions of many bloom-forming species will make interpretations and predictions about their responses to human influences more difficult. In Southeast Asia, for example, *Pyrodinium bahamense* var. *compressum* has predominated among harmful algae in causing major public health and economic impacts, yet the ecological impacts of these blooms are virtually unknown (Azanza and Taylor 2001).

5.5 Conceptual Frameworks to Advance Understanding

Harmful dinoflagellates are capable of exploiting "a bewildering array of ecological niches, survival strategies, and nutritional modes" (Cembella 1998, p.650). It is a daunting challenge to bring some semblance of order and predictive capability, overall, to their behavior and blooms. Nevertheless, conceptual frameworks can provide valuable, fresh insights and testable hypotheses to strengthen understanding of the ecology of harmful dinoflagellates.

Such efforts were greatly advanced by the classic work of Margalef (1978), who suggested unifying principals defining the niche of harmful planktonic, bloom-forming marine dinoflagellate assemblages (Fig. 5.2). "Margalef's Mandala" considers nutrient conditions and turbulence intensity (mixing) as key factors controlling morphology and physiology of phytoplankton "life forms," and their seasonal succession along a gradient from *r* to *K* growth strategies. The model predicts that harmful dinoflagellate blooms occur in calm, nutrient-rich waters, in apparent contradiction with empirical data for some bloom species; and it considers all "red tides" as similar, regardless of the species or bloom habitat. Reynolds' (1987) "life form Intaglio" refined Margalef's axes and proposed three major adaptive strategies for freshwater phytoplankton: (C) Colonists (invasive small, rapidly growing *r*-selected species, dominant in chemically disturbed habitats); (S), nutrient Stress-tolerant species (large, slowly growing, *K*-selected); and (R) Ruderal species (tolerant of shear/stress forces in physically disturbed waters). Reynolds' model allows species selection within a wide ecological space, except in "void" habitats where nutrients and light are continuously deficient.

In recent seminal work, Smayda and Reynolds (2001, 2003) constructed a conceptual model to re-evaluate habitat preferences of bloom species of harmful planktonic dinoflagellates along an onshore-offshore gradient of decreasing nutrients, reduced mixing, and deepened euphotic zone, including

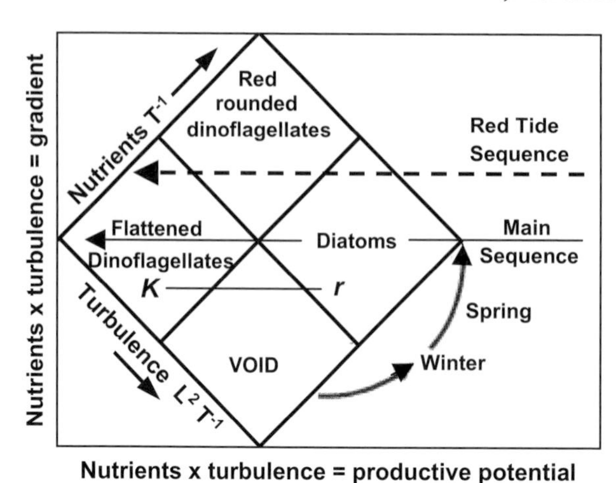

Fig. 5.2. Margalef's Mandala, including a trajectory for harmful planktonic dinoflagellate blooms ("red tides") considered collectively. From Margalef (1978)

consideration of life form traits (size, shape) and adaptive strategies. Smayda and Reynolds successfully applied Reynolds' model, earlier more so than Margalef's model, to harmful marine phytoplankton. Their analysis suggested that marine and freshwater phytoplankton species have similar adaptive strategies, and that the essential interaction in selection and succession of bloom species is more related to the degree of vertical microhabitat structural differentiation along the turbulence axis than a (presumed, in Margalef's model) correlation between degree of mixing and nutrient levels. Bloom dynamics were proposed to reflect two basic selection features, life-form and species-specific selection; life-form properties were viewed as overriding phylogenetic properties in selection; and selection of bloom species within a given life form was considered to follow "stochastic selection," characterized by a high degree of unpredictability. Harmful bloom-forming dinoflagellates sorted into nine categories along a "habitat template," with distinctive morphological and habitat preferences (Fig. 5.3). The model provides a framework for testing new hypotheses, a means of classifying harmful bloom-forming dinoflagellates on a functional basis, and a format for tracking and predicting compositional changes over seasonal mixing, vertical structure, and nutrient partitioning. Extension of this conceptual framework would assist in interpreting the spatial/temporal dynamics of benthic assemblages of harmful dinoflagellates as well.

Acknowledgements. The authors thank the editors for the invitation to write this chapter. JMB was funded by U.S. EPA and the North Carolina General Assembly.

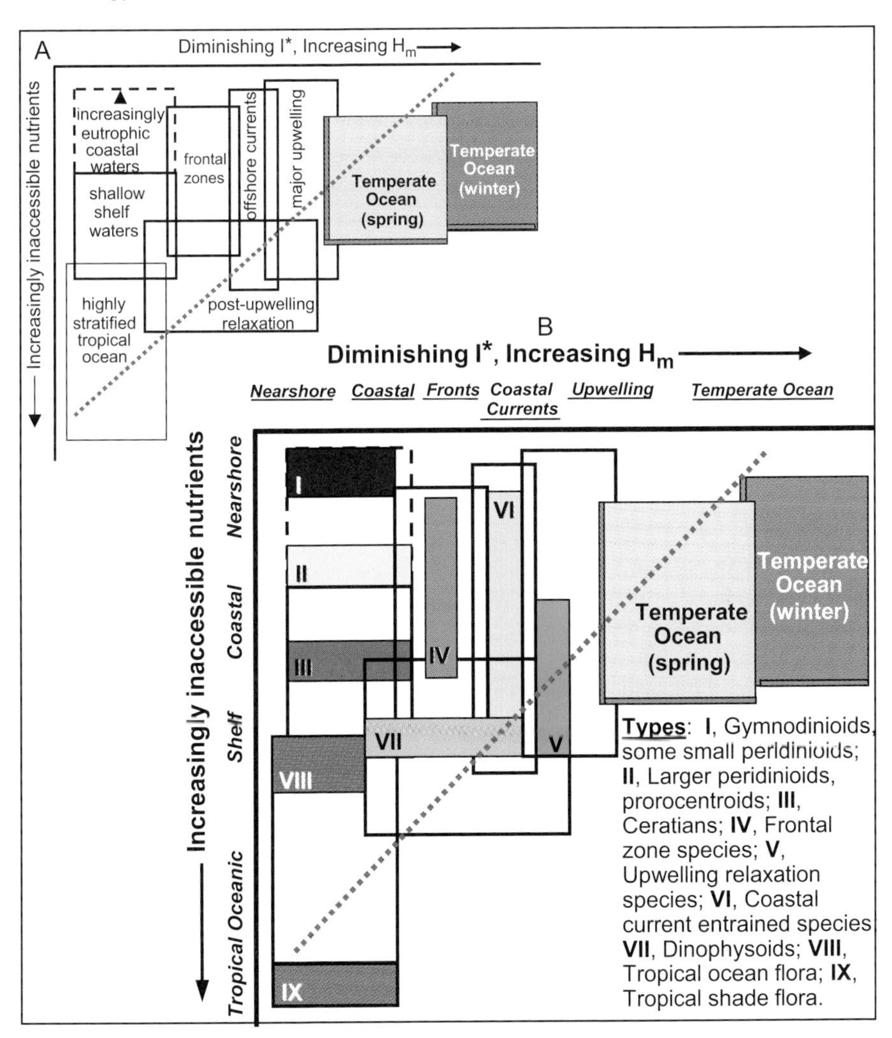

Fig. 5.3. Predominant dinoflagellate life form Types (I-IX) associated with the turbulence-nutrient matrix along an onshore-offshore continuum characterizing pelagic habitats (I^* the irradiance level received by cells; H_m depth of the mixed layer; diagonal line approximates the main *r*-to-*K* successional sequence depicted in Margalef et al. (1979). Overlap of the types does not always imply their contiguity (redrawn from Smayda and Reynolds 2003)

References

Azanza RV, Taylor FJRM (2001) Are *Pyrodinium* blooms in the Southeast Asian region recurring and spreading? A view at the end of the millennium. Ambio 30:356–364

Bates SS, Douglas DJ, Doucette GJ, Léger C (1995) Enhancement of domoic acid production by reintroducing bacteria to axenic cultures of the diatom *Pseudo-nitzschia multiseries*. Nat Toxins 3:428–435

Boyer GL, Brand LE (1998) Trace elements and harmful algal blooms. In: Anderson DM, Cembella AD, Hallegraeff GM (eds) Physiological ecology of harmful algal blooms. NATO ASI Series 41. Springer, Berlin Heidelberg New York, pp 489–508

Burkholder JM (1998) Implications of harmful microalgae and heterotrophic dinoflagellates in management of sustainable marine fisheries. Ecol Appl 8:S37–S62

Burkholder JM, Glasgow HB, Deamer-Melia NJ, Springer J, Parrow MW, Zhang C, Cancellieri P (2001) Species of the toxic *Pfiesteria* complex, and the importance of functional type in data interpretations. Environ Health Perspect 109:667–679

Burkholder JM, Gordon AS, Moeller PD, Law JM, Coyne KJ, Lewitus AJ, Ramsdell JS, Marshall HG, Deamer NJ, Cary SC, Kempton JW, Morton SL, Rublee PA (2005) Demonstration of toxicity to fish and mammalian cells by *Pfiesteria* species: comparison of assay methods and multiple strains. Proc Nat Acad Sci USA 102:3471–3476

Cachon J, Cachon M (1987) Parasitic dinoflagellates. In: Taylor FJR (ed) The biology of dinoflagellates. Botanical Monographs, vol 21. Blackwell Science Publications, Boston, pp 571–610

Carty S (2003) Dinoflagellates. In: Wehr JD, Sheath RG (eds) Freshwater algae of North America – ecology and classification. Academic Press, New York, pp 685–714

Cembella AD (1998a) Ecophysiology and metabolism of paralytic shellfish toxins in marine microalgae. In: Anderson DM, Cembella AD, Hallegraeff GM (eds) Physiological ecology of harmful algal blooms. NATO ASI Series 41. Springer, Berlin Heidelberg New York, pp 381–403

Cembella AD (1998b) Ecophysiological processes and mechanisms: towards common paradigms for harmful algal blooms. In: Anderson DM, Cembella AD, Hallegraeff GM (eds) Physiological ecology of harmful algal blooms. NATO ASI Series 41. Springer, Berlin Heidelberg New York, pp 649–652

Coats DW (1999) Parasitic life styles of marine dinoflagellates. J Eukaryot Microbiol 46:402–409

Cullen JJ, MacIntyre JG (1998) Behavior, physiology and the niche of depth-regulating phytoplankton. In: Anderson DM, Cembella AD, Hallegraeff GM (eds) Physiological ecology of harmful algal blooms. NATO ASI Series 41. Springer, Berlin Heidelberg New York, pp 559–580

Doblin MA, Blackburn SI, Hallegraeff GM (2000) Intraspecific variation in the selenium requirement of different geographic strains of the toxic dinoflagellate *Gymnodinium catenatum*. J Plankton Res 22:421–432

Doucette GJ, Kodama M, Franca S, Gallacher S (1998) Bacterial interactions with harmful algal bloom species: bloom ecology, toxigenesis, and cytology. In: Anderson DM, Cembella AD, Hallegraeff GM (eds) Physiological ecology of harmful algal blooms. NATO ASI Series 41. Springer, Berlin Heidelberg New York, pp 619–647

Eppley RW, Holm-Hansen O, Strickland JDH (1968) Some observations on the vertical migration of dinoflagellates. J Phycol 4:333–340

Estrada M, Berdalet E (1998) Effects of turbulence on phytoplankton. In: Anderson DM, Cembella AD, Hallegraeff GM (eds) Physiological ecology of harmful algal blooms. NATO ASI Series 41. Springer, Berlin Heidelberg New York, pp 601–618

Franks PJS (1997) Models of harmful algal blooms. Limnol Oceanogr 42:1273–1282

Gaines G, Elbrächter M (1987) Heterotrophic nutrition. In: Taylor FJR (ed) The biology of dinoflagellates. Botanical Monographs, vol 21. Blackwell Science Publications, Oxford, pp 224–268

Glibert PM, Seitzinger S, Heil CA, Burkholder JM, Parrow MW, Codispoti LA, Kelly V (2005) The role of eutrophication in the global proliferation of harmful algal blooms. Oceanography 18:198–209

Graham LE, Wilcox LW (2000) Algae. Prentice Hall, Upper Saddle River, NJ 640 pp

Granéli E, Carlsson P (1998) The ecological significance of phagotrophy in photosynthetic flagellates. In: Anderson DM, Cembella AD, Hallegraeff GM (eds) Physiological ecology of harmful algal blooms. NATO ASI Series 41. Springer, Berlin Heidelberg New York, pp 539–557

Hallegraeff GM (1993) A review of harmful algal blooms and their apparent global increase. Phycologia 32:79–99

Hallegraeff GM (2002) Aquaculturists' guide to harmful Australian microalgae, 2nd edn. School of Plant Science, Univ of Tasmania, Hobart, Tasmania, Australia, 136 pp

Hansen PJ, Calado AJ (1999) Phagotrophic mechanisms and prey selection in free-living dinoflagellates. J Eukaryot Microbiol 46:382–389

Havskum H, Hansen PJ, Berdalet E (2005) Effect of turbulence on sedimentation and net population growth of the dinoflagellate *Ceratium tripos* and interactions with its predator, *Fragilidium subglobosum*. Limnol Oceanogr 50:1543–1551

Hofmann E, Wrench PM, Sharples FP, Hiller RG, Welte W, Diederichs K (1996) Structural basis of light harvesting by carotenoids: Peridinin-chlorophyll-protein from *Amphidinium carterae*. Science 272:1788–1791

Jacobson DM (1999) A brief history of dinoflagellate feeding research. J Eukaryot Microbiol 46:376–381

James KJ, Moroney C, Roden C, Satake M, Yasumoto T, Lehane M, Furey A (2003) Ubiquitous 'benign' alga emerges as the cause of shellfisch contamination responsible for the human toxic syndrome, azaspiracid poisoning. Toxicon 41:145–151

Jeong HJ (1999) The ecological roles of heterotrophic dinoflagellates in marine planktonic community. J Eukaryot Microbiol 46:390–396

Kamykowski D, Yamazaki H, Yamazaki AK, Kirkpatrick GJ (1998) A comparison of how different orientation behaviors influence dinoflagellate trajectories and photoresponses in turbulent water columns. In: Anderson DM, Cembella AD, Hallegraeff GM (eds) Physiological ecology of harmful algal blooms. NATO ASI Series 41. Springer, Berlin Heidelberg New York, pp 581–599

Leggat W, Badger MR, Yellowlees D (1999) Evidence for an inorganic carbon-concentrating mechanism in the symbiotic dinoflagellate *Symbiodinium* sp. Plant Physiol 121:1247–1256

Lessard EJ, Swift E (1985) Species-specific grazing rates of heterotrophic dinoflagellates in oceanic waters, measured with a dual-label radioisotope technique. Mar Biol 87:289–296

Lewis J, Hallett R (1997) *Lingulodinium polyedrum* (*Gonyaulax polyedra*), a blooming dinoflagellate. Oceanogr Mar Biol Ann Rev 35:97–161

Lewitus AJ, Glasow HB, Burkholder JM (1999a) Kleptoplastidy in the toxic dinoflagellate, *Pfiesteria piscicida*. J Phycol 35:303–312

Lewitus AJ, Willis BM, Hayes KC, Burkholder JM, Glasgow HB, Glibert PM, Burke MK (1999b) Mixotrophy and nitrogen uptake by *Pfiesteria piscicida* (Dinophyceae). J Phycol 35:1430–1437

Lomas MW, Glibert PM (2000) Comparisons of nitrate uptake, storage, and reduction in marine diatoms and flagellates. J Phycol 36:903–913

Maestrini SY (1998) Bloom dynamics and ecophysiology of *Dinophysis* spp. In: Anderson DM, Cembella AD, Hallegraeff GM (eds) Physiological ecology of harmful algal blooms. NATO ASI Series 41. Springer, Berlin Heidelberg New York, pp 243–265

Mahoney J, Steimle FW Jr (1979) A mass mortality of marine animals associated with a bloom of *Ceratium tripos* in New York Bight. In: Taylor DL, Seliger HH (eds) Toxic dinoflagellate blooms. Elsevier, New York, pp 225–230

Margalef R (1978) Life-forms of phytoplankton as survival alternatives in an unstable environment. Oceanol Acta 1:493–509

Margalef R, Estrada M, Blasco D (1979) Functional morphology of organisms involved in red tides, as adapted to decaying turbulence. In: Taylor D, Seliger HH (eds) Toxic dinoflagellate blooms. Elsevier, New York, pp 88–94

Nishitani L, Erickson G, Chew KK (1985) Role of the parasitic dinoflagellate *Amoebophrya ceratii* in control of *Gonyaulax catenella* populations. In: Anderson DM, White AW, Baden DG (eds) Toxic dinoflagellates. Elsevier, New York, pp 225–230

Palmer JD (1996) Rubisco surprises in dinoflagellates. Plant Cell 8:343–345

Parrow MW, Burkholder JM (2003) Reproduction and sexuality in *Pfiesteria shumwayae* (Dinophyceae). J Phycol 39:697–711

Plumley FG (1997) Marine algal toxins: biochemistry, genetics, and molecular biology. Limnol Oceanogr 42:252–1264

Popovsky J, Pfiester LA (1990) Süßwasserflora von Mitteleuropa. Band 6: Dinophyceae (Dinoflagellida). Gustav Fischer Verlag, Jena, 272 pp

Rengefors K, Legrand C (2001) Toxicity in *Peridinium aciculiferum* – an adaptive strategy to outcompete other winter phytoplankton? Limnol Oceanogr 46:1990–1997

Reynolds CS (1987) Community organization in the freshwater plankton. Symp Br Ecol Soc 27:297–325

Smayda TJ (1997) Harmful algal blooms: their ecophysiology and general relevance to phytoplankton blooms in the sea. Limnol Oceanogr 42:1137–1153

Smayda TJ, Reynolds CS (2001) Community assembly in marine phytoplankton: application of recent models to harmful dinoflagellate blooms. J Plankton Res 23:447–461

Smayda TJ, Reynolds CS (2003) Strategies of marine dinoflagellate survival and some rules of assembly. J Sea Res 49:95–106

Stoecker DK (1999) Mixotrophy in dinoflagellates. J Eukaryot Microbiol 46:397–401

Sullivan JM, Swift E (2003) Effects of small-scale turbulence on net growth rate and size of ten species of marine dinoflagellates. J Phycol 39:38–94

Taylor FJR, Pahlinger U (1987) Ecology of dinoflagellates. In: Taylor FJR (ed) The biology of dinoflagellates. Botanical Monographs, vol 21. Blackwell Science Publications, Boston, pp 399–529

Walsh JJ, Penta B, Dieterle D, Bissett WP (2001) Predictive ecological modeling of harmful algal blooms. Hum Ecol Risk Assess 7:1369–1383

Yih W, Coats DW (2000) Infection of *Gymnodinium sanguineum* by the dinoflagellate *Amoebophrya* sp.: effect of nutrient environment on parasite generation time, reproduction, infectivity. J Eukaryot Microbiol 47:504–510

6 The Ecology of Harmful Flagellates Within Prymnesiophyceae and Raphidophyceae

B. EDVARDSEN and I. IMAI

6.1 Introduction

Some members of the classes Prymnesiophyceae and Raphidophyceae may form blooms resulting in fish kills and great economic losses. Harmful prymnesiophytes are found in the genera *Prymnesium*, *Chrysochromulina* and *Phaeocystis*, and harmful raphidophytes in the genera *Chattonella*, *Heterosigma* and *Fibrocapsa*. The aim of this chapter is to review the ecology and toxicity of harmful species within these genera.

6.2 Class Prymnesiophyceae (Division Haptophyta)

6.2.1 Taxonomy, Morphology and Life History

The predominantly marine division Haptophyta comprises at present ca. 300 species and includes mostly unicellular phototrophic organisms that typically have an appendage called haptonema, two smooth flagella and body scales that can be calcified or not. At present, about 60 species of *Chrysochromulina*, ten of *Prymnesium* and nine of *Phaeocystis* have been described. Their taxonomy and morphology were reviewed by Moestrup and Thomsen (2003). Phylogenetic analyses indicate that *Chrysochromulina* is paraphyletic and should be divided into two or more genera (e.g., Edvardsen et al. 2000).

A heteromorphic haplo-diploid life cycle is found in *Chrysochromulina polylepis*, *Prymnesium parvum* and *Phaeocystis* spp., as well as in several coccolithophorids (e.g., Houdan et al. 2004). In *C. polylepis*, three flagellated cell types (n or 2n) are joined in a life cycle (Edvardsen and Vaulot 1996 cited in Eschbach et al. 2005). The morphology, growth response to light and temperature, and toxin production may all be altered during the life cycle (Edvardsen

Ecological Studies, Vol. 189
Edna Granéli and Jefferson T. Turner (Eds.)
Ecology of Harmful Algae

and Paasche 1992; John et al. 2002). *P. parvum* and *P. patelliferum* were previously considered to be two separate species, but genetic and ploidy analyses showed that they have a common life cycle and the names were amended to *P. parvum* f. *parvum* and *P. parvum* f. *patelliferum* (Larsen 1999). The two forms differ in scale morphology, but not significantly in toxicity or growth preferences (see Larsen 1999). In *Phaeocystis globosa, P. pouchetii, P. antarctica* and *P. jahnii* a colonial stage with cells embedded in mucilage alternates with single celled stages. Five life cycle stages have been proposed for *P. globosa*: non-flagellated cells (solitary or colonial, 2n), macroflagellates (2n), mesoflagellates (n) and microflagellates (n) (e.g., Peperzak et al. 2000 and references therein).

6.2.2 Distribution and Abundance

Members of *Chrysochromulina* occur in all seas of the world, and most appear to have a worldwide distribution (Thomsen et al. 1994). Often, several *Chrysochromulina* species co-occur. More than 40 species have been recorded in Scandinavian coastal waters (Jensen 1998 cited in Moestrup and Thomsen 2003) and a similar number was found in Tasmanian waters (LeRoi and Hallegraeff 2004). *Chrysochromulina* normally occur in rather low cell numbers (10^3–10^5 cells L^{-1}, Thomsen et al. 1994), but concentrations up to 1–2x10^6 cells L^{-1} were recorded almost every year during a 14-year (1989–2002) survey in the Skagerrak, Norway (Dahl et al. 2005), and up to 6–9x10^6 cells L^{-1} between 1985 and 1987 in the Baltic Sea, Sweden (Hajdu et al. 1996). The peak season for *Chrysochromulina* spp. is from April to August in the Skagerrak and from May to October in the Baltic Sea. Blooms of *Chrysochromulina* species (ca. 3–100x10^6 cells L^{-1}) have up to the present only been reported from the North Atlantic and adjacent waters (Edvardsen and Paasche 1998). The toxic species *C. polylepis* appears to have a world-wide distribution, and blooms of this species were observed in 1988, 1989, 1994 and 1995, all in the Skagerrak or Kattegat.

Records of *Prymnesium* are restricted to temperate and tropical regions (Table 6.1). *Prymnesium parvum*, including the *patelliferum*-form, is by far the most common, and together with *P. calathiferum,* are the only species reported to form harmful blooms. Most records of *P. parvum* are from brackish inshore or coastal waters. Harmful *P. parvum* blooms have been reported from Europe, Israel, China, North America, Russia, Morocco and Australia (Guo et al. 1996; Edvardsen and Paasche 1998). Most blooms of *P. parvum* have occurred in lakes, ponds or lagoons limited in area, with standing water of low salinity (1–12 PSU) and temperatures between 12 and 30 °C. Rivers, lakes and reservoirs in Texas, with salinities up to 1 PSU have since 1981 experienced recurrent *P. parvum* blooms resulting in fish kills and extensive economic, ecological and recreational losses (www.tpwd.state.tx.us/hab/ga/).

Table 6.1. Distribution and toxicity to *Artemia* of *Prymnesium* species

Species*	Habitat	Distribution	Toxic
P. parvum	Brackish	Worldwide, temperate	Yes
P. annuliferum	Marine	France	Unknown
P. calathiferum	Marine	New Zealand, Viet Nam, E Africa, Martinique	Yes
P. faveolatum	Marine	Mediterranean Sea	Yes
P. nemamethecum	Marine, psammobic	S Africa, Australia, Denmark, Norway	Unknown
P. zebrinum	Marine	France	Yes

* Five additional species are reported in the literature, which have been studied under the light microscope only (*P. czosnowskii, P. gladiociliatum, P. minutum, P. papillarum* and *P. saltans*). Their status is uncertain and they may well be forms of *P. parvum*

The distribution and abundance of *Phaeocystis* was described by Moestrup and Thomsen (2003) and Schoemann et al. (2005). Flagellates of *Phaeocystis* spp. are cosmopolitan and may dominate the oceanic nanophytoplankton (Thomsen et al. 1994). Three *Phaeocystis* species produce blooms (>10^7 cells L^{-1}), which tend to be of the colonial stage: *P. antarctica, P. globosa* and *P. pouchetii*. *P. pouchetii* is a cold-water species, and occurs in arctic and temperate waters of the northern hemisphere. *P. globosa* occurs in temperate and tropical waters of both hemispheres, whereas *P. antarctica* is a cold-water species confined to Antarctic waters (e.g., Schoemann et al. 2005). Massive blooms of *Phaeocystis* have been observed in nutrient-rich, turbulent environments at all latitudes, but mainly in cold waters, and harmful effects of such blooms have been reported mainly from North Atlantic, New Zealand, and Antarctic waters (Moestrup 1994; Lancelot et al. 1998).

6.2.3 Autecology and Ecophysiology

The haptophytes *C. polylepis*, *P. parvum*, and *Phaeocystis* are all euryhaline and eurythermal, adapted to variable conditions. Maximum growth rates in cultures of *C. polylepis* and *P. parvum* are up to 1.5 div. per day (Edvardsen and Paasche 1998), which is an intermediate rate among microalgae. Optimal growth for *C. polylepis* from Norway was found at temperatures between 15 and 19 °C, salinities of 17–25 PSU and PFD of 100–200 μmol m^{-2} s^{-1}, conditions typical for temperate coastal waters during early summer (Edvardsen and Paasche 1992, 1998). *P. parvum* may grow at salinities between 0.8–100 PSU, and highest growth rates in cultures have usually been achieved at 10–20 PSU, at temperatures between 20–30 °C (Edvardsen and Paasche 1998). Cultures of *P. pouchetii* showed highest growth rate at 8 °C, but tolerated –2 to 14 °C. For *P. antarctica*, the growth optimum was at 4 °C, and it showed the same tolerance for temperature as *P. pouchetii*. *P. globosa* had a growth optimum of 15 °C but tolerated –0.6 to 22 °C (Baumann et al. 1994 cited in Schoemann et al. 2005).

Species of *Chrysochromulina* and *Prymnesium* have a requirement for the vitamins B_1 and B_{12}, the trace element selenium, and may take up dissolved organic material (Pintner and Provasoli 1968 cited in Jones et al. 1994). *P. parvum* and several *Chrysochromulina* species can ingest particles such as bacteria and microalgae, and thus are mixotrophic (Jones et al. 1994). The prey may be immobilized or killed by toxin(s) before ingestion (Tillmann 2003). A requirement for vitamin B_1 has been demonstrated for *Phaeocystis globosa* (Peperzak et al. 2000), but phagotrophy or osmotrophic uptake of organic carbon have not been observed in *Phaeocystis*. The colonial matrix of *Phaeocystis* consists of polysaccharids that can accumulate iron and manganese (Schoemann et al. 2005 and references therein).

6.2.4 Toxicity and Toxins

All *Prymnesium* species are suspected to be toxic to gill-breathing organisms (Table 6.1). *P. parvum* produces toxins with ichthyotoxic, cytotoxic, neurotoxic, antibacterial and allelopathic activity (e.g., Shilo 1971). The toxins act on biological membranes and the ichthyotoxic effect is assumed to be due to increased permeability in fish gills resulting in disturbed ion balance. Several different compounds have been ascribed the hemolytic effect of *P. parvum* such as proteolipids, glycolipids (hemolysins) and polyethers (prymnesin-1 and -2, e.g., Moestrup 1994; Morohashi et al. 2001). Factors found to activate or promote toxicity in *P. parvum* are phosphorus- and nitrogen-limited growth, the presence of a cofactor (e.g., divalent cations and streptomycin) and of cationic polyamines, light and strong aeration (e.g., Shilo 1971). The mechanisms underlying activation or promotion of toxicity, however, remain obscure.

Although several *Chrysochromulina* species have demonstrated harmful effects in nature or in the laboratory (Edvardsen and Paasche 1998; Moestrup and Thomsen 2003), only *C. polylepis* and *C. leadbeateri* have shown sustained toxicity. Several blooms of *Chrysochromulina* spp. have caused fish kills (Edvardsen and Paasche 1998), but the exceptional bloom of *C. polylepis* in the Skagerrak and Kattegat in 1988 also killed large populations of benthic organisms (Gjøsæter et al. 2000 cited in Dahl et al. 2005). The toxic effects of *C. polylepis* are similar to those of *P. parvum*, but chemical characterization of *C. polylepis* toxin(s) is still lacking. A glycolipid and a fatty acid from extracts of *C. polylepis* caused hemolytic activity (Yasumoto et al. 1990 cited in John et al. 2002). However, John et al. (2002) could not detect differences in lipid and fatty acid composition between a toxic and apparently non-toxic clone of *C. polylepis*. Simonsen and Moestrup (1997) found six hemolytic compounds in a crude extract of *C. polylepis*, but only one was toxic to *Artemia*. Seven additional haptophyte non-toxic strains showed hemolytic activity. This demonstrates that hemolytic activity is not always a good proxy for toxicity. Toxicity in *C. polylepis* increases under phosphorus- and nitrogen-deficiency and appears to be dependent on pH, salinity (Edvardsen and Paasche 1998, Chap. 18 in this book), cell cycle and light (Eschbach et al. 2005).

As with several other haptophytes, *Phaeocystis* produces dimethylsulfoniopropionate (DMSP) and converts it into dimethylsulfide (DMS) and acrylic acid. Blooms may affect global climate (DMS contributes to acidity and cloud condensation) and biogeochemical cycles of C and S. High-density blooms of *Phaeocystis* may also affect human activities such as fisheries (through net clogging and alteration of fish taste), fish farming, shellfish production (growth and reproduction) and recreational use of beaches through odorous foam (Schoemann et al. 2005 and references therein). Acrylic acid has an antibacterial activity, but this or DMSP/DMS are not considered toxic. However, toxic activity of *P. pouchetii* to cod larvae has been demonstrated. A polyunsaturated aldehyde was isolated from a culture filtrate of *P. pouchetii* and proposed to have caused the observed cytotoxicity (Hansen et al. 2004). *P. pouchetii* is thus far the only species in the genus known to be toxic, but *P. globosa* and *P. antarctica* are also considered potentially harmful.

6.2.5 Ecological Strategies

Members of *Chrysochromulina* and *Prymnesium* may produce and excrete compounds that reduce grazing from zooplankton (e.g., John et al. 2002; Granéli and Johansson 2003) and reduce or obstruct growth in other phytoplankton (allelopathy, e.g., Legrand et al. 2003). When toxicity is low protozoan or mesozooplankton grazing are assumed to control populations of these haptophytes. However, when sufficient amounts of toxins are produced and released into the water, they may act as a chemical defense to repel grazers (Granéli and Johansson 2003; Tillmann 2003).

Fourteen-year time-series data from southern Norway indicated that high spring N:P ratios, freshwater influence, stratified conditions with a low-salinity surface layer, together with low phosphate concentrations during summer were associated with favorable conditions for *Chrysochromulina* spp. (Dahl et al. 2005). High N:P ratios (>20:1) may eventually lead to phosphorous deficiency, which promote toxicity in *C. polylepis*. Efficient nutrient uptake due to small size, motility, mixotrophic ability, reduced grazing, and increased availability of nutrients due to reduced competition may promote its dominance.

Fish kills due to blooms of *Prymnesium* have usually occurred only at very high algal concentrations (>50–100x10^6 cells L^{-1}), requiring high amounts of nutrients, and many of the affected waters are clearly eutrophic (Edvardsen and Paasche 1998). In order to reduce harmful effects of *Prymnesium* blooms, it is recommended to reduce the effluents of both phosphorus and nitrogen and prevent conditions that increase toxin production. In fish ponds, measures to keep salinity below 1–2 PSU are also recommended (Guo et al. 1996).

The success of *Phaeocystis* has been attributed to the alternation between the colonial and flagellated stages. The matrix in the colonial stage may protect the embedded cells from grazers, pathogen infection, etc., while keeping the cells buoyant in turbulent waters and high light conditions, and may serve as an energy and nutrient reservoir (Schoemann et al. 2005, and references therein). The flagellated cells appear to be adapted to oligotrophic environments, and may allow for genetic recombination. The endogenous and environmental factors that direct the transition between life cycle stages of haptophytes in nature are poorly known, but this information is needed in order to understand bloom formation and survival.

6.3 Class Raphidophyceae (Division Heterokontophyta)

6.3.1 Taxonomy, Morphology and Life History

Marine species of Raphidophyceae belong to the genera *Chattonella*, *Fibrocapsa*, *Heterosigma*, *Haramonas*, *Olisthodiscus* and *Oltmannsia*. The first three contain notorious fish-killing species such as *C. antiqua*, *C. marina*, *C. ovata*, *C. verruculosa*, *F. japonica* and *H. akashiwo* (Fukuyo 2004; Hiroishi et al. 2005). These are unicellular, naked flagellates possessing a forwardly directed flagellum with tubular hairs and a smooth trailing flagellum. The cells are golden-brown to brown, somewhat dorsi-ventrally flattened and contain numerous ejectosomes, trichocysts, or mucocysts that readily discharge. Most pelagic blooms attributed to the sand-dwelling species *O. luteus* are almost certainly of *H. akashiwo* (Smayda 1998). Analyses of the D1/D2 domains of LSU rDNA showed that *C. antiqua*, *C. marina* and *C. ovata* cannot be separated into distinct subgroups (Hirashita et al. 2000; Sako pers. comm.), and that *C. verruculosa* not is a raphidophyte, but belongs to the class Dictyochophyceae (Edvardsen et al. unpubl.). There is thus a strong need for a taxonomic revision of *Chattonella* spp.

Raphidophyte species normally multiply with longitudinal binary cell fission. *C. antiqua* and *C. marina* generally divide once during a dark period, and *H. akashiwo* twice or more. Cysts have been identified in *C. antiqua*, *C. marina*, *C. ovata*, *F. japonica* and *H. akashiwo* from sediments of coastal waters of Japan (e.g., Imai et al. 1993, 1998; Yamaguchi et al. unpubl.). There is no distinct difference in morphology between the cysts of *C. antiqua* and *C. marina* (Imai et al. 1998). A stage of "over-summering" cysts is suspected in *C. verruculosa* because it is unable to survive at temperatures ≥ 25 °C (Yamaguchi et al. 1997).

A diplontic life cycle was discovered in *C. antiqua* and *C. marina* when the nuclear DNA content was determined in various life cycle stages (Yamaguchi and Imai 1994 cited in Imai et al. 1998). Vegetative cells of both species were found to be diploid and pre-encystment small cells and cysts were haploid. In *H. akashiwo*, the DNA content of cysts was smaller than that of vegetative cells (Imai et al. 1993), suggesting a diplontic life cycle also in this species.

6.3.2 Distribution and Abundance

The distributions of *Chattonella, Fibrocapsa* and *Heterosigma* are shown in Fig. 6.1. The first massive bloom of *Chattonella* (*C. marina*) accompanied by mass mortality of fish was reported from Malabar Coast, India (Subrahmanyan 1954 cited in Imai et al. 1998). Red tides of *Chattonella* causing massive fish kills have been recorded thereafter in Japan, China, USA (Florida), and South Australia. *Chattonella* spp. has also been observed in e.g., Southeast

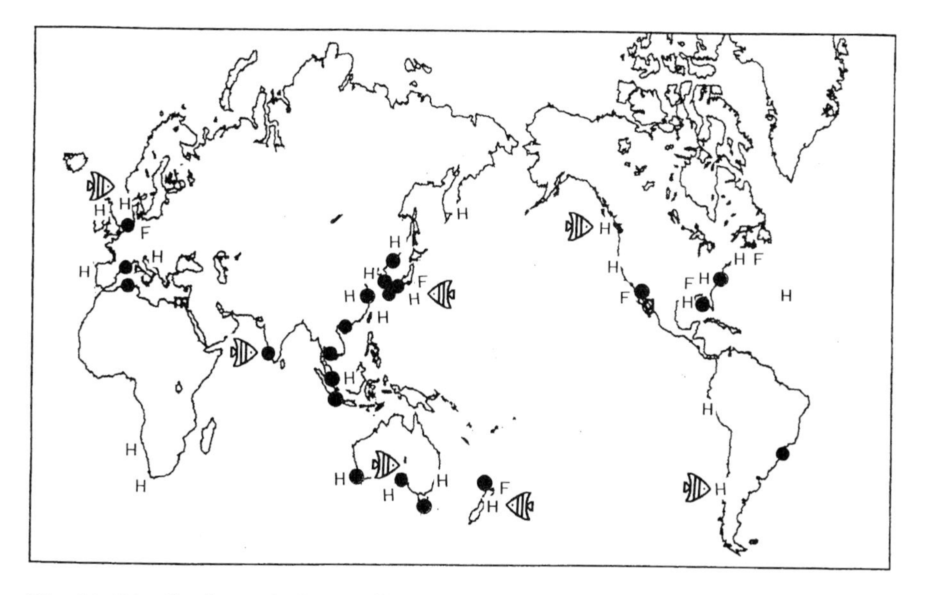

Fig. 6.1. Distributions of *Chattonella* spp. (*C. antiqua, C. marina, C. ovata,* and *C. subsalsa,* closed circles), *Heterosigma akashiwo* (*H*) and *Fibrocapsa japonica* (*F*). Fish icons indicate sites where mass mortalities of farmed fish have occurred except for India. Modified from Smayda (1998)

Asia, New Zealand, Brazil, and Europe (North Sea). Red tides of *H. akashiwo* accompanied by fish kills of salmon and yellowtail have occurred in e.g., Japan, Canada (British Columbia), New Zealand, Chile, and Scotland. The distribution of this species is world-wide, excluding arctic areas. Blooms of *F. japonica* causing fish kills have been observed in Japan, and it has been recorded in the North Sea, New Zealand, and the USA. The highest reported cell densities are 10^4 cells mL^{-1} for *Chattonella* spp. and *F. japonica*, and 10^5 cells mL^{-1} for *H. akashiwo*. *C. antiqua* and *C. marina* may kill fish at cell densities of 10^2 cells mL^{-1}. A noteworthy feature of raphidophyte blooms is the frequent co-occurrence of *H. akashiwo*, *F. japonica* and one or more *Chattonella* spp. (Smayda 1998).

6.3.3 Autecology and Ecophysiology

Important factors affecting the growth of raphidophytes are light intensity, temperature, salinity and nutrients. *C. antiqua* and *C. marina* may grow at light intensities ≥ 30 µmol m^{-2} s^{-1}, and obtain growth saturation at 110 µmol m^{-2} s^{-1}. In Japanese coastal waters such as the Seto Inland Sea, *C. antiqua* and *C. marina* are common during summer. They may grow at temperatures of 15–30 °C and salinities 10–35 PSU, and optimal conditions are at 25 °C and 25 PSU for *C. antiqua*, and at 25 °C and 20 PSU for *C. marina* (Yamaguchi et al. 1991). *C. antiqua* and *C. marina* cannot survive temperatures below 10 °C, and over-winter as cysts. *C. antiqua* uses nitrate, ammonium and urea, but not amino acids (glycine, alanine and glutamate) as nitrogen sources, and phosphate as the phosphorus source. Iron and vitamin B_{12} are essential (e.g., Nishijima and Hata 1986).

H. akashiwo can grow at low light intensities (≥ 9 µmol m^{-2} s^{-1}), and obtains growth saturation at 108 µmol m^{-2} s^{-1} (Smayda 1998). *H. akashiwo* is eurythermal and euryhaline and can grow in the salinity range of 2–50 PSU and a temperature range of 5–30 °C, and grows optimally between 9 and 31 PSU and 15–26 °C (Smayda 1998). Red tides of *H. akashiwo* have occurred at temperatures ≥ 15 °C. As micronutrients, Fe, Mn, and vitamin B_{12} are necessary (Smayda 1998). Cells of *H. akashiwo* and *Chattonella* spp. are enveloped by glycocalyx. The function of this mucus cover is unclear, but it is suggested to have an affinity to iron (Honjo 1994 in Smayda 1998).

For *F. japonica* (Iwasaki 1973), reported growth enhancements by Fe, Mn, purines, pyrimidines, plant hormones and vitamin B_{12} and a preference for ammonium and urea over nitrate as a nitrogen source. Investigations on the effects of light intensity, temperature, salinity on the growth of *F. japonica* are scarce.

6.3.4 Toxicity

The mechanism by which *Chattonella* spp. kill fish remains unclear, but suffocation due to gill tissue damage is the ultimate cause of fish death. Neurotoxins and free fatty acids have been suggested to be the responsible compounds. A recent hypothesis is that harmful raphidophytes generate reactive oxygen species (ROS, e.g., superoxide), which are responsible for the gill tissue injury and mucus production that leads to fish death (e.g., Oda et al. 1997). Marshall et al. (2003) proposed a synergistic role of ROS and free fatty acids during fish death.

Heterosigma akashiwo has also been reported to produce ROS. Twiner et al. (2005) reported that extracellular organic compounds from cultures of *H. akashiwo* acutely increase the level of cytosolic calcium by inhibiting the plasma membrane Ca^{2+}-ATPase transporter and ultimately induce apoptotic cell death. Hence, these excreted organic compounds may play a significant role in the ichthyotoxic behavior of *H. akashiwo*.

6.3.5 Ecological Strategies

Cyst formation has been observed in cultures of *C. marina* and *C. antiqua* (Imai et al. 1998 and references therein), but germination has been confirmed in *C. marina* only. Nitrogen limitation induced the formation of pre-encystment small cells, and cyst formation was successfully completed under low light ($\leq$ 15 μmol m^{-2} s^{-1}) or darkness. *C. ovata* also formed pre-encystment small cells in culture, but the completion of encystment has been unsuccessful. In *H. akashiwo*, encystment was observed in a natural population at the final stage of a red tide (Itakura et al. 1996). Nutrient limitation might be an inducing factor. Cyst formation has been unsuccessful in cultures of *H. akashiwo*, but Tomas (1978 cited in Smayda 1998) reported on formation of non-motile vegetative cells agglutinated into plasmodial masses with tolerance to low temperature. There is no information on cyst formation in cultures of *F. japonica*.

The cysts settle to the bottom sediments to over-winter and thereby ensure the persistent existence in the same area, and the germination of cysts provides the inoculum for blooms to overlying waters in summer. The physiology of cysts of *Chattonella* is significantly affected by seasonal fluctuations of temperature in temperate coastal areas such as the Seto Inland Sea (Imai et al. 1998).

The annual life cycle of *Chattonella* has been described by Imai et al. (1998). In the Seto Inland Sea, vegetative cells of *C. antiqua* and *C. marina* generally appear from June to September. These vegetative cells originate from the germination of cysts in the bottom sediments in early summer when the bottom temperature reaches an adequate level of ca. 20 °C. The cells multiply asexually during the summer and produce pre-encystment small cells at unfavorable conditions such as at nutrient depletion. The small cells then sink to the sea bottom, where the encystment is completed after the adhesion to solid surfaces such as diatom frustules and sand grains. The cysts spend a period of spontaneous dormancy there until the following spring, and never germinate in autumn. In spring, most cysts complete the period of spontaneous dormancy and acquire the ability to germinate. However, they must first undergo a post dormancy period (enforced dormancy) because the temperature is too low for germination. Vegetative cells thereafter appear in the overlying water column by excystment. Most of the cyst populations, however, lie buried and remain in the sediments without germination, and survive from one year to another via a secondary dormancy. As *C. antiqua* and *C. marina* appear only in the summer, they spend most of their life in sediments as cysts. In conclusion, the life cycle of *C. antiqua* and *C. marina* seems to be well adapted to the seasonal temperature fluctuations in temperate seas and to shallow waters such as the Seto Inland Sea.

In *H. akashiwo*, vegetative cells can overwinter in the water column, but *H. akashiwo* also has benthic cysts. The germination of cysts was not observed at 5 °C, was low at 10 °C, and was optimal at ≥ 15 °C (Imai and Itakura 1999). Consequently, temperature is a crucial factor at excystment and initial growth just after germination. The timing of the spring bloom of *H. akashiwo* shows great regularity in temperate coastal waters (Smayda 1998). This suggests that the seed population are cells from the excystment, which are greatly affected by bottom temperature, and that population growth is affected by surface temperature (favorable at ≥ 18 °C). After the initial growth, allelopathic activity of *H. akashiwo* against competing phytoplankton and grazers is probably important (Smayda 1998). During periods of nutrient depletion, diel vertical migration presumably facilitates nutrient retrieval and maintenance of *H. akashiwo* blooms in shallow coastal areas such as fishing ports and river mouths (Yamochi and Abe 1984 cited in Smayda 1998).

Cysts of *F. japonica* have dormancy periods of 2–3 months in sediments. The role of cysts for bloom dynamics is unclear because of the scarcity of ecophysiological studies of this species.

Harmful raphidophytes are usually found in coastal waters of subtropical and temperate areas. They have an adaptive survival strategy by virtue of over-wintering cysts. During the last decade, raphidophytes have established populations in northern regions such as the North Sea (Nehring 1998). Since cysts enable these species to persist in the same area once established, human activities such as shipping and transfer of marine organisms for aquaculture should undertake necessary measures to avoid dispersal of harmful raphidophytes. Comparative studies on the ecophysiology of vegetative cells and cysts of newly isolated strains from various geographic areas are also essential.

References

Dahl E, Bagøien E, Edvardsen B, Stenseth N-C (2005) The dynamics of *Chrysochromulina* species in the Skagerrak in relation to environmental conditions. J Sea Res 54:15–24
Edvardsen B, Paasche E (1992) Two motile stages of *Chrysochromulina polylepis* (Prymnesiophyceae): morphology, growth and toxicity. J Phycol 28:104–114
Edvardsen B, Paasche E (1998) Bloom dynamics and physiology of *Prymnesium* and *Chrysochromulina*. In: Anderson DM, Cembella AD, Hallegraeff GM (eds) Physiological ecology of harmful algal blooms. Springer, Berlin Heidelberg New York, pp 193–208
Edvardsen B, Eikrem W, Green JC, Andersen RA, Moon-Van Der Staay SY, Medlin LK (2000) Phylogenetic reconstructions of the Haptophyta inferred from 18S ribosomal DNA sequences and available morphological data. Phycologia 39:19–35
Eschbach E, John U, Reckermann M, Cembella AD, Edvardsen B, Medlin LK (2005) Cell cycle dependent expression of toxicity by the ichthyotoxic prymnesiophyte *Chrysochromulina polylepis*. Aquat Microb Ecol 39:85–95
Fukuyo Y (2004) Morphology of red-tide organisms and their taxonomic difficulty. In: Okaichi T (ed) Red tides. TERRAPUB/Kluwer, Tokyo, pp 67–84
Granéli E, Johansson N (2003) Effects of the toxic haptophyte *Prymnesium parvum* on the survival and feeding of a ciliate: the influence of different nutrient conditions. Mar Ecol Prog Ser 254:49–56
Guo M, Harrison PJ, Taylor FJR (1996) Fish kills related to *Prymnesium parvum* N. Carter (Haptophyta) in the People's Republic of China. J Appl Phycol 8:111–117
Hajdu S, Larsson U, Moestrup Ø (1996) Seasonal dynamics of *Chrysochromulina* species (Prymnesiophyceae) in a coastal area and a nutrient-enriched inlet of the northern Baltic proper. Bot Mar 39:281–295
Hansen E, Ernstsen A, Eilertsen HC (2004) Isolation and characterisation of a cytotoxic polyunsaturated aldehyde from the marine phytoplankter *Phaeocystis pouchetii* (Hariot) Lagerheim. Toxicology 199:207–217
Hirashita T, Ichimi K, Montani S, Nomura M, Tajima S (2000) Molecular analysis of ribosomal RNA gene of red tide algae obtained from the Seto Inland Sea. Mar Biotechnol 2:267–273
Hiroishi S, Okada H, Imai I, Yoshida T (2005) High toxicity of the novel bloom-forming species *Chattonella ovata* (Raphidophyceae) to cultured fish. Harmful Algae 4:783–787
Houdan A, Billard C, Marie D, Not F, Sáez AG, Young JR, Probert I (2004) Holococcolithophore-heterococcolithophore (Haptophyta) life cycles: flow cytometric analysis of relative ploidy levels. Syst Biodiv 1:453–465
Imai I, Itakura S (1999) Importance of cysts in the population dynamics of the red tide flagellate *Heterosigma akashiwo* (Raphidophyceae). Mar Biol 133:755–762
Imai I, Itoh K (1987) Annual life cycle of *Chattonella* spp., causative flagellates of noxious red tides in the Inland Sea of Japan. Mar Biol 94:287–292
Imai I, Itakura S, Itoh K (1993) Cysts of the red tide flagellate *Heterosigma akashiwo*, Raphidophyceae, found in bottom sediments of northern Hiroshima Bay, Japan. Nippon Suisan Gakk 59:1669–1673
Imai I, Yamaguchi M, Watanabe M (1998) Ecophysiology, life cycle, and bloom dynamics of *Chattonella* in the Seto Inland Sea, Japan. In: Anderson DM, Cembella AD, Hallegraeff GM (eds) Physiological ecology of harmful algal blooms. Springer, Berlin Heidelberg New York, pp 95–112
Itakura S, Nagasaki K, Yamaguchi M, Imai I (1996) Cyst formation in the red tide flagellate *Heterosigma akashiwo* (Raphidophyceae). J Plankton Res 18:1975–1979

Iwasaki H (1973) The physiological characteristics of neritic red tide flagellates. Bull Plankton Soc Japan 19:104–114

John U, Tillmann U, Medlin LK (2002) A comparative approach to study inhibition of grazing and lipid composition of a toxic and non-toxic clone of *Chrysochromulina polylepis* (Prymnesiophyceae). Harmful Algae 1:45–57

Jones HLJ, Leadbeater BSC, Green JC (1994) Mixotrophy in haptophytes. In: Green JC, Leadbeater BSC (eds) The haptophyte algae. Clarendon Press, Oxford, pp 247–263

Lancelot C, Keller MD, Rousseau V, Smith WO, Mathot S (1998) Autecology of the marine haptophyte *Phaeocystis* sp. In: Anderson DM, Cembella AD, Hallegraeff GM (eds) Physiological ecology of harmful algal blooms. Springer, Berlin Heidelberg New York, pp 209–224

Larsen A (1999) *Prymnesium parvum* and *P. patelliferum* (Haptophyta) - one species. Phycologia 38:541–543

Legrand C, Rengefors K, Fistarol GO, Granéli E (2003) Allelopathy in phytoplankton – biochemical, ecological and evolutionary aspects. Phycologia 42:406–419

LeRoi J-M, Hallegraeff GM (2004) Scale-bearing nanoflagellates from southern Tasmanian coastal waters, Australia. I. Species of the genus *Chrysochromulina* (Haptophyta). Bot Mar 47:73–102

Marshall JA, Nichols PD, Hamilton B, Lewis RJ, Hallegraeff GM (2003) Ichthyotoxicity of *Chattonella marina* (Raphidophyceae) to damselfish (*Acanthochromis polycanthus*): the synergistic role of reactive oxygen species and free fatty acids. Harmful Algae 2:273–281

Moestrup Ø (1994) Economic aspects: 'blooms', nuisance species, and toxins. In: Green JC, Leadbeater BSC (eds) The haptophyte algae. Clarendon Press, Oxford, pp 265–285

Moestrup Ø, Thomsen HA (2003) Taxonomy of toxic haptophytes (prymnesiophytes). In: Hallegraeff GM, Anderson DM, Cembella AD (eds) Manual on harmful marine microalgae. UNESCO Publishing, Paris, pp 433–463

Morohashi A, Satake M, Oshima Y, Igarashi T, Yasumoto T (2001) Absolute configuration at C14 and C85 in prymnesin-2, a potent hemolytic and ichthyotoxic glycoside isolated from the red tide alga *Prymnesium parvum*. Chirality 13:601–605

Nehring S (1998) Establishment of thermophilic phytoplankton species in the North Sea: biological indicators of climate changes? ICES J Mar Sci 55:818–823

Nishijima T, Hata Y (1986) Physiological ecology of *Chattonella antiqua* (Hada) Ono on B group vitamin requirements. Bull Jpn Soc Sci Fish 52:181–186

Oda T, Nakamura A, Shikayama M, Kawano I, Ishimatsu A, Muramatsu T (1997) Generation of reactive oxygen species by raphidophycean phytoplankton. Biosci Biotech Biochem 61:1658–1662

Peperzak L, Colijn F, Vrieling EG, Gieskes WWC, Peeters JCH (2000) Observations of flagellates in colonies of *Phaeocystis globosa* (Prymnesiophyceae); a hypothesis for their position in the life cycle. J Plankton Res 22:2181–2203

Schoemann V, Becquevort S, Stefels J, Rousseau V, Lancelot C (2005) *Phaeocystis* blooms in the global ocean and their controlling mechanisms: a review. J Sea Res 53:43–66

Shilo M (1971) Toxins of Chrysophyceae. In: Kadis S, Ciegler A, Ajl S (eds) Microbial toxins. Academic Press, New York, pp 67–103

Simonsen S, Moestrup Ø (1997) Toxicity tests in eight species of *Chrysochromulina* (Haptophyta). Can J Bot 75:129–136

Smayda TJ (1998) Ecophysiology and bloom dynamics of *Heterosigma akashiwo* (Raphidophyceae). In: Anderson DM, Cembella AD, Hallegraeff GM (eds) Physiological ecology of harmful algal blooms. Springer, Berlin Heidelberg New York, pp 113–131

Tillmann U (2003) Kill and eat your predator: a winning strategy of the planktonic flagellate *Prymnesium parvum*. Aquat Microb Ecol 32:73–84

Thomsen HA, Buck KR, Chavez FP (1994) Haptophytes as components of marine phytoplankton. In: Green JC, Leadbeater BSC (eds) The haptophyte algae. Clarendon Press, Oxford, pp 187–208

Twiner MJ, Chidiac P, Dixon SJ, Trick CG (2005) Extracellular organic compounds from the ichthyotoxic red tide alga *Heterosigma akashiwo* elevate cytosolic calcium and induce apoptosis in Sf9 cells. Harmful Algae 4:789–800

Yamaguchi M, Imai I, Honjo T (1991) Effects of temperature, salinity and irradiance on the growth rates of the noxious red tide flagellates *Chattonella antiqua* and *C. marina* (Raphidophyceae). Nippon Suisan Gakk 57:1277–1284

Yamaguchi M, Itakura S, Nagasaki K, Matsuyama Y, Uchida T, Imai I (1997) Effects of temperature and salinity on the growth of the red tide flagellates *Heterocapsa circularisquama* (Dinophyceae) and *Chattonella verruculosa* (Raphidophyceae). J Plankton Res 19:1167–1174

7 The Ecology of Harmful Diatoms

S.S. Bates and V.L. Trainer

7.1 Introduction

Diatoms represent one of the most important groups within marine phytoplankton and are characterized by having a siliceous cell wall (frustule). They contribute up to 45 % of the total primary production in the ocean (Mann 1999), or 20–25 % globally (Werner 1977). Diatoms form the base of the food web in many marine ecosystems and are major players in the biogeochemical cycling of C, N, P, Si, and biologically required trace metals (Sarthou et al. 2005). Their success implies that they have highly efficient and adaptable survival mechanisms and growth strategies. A key to this ecological success may lie in their use of Si to form a frustule, which requires less energy to synthesize relative to organic cell walls (Raven 1983). Diatoms may thus take advantage of available Si, and given favorable light, may grow rapidly and dominate the phytoplankton assemblage, forming a "bloom". As such, they are often classed as opportunistic r-strategists, although some stages of their life cycle may be more like K-strategists (Fryxell and Villac 1999). The success of some diatom species may also be explained by their ability to form long chains, which are difficult or impossible for some grazers to ingest. Their frustules also provide mechanical protection against some classes of grazers because exceptional force is required to disrupt them (Hamm et al. 2003). In addition to this mechanical defense, some diatoms produce unsaturated aldehydes that significantly reduce the reproductive success and hence the population growth of zooplankton and other invertebrates (Ianora et al. 2003). In contrast to other smaller phytoplankton, diatoms require nutrient-rich conditions for growth, as well as turbulence to keep them in suspension. They are therefore often found in coastal regions, where their impacts on humans and marine food webs are more often observed.

Most diatoms are considered benign, but some are known to cause harm either by physical means, by causing oxygen depletion, or by the production of a phycotoxin. The first diatom species found to produce a phycotoxin belongs to the genus *Pseudo-nitzschia*; the remainder of this chapter will focus on the

Ecological Studies, Vol. 189
Edna Granéli and Jefferson T. Turner (Eds.)
Ecology of Harmful Algae

ecology of this pennate diatom. For a description of harmful non-toxic diatoms, see Hasle and Fryxell (1995), and Fryxell and Hasle (2003); Ochoa et al. (2002) list harmful diatoms from Mexico.

7.2 Toxin-Producing Diatoms, Genus *Pseudo-nitzschia*

In 1987, the pennate diatom *Pseudo-nitzschia multiseries* (then called *Nitzschia pungens* f. *multiseries*) was identified as the source of the neurotoxin domoic acid (DA) that poisoned humans in eastern Canada (reviewed in Bates et al. 1998). Previously, diatoms were not thought to produce phycotoxins. However, since this first event in Canada, other species of *Pseudo-nitzschia*, thus far totaling 11–12 (depending on the inclusion of *P. pseudodelicatissima*), have become problematic in other parts of the world due to the production of DA. These species include (see Bates et al. 1998; Bates 2000; Moestrup 2004): *P. australis*, *P. calliantha* (Lundholm et al. 2003), *P. cuspidata* (Bill et al. 2005), *P. delicatissima*, *P. fraudulenta*, *P. galaxiae* (Cerino et al. 2005), *P. multiseries*, *P. multistriata*, *P. pseudodelicatissima*, *P. pungens*, *P. seriata*, and *P. turgidula*. Interestingly, all toxigenic species are primarily coastal, although some may be found up to 150 km offshore. Hasle (2002) tentatively concluded that most DA-producing *Pseudo-nitzschia* species, with the exception of *P. seriata*, which is restricted to cold waters of the North Atlantic Ocean, are cosmopolites (see Chap. 3).

On the west coast of North America, the major DA producers are *P. australis*, *P. multiseries*, and *P.* cf. *pseudodelicatissima* (e.g., Adams et al. 2000; Stehr et al. 2002); the latter may have been misidentified and may actually be *P. cuspidata* (cf. Lundholm et al. 2003), which is now a confirmed DA producer (Bill et al. 2005). The *Pseudo-nitzschia* species that contaminated molluscan shellfish in the Bay of Fundy, eastern Canada in 1989 and 1995, was reported as *P. pseudodelicatissima* (see Bates et al. 1998). However, Lundholm et al. (2003) re-examined the field material and identified the cells as *P. calliantha* sp. nov. The question of which *Pseudo-nitzschia* species (*P. pseudodelicatissima* or *P. calliantha*) is the source of the toxin in the Bay of Fundy is still under debate (cf. Kaczmarska et al. 2005b). However, high numbers of non-toxic *P. calliantha* were found in bays of Prince Edward Island, eastern Canada in 2001 and 2002 (Bates et al. unpubl.). In 2002, an unusual spring closure of most of the southern Gulf of St. Lawrence was caused by toxic *P. seriata* (Bates et al. 2002). In Europe, the problematic *Pseudo-nitzschia* species are *P. australis*, *P. seriata*, and *P. multiseries*. In New Zealand, *P. australis* is the main source of DA, although other toxigenic species are present (Rhodes et al. 1998).

Note that all of the *Pseudo-nitzschia* species shown to be toxigenic, with the exception of *P. multiseries*, also have strains that do not produce DA at detectable levels (Bates et al. 1998). Coastal species that have not been shown to produce DA include *P. americana* (Villac et al. 1993), *P. brasiliana* (Lundholm et al. 2002b), and *P.* cf. *subpacifica* (Lundholm et al. 2002a); several other species have not yet been tested. Genetic studies are needed to clarify whether all *Pseudo-nitzschia* species are capable of DA production. It should also be noted that toxin production has been reported for *Nitzschia navis-varingica* (Kotaki et al. 2000, 2004), suggesting that the ability to produce DA may include other genera, as well as other *Pseudo-nitzschia* species thus far not shown to be toxigenic.

7.3 Domoic Acid in the Marine Food Web

Domoic acid has frequent, recurrent impacts on many levels of the food web in certain coastal areas and minimal impacts in others. This may be due to a combination of factors, including the variability of toxin production among *Pseudo-nitzschia* strains, differences in shellfish retention or release of toxin, sensitivity and resistance of exposed organisms to ingested toxins, and composition of the food webs in each region. DA is available to pelagic and benthic organisms that filter feed directly on toxic *Pseudo-nitzschia* cells or on "marine snow" containing flocculated intertwined chains of *Pseudo-nitzschia* (e.g., Trainer et al. 1998), and to fish, birds and mammals that feed on contaminated food at higher trophic levels (see Chap. 22). Molluscan shellfish are the most common vector for DA transfer. However, other vectors continue to be found, implicating DA as an important agent for disrupting marine food webs. DA can be passed up the food web via krill (Bargu et al. 2002, 2003; Lefebvre al. 2002a; Bargu and Silver 2003), copepods (Lincoln et al. 2001; Tester et al. 2001; Maneiro et al. 2005), crabs (e.g., Wekell et al. 1994; Costa et al. 2003), other benthic organisms (Goldberg 2003), cephalopods (Costa et al. 2004, 2005), and fish (Lefebvre et al. 1999, 2001, 2002a, 2002b; Vale and Sampayo 2001; Costa and Garrido 2004; Busse et al. 2006). The latter has led to notable mortalities of marine birds (Sierra-Beltrán et al. 1997) and marine mammals (Lefebvre et al. 1999, 2002b; Scholin et al. 2000; Kreuder et al. 2003). Cellular toxicity may vary greatly, depending on the physiological condition of the *Pseudo-nitzschia* cells (Bates et al. 1998); therefore, it is difficult to predict toxin transfer based solely on cell concentrations.

It is interesting that examples of toxigenic *Pseudo-nitzschia* blooms in which DA is found at several levels of the food web appear primarily in upwelling regions, i.e., off the west coasts of the USA, Spain, Portugal, and in Chile. These regions are conducive to blooms of several toxic *Pseudo-nitzschia* species, but especially of *P. australis*, which can contain high levels of DA because of its large cell size (e.g., Cusack et al. 2002). Recently, the presence of DA in phytoplankton and planktivorous fish (pilchard) samples associated with a wildlife mortality event off the Namibian coast, also an upwelling area, was confirmed by liquid chromatography-mass spectrometry/mass spectrometry (LC-MS/MS) (DC Louw, B Currie, GJ Doucette pers. comm.). In contrast, molluscan shellfish continue to be the primary vector on the Canadian east coast, and DA has so far not been found at any other trophic level; no resulting mortalities of sea birds or marine mammals have been observed. This is curious because comparable links in the food web (e.g., herring, seals) are present. An exception may be in the Bay of Fundy, where LC-MS/MS has confirmed the presence of DA in North Atlantic right whales; the vector is still being sought (GJ Doucette, RM Rolland pers. comm.).

7.4 Physiological Ecology of *Pseudo-nitzschia* spp.

Laboratory studies with cultured isolates of *Pseudo-nitzschia* in natural and artificial seawater media have given us clues about environmental factors that may control cell growth and DA production (see Chap. 18). These physiological studies (up to about 1997) have been reviewed by Bates (1998). Briefly, DA production is non-detectable or minimal during exponential growth in batch culture, and increases during the stationary phase as cell division slows and then ceases due to stress. Pan et al. (1998) argued that the preferential need for cellular energy (ATP) limited DA biosynthesis during exponential growth when metabolic energy is used for primary metabolism. During the stationary phase, photosynthesis continues to produce ATP, which hence becomes increasingly available for DA biosynthesis. Early studies with *P. multiseries* consistently demonstrated that DA production was correlated with Si or P limitation, both in batch and in chemostat cultures. This same pattern has also been shown for *P. australis* (Cusack et al. 2002) and *P. seriata* (Bates et al. 2002; Fehling et al. 2004). An exception is *P.* cf. *pseudodelicatissima* (which may be identified as either *P. pseudodelicatissima* or *P. cuspidata*) (Lundholm et al. 2003) from the Gulf of Mexico. The highest DA production rates were during the early exponential phase, with no net production during the stationary phase (Pan et al. 2001). Other *Pseudo-nitzschia* species require study to determine if there are truly different patterns of DA production.

Recent laboratory studies with *P. multiseries* and *P. australis* have revealed that DA production is also associated with stress due to limitation by Fe and/or Cu, as well as to excess Cu (see Chap. 16). The presence of three carboxyl groups in the chemical structure of DA suggests that it could chelate trace metals (Bates et al. 2001), as was demonstrated by Rue and Bruland (2001). Fe- and Cu-stressed *P. multiseries* and *P. australis* cells produce increasing amounts of dissolved and particulate DA during the exponential phase (Rue and Bruland 2001; Maldonado et al. 2002; Wells et al. 2005). In addition, dissolved DA reduces Cu toxicity in cultured *P. multiseries* and *P. australis* (Maldonado et al. 2002; Ladizinsky 2003), and high Cu concentrations increase DA production by *P. australis* during stationary phase (Rhodes et al. 2004). Cu chelation by DA may play a role in a Cu-reliant high-affinity Fe acquisition system, which would potentially provide toxigenic *Pseudo-nitzschia* species with a competitive advantage in areas where Fe is limiting (Wells et al. 2005). Both laboratory and field evidence indicate that dissolved DA enhances the rate of Fe uptake (Maldonado et al. 2002; Wells et al. 2005). Given this potential role of DA, it is surprising that none of the open ocean *Pseudo-nitzschia* species (i.e., *P. granii*, *P.* cf. *fraudulenta*, *P.* cf. *heimii*, *P.* cf. *inflatula*, *P. turgidula*) isolated from "high-nutrient, low-chlorophyll" Fe-limited waters of the NE subarctic Pacific produced detectable amounts of DA when Si-starved (Marchetti 2005); different conditions may be required to stimulate DA production in oceanic *Pseudo-nitzschia* species. In contrast to the above results showing increased DA production in Fe-stressed cells, Bates et al. (2001) found decreased DA production during the stationary phase when *P. multiseries* was grown in artificial seawater with decreasing amounts of added Fe. Differences in initial nutrient levels and in the time required to acclimate to low Fe stress may help to explain these disparities.

Studies (see Bates 1998; Kotaki et al. 2000; Chap. 19) have also shown that bacteria play an important role in enhancing DA production by *P. multiseries*; there is still no conclusive proof of autonomous production of DA by bacteria. These bacteria may be attached (Kaczmarska et al. 2005a) or free-living. Certain bacteria may provoke *Pseudo-nitzschia* to produce DA. One hypothesis (Osada and Stewart 1997) is that some bacteria (e.g., *Alteromonas* spp.) produce chelating agents (e.g., gluconic acid) that remove essential trace metals from use by the *P. multiseries* cells. To counter this, the diatom may produce its own chelator, i.e., DA. This hypothesis links the potential role of DA as a chelator with the observed stimulation of DA production by bacteria. Ultimately, field studies are required to tie together the various findings of laboratory studies.

Understanding the influence of other biotic and abiotic factors is necessary to help predict blooms and their toxicity. Photoperiod differentially affects the growth of *P. delicatissima* and *P. seriata* and the toxicity of *P. seriata* (Fehling et al. 2005), as well as the sexual reproduction of *P. multiseries* (Hiltz et al. 2000). The timing of sexual reproduction is important, as the cells cannot increase in number while undergoing gametogenesis, and cell toxicity may change with the sexual stage (Davidovich and Bates 1998; Bates et al. 1998). Elevated pH, as can be found during intense blooms, also enhances DA production by *P. multiseries* when growth rates decrease with increasing pH (Lundholm et al. 2004). *Pseudo-nitzschia* species are euryhaline (see also Bates et al. 1998), able to grow in culture from ca. 6–45 PSU and observed at salinities from 1 to ~35 PSU in Louisiana-Texas coastal waters (Thessen et al. 2005); on the other hand, these authors also found that *P. delicatissima*, *P.* cf. *pseudodelicatissima* and *P. multi-*

series have distinct salinity preferences for growth. The form of N may influence DA production. Nitrate- or ammonium-grown *P. australis* cultures in exponential growth produce equivalent amounts of DA, whereas DA production is enhanced in cultures growing on urea as their sole N source, while their growth rate is reduced (Cochlan et al. 2005). Regarding biotic factors, it is curious that *P. multiseries* lacks allelopathic effects (Lundholm et al. 2005; see also Bates 1998; Chap. 15), given that almost monospecific blooms may last for extended periods. Parasitic fungi and viruses may also play an important role in *Pseudo-nitzschia* bloom dynamics (see Bates et al. 1998).

An understanding of the hydrographic environments in which *Pseudo-nitzschia* spp. thrive will aid in bloom prediction (see Chap. 10). One approach is to study retentive zones where phytoplankton, including HAB species, accumulate because of unique chemical, biological and physical characteristics. Field surveys have shown that toxigenic *Pseudo-nitzschia* spp. are found at certain seasonally retentive sites, e.g., the Juan de Fuca eddy region (Washington State), Heceta Bank (Oregon), and Point Conception (California) (Trainer et al. 2001). Toxic cells can be reliably found during summer months in the Juan de Fuca eddy region, a "natural laboratory" where ecological studies can be carried out with field populations to determine environmental factors that enhance or diminish DA-producing capabilities (Trainer et al. 2002; Marchetti et al. 2004).

Because *Pseudo-nitzschia* spp. are planktonic, their movement depends greatly on the surrounding ocean physics (e.g., Horner et al. 2000). Topographical features (e.g., canyons, shallow shelves, sills) influence both nutrient flux and phytoplankton placement in retentive regions. The coupling of physical and biological processes has concentrated *Pseudo-nitzschia* cells into layers from several meters (Ryan et al. 2005) to less than a meter (Rines et al. 2002) thick; these may be missed by normal sampling techniques. Apparently healthy cells in deep layers may be transported long distances, thus providing an inoculum to distant surface waters, resulting in an unexpected bloom. Another form of hidden flora is *P. pseudodelicatissima* cells intermingled within colonies of the diatom *Chaetoceros socialis* (Rines et al. 2002). This close association suggests a chemically mediated interaction, and may provide a competitive advantage by offering a microenvironment different from that of the water column.

7.5 Molecular Tools for Studying *Pseudo-nitzschia*

Molecular techniques are currently being applied to *Pseudo-nitzschia* spp. for identification and quantification (see Bates et al. 1998); elucidating phylogenetic relationships (Lundholm and Moestrup 2002; Lundholm et al. 2002a, 2002b, 2003; Orsini et al. 2002, 2004); discriminating among populations of the same species (Orsini et al. 2004; Evans and Hayes 2004; Evans et al. 2004, 2005); and for gene discovery and expression profiling (Boissonneault 2004).

Studies of mating compatibility among strains of presumably the same species of *Pseudo-nitzschia* are augmenting information gained by genetic and classical morphology studies (Davidovich and Bates 1998; Evans et al. 2004; Amato et al. 2005). Such research is just beginning to confirm the existence of cryptic intraspecific diversity within *Pseudo-nitzschia* species. These studies may also help to explain the existence of the great physiological variability, including toxin production (e.g., Kudela et al. 2004), among different *Pseudo-nitzschia* strains of the same species. The investigation of intraspecific genetic diversity (e.g., Evans et al. 2004) will help us to understand how individual cells within a population can respond differentially to changing environmental conditions.

Identifying and characterizing genes that are related to DA biosynthesis will be valuable for further understanding *Pseudo-nitzschia* physiology and ecology. Two approaches are being taken to elucidate these genes. Boissonneault (2004) designed a cDNA microarray to screen for genes whose expression patterns were correlated with DA production in *P. multiseries*. Expression analysis of 5,372 cDNAs revealed 12 transcripts that were up-regulated during toxin production in stationary phase; among them were several that may be directly involved in DA metabolism. This study demonstrates the potential of applying cDNA microarray technology to investigate the control of toxin production in *P. multiseries*. It has also currently generated sequence data for 2,552 cDNAs, providing a database of actively expressed gene tags that may be used as markers or for further characterizing specific biological functions within *Pseudo-nitzschia*.

A second approach is using subtraction hybridization techniques on Si-replete and deplete cultures of *P. australis* to identify genes that are involved in DA production (B. Jenkins, E. Ostlund, V. Armbrust pers. comm.). Sequence analysis of the subtracted libraries has revealed genes implicated in amino acid transport and biosynthesis. The data provided by both of these described approaches, the completion of the first diatom genome project using *Thalassiosira pseudonana* (Armbrust et al. 2004), and the recent start of the whole genome sequencing of *Pseudo-Nitzschia* multiseries (carried out by E.V. Armbrust, B. Jenkins, and S. Bates), will greatly assist in characterizing *Pseudo-nitzschia* genes involved in cell growth and physiology, including DA production.

7.6 Conclusions and Directions for Future Research

Field sites where toxic diatoms can reliably be found provide natural laboratories where researchers can determine the role of environmental factors in influencing toxin production through manipulation of healthy populations of cells. These sites will enhance our understanding of why certain coastlines are plagued by recurring toxigenic blooms, whereas others are not. The establishment of long-term monitoring programs in coastal regions is essential to our understanding of the survival mechanisms of toxic and harmful algae amidst the larger complex of phytoplankton species. It has been suggested that the increased incidence and intensity of HABs may be linked to regime shifts, manifested as changes in strength of the North Pacific and North Atlantic pressure systems (Hayes et al. 2001). Global factors also affect the plankton; e.g., high sea surface temperatures increase photosynthesis and metabolism and may contribute to the growth of tropical and temperate species in higher northern and southern latitudes. Only through the collection of comprehensive data sets that address specifically the mechanisms of survival of these highly successful diatoms, will such questions be answered. Further understanding of the factors that control growth and DA production (e.g., Fe limitation, bacteria) will be assisted by the current development of molecular tools.

Acknowledgements. We thank K.R. Boissonneault, W.P. Cochlan, G.J. Doucette, K. Lefebvre, N. Lundholm, A. Marchetti, and M. Wells for their constructive comments.

References

Adams NG, Lesoing M, Trainer VL (2000) Environmental conditions associated with domoic acid in razor clams on the Washington coast. J Shellfish Res 19:1007–1015

Amato A, Orsini L, D'Alelio D, Montresor M (2005) Life cycle, size reduction patterns, and ultrastructure of the pennate planktonic diatom *Pseudo-nitzschia delicatissima* (Bacillariophyceae). J Phycol 41:542–556

Armbrust EV, Berges JA, Bowler C et al (2004) The genome of the diatom *Thalassiosira pseudonana*: ecology, evolution, and metabolism. Science 306:79–86

Bargu S, Silver MW (2003) Field evidence of krill grazing on the toxic diatom genus *Pseudo-nitzschia* in Monterey Bay, California. Bull Mar Sci 72:629–638

Bargu S, Powell CL, Coale SL, Busman M, Doucette GJ, Silver MW (2002) Krill: a potential vector for domoic acid in marine food webs. Mar Ecol Prog Ser 237:209–216

Bargu S, Barinovic B, Mansergh S, Silver MW (2003) Feeding responses of krill to the toxin-producing diatom *Pseudo-nitzschia*. J Exp Mar Biol Ecol 284:87–104

Bates SS (1998) Ecophysiology and metabolism of ASP toxin production. In: Anderson DM, Cembella AD, Hallegraeff GM (eds) Physiological ecology of harmful algal blooms. NATO ASI Series 41. Springer, Berlin Heidelberg New York, pp 405–426

Bates SS (2000) Domoic-acid-producing diatoms: another genus added! J Phycol 36:978–983

Bates SS, Garrison DL, Horner RA (1998) Bloom dynamics and physiology of domoic-acid-producing *Pseudo-nitzschia* species. In: Anderson DM, Cembella AD, Hallegraeff GM (eds) Physiological ecology of harmful algal blooms. NATO ASI Series 41. Springer, Berlin Heidelberg New York, pp 267–292

Bates SS, Léger C, Satchwell M, Boyer GL (2001) The effects of iron on domoic acid production by *Pseudo-nitzschia multiseries*. In: Hallegraeff GM, Blackburn SI, Bolch CJ, Lewis RJ (eds) Harmful algal blooms 2000. IOC-UNESCO, Paris, pp 320–323

Bates SS, Léger C et al (2002) Domoic acid production by the diatom *Pseudo-nitzschia seriata* causes spring closures of shellfish harvesting for the first time in the Gulf of St. Lawrence, eastern Canada. Xth Int Conf on Harmful Algae, St. Pete Beach, FL (Abstract) p 23

Bill BD, Lundholm N, Connell L, Baugh KA, Trainer VL (2005) Domoic acid in *Pseudo-nitzschia cuspidata* in Washington State coastal waters. 3rd Symp on Harmful Marine Algae in the U.S., Monterey, CA (Abstract) p 77

Boissonneault KR (2004) Gene discovery and expression profiling in the toxin-producing marine diatom, *Pseudo-nitzschia multiseries* (Hasle) Hasle. PhD Thesis, Massachusetts Inst Technol, Cambridge, MA, USA, 181 pp

Busse LB, Venrick EL, Antrobus R, Miller PE, Vigilant V, Silver MW, Mengelt C, Mydlarz L, Prezelin BB (2006) Domoic acid in phytoplankton and fish in San Diego, CA, USA. Harmful Algae 5:91–101

Cerino F, Orsini L, Sarno D, Dell'Aversano C, Tartaglione L, Zingone A (2005) The alternation of different morphotypes in the seasonal cycle of the toxic diatom *Pseudo-nitzschia galaxiae*. Harmful Algae 4:33–38

Cochlan WP, Herdon J, Ladizinsky L, Kudela RM (2005) Nitrogen uptake by the toxigenic diatom *Pseudo-nitzschia australis*. GEOHAB Open Science Meeting: HABs and Eutrophication, Baltimore, MD (Abstract) pp 27–28

Costa PR, Garrido S (2004) Domoic acid accumulation in the sardine *Sardina pilchardus* and its relationship to *Pseudo-nitzschia* diatom ingestion. Mar Ecol Prog Ser 284:261–268

Costa PR, Rodrigues SM, Botelho MJ, Sampayo MA (2003) A potential vector of domoic acid: the swimming crab *Polybius henslowii* Leach (Decapoda-brachyura). Toxicon 42:135–141

Costa PR, Rosa R, Sampayo MAM (2004) Tissue distribution of the amnesic shellfish toxin, domoic acid, in *Octopus vulgaris* from the Portuguese coast. Mar Biol 144:971–976

Costa PR, Rosa R, Duarte-Silva A, Brotas V, Sampayo MA (2005) Accumulation, transformation and tissue distribution of domoic acid, the amnesic shellfish poisoning toxin, in the common cuttlefish, *Sepia officinalis*. Aquat Toxicol 74:82–91

Cusack CK, Bates SS, Quilliam MA, Patching JW, Raine R (2002) Confirmation of domoic acid production by *Pseudo-nitzschia australis* (Bacillariophyceae) isolated from Irish waters. J Phycol 38:1106–1112

Davidovich NA, Bates SS (1998) Sexual reproduction in the pennate diatoms *Pseudo-nitzschia multiseries* and *P. pseudodelicatissima* (Bacillariophyceae). J Phycol 34:126–137

Evans KM, Hayes PK (2004) Microsatellite markers for the cosmopolitan marine diatom *Pseudo-nitzschia pungens*. Mol Ecol Notes 4:125–126

Evans KM, Bates SS, Medlin LK, Hayes PK (2004) Microsatellite marker development and genetic variation in the toxic marine diatom *Pseudo-nitzschia multiseries* (Bacillariophyceae). J Phycol 40:911–920

Evans KM, Kühn SF, Hayes PK (2005) High levels of genetic diversity and low levels of genetic differentiation in North Sea *Pseudo-nitzschia pungens* (Bacillariophyceae) populations. J Phycol 41:506–514

Fehling J, Davidson K, Bolch CJ, Bates SS (2004) Growth and domoic acid production by *Pseudo-nitzschia seriata* (Bacillariophyceae) under phosphate and silicate limitation. J Phycol 40:674–683

Fehling J, Davidson K, Bates SS (2005) Growth dynamics of non-toxic *Pseudo-nitzschia delicatissima* and toxic *P. seriata* (Bacillariophyceae) under simulated spring and summer photoperiods. Harmful Algae 4:763–769

Fryxell GA, Hasle GR (2003) Taxonomy of harmful diatoms. In: Hallegraeff GM, Anderson DM, Cembella AD (eds) Manual on harmful marine microalgae. UNESCO Publishing, Paris, pp 465–509

Fryxell GA, Villac MC (1999) Toxic and harmful marine diatoms. In: Stoermer EF, Smol JP (eds) The diatoms. Applications for the environmental and earth sciences. Cambridge Univ Press, Cambridge, pp 419–428

Goldberg JD (2003) Domoic acid in the benthic food web of Monterey Bay, California. MSc Thesis, California State Univ, Monterey Bay, CA, USA, 33 pp

Hamm CE, Merkel R, Springer O, Jurkojc P, Maier C, Prechtel K, Smetacek V (2003) Architecture and material properties of diatom shells provide effective mechanical protection. Nature 421:841–843

Hasle GR (2002) Are most of the domoic acid-producing species of the diatom genus *Pseudo-nitzschia* cosmopolites? Harmful Algae 1:137–146

Hasle GR, Fryxell GA (1995) Taxonomy of diatoms. In: Hallegraeff GM, Anderson DM, Cembella AD (eds) IOC manual on harmful marine microalgae, IOC Manuals and Guides 33. UNESCO, Paris, pp 341–366

Hayes ML, Bonaventura J, Mitchell TP, Prospero JM, Shinn EA, Van Dolah F, Barber RT (2001) How are climate and marine biological outbreaks functionally linked? Hydrobiologia 460:213–220

Hiltz MF, Bates SS, Kaczmarska I (2000) Effect of light:dark cycles and cell apical length on the sexual reproduction of *Pseudo-nitzschia multiseries* (Bacillariophyceae) in culture. Phycologia 39:59–66

Horner RA, Hickey BM, Postel JR (2000) *Pseudo-nitzschia* blooms and physical oceanography off Washington State, USA. S Afr J Mar Sci 22:299–308

Ianora A, Poulet SA, Miralto A (2003) The effects of diatoms on copepod reproduction: a review. Phycologia 42:351–363

Kaczmarska I, Ehrman JM, Bates SS, Green DH, Léger C, Harris J (2005a) Diversity and distribution of epibiotic bacteria on *Pseudo-nitzschia multiseries* in culture, and comparison with those on diatoms in native seawater. Harmful Algae 4:725–741

Kaczmarska I, LeGresley MM, Martin JL, Ehrman J (2005b) Diversity of the diatom genus *Pseudo-nitzschia* Peragallo in the Quoddy Region of the Bay of Fundy, Canada. Harmful Algae 4:1–19

Kotaki Y, Koike K, Yoshida M, Thuoc CV, Huyen NTM, Hoi NC, Fukuyo Y, Kodama M (2000) Domoic acid production in *Nitzschia* sp. isolated from a shrimp-culture pond in Do Son, Vietnam. J Phycol 36:1057–1060

Kotaki Y, Lundholm N, Onodera H, Kobayashi K, Bajarias FFA, Furio E, Iwataki M, Fukuyo Y, Kodama M (2004) Wide distribution of *Nitzschia navis-varingica*, a new domoic acid-producing benthic diatom found in Vietnam. Fish Sci 70:28–32

Kreuder C, Miller MA, Jessup DA, Lowenstine LJ, Harris MD, Ames JA, Carpenter TE, Conrad PA, Mazet JAK (2003) Patterns of mortality in southern sea otters (*Enhydra lutris nereis*) from 1998–2001. J Wildlife Dis 39:495–509

Kudela R, Roberts A, Armstrong, M (2004) Laboratory analyses of nutrient stress and toxin accumulation in *Pseudo-nitzschia* species from Monterey Bay, California. In: Steidinger KA, Landsberg JH, Tomas CR, Vargo GA (eds) Harmful algae 2002. Florida Fish and Wildlife Conserv Comm, Florida Inst of Oceanogr, IOC-UNESCO, St. Petersburg, USA, pp 136–138

Ladizinsky NL (2003) The influences of dissolved copper on the production of domoic acid by *Pseudo-nitzschia* species in Monterey Bay, California: laboratory experiments and field observations. MSc Thesis, California State Univ, Monterey Bay, CA, USA, 68 pp

Lefebvre KA, Powell CL, Busman M, Doucette GJ, Moeller PDR, Silver JB, Miller PE, Hughes MP, Singaram S, Silver MW, Tjeerdema RS (1999) Detection of domoic acid in northern anchovies and California sea lions associated with an unusual mortality event. Natural Toxins 7:85–92

Lefebvre KA, Dovel SL, Silver MW (2001) Tissue distribution and neurotoxic effects of domoic acid in a prominent vector species, the northern anchovy *Engraulis mordax*. Mar Biol 138:693–700

Lefebvre KA, Bargu S, Kieckhefer T, Silver MW (2002a) From sandabs to blue whales: the pervasiveness of domoic acid. Toxicon 40:971–977

Lefebvre KA, Silver MW, Coale SL, Tjeerdema RS (2002b) Domoic acid in planktivorous fish in relation to toxic *Pseudo-nitzschia* cell densities. Mar Biol 140:625–631

Lincoln JA, Turner JT, Bates SS, Léger C, Gauthier DA (2001) Feeding, egg production, and egg hatching success of the copepods *Acartia tonsa* and *Temora longicornis* on diets of the toxic diatom *Pseudo-nitzschia multiseries* and the non-toxic diatom *Pseudo-nitzschia pungens*. Hydrobiologia 453/454:107–120

Lundholm N, Moestrup Ø (2002) The marine diatom *Pseudo-nitzschia galaxiae* sp. nov. (Bacillariophyceae): morphology and phylogenetic relationships. Phycologia 41:594–605

Lundholm N, Daugbjerg N, Moestrup Ø (2002a) Phylogeny of the Bacillariaceae with emphasis on the genus *Pseudo-nitzschia* (Bacillariophyceae) based on partial LSU rDNA. Eur J Phycol 37:115–134

Lundholm N, Hasle GR, Fryxell GA, Hargraves PE (2002b) Morphology, phylogeny and taxonomy of species within the *Pseudo-nitzschia americana* complex (Bacillariophyceae) with descriptions of two new species, *Pseudo-nitzschia brasiliana* and *Pseudo-nitzschia linea*. Phycologia 41:480–497

Lundholm N, Moestrup Ø, Hasle GR, Hoef-Emden K (2003) A study of the *P. pseudodelicatissima/cuspidata* complex (Bacillariophyceae): what is *P. pseudodelicatissima*? J Phycol 39:797–813

Lundholm N, Hansen PJ, Kotaki Y (2004) Effect of pH on growth and domoic acid production by potentially toxic diatoms of the genera *Pseudo-nitzschia* and *Nitzschia*. Mar Ecol Prog Ser 273:1–15

Lundholm N, Hansen PJ, Kotaki Y (2005) Lack of allelopathic effects of the domoic acid-producing marine diatom *Pseudo-nitzschia multiseries*. Mar Ecol Prog Ser 288:21–33

Maldonado MT, Hughes MP, Rue EL, Wells ML (2002) The effect of Fe and Cu on growth and domoic acid production by *Pseudo-nitzschia multiseries* and *Pseudo-nitzschia australis*. Limnol Oceanogr 47:515–526

Maneiro I, Iglesias P, Guisande C, Riveiro I, Barreiro A, Zervoudaki S, Granéli E (2005) Fate of domoic acid ingested by the copepod *Acartia clausi*. Mar Biol 148:123–130

Mann DG (1999) The species concept in diatoms. Phycologia 38:437–495

Marchetti A (2005) Ecophysiological aspects of iron nutrition and domoic acid production in oceanic and coastal diatoms of the genus *Pseudo-nitzschia*. PhD Thesis. University of British Columbia, Vancouver, BC, Canada, 220 pp

Marchetti A, Trainer VL, Harrison PJ (2004) Environmental conditions and phytoplankton dynamics associated with *Pseudo-nitzschia* abundance and domoic acid in the Juan de Fuca eddy. Mar Ecol Prog Ser 281:1–12

Moestrup Ø (ed) (2004) IOC taxonomic reference list of toxic plankton algae. Intergov Oceanogr Comm of UNESCO, http://www.bi.ku.dk/ioc/default.asp

Ochoa JL, Hernández-Becerril DU et al (2002) Marine biotoxins and harmful algal blooms in Mexico's Pacific littoral. In: Taylor FJR, Trainer VL (eds) Harmful algal blooms in the PICES region of the North Pacific. PICES Scientific Report no. 23, Sidney, BC, Canada, pp 119–128

Orsini L, Sarno D, Procaccini G, Poletti R, Dahlmann J, Montresor M (2002) Toxic *Pseudo-nitzschia multistriata* (Bacillariophyceae) from the Gulf of Naples: morphology, toxin analysis and phylogenetic relationships with other *Pseudo-nitzschia* species. Eur J Phycol 37:247–257

Orsini L, Procaccini G, Sarno D, Montresor M (2004) Multiple rDNA ITS-types within the diatom *Pseudo-nitzschia delicatissima* (Bacillariophyceae) and their relative abundances across a spring bloom in the Gulf of Naples. Mar Ecol Prog Ser 271:87–98

Osada M, Stewart JE (1997) Gluconic acid/gluconolactone: physiological influences on domoic acid production by bacteria associated with *Pseudo-nitzschia multiseries*. Aquat Microb Ecol 12:203–209

Pan Y, Bates SS, Cembella AD (1998) Environmental stress and domoic acid production by *Pseudo-nitzschia*: a physiological perspective. Natural Toxins 6:127–135

Pan Y, Parsons ML, Busman M, Moeller PDR, Dortch Q, Powell CL, Doucette GJ (2001) *Pseudo-nitzschia* sp. cf. *pseudodelicatissima* – a confirmed producer of domoic acid from the northern Gulf of Mexico. Mar Ecol Prog Ser 220:83–92

Raven JA (1983) The transport and function of silicon in plants. Biol Rev 58:179–207

Rhodes L, Scholin C, Garthwaite I, Haywood A, Thomas A (1998) Domoic acid producing *Pseudo-nitzschia* species educed by whole cell DNA probe-based and immunochemical assays. In: Reguera B, Blanco J, Fernández ML, Wyatt T (eds) Harmful algae. Xunta de Galicia and IOC-UNESCO, Paris, pp 274–277

Rhodes L, Holland P, Adamson J, Selwood A, McNabb P (2004) Mass culture of New Zealand isolates of *Pseudo-nitzschia australis* for production of a new isomer of domoic acid. In: Steidinger KA, Landsberg JH, Tomas CR, Vargo GA (eds) Harmful algae 2002. Florida Fish and Wildlife Conserv Comm, Florida Inst of Oceanogr, IOC-UNESCO, St. Petersburg, USA, pp 125–127

Rines JEB, Donaghay PL, Dekshenieks MM, Sullivan JM, Twardowski MS (2002) Thin layers and camouflage: hidden *Pseudo-nitzschia* spp. (Bacillariophyceae) populations in a fjord in the San Juan Islands, Washington, USA. Mar Ecol Prog Ser 225:123–137

Rue E, Bruland K (2001) Domoic acid binds iron and copper: a possible role for the toxin produced by the marine diatom *Pseudo-nitzschia*. Mar Chem 76:127–134

Ryan JP, Chavez FP, Bellingham JG (2005) Physical-biological coupling in Monterey Bay, California: topographic influences on phytoplankton ecology. Mar Ecol Prog Ser 287:23–32

Sarthou G, Timmermans KR, Blain S, Tréguer P (2005) Growth physiology and fate of diatoms in the ocean: a review. J Sea Res 53:25–42

Scholin CA, Gulland F et al (2000) Mortality of sea lions along the central California coast linked to a toxic diatom bloom. Nature 403:80–84

Sierra Beltrán A, Palafox-Uribe M, Grajales-Montiel J, Cruz-Villacorta A, Ochoa JL (1997) Sea bird mortality at Cabo San Lucas, Mexico: evidence that toxic diatom blooms are spreading. Toxicon 35:447–453

Stehr CM, Connell L, Baugh KA, Bill BD, Adams NG, Trainer VL (2002) Morphological, toxicological, and genetic differences among *Pseudo-nitzschia* (Bacillariophyceae) species in inland embayments and outer coastal waters of Washington State, USA. J Phycol 38:55–65

Tester PA, Pan Y, Doucette GJ (2001) Accumulation of domoic acid activity in copepods. In: Hallegraeff GM, Blackburn SI, Bolch CJ, Lewis RJ (eds) Harmful algal blooms 2000. IOC-UNESCO, Paris, pp 418–421

Thessen AE, Dortch Q, Parsons ML, Morrison W (2005) Effect of salinity on *Pseudo-nitzschia* species (Bacillariophyceae) growth and distribution. J Phycol 41:21–29

Trainer VL, Adams NG, Bill BD, Anulacion BF, Wekell JC (1998) Concentration and dispersal of a *Pseudo-nitzschia* bloom in Penn Cove, Washington, USA. Natural Toxins 6:113–126

Trainer VL, Adams NG, Wekell JC (2001) Domoic acid-producing *Pseudo-nitzschia* species off the U.S. west coast associated with toxification events. In: Hallegraeff GM,

Blackburn SI, Bolch CJ, Lewis RJ (eds) Harmful algal blooms 2000. IOC-UNESCO, Paris, pp 46–48
Trainer VL, Hickey BM, Horner RA (2002) Biological and physical dynamics of domoic acid production off the Washington coast. Limnol Oceanogr 47:1438–1446
Vale P, Sampayo MA (2001) Domoic acid in Portuguese shellfish and fish. Toxicon 39:893–904
Villac MC, Roelke DL, Chavez FP, Cifuentes LA, Fryxell GA (1993) *Pseudo-nitzschia australis* Frenguelli and related species from the west coast of the USA: occurrence and domoic acid production. J Shellfish Res 12:457–465
Wekell JC, Gauglitz EJ Jr, Barnett HJ, Hatfield CL, Simons D, Ayres D (1994) Occurrence of domoic acid in Washington state razor clams (*Siliqua patula*) during 1991–1993. Natural Toxins 2:197–205
Wells ML, Trick CG, Cochlan WP, Hughes MP, Trainer VL (2005) Domoic acid: the synergy of iron, copper, and the toxicity of diatoms. Limnol Oceanogr 50:1908–1917
Werner D (1977) Introduction with a note on taxonomy. In: Werner D (ed) The biology of diatoms. University of California Press, Berkeley, pp 1–17

8 Ecology of Harmful Cyanobacteria

H.W. Paerl and R.S. Fulton III

8.1 Introduction

Cyanobacteria (blue-green algae) are the Earth's oldest known (>2.5 billion years) oxygenic phototrophs, yet today still exhibit remarkable adaptations and diversification in response to evolutionary change. This physiologically and ecologically diverse microbial group has survived and adapted to the range of geochemical changes marking the evolution of the Earth's biosphere. Recently, cyanobacteria have exploited human alterations of aquatic environments, most notably nutrient-enhanced primary production, or eutrophication (Fogg 1969). Worldwide, cyanobacterial blooms are visible, well-documented, widespread indicators of freshwater, brackish and marine eutrophication. Toxic blooms pose serious water quality, fisheries resource, animal, and human health problems (Chorus and Bartram 1999; Huisman and Visser 2005), including foul odors and tastes, hypoxia, and fish kills. Nutrient cycling and food web dynamics may be altered, and blooms can contribute large amounts of "new" C and N as fuel for hypoxia and anoxia (Horne 1977).

Harmful cyanobacterial blooms (CyanoHABs) have been present for centuries in eutrophic inland, estuarine, and coastal waters (Fogg 1969; Paerl 1988). Proliferation into more recently eutrophying waters is also underway. Examples include expansion of planktonic N_2-fixing and non-N_2-fixing genera (Table 8.1) in lacustrine, riverine, estuarine, and coastal habitats worldwide (Paerl et al. 2001). Aquaculture operations are also susceptible to cyanoHAB invasions. Molecular analyses indicate that the diversity of cyanoHABs at genera, species and strain levels is much greater than observed microscopically (Dyble et al. 2002) (Fig. 8.1). Using N_2-fixing gene (*nifH*) sequence analysis techniques, MacGregor et al. (2001) recently detected the notorious Baltic Sea N_2-fixing bloom former *Nodularia* spp. in oligo-mesotrophic Lake Michigan.

Benthic filamentous and coccoid genera (Table 8.1) can also undergo explosive growth in diverse habitats ranging in salinity. Blooms of the non-

Ecological Studies, Vol. 189
Edna Granéli and Jefferson T. Turner (Eds.)
Ecology of Harmful Algae

Table 8.1. Non-N_2-fixing and N_2-fixing filamentous non-heterocystous (*FNH*), coccoid (*C*), and heterocystous filamentous (*HF*) cyanoHAB genera

Genus	Habitat	Freshwater(F)/ Estuarine(E)/Marine(M)
Non-N_2-Fixing		
Microcystis (C)	Planktonic	F, E
Oscillatoria (FNH)	Planktonic/benthic	F, E, M
Planktothrix (FNH)	Planktonic	F, E
N_2-Fixing		
Anabaena (HF)	Planktonic/benthic	F, E, M
Aphanizomenon (HF)	Planktonic	F, E
Calothrix (HF)	Benthic	F, E, M
Cylindrospermopsis (HF)	Planktonic	F
Lyngbya (FNH)	Planktonic/benthic	F, E, M
Nodularia (HF)	Planktonic	F, E, M
Nostoc (HF)	Planktonic/benthic	F, E, M
Scytonema (HF)	Planktonic/benthic	F, E, M

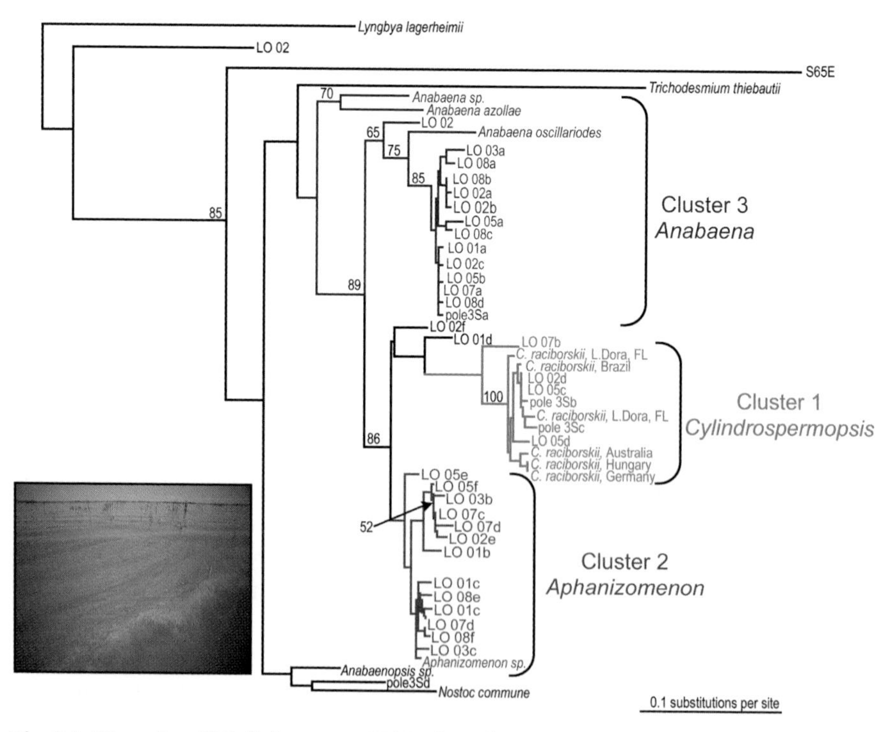

Fig. 8.1. Diversity of N_2 fixing cyanoHABs, based on sequence analysis of the N_2-fixing gene *nifH*, from Lake Okeechobee, Florida. Only six heterocystous cyanoHAB species were identified microscopically. Bootstrap values of >50 % are given next to the corresponding nodes. *Desulfovibrio gigas* (GenBank accession #u68183) was used as the outgroup. From Dyble et al. (2002)

heterocystous filamentous genera *Lyngbya* and *Oscillatoria* are becoming more numerous and persistent in nutrient-enriched springs, rivers, reservoirs, estuarine and coastal embayments, and reefs (Speziale and Dyck 1992; Lapointe 1997; Dennison et al. 1999). Benthic CyanoHABs are indicative of eutrophication and decreased grazing by herbivorous fishes due to overfishing. Their adverse impacts include smothering seagrass communities, corals and mudflats, negatively affecting infauna (Dennison et al. 1999; Paul et al. 2001). Some *Lyngbya* strains produce toxic secondary metabolites (Paul et al. 2001), including aplysiatoxins, debromoaplysiatoxin, and lyngbyatoxin A, metabolites of *L. majuscula*, causative agents of the acute dermal lesions following contact with this cyanoHAB. These compounds are protein kinase C activators and potent tumor promoters *in vivo*, issues of broad public health concern (Mynderse et al. 1977 cited in Paul et al. 2001). *Lyngbya* strains also fix N_2 (Jones 1990), which explains why P may stimulate growth more than N additions (Phlips et al. 1991). These strains thrive in N-deprived waters where N_2 fixation is advantageous, and they bloom in response to land-derived nutrients (N, P, Fe), and organic matter (Dennison et al. 1999).

8.2 Environmental Factors Controlling CyanoHABs

Interactive physical, chemical and biotic factors involved in the control of cyanoHAB growth and dominance are shown in Fig. 8.2.

8.2.1 Nutrients

Among the macronutrients required for aquatic plant growth, nitrogen (N) and phosphorus (P) are often in shortest supply relative to need. In most freshwater ecosystems, excessive P loading has been linked to eutrophication. P-driven eutrophication may favor the development of either N_2-fixing or non-N_2-fixing cyanobacteria, especially if the affected water body exhibits relatively long residence times (low rates of flushing), high surface water temperatures (>20 °C), and strong vertical stratification (Paerl 1988). N_2-fixing cyanobacterial dominance depends on several factors, most notably the availability of biologically utilizable nitrogen relative to phosphorus. Systems having low molar ratios of total and soluble N to P (<15) appear most susceptible to cyanobacterial dominance (Smith 1990). In some instances, the "N:P rule" does not apply. For example, Downing et al. (2001) showed that increased nutrient concentrations and loadings of N and P were better predictors of CyanoHAB dominance. Other exceptions include; (1) highly eutrophic systems in which *both* N and P inputs may exceed the assimilative capacity of the phytoplankton, and (2) well-flushed systems, in which the flushing rate

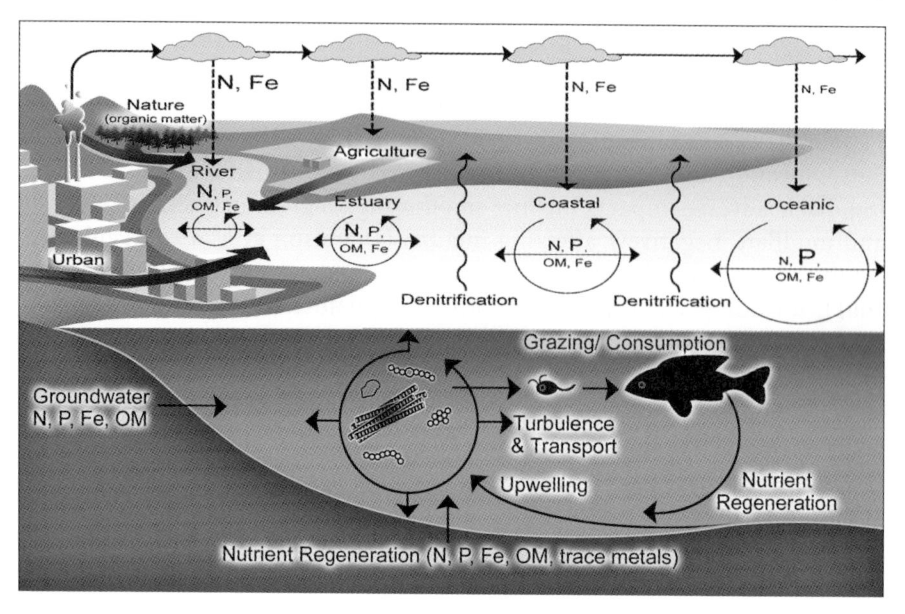

Fig. 8.2. Illustration of the interactive physical, chemical and biotic variables controlling CyanoHAB expansion across the freshwater–marine interface

exceeds the maximal cyanobacterial growth rates, typically >1 d^{-1}. In N- and P-enriched systems, N:P ratios may be >20, but since both N and P may be supplied in excess, factors other than nutrient availability (e.g., irradiance, vertical mixing, residence time, salinity, micronutrients) may dictate algal community dynamics. Here, N_2 fixation confers no advantage, and non-N_2-fixing taxa will predominate.

Moderately N- and P-enriched waters can support diazotrophic and non-diazotrophic species sequentially. During the wet spring months, non-point source runoff, which tends to be high in N, dominates. Later, during drier summer months external N loads decrease, and P-rich point source inputs and internal nutrient cycling dominate. Resultant relative P enrichment (i.e., declining N:P ratios) will select for N_2-fixing species (Paerl 1988). Once N_2 fixers are established, non-diazotrophic species may coexist, because they are able to utilize fixed N produced and released by N_2 fixing species (Paerl 1990). Co-occurring diazotrophic and N-requiring bloom species are capable of buoyancy regulation, and thus a near-surface existence, in turbid productive waters. Mixed assemblages can persist as bloom "consortia" throughout the summer and fall months (Paerl et al. 2001), until unfavorable physical conditions including rapid cooling (<15 °C) and turnover take place. Because these assemblages have complex nutritional requirements, efforts aimed at reducing cyanobacterial dominance by manipulating N:P ratios have met with mixed results.

P input constraints are often the preferred, least costly approach. In certain cases, P cutbacks can be effective on their own (without parallel N removal), because; (1) they may reduce total P availability enough to control growth of *all* bloom taxa, and (2) they may increase N:P ratios enough to provide eukaryotic algae with a competitive advantage over cyanobacteria. Examples include; Lakes Washington and Erie, USA, and the Baltic Sea embayments and fjords of Sweden, where cyanobacterial blooms have been controlled by both point and non-point source P reductions (Edmondson and Lehman 1981; Elmgren and Larsson 2001).

Nitrogen-limited (i.e., P-sufficient) estuarine and coastal waters have also experienced a recent upsurge in algal blooms. This has been attributed to increases in anthropogenic N loading (Paerl 1988). In the Neuse River Estuary, NC, USA, deteriorating water quality due to excessive N loading has led to an N input "cap" and an overall 30 % reduction in N loading (Paerl et al. 2004). Reductions in N loading lead to lower N:P ratios, potentially selecting for N_2-fixing cyanobacteria that could circumvent N-limitation imposed by N reductions. During the summer of 1997, Piehler et al. (2002) observed N_2-fixing *Anabaena* strains in mesohaline segments of this estuary. In a parallel laboratory study (Moisander et al. 2002a), two toxic Baltic Sea *Nodularia* strains and native *Anabaenopsis* and *Anabaena* thrived in Neuse River Estuary water over a range of salinities, demonstrating the potential for estuarine cyanoHAB expansion (Fig. 8.3).

While diazotrophs have an advantage of subsistence on atmospheric N_2, they also proliferate on combined inorganic and organic N sources. In Florida's St. Johns River system, growth of the invasive, toxic, heterocystous cyanoHAB *Cylindrospermopsis raciborskii* can be stimulated by nitrate and/or ammonium, for which it competes well with eukaryotic phytoplankton during runoff-rich springtime (Fig. 8.4). During the drier summer months when N is chronically limiting, it switches to N_2 fixation as its main N source. This nutritional flexibility provides *C raciborskii* a competitive advantage in response to seasonal N loading.

Iron (Fe) and trace metal micronutrients are essential for cyanobacterial growth. Iron is a cofactor of enzymes involved in photosynthesis, electron transport, energy transfer, N (NO_3^- and NO_2^-) assimilation and N_2 fixation. Iron limitation of cyanobacterial growth has been demonstrated in freshwater (Paerl 1990) and marine ecosystems (Paerl et al. 1994 in Paerl and Zehr 2000). In oceanic and coastal regions removed from terrigenous or atmospheric Fe inputs, diazotrophic growth may be mediated by Fe availability. Populations of the bloom-forming N_2-fixing cyanobacteria *Trichodesmium* exhibited Fe-limited CO_2 and N_2 fixation in the W. Atlantic Ocean off the North Carolina coast (Paerl and Zehr 2000).

Bloom-forming cyanobacteria produce potent siderophore (hydroxamate) chelators capable of sequestering iron at low ambient concentrations (Murphy et al. 1976 in Paerl 1990). This may provide a competitive advantage over

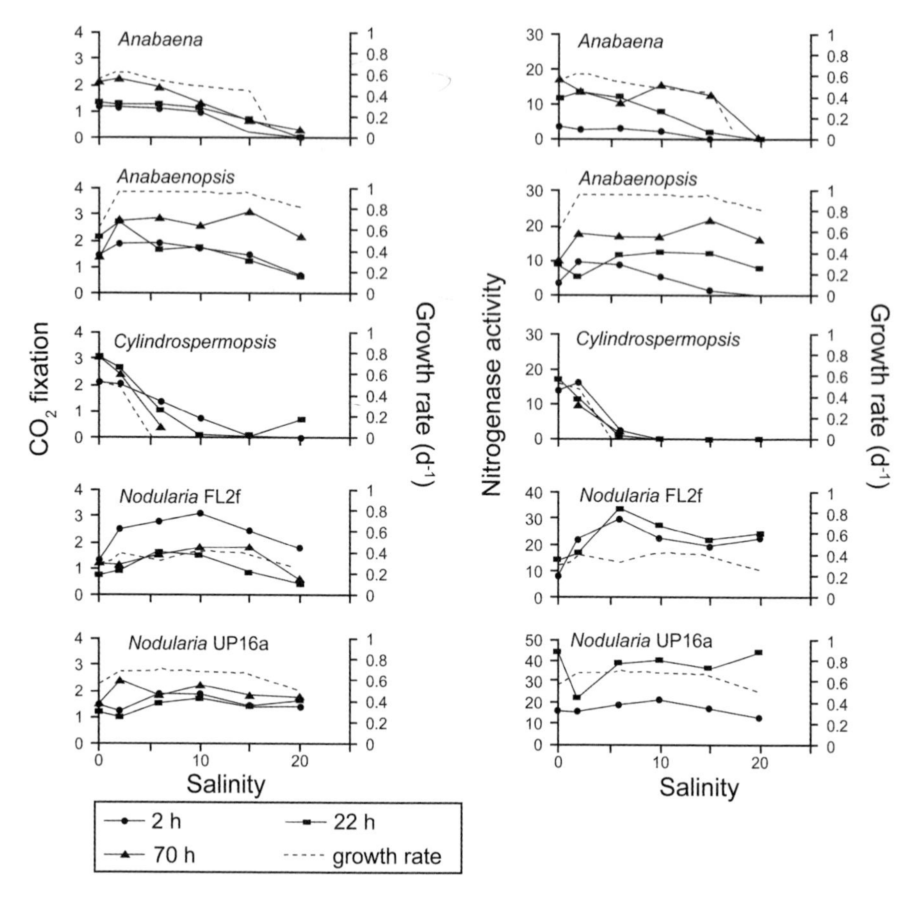

Fig. 8.3. Experimentally determined effects of salinity on CO_2 fixation (mgC mg Chl a^{-1} h^{-1}), nitrogenase activity (µmol C_2H_4 mg Chl a^{-1} h^{-1}), and growth of cyanoHABs found at the freshwater-marine interface. Salinity is in practical salinity units (PSU). Isolates of cyanoHABs used were obtained from the Baltic Sea, Neuse River Estuary (North Carolina), and lakes in Central Florida that drain into the St. Johns River and Estuary. Figure adapted from Moisander et al. (2002a)

eukaryotic phytoplankton when Fe availability is restricted. Bloom species may themselves affect Fe availability by mediating ecosystem productivity (i.e., organic matter production), bottom water hypoxia and anoxia, which can lead to the liberation of significant amounts of Fe, as soluble Fe^{2+}. Since most cyanoHABs are tolerant to sulfide that often accompanies hypoxia, these species are free to migrate into the Fe-rich hypolimnion to replenish Fe supplies.

Cyanobacteria require a suite of trace metals for metabolism, growth and reproduction. Manganese, cobalt, copper, molybdenum and zinc are most frequently mentioned. Photosynthesis and N_2 fixation require these metals for

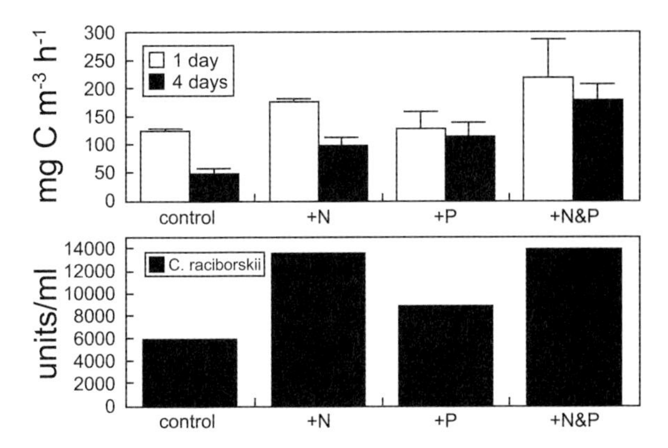

Fig. 8.4. Photosynthetic activity and growth response of Lake George, Florida, natural phytoplankton community to the additions of nitrogen (as nitrate at 20 mM N), phosphate (at 10 mM P) and N and P combined, in an in situ bioassay. Conducted during a period (May 2002) of dominance by the heterocystous cyanoHAB *Cylindrospermopsis raciborskii*. *C. raciborskii* accounted for approximately 70 % of cyanobacterial biomass and 40 % of total phytoplankton biomass during the bioassay period. Photosynthetic and growth responses were assessed after 24 and 72-h periods of nutrient enrichment, relative to untreated controls. Bioassays were conducted in 4-L transparent, polyethylene Cubitainers that were suspended in the water under natural irradiance and temperature conditions. Total phytoplankton community photosynthetic activity response was shown in the upper frame, while *C. raciborskii* response was specifically assessed using counts of filaments (normalized for length) shown in the lower frame. Results indicate that *C. raciborskii* growth was stimulated under individual N and P enrichment as well as combined N and P enrichment conditions, underscoring the broad competitive ability of the cyanoHAB to exploit nutrient enriched conditions

enzyme synthesis and function. Molybdenum (Mo) is a cofactor in the N_2-fixing enzyme nitrogenase. Howarth and Cole (1985 in Paerl and Zehr 2000) proposed that the relatively high (>20 mM) concentrations of sulfate (SO_4^{2-}), a structural analogue of the most common form of Mo found in seawater, molybdate (MoO_4^{2-}), could competitively (via the uptake process) inhibit N_2 fixation, thereby controlling this process. However, MoO_4^{2-} is highly soluble in seawater with concentrations in the order of ≈100 μM. Paulsen et al. (1991 in Paerl and Zehr 2000) showed that despite the potential for SO_4^{2-} competition, Mo was available at concentrations much lower than 100 μM. In W. Atlantic Ocean waters, N_2-fixing potentials were unaffected by this competition (Paulsen et al. 1991 in Paerl and Zehr 2000). It appears that cellular Mo requirements for N_2 fixation are met by sufficient uptake and storage. In addition, "alternative" non Mo-requiring nitrogenases exist in bacterial and cyanobacterial diazotrophs (Paerl and Zehr 2000). If such microbes are broadly distributed, they would have a mechanism circumventing Mo limitation.

Early studies (cf. Fogg 1969) cite dissolved organic matter (DOM) as a factor potentially controlling cyanobacterial blooms. The hypothesized mechanism for DOM-stimulated cyanobacterial growth is that DOM „conditions“ the water for cyanobacteria, possibly by inducing nutrient assimilatory enzymes and heterotrophy, or acting as nutrient (Fe and other trace metal) chelators, and providing a source of energy and nutrition for closely associated heterotrophic bacteria, which may form synergistic interactions with cyanobacteria (Paerl and Pinckney 1996). However, as pointed out by Fogg (1969), elevated DOM may be a result (due to DOM excretion, bacterial and viral lysis, and zooplankton „sloppy feeding“), rather than a cause of cyanobacterial blooms. Lastly, terrigenous organic substances discharged to coastal waters can regulate phytoplankton growth and composition. These substances interact with N and P availability to determine growth and dominance potentials of the N_2-fixing cyanoHAB *Nodularia spumigena* in the Baltic Sea (Panosso and Granéli 2000).

Cyanobacterial bloom taxa exhibit excellent CO_2 uptake kinetics, which are advantageous under dissolved inorganic carbon (DIC) limited conditions that can occur during blooms (Paerl 1988). This, combined with the ability to form buoyant surface blooms to obtain atmospheric CO_2 under DIC-limited conditions, provides a competitive advantage over eukaryotic algal taxa in nutrient-enriched, productive waters (Paerl 1988).

8.2.2 Physical-Chemical Factors: Salinity and Turbulence

Salinity, specifically ionic composition and strength, can inhibit growth among freshwater cyanoHABs (e.g., *Microcystis*, *Cylindrospermopsis*). However, salinity, per se, is not necessarily an across-the-board barrier to cyanobacterial growth and proliferation, as witnessed by the presence of diverse epiphytic and epibenthic N_2-fixing (e.g., *Calothrix, Lyngbya, Nostoc, Scytonema*) and non-N_2-fixing genera (e.g., *Oscillatoria*) in estuarine and coastal ecosystems (Potts 1994). Species of some benthic cyanoHAB genera, including *Lyngbya* and *Oscillatoria* are well adapted to freshwater or saline (including hypersaline) conditions. N_2- and CO_2-fixing activities in some freshwater cyanoHABs may be inhibited at increasing salt concentrations (Moisander et al. 2002a) (Fig. 8.3). Some indigenous populations can adapt to varying salinities by production of compatible osmolytes (Reed and Stewart 1985 in Paerl and Zehr 2000). Moisander and Paerl (2000) showed that Baltic Sea *Nodularia* were capable of diazotrophic growth under a wide range of salinities (0 to over 30 PSU).

8.2.3 Salinity and Turbulence

Turbulence over a range of scales (cellular to ecosystem) plays an important regulatory role in cyanoHAB bloom dynamics (Reynolds 1987; Kononen et al.

1996). CyanoHABs prefer calm, vertically stratified conditions, given adequate nutrient supplies (Paerl and Millie 1996). Following wind- or flow-induced destratification, cyanoHABs may lose their competitive advantage, and if such conditions persist, blooms can rapidly degrade or "crash". However, when intermittent weak stratification occurs during favorable growth periods (summer), cyanoHABs can quickly reemerge. Shallow, periodically mixed eutrophic lakes, the brackish Baltic Sea, reservoir and lagoonal ecosystems exhibit these tendencies. Ibelings et al. (1991) showed that periods of gentle intermittent mixing are preferred over consistently calm conditions, where colonial buoyant CyanoHABs like *Microcystis* can benefit from nutrient exchange during mixing, but then rapidly respond to an increase in water column stability by floating to the illuminated near-surface mixed layer.

Turbulence also affects phytoplankton growth rates and structural integrity. Among cyanobacteria, non-disruptive, low-level turbulence can promote localized „phycosphere" nutrient cycling, alleviate certain forms of nutrient limitation (DIC, PO_4^{3-}, trace metals), and enhance growth. Gently stirred cultures of cyanoHABs frequently show higher growth than static cultures. Increases in turbulence either as stirring or shaking or more well-defined small-scale shear can, however, inhibit CO_2- and N_2-fixing activities and growth, with excessive turbulence causing disaggregation, cell and filament damage, and rapid death among diverse colonial genera in culture and in nature (Paerl 1990; Moisander et al. 2002b). Optimal CO_2, N_2 fixation and growth of these genera often rely on mutually beneficial microbial consortial interactions with host cyanobacteria. Turbulence can disrupt consortia and negatively affect growth (Paerl and Pinckney 1996).

Many bloom genera can regulate buoyancy by varying intracellular gas vesicles, a feature ensuring optimal (for growth) vertical orientation with respect to light and nutrient regimes in the water column (Reynolds 1987). In highly turbulent waters, the ability to maintain optimal vertical positioning can be overcome by mixing (Reynolds 1987), which may negatively affect growth and long-term competition for nutrient and radiant energy resources with eukaryotic phytoplankton. The turbulent conditions characterizing wind-exposed, well-mixed estuarine/coastal surface waters constrain N_2 and CO_2 fixation in CyanoHABs, and hence represent a potential barrier to their expansion (Paerl et al. 1995, in Paerl et al. 2001).

Despite their chronic N-limited condition and increasing nutrient loads, many estuarine and coastal waters do not show significant cyanobacterial dominance or blooms. Do physical constraints play a role in what appears to be an available niche for expansion from a nutrient perspective?

While non-N_2-fixing cyanoHAB taxa (e.g., *Microcystis* and *Oscillatoria* spp) are severely constrained in most N-limited, P-sufficient estuarine and coastal systems, diazotrophic taxa can, at times, proliferate in these waters; most often as endosymbionts in diatoms (*Rhizoselenia-Richelia*), or in benthic habitats. Adequate genetic diversity exists for potential colonization of diazotrophic cyanobacteria in estuaries (Paerl 2000; Paerl and Zehr 2000). Zooplankton and/or fish grazing do not appear to be the dominant mechanism explaining the scarcity of cyanobacteria in estuaries (Paerl et al. 2001). Turbulence has been proposed as a possible control in exposed, wind-mixed estuarine and coastal surface waters. This hypothesis has been tested in the laboratory using Couette chambers, capable of duplicating the shear rates experienced in a shallow wind-exposed (10–15 kt), N-limited estuary, the Neuse River Estuary, NC (Moisander et al. 2002b). These shear rates suppressed N_2 fixation and CO_2 fixation rates of *Anabaena* and *Nodularia* spp., relative to calm (low-shear) conditions. Persistent wind and shear in estuarine, coastal and oceanic waters may indeed be a constraint to cyanoHAB expansion.

8.3 CyanoHAB Interactions with Micro/Macroorganisms

CyanoHABs form close associations with other microorganisms in nature and in culture (Paerl and Kellar 1978 in Paerl 1990). Associated microorganisms include eubacteria, fungi, phytoflagellates, and protozoans (Paerl 1982 in Paerl 1990). These associations occur during all bloom phases, but the intensity and specificity of microbial epiphytism vary. During the initiation of *Anabaena oscillarioides* blooms, when high biomass-specific rates of CO_2 and N_2 fixation were observed, an association between the heterotrophic bacterium *Pseudomonas aeruginosa* and heterocysts prevailed (Gallucci and Paerl 1983 in Paerl 1990). The intensity and frequency of *P. aeruginosa* attachment to heterocysts exhibited a diel pattern, and *P. aeruginosa* was chemotactically attracted to *A. oscillarioides* heterocysts and N-containing compounds, including amino acids (Gallucci and Paerl 1983 in Paerl 1990).

Cyanobacteria excrete organic compounds, including organic and amino acids, peptides, alkaloids, carbohydrates and lipopolysaccharides. Excretion products both attract (chemotactically) and support the growth of phycosphere-associated bacteria (Paerl and Gallucci 1985). $^{15}N_2$ fixed by host *A. oscillarioides* was rapidly transferred to heterocyst-associated *P. aeruginosa* (Paerl 1984). Axenic isolates of *A. oscillarioides* showed optimal growth and N_2 fixation rates when inoculated with *P. aeruginosa* (Paerl and Kellar 1978), which specifically recolonized heterocysts.

Among *Oscillatoria* and *Lyngbya*, bacterized strains revealed higher growth rates and were easier to maintain in culture than axenic strains (Paerl 1982 in Paerl et al. 2001). Mutually beneficial mechanisms include exchange of nutrients, metabolites and gases as well as detoxifying roles of associated bacteria (Paerl 1986). Cyanobacterial toxins may mediate both antagonistic (protozoan and metazoan grazing, viral and bacterial lysis), and mutually beneficial consortial associations with heterotrophic bacteria (Paerl and Millie

1996). In this context, "toxins" are chemical mediators of microbial interactions and bloom dynamics.

Effects of cyanoHABs on higher trophic levels may occur through direct exposure to cyanotoxins, trophic transfer of toxins, or altered food chain structure (Paerl et al. 2001). CyanoHABs are often associated with decreases in abundance of large cladocerans and increases in importance of smaller cladocerans, rotifers and copepods (Leonard and Paerl 2005). Decreases in large cladocerans may be due to increased predation by planktivorous fish, or due to direct negative effects by cyanobacteria (Ghadouani et al. 2003). Inhibitory effects of cyanobacteria on grazing zooplankton may be due to interference with feeding as a result of awkward size or shape of filaments or colonies (Fulton and Paerl 1987 in Paerl et al. 2001; DeMott et al. 2001), chemical factors (toxicity, poor taste, poor nutritional value), or high abundances (i.e., blooms), which displace more nutritious non-bloom-forming algae or limit the ability of herbivores to utilize coexisting algae.

Cyanobacterial toxins can cause rapid mortality of herbivorous zooplankton (e.g., Fulton 1988; Rohrlack et al. 1999, both in Paerl et al. 2001), as well as long-term chronic effects on zooplankton growth and reproduction (e.g., Lürling 2003). While purified microcystin and anatoxin have toxic effects on zooplankton (Gilbert 1994 in Paerl et al. 2001), other compounds from cyanobacteria not generally considered to be "toxins" may be more inhibitory to zooplankton than known toxins (Rohrlack et al. 1999 in Paerl et al. 2001; Lürling 2003). Generally, toxic effects of "healthy" cyanobacteria on zooplankton appear to be due to consumption of cells containing endotoxins (Paerl et al. 2001; Rohrlack et al. 2001). However, filtrates or purified toxins from cyanobacteria can also inhibit herbivorous zooplankton (Paerl et al. 2001). Release of endotoxins from cyanobacteria during senescence or following treatment of blooms by algicides could have a strong impact on zooplankton communities.

Even when there is no evidence of cyanobacterial toxicity, unpalatability (Rohrlack et al. 1999 in Paerl et al. 2001) and poor nutritional value can adversely affect herbivorous zooplankton. Factors contributing to poor nutritional value include indigestibility and deficiency of highly unsaturated fatty acids (Ferrão-Filho et al. 2000). Although cyanobacteria are usually poor sole food sources for zooplankton, they can complement other food sources in mixed diets (Koski et al. 2002), suggesting that non-toxic cyanobacteria can contribute to zooplankton nutrition even when they do not dominate the phytoplankton community.

Microcystis aeruginosa can inhibit grazing rates, be poorly digested, or of poor nutritional value for herbivorous fish (Kamjunke et al. 2002). Fish kills associated with cyanobacterial blooms often result from oxygen depletion, but cyanobacterial toxins can also kill fish directly (Rodger et al. 1994 in Paerl et al. 2001). Conflicting evidence exists regarding effects of *Microcystis* on feeding rates and food selection by the zebra mussel, *Dreissena* (Vanderploeg et al. 2001; Pires et al. 2005). Cyanotoxins can accumulate in consumers, including freshwater clams, crayfish, zooplankton, gastropods, and fish (Paerl et al. 2001; Ibelings et al. 2005).

Although cyanobacterial blooms produce large biomass, their toxic and other inhibitory effects on grazers could produce alterations in aquatic food web structure that would reduce productivity of higher trophic levels. A shift in zooplankton composition from large cladocerans to smaller or more evasive cladocerans, rotifers and copepods may reduce the feeding success of young-of-the-year and other planktivorous fish. If cyanobacterial blooms exclude more nutritious algae, even zooplankton relatively resistant to cyanobacterial toxins may decrease in abundance and this could potentially lead to fish recruitment failures. Senescence or herbicidal treatment of blooms may release dissolved toxins, affecting organisms that would not normally consume cyanobacteria. Much of the biomass from these blooms probably enters microbial or detrital trophic pathways, with additional steps in the food chain reducing the potential productivity of fish and other commercially important species.

8.4 CyanoHAB Management

Ecosystem level, physical, chemical and biotic regulatory variables often co-occur and may interact synergistically and antagonistically to control N_2 fixation, photosynthesis and growth potentials of cyanoHABs (Paerl and Millie 1996) (Fig. 8.2). Potentially effective, achievable means of controlling bloom expansion include; (1) applications of algacides, the most common of which is copper sulfate, (2) nutrient input reductions and ratio manipulation, (3) disrupting vertical stratification, through mechanically or hydrologically induced vertical mixing, (4) reducing retention time (increasing flushing) of bloom-impacted waters, and (5) biological manipulation. Option 1 has been used with considerable short-term success in small impoundments, including ponds and small reservoirs. This approach is not advised for larger ecosystems or any waters to be used for fishing, drinking water and other animal or human purposes. Likewise, in small water bodies, option (3) may be feasible. If abundant water supplies (i.e., upstream reservoirs) are available for hydrodynamic manipulative purposes, option (4) may be possible. Biological manipulation includes approaches to increase grazing pressure on cyanoHABs and reduce recycling of nutrients. Biomanipulation approaches include introducing benthic filter feeders or fish (e.g., *Tilapia*) capable of directly consuming cyanobacteria from the water column, or introduction of lytic bacteria and viruses. Questions have been raised about the long-term efficacy of curtailing cyanobacterial blooms by increasing grazing pressure, because this may lead to dominance by ungrazable or toxic strains

(Ghadouani et al. 2003). Presently, biomanipulation is viewed as one component of an integrated approach to water quality management when nutrient reductions alone are insufficient to restore water quality (Elser 1999). Overall, nutrient input controls remain the most effective and feasible long-term option.

Acknowledgements. We appreciate the assistance of A. and J. Joyner, J. Dyble, K. Kononen, P. Moisander, L. Valdes, J. Leonard, and B. Peierls. Support came from the National Science Foundation (DEB 9815495, OCE 9905723), the US Department of Agriculture NRI Project 00-35101-9981, U.S. EPA STAR Projects R82-5243-010 and R82867701, NOAA/North Carolina Sea Grant Program R/MER-43, the St. Johns Water Management District, Florida, and the Florida Department of Health.

References

Chorus I, Bartram J (eds) (1999) Toxic cyanobacteria in water. E&F Spon, London, 416 pp

Demott WR, Gulati RD, Van Donk E (2001) *Daphnia* food limitation in three hypereutrophic Dutch lakes: Evidence for exclusion of large-bodied species by interfering filaments of cyanobacteria. Limnol Oceanogr 46:2054–2060

Dennison WC, O'Neil JM, Duffy EJ, Oliver PE, Shaw GR (1999) Blooms of the cyanobacterium *Lyngbya majuscula* in coastal waters of Queensland, Australia. In: Charpy L, Larkum A (eds) Marine cyanobacteria, Bulletin de l'Institut Oceanographique, Monaco, pp 501–507

Downing JA, Watson SB, McCauley E (2001) Predicting cyanobacteria dominance in lakes. Can J Fish Aquat Sci 58:1905–1908

Dyble JH, Paerl HW, Neilan BA (2002) Genetic characterization of *Cylindrospermopsis raciborskii* (Cyanobacteria) isolates from diverse geographic origins based on *nifH* and *cpcBA*-IGS nucleotide sequence analysis. Appl Environ Microbiol 68:2567–2571

Edmondson WT, Lehman JT (1981) The effect of changes in the nutrient income and conditions of Lake Washington. Limnol Oceanogr 26:1–29

Elser JJ (1999) The pathway to noxious cyanobacteria blooms in lakes: The food web as the final turn. Freshwater Biol 42:537–543

Elmgren R, Larsson U (2001) Nitrogen and the Baltic Sea: Managing nitrogen in relation to phosphorus. The Scientific World 1371–377

Ferrão-Filho AS, Azevedo SMFO, DeMott WR (2000) Effects of toxic and non-toxic cyanobacteria on the life history of tropical and temperate cladocerans. Freshw Biol 45:1–19

Fogg GE (1969) The physiology of an algal nuisance. Proc R Soc Lond B. 173:175–189

Ghadouani A, Pinel-Alloul B, Prepas EE (2003) Effects of experimentally induced cyanobacterial blooms on crustacean zooplankton communities. Freshw Biol 48:363–381

Horne AJ (1977) Nitrogen fixation – a review of this phenomenon as a polluting process. Progr Wat Technol 8:359–372

Huisman JM, Visser PM (2005) Harmful cyanobacteria. Aquatic ecology series 3. Springer, Berlin Heidelberg New York, 243 pp

Ibelings BW, Mur LR, Walsby AE (1991) Diurnal changes in buoyancy and vertical distribution in populations of *Microcystis* in two shallow lakes. J Plankton Res 13:419–436

Ibelings BW, Bruning K, De Jonge J, Wolfstein K, Pires LM, Postma J, Burger T (2005) Distribution of microcystins in a lake food web: no evidence for biomagnification. Microb Ecol 49:487–500

Jones K (1990) Aerobic nitrogen fixation by *Lyngbya* sp., a marine tropical cyanobacterium. Brit Phycol J 25:287–289

Kamjunke N, Mendonca R, Hardewig I, Mehner T (2002) Assimilation of different cyanobacteria as food and the consequences for internal energy stores of juvenile roach. J Fish Biol 60:731–738

Kononen K, Kuparinen J, Mäkelä J, Laanemets J, Pavelson J, Nõmmann S (1996) Initiation of cyanobacterial blooms in a frontal region at the entrance to the Gulf of Finland, Baltic Sea. Limnol Oceanogr 41:98–112

Koski M, Schmidt K, Engström-Öst J, Vittasalo M, Jónasdóttir S, Repka S, Sivonen K (2002) Calanoid copepods feed and produce eggs in the presence of toxic cyanobacteria *Nodularia spumigena*. Limnol Oceanogr 47:878–885

Lapointe BE (1997) Nutrient thresholds for bottom-up control of macroalgal blooms on coral reefs in Jamaica and southeast Florida. Limnol Oceanogr 42:312–323

Leonard JA, Paerl HW (2005) Zooplankton community structure, microzooplankton grazing impact, and seston energy content in the St. Johns river system, FL as influenced by the toxic cyanobacterium *Cylindrospermopsis raciborskii*. Hydrobiologia 537:89–97

Lürling M (2003) *Daphnia* growth on microcystin-producing and microcystin-free *Microcystis aeruginosa* in different mixtures with the green alga *Scenedesmus obliquus*. Limnol Oceanogr 48:2214–2220

MacGregor BJ, Van Mooy B, Baker BJ, Mellon M, Moisander PH, Paerl HW, Zehr J, Hollander D, Stahl DA (2001) Microbiological, molecular biological and stable isotopic evidence for N_2 fixation in the open waters of Lake Michigan. Environ Microb 3:205–219

Moisander PH, Paerl HW (2000) Growth, primary productivity and nitrogen fixation potential of *Nodularia* spp. (Cyanophyceae) in water from a subtropical estuary in the United States. J Phycol 36:645–658

Moisander PH, McClinton III E, Paerl HW (2002a) Salinity effects on growth, photosynthetic parameters and nitrogenase activity in estuarine planktonic cyanobacteria. Microb Ecol 43:432–442

Moisander PH, Hench JL, Kononen K, Paerl HW (2002b) Small-scale shear effects on heterocystous cyanobacteria. Limnol Oceanogr 47:108–119

Paerl HW (1984) Transfer of N_2 and CO_2 fixation products from *Anabaena oscillarioides* to associated bacteria during inorganic carbon sufficiency and deficiency. J Phycol 20:600–608

Paerl HW (1986) Growth and reproductive strategies of freshwater blue-green algae (cyanobacteria), pp 261–317, In: Sandgren CD (ed) Growth and reproductive strategies of freshwater phytoplankton. Cambridge Univ Press, Cambridge 442 pp

Paerl HW (1988) Nuisance phytoplankton blooms in coastal, estuarine, and inland waters. Limnol Oceanogr 33:823–847

Paerl HW (1990) Physiological ecology and regulation of N_2 fixation in natural waters. Adv Microbiol Ecol 11:305–344

Paerl HW, Gallucci KK (1985) Role of chemotaxis in establishing a specific nitrogen-fixing cyanobacterial-bacterial association. Science 227:647–649

Paerl HW, Millie DF (1996) Physiological ecology of toxic cyanobacteria. Phycologia 35:160–167

Paerl HW, Pinckney JL (1996) Microbial consortia: Their roles in aquatic production and biogeochemical cycling. Microb Ecol 31:225–247

Paerl HW, Zehr JP (2000) Nitrogen fixation. In: Kirchman D (ed) Microbial ecology of the oceans. Academic Press, New York, pp 387–426

Paerl HW, Fulton RS, Moisander PH, Dyble J (2001) Harmful freshwater algal blooms, with an emphasis on cyanobacteria. The Scientific World 1:76–113

Paerl HW, Valdes LM, Piehler MF, Lebo ME (2004) Solving problems resulting from solutions: the evolution of a dual nutrient management strategy for the eutrophying Neuse River Estuary, NC, USA. Environ Sci Tech 38:3068–3073

Panosso R, Granéli E (2000) Effects of dissolved organic matter on the growth of Nodularia spumigena (Cyanophyceae) cultivated under N and P deficiency. Mar Biol 136:331–336

Paul VJ, Thacker RW, Cruz-Rivera E (2001) Chemical mediation of seaweed-herbivore interactions: ecological and evolutionary perspectives. In: McClintock J, Baker B (eds) Marine chemical ecology. CRC Press, New York, pp 227–265

Phlips EJ, Ihnat J, Conroy M (1991) Nitrogen fixation by the benthic freshwater cyanobacterium *Lyngbya wollei*. Hydrobiologia 234:59–64

Piehler MF, Dyble J, Moisander PH, Pinckney JL, Paerl HW (2002) Effects of modified nutrient concentrations and ratios on the structure and function of the native phytoplankton community in the Neuse River Estuary, North Carolina USA. Aquat Ecol 36:371–385

Pires LMD, Bontes BM, Van Donk E, Ibelins B (2005) Grazing on colonial and filamentous, toxic and non-toxic cyanobacteria by the zebra mussel *Dreissena polymorpha*. J Plankton Res 27:331–339

Potts M (1994) Desiccation tolerance of prokaryotes. Microbiol Rev 58:755–805

Reynolds CS (1987) Cyanobacterial water blooms. Adv Bot Res 13:67–143

Rohrlack T, Dittman E, Börner T, Christoffersen K (2001) Effects of cell-bound microcystins on survival and feeding of *Daphnia* spp. Appl Environ Microbiol 67:3523–3529

Smith VH (1990) Nitrogen, phosphorus and nitrogen fixation in lacustrine and estuarine ecosystems. Limnol Oceanogr 35:1852–1859

Speziale BJ, Dyck LA (1992) *Lyngbya* infestations: comparative taxonomy of *Lyngbya wollei* comb. nov. (cyanobacteria). J Phycol 28:693–706

Vanderploeg HA, Liebig JR, Carmichael WW, Agy MA, Johengen TH, Fahnenstiel GL, Nalepa TF (2001) Zebra mussel selective filtration promoted toxic *Microcystis* blooms in Saginaw Bay (Lake Huron) and Lake Erie. Can J Fish Aquat Sci 58:1208–1221

9 Brown Tides

C. J. Gobler and W. G. Sunda

9.1 Background

Brown tides are caused by the pelagophytes *Aureococcus anophagefferens* Hargraves et Sieburth and *Aureoumbra lagunensis* DeYoe et Stockwell. *Aureococcus* has caused destructive 'brown tide' blooms in northeast and mid-Atlantic US estuaries for two decades (Nuzzi and Waters 2004), and more recently in South Africa (Probyn et al. 2001). *Aureoumbra* has caused similar blooms in coastal bays in and around the Laguna Madre, Texas, for the past 15 years (Buskey et al. 1997; Villareal et al. 2004). Neither species was known to science prior to the first documented bloom events, but both have since gained notoriety for their ability to disrupt and damage the coastal ecosystems in which they occur.

Although *Aureococcus* and *Aureoumbra* are genetically distinct organisms (DeYoe et al. 1995), they share many similarities. Both contain the unique pigment 19'-butanoyloxyfucoxanthin and are small (4–5 µm for *Aureoumbra* and 2–3 µm for *Aureococcus*), spherical, non-motile cells with simple life cycles (DeYoe et al. 1997). They were originally assigned to the class Chrysophyceae (Sieburth et al. 1988; Buskey and Stockwell 1993), but later examination of their pigments, physiology, 18S rRNA sequences, and morphology led to their formal placement in the class Pelagophyceae (DeYoe et al. 1997). Both contain the pelagophyte-specific sterol, (Z)-24-propylidenecholesterol, which further supports this classification (Giner et al. 2001). While they are sufficiently different genetically to be placed in separate genera, the genetic diversity within each species is poorly described. There has been only one study of intra-species diversity in brown tides, Bailey and Andersen (1999), which examined 14 clones of *Aureococcus* isolated over a 12-year period (1986–1998) from three New York (NY) and New Jersey (NJ) estuaries. Bailey and Andersen found identical sequences for the small subunit rRNA gene, the small subunit RUBISCO gene, and the non-encoding spacer regions of RUBISCO, suggesting little genetic diversity among *Aureococcus* blooms. Despite these genetic similarities, individual *Aureococcus* clones differ in their

Ecological Studies, Vol. 189
Edna Granéli and Jefferson T. Turner (Eds.)
Ecology of Harmful Algae

impacts on shellfish (Bricelj et al. 2001) and their susceptibility to viral lysis (Gobler et al. 2004a). However, it is unknown if these differences have a genetic basis.

The first *Aureococcus* blooms occurred simultaneously in the summer of 1985 in several estuaries in the northeastern USA, including Narragansett Bay, Rhode Island, Great South Bay and the Peconic Estuary on Long Island, NY, and putatively in Barnegat Bay, NJ (Cosper et al. 1989; Sieburth et al. 1988; Olsen 1989). Blooms returned to Long Island bays in 1986–88 and have since occurred sporadically in these waters. They have not occurred in the Peconic Estuary since 1995, although low cell densities are still present (up to 10^4 ml^{-1}; Nuzzi and Waters 2004). While *Aureococcus* has been observed in Narragansett Bay in subsequent years, it has not formed a significant bloom there since 1985 (Sieburth et al. 1988). In recent years, blooms have expanded south along the US east coast into bays in New Jersey (Barnegat Bay; Gastrich et al. 2004), Delaware (Little Assawoman Bay; Popels et al. 2003), Maryland, and Virginia (Chincoteague Bay; Trice et al. 2004). *Aureococcus* blooms have also recently occurred in Saldanha Bay, South Africa (Probyn et al. 2001). Low abundances of *Aureococcus* cells have been observed along the entire eastern seaboard of the USA from Maine to Florida, indicating the potential for these blooms to continue to spread north and south of their current range (Anderson et al. 1993; Popels et al. 2003).

The initial *Aureoumbra* bloom in the Laguna Madre and Baffin Bay, Texas began in January 1990 and persisted for nearly 8 years, representing the longest continuous HAB event ever recorded (Buskey et al. 2001). During this time, *Aureoumbra* comprised most of the algal biomass and generally maintained cell densities from $0.5–5x10^6$ ml^{-1} (Buskey and Stockwell 1993; Buskey et al. 1997). The bloom terminated in the fall of 1997, but re-emerged during the summer and fall of 1998 (Buskey et al. 2001). Since then it has occurred intermittently in the Laguna Madre system. Low concentrations of *Aureoumbra* cells have also been found in coastal bays across Florida, Texas, and Mexico (Villareal et al. 2004).

During the 1985–1987 *Aureococcus* blooms in eastern and southern Long Island bays, high cell concentrations ($>10^6$ ml^{-1}) substantially increased light attenuation, which caused a large-scale die-off of seagrass beds of *Zostera marina*, a critical habitat for scallops, larval fish, and other species (Dennison 1988). The bloom caused mass mortality and recruitment failure in populations of *Argopecten irradians*, which resulted in the collapse of the multi-million dollar scallop industry in eastern Long Island (Bricelj et al. 1989). The *Aureococcus* blooms also appear to have negatively impacted populations of clams (*Mercenaria mercenaria*) in Great South Bay, which was formerly the largest clam fishery in the state of New York. Subsequent research has established that *Aureococcus* adversely impacts the growth and survival of many algal grazers, including juvenile and adult hard clams (*M. mercenaria*; Bricelj et al. 2001; Greenfield and Lonsdale 2002), larval and adult bay scallops (*A.

irradians; Bricelj et al. 1989; Gallager et al. 1989), adult blue mussels (*Mytilus edulis*; Bricelj et al. 1989), and micro- and mesozooplankton (Lonsdale et al. 1996, Caron et al. 2004). Although a toxin has never been isolated from *Aureococcus*, there is evidence for toxic activity, most likely within the extracellular polysaccharide (EPS) sheath surrounding the cells (Sieburth et al. 1988). This putative toxin deters feeding in bivalves by causing the cessation of cilia movement (Gainey and Shumway 1991). In addition, it has also been argued that *Aureococcus* may be a poor source of nutrition for zooplankton (Lonsdale et al. 1996, Caron et al. 2004).

The ecological impacts of *Aureoumbra* appear to have been similar to those of *Aureococcus*, although the economic impacts have been lower because of a less-developed shellfish industry in and around Laguna Madre. As with *Aureococcus*, the *Aureoumbra* bloom caused a substantial increase in light attenuation, and the resulting shading of the bottom has decreased the abundance of the once extensive sea grass beds (Onuf 1996). There has also been a decrease in the biomass and diversity of benthic invertebrates in the Laguna Madre (Ward et al. 2000). The dominant clam, *Mulinia lateralis*, virtually disappeared after the onset of the brown tide bloom (Montagna et al. 1993), and the dominate polychaete, *Streblospio benedicti*, an important grazer of phytoplankton, has declined in abundance by two orders of magnitude (Buskey et al. 1997). However, population decreases in both species may have begun during the period of high salinity and freezing temperatures prior to the bloom (Buskey et al. 1997). The brown tide bloom also was associated with substantial decreases in the grazing activity, growth, and egg release rates in mesozooplankton (e.g., *Acartia tonsa*), and decreases in the abundance and grazing rates of microzooplankton (Buskey and Stockwell 1993). Like *Aureococcus*, an EPS polymer sheath covers the cell surface of *Aureoumbra*. In feeding experiments with *Aureoumbra*, the presence of high levels of this exopolymer decreased rates of grazing and/or growth in three protozoan species and altered protozoan motility (Liu and Buskey 2000). Liu and Buskey speculated that these effects on grazing and motility may have been caused by the adherence of the exopolymer to cilia on the surface of the protozoans. These cilia are essential for protozoan feeding and motility. Alternatively, the adverse effects may be caused by an unidentified toxin within the polysaccharide sheath, as has been suggested for *Aureococcus*.

9.2 Nutrients and Physical Factors

Although harmful algal blooms in coastal waters have been commonly attributed to nutrient loading, the role of nutrients in the occurrence of brown tides appears to be more complex than a simple nutrient stimulation of brown tide growth. An examination of spatial and temporal patterns of concentrations of

Aureococcus cells and inorganic nutrients indicated that blooms occurred when inorganic nutrient levels were low (Cosper et al. 1989; LaRoche et al. 1997). Moreover, nitrate additions during mesocosm and bottle experiments have consistently yielded reduced *Aureococcus* cell densities relative to those of competing algae (Keller and Rice 1989; Gobler and Sanudo-Wilhelmy 2001a; Gobler et al. 2002, 2004b, 2004 c). The nitrogen-uptake characteristics of *Aureococcus* (low K_s and V_{max} for ammonium) suggest that this species is well adapted to low-nutrient environments. The ability of *Aureococcus* to attain high biomass levels when inorganic nutrient concentrations are low is partly linked to its ability to utilize organic forms of C, N, and P. Cultures (both axenic and non-axenic) and field populations of *Aureococcus* have been shown to obtain nitrogen from a variety of organic compounds, including urea, amino acids, proteins, chitobiose, and acetamide (Berg et al. 2002; Mulholland et al. 2002), and to assimilate organic carbon from glucose and amino acids (Mulholland et al. 2002). Experimental additions of DOM (glucose, amino acids, DOM from marcroalgae) have enhanced the growth and relative abundance of *Aureococcus* during field experiments (Gobler and Sanudo-Wilhelmy 2001a). Dissolved organic nitrogen (DON) and carbon (DOC) are often elevated during bloom initiation (LaRoche et al. 1997; Gobler et al. 2002, 2004b) and reductions in DOC and DON concentrations have been noted with bloom development, a pattern consistent with utilization of these substances by *Aureococcus* (LaRoche et al. 1997; Gobler et al. 2004b). While DON provides an important N source to blooms, heterotrophic C consumption may partly circumvent the need for photosynthetic carbon fixation. This mixotrophy may give *Aureococcus* a competitive advantage over strictly autotrophic algae under severe shading during blooms.

Blooms of *Aureococcus* often occur after 'pre-blooms' of other algae, which both draw down inorganic nutrients to low levels and cause gradual increases in more biologically refractory DON and DOC pools (Gobler and Sanudo-Wilhelmy 2001b). Nutrient remineralization processes during, and following these preblooms can result in enhanced DOM levels, which can serve as a source of DON and DOC to subsequent *Aureococcus* blooms (LaRoche et al. 1997, Gobler and Sañudo-Wilhelmy 2001b). Since such blooms usually occur in shallow bays, fluxes from the benthos are important sources of inorganic and organic nutrients (Lomas et al. 2004). Whether sediments serve as a net source or sink for nutrients often depends on the degree to which benthic primary producers (microalgae, sea grass, or macroalgae) utilize nutrients regenerated in sediments (MacIntyre et al. 2004). Brown-tide shading of the bottom causes a net loss of these benthic phototrophs, and this loss provides a source of nutrients and DOM via plant matter decay and recycling processes. This loss also increases net benthic flux of nutrients to the water column due to decreased utilization by benthic phototrophs. Both processes should promote the development of brown tide blooms (MacIntyre et al. 2004).

Considerably less is known about the role of nutrients in the development and persistence of *Aureoumbra* blooms. The initial brown tide bloom in the Laguna Madre and Baffin Bay, Texas developed after an extremely dry period and a rare freeze in December 1989. This freeze caused extensive mortalities of fish and other organisms, and a subsequent large pulse of regenerated ammonium (up to 25 μM; Buskey et al. 1996). The ammonium pulse stimulated the growth of *Aureoumbra* to bloom levels (~10^6 cells L^{-1}) in the upper reaches of Baffin Bay during January 1990, but the bloom did not spread to lower Baffin Bay or the adjacent Laguna Madre until June of that year (Buskey et al. 1997). Thereafter, ammonia concentrations plummeted to very low levels (<1 μM) in the Laguna Madre system and remained low during the following year (Buskey et al. 1996). The decrease in ammonia in the Laguna Madre may be attributed to the high demand created by the high *Aureoumbra* biomass. However, there is evidence that this species is also well adapted to low concentrations of available nutrients. Chemostat studies indicate that *Aureoumbra* can grow at extremely low cellular P:C ratios, and thus may be well adapted to low-phosphorus environments (Liu et al. 2001). Moreover, its small cell size, like that of *Aureococcus*, results in a high surface to volume ratio and a thin surface diffusive boundary layer, which should further favor growth of this species at low-nutrient concentrations (Raven and Kubler 2002). *Aureoumbra* utilizes ammonia or urea as nitrogen sources, but is unable to grow on nitrate, usually the most available nitrogen source under high-nutrient conditions (DeYoe and Suttle 1994). In a mesocosm study in the Laguna Madre in the summer of 1999, the addition of 40 μM ammonia to water containing 0.4 μM of that nutrient had no effect on the net growth rate of *Aureoumbra*, suggesting that brown tide growth was not limited by nitrogen availability (Buskey et al. 2003). There have been no studies of the ability of *Aureoumbra* to utilize organic sources of nitrogen, but it is likely that this putative low-nutrient species, like *Aureococcus* and other low-nutrient species, is able to grow on a variety of organic nitrogen substrates. The mixotrophic capabilities in *Aureoumbra* have not been examined.

Physical factors (salinity, temperature, light, and water residence times) can also influence brown tide blooms. Both *Aureococcus* and *Aureoumbra* blooms are associated with high chlorophyll-*a* (chl-*a*) concentrations (30–60 μg L^{-1}) and severe light attenuation. *Aureococcus* is able to maintain a near-maximum growth rate (at 20 °C) at a light intensity of 100 μmol photons m^{-2} s^{-1} (~4% of noon solar irradiance; Milligan and Cosper 1997; MacIntyre et al. 2004) while *Aureoumbra* is able to do the same at a slightly higher irradiance (150 μmol photons m^{-2} s^{-1}; Buskey et al. 1998). Curiously, the accessory pigments in *Aureococcus* make it best adapted for growth at low levels of blue light, as found in the deep chlorophyll maximum in the open ocean where pelagophytes are commonly found, and less-well-adapted for growth at low intensities of the longer wavelength visible light that occur during brown tide blooms (Yentsch et al. 1989). Based on these observations,

Yentsch et al. suggested that *Aureococcus* might be an expatriate open-ocean species. The ability of the two brown-tide species to grow maximally at low light intensities likely contributes to their ability to compete well with non-HAB species during blooms (Milligan and Cosper 1997; MacIntyre et al. 2004).

Aureococcus grows optimally at salinities > 24 PSU and blooms of this species tend to occur during dry years with elevated salinities (Cosper et al. 1989; LaRoche et al. 1997). However, this phenomenon is more likely related to the reduced inorganic nutrient loading from reduced groundwater inputs, rather than salinity per se since the salinities of most bays that experience brown tide blooms rarely fall below 24 PSU (Cosper et al. 1989; LaRoche et al. 1997; Gobler and Sañudo-Wilhelmy 2001b).

Aureoumbra can grow over a wide range of salinities (10–90 PSU) and maintains near optimal growth at salinities of 20–70 PSU at 25 °C (Buskey et al. 1998). This ability to acclimate to a wide salinity range and to grow well at very high salinities appears to have contributed to the initiation of the brown tide bloom in Laguna Madre (Buskey et al. 1998). The Laguna Madre estuary is a shallow (1.2 m average depth) hypersaline lagoon within a semi-arid region of south Texas. Salinities can vary considerably, depending on rainfall, and the initial *Aureoumbra* bloom occurred after a prolonged dry period. It was initiated in the hypersaline upper reaches of Baffin Bay in January 1990, where maximum salinities reached 70 PSU. This salinity was within the range for near-optimal growth of *Aureoumbra*, but well above the optimal range for most algal grazers and competing algal species (Buskey et al. 1998). During the period just before the bloom there was a significant negative correlation between microzooplankton abundance and salinity, with very low abundances (<10 ml^{-1}) at the highest salinities. The low grazing pressure on *Aureoumbra* associated with the high salinities was likely a contributing factor in bloom initiation (Buskey et al. 1998).

The wide salinity tolerance of *Aureoumbra* also likely contributed to the unprecedented persistence of the bloom. The bloom temporally terminated in the fall of 1997, after intense rains decreased average salinities in Baffin Bay from ~50 to 15 PSU, below the range for optimal *Aureoumbra* growth (Buskey et al. 2001). Thus, the demise of the bloom may have been partly caused by a decrease in brown tide growth and a concomitant increase in the growth rate of competing phytoplankton associated with the lower salinities. In addition, an influx of nutrients accompanying the large freshwater input stimulated the growth of other algae such as the diatom *Rhizosolenia* sp. (Buskey et al. 1998).

Aureococcus blooms under a wide range of temperatures (0–25 °C) and shows optimal growth at ~20 °C (Cosper et al. 1989; Gobler et al. 2002). Blooms typically initiate during late May to early June as temperatures approach 15–20 °C, and decline when temperatures exceed 25 °C (Nuzzi and Waters 2004). However, *Aureococcus* blooms have been observed to persist and even initiate during the fall when temperatures decline, and can persist during winter when temperatures reach freezing levels (Nuzzi and Waters 2004; Gobler et al. 2002).

Temperature may also influence *Aureoumbra* bloom dynamics. Temperatures in the Laguna Madre usually range from 10–15 °C in the winter to ~30 °C in summer, within a favorable range for *Aureoumbra* (Buskey et al. 1996, 1998). In culture studies, the specific growth rate of *Aureoumbra* increased with temperature within the seasonal range, with a rate of 0.25 d^{-1} at 15 °C, 0.62 d^{-1} at 25 °C, and 0.8 d^{-1} at 30 °C (Buskey et al. 1998; Sunda and Hardison, unpubl. data). Thus unlike *Aureococcus*, the highest summer temperatures in the Laguna Madre system appear to favor rapid growth of *Aureoumbra*. The ability to grow over a wide range of temperatures likely contributed to the persistence of the Texas brown tide bloom.

While bays that host frequent *Aureococcus* blooms are known to have high salinities and are relatively shallow, an additional attribute of these bays is their long residence times (up to 100 days), which are associated with low freshwater inputs and low rates of mixing with coastal seawater (Wilson 1995). Similarly, estimates of residence times for the upper Laguna Madre and Baffin Bay, Texas, which host *Aureoumbra* blooms, can range from 300 days to several years (Buskey et al. 1998, 2001). While long residence times clearly permit the accumulation of high algal biomass, they would not necessarily give a competitive advantage to one algal species over another. However, the high biomass and low inputs of external nutrients in these systems create large algal nutrient demand:supply ratios, and hence low concentrations of available inorganic nutrients. Moreover, intense nutrient recycling within these shallow, long-residence time systems promotes the accumulation of DOM (Lomas et al. 2004). Therefore, although the long residence times of these shallow estuaries do not themselves directly stimulate brown tide blooms, the resulting low concentrations of available nutrients and low ratios of inorganic to organic pools of N and P likely favor the net growth of brown tide species.

9.3 Sources of Cell Mortality

Because of their small size, *Aureococcus* and *Aureoumbra* should be efficiently grazed by microzooplankton, whose rapid growth rates should prevent blooms from developing (Sherr and Sherr 2002). However, grazing rates on these species are typically low due to unpalatability, physical interference with grazing, or toxicity. These low rates of grazing mortality have been proposed to be important factors in the initiation and persistence of brown tide blooms (Buskey et al. 1997; Gobler et al. 2002). While the zooplankton are able to actively graze during *Aureococcus* blooms, specific grazing rates on *Aureococcus* can be considerably less (e.g., 30 %; Gobler et al. 2002) than those for competing algae, and lower than concurrent specific growth rates for *Aureococcus*

(Gobler et al. 2002, 2004 c; Caron et al. 2004). Reduced grazing may be partly facilitated by the EPS layer of *Aureococcus*. Another mechanism that may decrease grazing on *Aureococcus* is a microbial trophic cascade in which enhanced grazing by larger zooplankton reduces microzooplankton populations, which in turn reduces grazing on brown tide cells (Sieracki et al. 2004). It appears that this mechanism does not promote Texas brown tides (Buskey et al. 2003).

Aureococcus blooms have waned in their intensity and frequency in some US estuaries (e.g., Great South Bay, NY) and have vanished from others (Peconic Estuary, NY). These systems do not appear to have experienced changes in their chemical or physical characteristics relative to periods when intense blooms first occurred (Nuzzi and Waters 2004). Zooplankton communities can evolve resistance to harmful algae over time via natural selection processes (Hairston et al. 2002). The selective adverse impact of blooms on more sensitive grazer species and more sensitive phenotypes within populations (Caron et al. 2004) should eventually lead to the establishment of grazer communities that are better adapted to co-exist with and consume *Aureococcus*. Recent studies have shown that some NY estuaries, which formerly hosted zooplankton communities that consumed *Aureococcus* at low rates during massive blooms, now have communities that actively graze *Aureococcus* at similar rates to those for other species when brown tide densities are low (10^4 cells ml^{-1}; Deonarine 2005). As such, the recent abatement of brown tides in some Long Island bays (e.g., Peconic Estuary and West Neck Bay; Nuzzi and Waters 2004) may reflect a shift in grazing communities toward an increased ability to feed and grow on *Aureococcus*.

The occurrence of brown tides in Long Island bays may be partly related to the dramatic decline in shellfish populations from over-harvesting in recent decades, and the subsequent loss of this important source of algal grazing mortality. Such a loss of these benthic filter feeders may have shifted planktonic grazing pressure toward zooplankton, which may be more efficient than bivalves at food selection and rejection, including avoidance of *Aureococcus* (Gobler et al. 2002). In support of this hypothesis, mesocosms containing high but environmentally realistic concentrations of hard clams (*M. mercenaria*) maintained non-bloom densities of *Aureococcus* ($\sim 10^3$ ml^{-1}), while identical mesocosms with no or few clams developed dense brown tide blooms ($>10^5$ ml^{-1}) (Cerrato et al. 2004). *Aureococcus* inhibits *M. mercenaria* grazing rates at densities above 4×10^3 cells ml^{-1}. Thus, there may be a threshold effect whereby the clams consume *Aureococcus* at low cell densities, and grazing retards bloom development, but clams are inhibited at high cell concentrations, promoting brown tide blooms (Bricelj et al. 2001).

Reduced grazing pressure appears to have been an important factor in both the initiation and persistence of the eight-year Texas brown tide bloom (Buskey and Stockwell 1993; Buskey et al. 1997). The bloom was initiated in the hypersaline upper reaches of Baffin Bay and was preceded by a period of unusually high salinities (up to 70 PSU), which greatly reduced the abundance of microzooplankton, and important benthic grazers such as the bivalve *Mulinia lateralis* (Buskey et al. 1997). The spread of the bloom to lower Baffin Bay and the Laguna Madre in June 1990 was accompanied by a large decline in both the abundance of mesozooplankton (e.g., *Acartia tonsa*), and in grazing rates of microzooplankton (Buskey and Stockwell 1993). Gut contents, growth, and egg laying in adult *Acartia* were severely reduced during the bloom, suggesting that the zooplankton were either unable to feed upon, or actively avoided *Aureoumbra* cells. Subsequent laboratory experiments verified these findings for adult *Acartia* and showed that the presence of *Aureoumbra* increased the mortality of *Acartia* nauplii relative to unfed (starved) nauplii (Buskey and Hyatt 1995). Increased concentrations of *Aureoumbra* also caused a dose-dependent increase in mortality in two microzooplankton species (the ciliate *Strombidinopsis* sp. and the rotifer *Brachionus plicatilus*). A third species was unable to grow on *Aureoumbra* and three others had reduced growth rates compared to cultures fed on equal concentrations of control algae (Buskey and Hyatt 1995). The increased mortality observed in some zooplankton species suggests the presence of a toxin, but no toxin has yet been identified (Liu and Buskey 2000). The putative toxicity of *Aureoumbra* or an inability of zooplankton to feed and grow on this algal food source likely contributed substantially to the development and persistence of the Texas brown tide bloom (Buskey and Stockwell 1993).

Viruses may influence the dynamics of *Aureococcus* blooms. Electron micrographs of the first observed brown tide event in Rhode Island and subsequent blooms in New Jersey and New York revealed the presence of intracellular, icosahedral virus-like particles in *Aureococcus* cells (Sieburth et al. 1988; Gastrich et al. 2002), suggesting that brown tide populations may experience viral infection and lysis. Virus densities during *Aureococcus* blooms are generally elevated compared to most estuarine environments (Gobler et al. 2004a) and field studies have observed a high percentage (~40 %) of virally infected *Aureococcus* cells as blooms end. Thus, viruses may be an important source of mortality during bloom termination (Gastrich et al. 2004). *Aureococcus*-specific viruses capable of completely lysing brown tide cultures have been isolated from bloom waters in Long Island estuaries (Gobler et al. 2004 c). However, these viruses infect only a portion of available clones of *Aureococcus* (Gobler et al. 2004 c). Hence, viruses may have a greater impact on bloom diversity than biomass. Thus far, only a small number of viral strains have been tested, and there may be others yet to be isolated, which can infect the *Aureococcus* clones resistant to the currently available virus strains. To date, there is no evidence for viruses that infect *Aureoumbra*.

References

Anderson DM, Keafer BA, Kulis DM, Waters RM, Nuzzi R (1993) An immunofluorescent survey of the brown tide chrysophyte *Aureococcus anophagefferens* along the northeast coast of the United States. J Plankton Res 15:563–580

Bailey JC, Andersen RA (1999) Analysis of clonal cultures of the brown tide algae *Aureococcus* and *Aureoumbra* (Pelagophyceae) using 18S rRNA, rbcL, and rubisco spacer sequences. J Phycol 35:570–574

Berg GM, Repeta DJ, LaRoche J (2002) Dissolved organic nitrogen hydrolysis rates in axenic cultures of *Aureococcus anophagefferens* (Pelagophyceae): comparison with heterotrophic bacteria. Appl Environ Microbiol 68:401–404

Bricelj VM, Fisher NS, Guckert JB, Chu F-LE (1989) Lipid composition and nutritional value of the brown tide alga *Aureococcus anophagefferens*. Coastal Estuarine Studies 35:85–100

Bricelj VM, MacQuarrie SP, Schaffner RA (2001) Differential effects of *Aureococcus anophagefferens* isolates („brown tide") in unialgal and mixed suspensions on bivalve feeding. Mar Biol 139:605–615

Buskey EJ, DeYoe H, Jochem FJ, Villareal TA (2003) Effects of mesozooplankton removal and ammonium addition on planktonic trophic structure during a bloom of the Texas 'brown tide': a mesocosm study. J Plankton Res 25:215–228

Buskey EJ, Hyatt CJ (1995) Effects of the Texas (USA) „brown tide" alga on planktonic grazers. Mar Ecol Prog Ser 126:285–292

Buskey EJ, Liu HB, Collumb C, Bersano JGF (2001) The decline and recovery of a persistent Texas brown tide algal bloom in the Laguna Madre (Texas, USA). Estuaries 24:337–346

Buskey EJ, Montagna PA, Amos AF, Whitledge TE (1997) Disruption of grazer populations as a contributing factor to the initiation of the Texas brown tide algal bloom. Limnol Oceanogr 42:1215–1222

Buskey EJ, Stewart S, Peterson J, Collumb C (1996) Current status and historical trends of brown tide and red tide phytoplankton blooms in the Corpus Christi Bay Nat Estuary Progr study area. Texas Nat Resource Cons Comm, Austin, TX, USA, 65 pp

Buskey EJ, Stockwell DA (1993) Effects of a persistent „brown tide" on zooplankton populations in the Laguna Madre of south Texas. In: Smayda TJ, Shimizu Y (eds) Toxic phytoplankton blooms in the sea. Elsevier, Amsterdam, pp 659–666

Buskey EJ, Wysor B, Hyatt C (1998) The role of hypersalinity in the persistence of the Texas 'brown tide' in the Laguna Madre. J Plankton Res 20:1553–1565

Caron DA, Gobler CJ, Lonsdale DJ, Buck NJ, Cerrato RM, Schaffner RA, Rose JM, Taylor GT, Boissonneault KR, Mehran R (2004) Microbial herbivory on the brown tide alga, *Aureococcus anophagefferens*: results from natural ecosystems, mesocosms and laboratory experiments. Harmful Algae 3:439–457

Cerrato RM, Caron DA, Lonsdale DJ, Rose JM, Schaffner RA (2004) An experimental approach to examine the effect of the hard clam *Mercenaria mercenaria* on the development of blooms of the brown tide alga, *Aureococcus anophagefferens*. Mar Ecol Prog Ser 281:93–108

Cosper EM, Dennison W, Milligan A (1989) An examination of environmental factors important to initiating and sustaining „brown tide" blooms. Coastal Est Studies 35:317–340

Dennison WC (1988) „Brown tide" algal blooms shade out eelgrass. J Shellfish Res 7:155

Deonarine SN (2005) Zooplankton grazing on *Aureococcus anophagefferens*: impacts on the ecology of brown tides in U.S. mid-Atlantic region estuaries. MSC Thesis, Stony Brook Univ, 57 pp

DeYoe HR, Chan AM, Suttle CA (1995) Phylogeny of *Aureococcus anophagefferens* and a morphologically similar bloom-forming alga from Texas as determined by 18S ribosomal RNA sequence analysis. J Phycol 31:413–418

DeYoe HR, Suttle CA (1994) The inability of the Texas „brown tide" alga to use nitrate and the role of nitrogen in the initiation of a persistent bloom of this organism. J Phycol 30:800–806

DeYoe HR, Stockwell DA, Bidigare RR, Latasa M, Johnson PW, Hargraves PE, Suttle CA (1997) Description and characterization of the algal species *Aureoumbra lagunensis* gen. et sp. nov. and referral of *Aureoumbra* and *Aureococcus* to the Pelagophyceae. J Phycol 33:1042–1048

Gainey LF, Shumway SE (1991) The physiological effect of *Aureococcus anophagefferens* („brown tide") on the lateral cilia of bivalve mollusks. Biol Bull 181:298–306

Gallager SM, Stoecker DK, Bricelj VM (1989) Effects of the brown tide algae on growth, feeding physiology and locomotory behavior of scallop larvae (*Argopecten irradians*). Coastal Estuarine Studies 35:511–542

Gastrich MD, Anderson OR, Cosper EM (2002) Viral-like particles (VLPS) in the alga, *Aureococcus anophagefferens* (Pelagophyceae), during 1999–2000 brown tide blooms in Little Egg Harbor, New Jersey. Estuaries 25:938–943

Gastrich MD, Bell JL, Gobler CJ, Anderson OR, Wilhelm SW (2004) Viruses as potential regulators of regional brown tide blooms caused by the alga, *Aureococcus anophagefferens:* a comparison of bloom years 1999–2000 and 2002. Estuaries 27:112–119

Giner JL, Li XY, Boyer GL (2001) Sterol composition of *Aureoumbra lagunensis*, the Texas brown tide alga. Phytochemistry 57:787–789

Gobler CJ, Sañudo-Wilhelmy SA (2001a) Effects of organic carbon, organic nitrogen, inorganic nutrients, and iron additions on the growth of phytoplankton and bacteria during a brown tide bloom. Mar Ecol Prog Ser 209:19–34

Gobler CJ, Sañudo-Wilhelmy SA (2001b) Temporal variability of groundwater seepage and brown tide blooms in a Long Island embayment. Mar Ecol Prog Ser 217:299–309

Gobler CJ, Renaghan MJ, Buck NJ (2002) Impacts of nutrients and grazing mortality on the abundance of *Aureococcus anophagefferens* during a New York brown tide bloom. Limnol Oceanogr 47:129–141

Gobler CJ, Deonarine S, Leigh-Bell J, Gastrich MD, Anderson OR, Wilhelm SW (2004a) Ecology of phytoplankton communities dominated by *Aureococcus anophagefferens*: the role of viruses, nutrients, and microzooplankton grazing. Harmful Algae 3:471–483

Gobler CJ, Boneillo GE, Debenham C, Caron DA (2004b) Nutrient limitation, organic matter cycling, and plankton dynamics during an *Aureococcus anophagefferens* bloom in Great South Bay, NY. Aquat Microb Ecol 35:31–43

Gobler CJ, Pererya G, Krause J, Maurer K, Gastrich MD, Anderson OR, Wilhelm SW (2004 c) Impacts of viruses isolated from New York waters on growth of the brown tide alga, *Aureococcus anophagefferens:* a field and laboratory assessment. Harmful Algae 3:209–210

Greenfield DI, Lonsdale DJ (2002) Mortality and growth of juvenile hard clams *Mercenaria mercenaria* during brown tide. Mar Biol 141:1045–1050

Hairston NG, Holtmeier CL, Lampert W, Weider LJ, Post DM, Fisher JM, Caceres CE, Gaedke U (2002) Natural selection for grazer resistance to toxic cyanobacteria: evolution of phenotypic plasticity? Evolution 55:2203–2214

Keller AA, Rice RL (1989) Effects of nutrient enrichment on natural populations of the brown tide phytoplankton *Aureococcus anophagefferens* (Chrysophyceae). J Phycol 25:636–646

LaRoche J, Nuzzi R, Waters R, Wyman K, Falkowski PG, Wallace DWR (1997) Brown tide blooms in Long Island's coastal waters linked to interannual variability in groundwater flow. Glob Change Biol 3:397–410

Liu H, Buskey EJ (2000) The exopolymer secretions (EPS) layer surrounding *Aureoumbra lagunensis* cells affects growth, grazing, and behavior of protozoa. Limnol Oceanogr 45:1187–1191

Liu H, Laws EA, Villareal TA, Buskey EJ (2001) Nutrient-limited growth of *Aureoumbra lagunensis* (Pelagophyceae), with implications for its capability to outgrow other phytoplankton species in phosphate-limited environments. J Phycol 37:500–508

Lomas MW, Kana TM, MacIntyre HL, Cornwell JC, Nuzzi R, Waters R (2004) Interannual variability of *Aureococcus anophagefferens* in Quantuck Bay, Long Island: natural test of the DON hypothesis. Harmful Algae 3:389–402

Lonsdale DJ, Cosper EM, Kim W-S, Doall MH, Divadeenam A, Jonasdottir SH (1996) Food web interactions in the plankton of Long Island bays, with preliminary observations on brown tide effects. Mar Ecol Prog Ser 134:247–263

MacIntyre HL, Lomas MW, Cornwell J, Suggett DJ, Gobler CJ, Koch EW, Kana TM (2004) Mediation of benthic-pelagic coupling by microphytobenthos: an energy- and material-based model for initiation of blooms of *Aureococcus anophagefferens.* Harmful Algae 3:403–437

Milligan AJ, Cosper EM (1997) Growth and photosynthesis of the 'brown tide' microalga *Aureococcus anophagefferens* in subsaturating constant and fluctuating irradiance. Mar Ecol Prog Ser 153:67–75

Montagna PA, Stockwell DA, Kalke RD (1993) Dwarf surf clam *Mulinea lateralis* (Say, 1822) populations and feeding during the Texas brown tide. J Shellfish Res 12:833–842

Mulholland MR, Gobler CJ, Lee C (2002) Peptide hydrolysis, amino acid oxidation, and nitrogen uptake in communities seasonally dominated by *Aureococcus anophagefferens.* Limnol Oceanogr 47:1094–1108

Nuzzi R, Waters RA (2004) Long-term perspective on the dynamics of brown tide blooms in Long Island coastal bays. Harmful Algae 3:279–293

Olsen PS (1989) Development and distribution of a brown-water algal bloom in Barnegat Bay, New Jersey with perspective on resources and other red tides in the region. Coastal Est Studies 35:189–212

Onuf CP (1996) Seagrass responses to long-term light reduction by brown tide in upper Laguna Madre, Texas: distribution and biomass patterns. Mar Ecol Prog Ser 138:219–231

Popels LC, Coyne KJ, Forbes R, Pustizzi F, Gobler CJ, Cary SC, Hutchins DA (2003) The use of quantitative polymerase chain reaction for the detection and enumeration of the harmful alga *Aureococcus anophagefferens* in environmental samples along the United States east coast. Limnol Oceanogr Methods 1:92–102

Probyn T, Pitcher G, Pienaar R, Nuzzi R (2001) Brown tides and mariculture in Saldanha Bay, South Africa. Mar Poll Bull 42:405–408

Raven JA, Kubler JE (2002) New light on the scaling of metabolic rate with the size of algae. J Phycol 38:11–16

Sherr EB, Sherr BF (2002) Significance of predation by protists in aquatic microbial food webs. Antonie van Leeuwenhoek 81:293–308

Sieburth JM, Johnson PW, Hargraves PE (1988) Ultrastructure and ecology of *Aureococcus anophagefferens* gen. et sp. nov. (Chrysophyceae): the dominant picoplankter during a bloom in Narragansett Bay, Rhode Island, summer 1985. J Phycol 24:416–425

Sieracki ME, Gobler CJ, Cucci T, Their E, Hobson I (2004) Pico- and nanoplankton dynamics during bloom initiation of *Aureococcus* in a Long Island, NY bay. Harmful Algae 3:459–470

Trice TM, Glibert PM, Lea C, Van Heukelem L (2004) HPLC pigment records provide evidence of past blooms of *Aureococcus anophagefferens* in the Coastal Bays of Maryland and Virginia, USA. Harmful Algae 3:295–304

Villareal TA, Chirichella T, Buskey EJ (2004) Regional distribution of the Texas Brown Tide (*Aureoumbra lagunensis*) in the Gulf of Mexico. In: Steidinger KA, Landsberg JH, Tomas CR, Vargo GA (eds) Harmful algae 2002. Florida Fish and Wildlife Conservation Comm, Fl Inst Oceanogr, IOC-UNESCO, St. Petersburg, USA, pp 374–376
Ward LA, Montagna PA, Kalke RD, Buskey EJ (2000) Sublethal effects of Texas brown tide on *Streblospio benedicti* (Polychaeta) larvae. J Exp Mar Biol Ecol 248:121–129
Wilson R (1995) Aspects of tidal and subtidal flushing within the Peconic Bays Estuary. In: McElroy A (ed) Proc of the Brown Tide Summit, NYSGI-W-95-001, New York Sea Grant Inst, NY, pp 53–56
Yentsch CS, Phinney DA, Shapiro LP (1989) Absorption and fluorescent characteristics of the brown tide chrysophyte: its role on light reduction in coastal marine environments. Coastal Est Studies 35:77–83

Part C
The Ecology and Physiology of Harmful Algae

10 Harmful Algal Bloom Dynamics in Relation to Physical Processes

F.G. FIGUEIRAS, G.C. PITCHER, and M. ESTRADA

10.1 Introduction

The term harmful algal bloom (HAB) has been applied to a diverse range of phytoplankton to which harmful impacts have been attributed. These algae belong to a wide variety of phylogenetic groups. The dinoflagellates, raphidophytes, and prymnesiophytes have historically been considered the primary cause of HABs, but other groups of phytoplankton, such as the diatoms and cyanobacteria now also include taxa known to be harmful.

HAB dynamics are influenced at all spatial-temporal scales relevant to phytoplankton physiology, ecology, and distribution; from the level of short-lived sub-cellular physiological processes to global biogeographical ranges and long-term fluctuations of natural populations (Donaghay and Osborn 1997; Zingone and Wyatt 2005; Franks 2006). Harmful algal blooms should not be considered marginal biological phenomena. The view that HABs are irregular, unpredictable events is misleading, when in fact many harmful species constitute normal components of the seasonal succession of phytoplankton driven by exogenous forces, which contribute to the establishment of habitat templates favoring one or another phylogenetic or functional groups.

The upper ocean is physically forced by air–sea interactions, land and rivers, and the ocean interior. Winds force various physical processes, which are not mutually independent, such as waves, currents and turbulence. Coherent flow structures and convection further complicate upper ocean dynamics (Yamazaki et al. 2002). Hence, the upper ocean is the site of a myriad of interacting dynamic physical processes, all of which influence the functioning and distribution of organisms both directly and indirectly. Direct effects include those of turbulence in altering algal growth rates (Berdalet 1992; Juhl and Latz 2002; Sullivan et al. 2003), cell shape (Zirbel et al. 2000), and the transport of substances in and out of cells (Karp-Boss et al. 1996). The establishment of aggregations required for sexual exchange or the ability of grazers to capture

Ecological Studies, Vol. 189
Edna Granéli and Jefferson T. Turner (Eds.)
Ecology of Harmful Algae

cells is also directly influenced by turbulence in determining encounter rates and perceptive abilities (Yamazaki et al. 2002). Indirect physical effects include the influence of mixing patterns on the distribution of temperature and nutrients, and the transport of phytoplankton in or out of the euphotic zone.

This chapter focuses on the indirect effects of water motion and turbulence on HAB development, as mediated through transport, and provides examples of various ecosystem types in which these processes are operative.

10.2 Physical Constraints: From Diffusion to Advection

Phytoplankton inhabits a heterogeneous environment in which diffusion ($K_{x,y,z}$) and advection ($V_{x,y,z}$) are the main physical processes influencing population development. In this environment, the rate of change of any harmful algal population at a particular location (dN/dt) can be defined as a function of these two physical fluxes and growth (μ):

$$dN/dt = K_x\partial^2 N/\partial x^2 + K_y\partial^2 N/\partial y^2 + K_z\partial^2 N/\partial z^2 \\ -V_x\partial N/\partial x - V_y\partial N/\partial y - V_z\partial N/\partial z + \mu N \quad (10.1)$$

Diffusion tends to reduce the spatial gradients of a population by mixing across gradients. Mixing tends to decrease the density of local population maxima, and increase local population minima by mixing from regions of high concentration to low. While vertical mixing tends to be many orders of magnitude weaker than horizontal mixing, the strong vertical gradients of many physical and chemical properties make vertical mixing of fundamental importance to biological phenomena.

Advection is driven by physical velocities, which move spatial gradients of a population in *x*, *y* and *z* directions. The importance of these terms depends on the relative strengths of these velocities and of the spatial gradients of the populations. Swimming can be formulated as a term for vertical advection (Margalef 1978) that may counter physically driven losses. Organism motility and its behavioral control (Kamykowski et al. 1998; Karp-Boss et al. 2000; Sullivan et al. 2003) interact with water circulation patterns and play a key role in favoring retention (Seliger et al. 1970; Anderson and Stolzenbach 1985) or accumulation (Tyler and Seliger 1981; Fraga et al. 1988; Figueiras et al. 1995) of motile phytoplankton. Swimming alters population gradients and leads to surface or subsurface accumulations of phytoplankton, especially in environments with downward water motions such as negative estuaries, convergent fronts or Langmuir circulation cells (Holligan 1979; Kamykowski 1981). The superior swimming ability of chain-forming dinoflagellates, like *Gymnodinium catenatum* or *Alexandrium catenella* (Fraga et al. 1989) or of ciliates such as *Mesodinium rubrum*, enables them to balance relatively strong dispersing currents and to form blooms in high-energy frontal zones.

10.3 Life-Forms

Determination of the particular characteristics and adaptations that enable a harmful species to bloom is complicated by the considerable physiological and phylogenetic variety presented by these microalgae. Physiological factors increasing the net-growth-rate parameter (μ), such as a rapid intrinsic division rate or allelopathic defenses against predation, are important in determining the bloom-forming capability of some species. However, these factors are in many instances overridden by the ecological response to hydrodynamic forcing. Margalef (1978) considered the available external energy, controlling water movement and nutrient supply, the key factor determining the selection of phytoplankton species. Populations selected under recurrent patterns of environmental conditions tend to share sets of *characteristics*, or *adaptation syndromes*, which provide a basis for the classification of phytoplankton life-forms, as depicted in the *phytoplankton mandala* of Margalef (Fig. 10.1a), in which major taxonomic groupings are presented as *proxies* for phytoplankton life-forms. Diatoms tend to dominate in high-nutrient, relatively high-turbulence conditions, while (*flat*) dinoflagellates are better adapted to survival

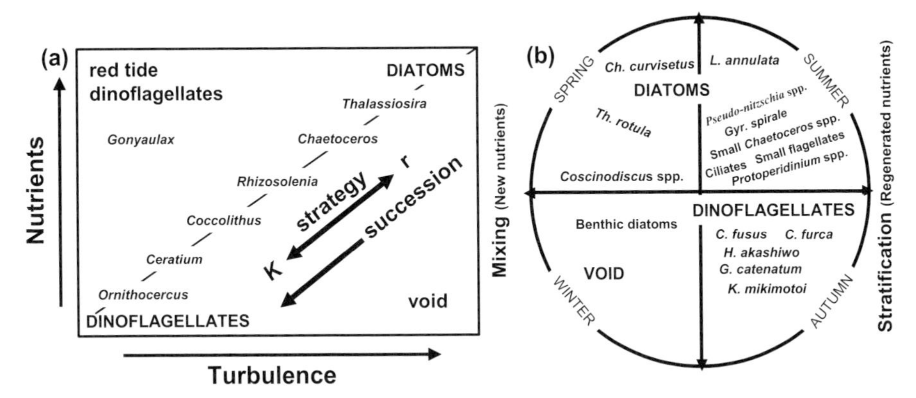

Fig. 10.1. Phytoplankton succession and life-forms in an ecological space defined by turbulence and nutrient concentration, as depicted by the mandala of Margalef (**a**) and annual cycle of phytoplankton abundance in The Rías Baixas of Galicia (**b**) (see Fig. 10.4a for location) according to a mixing-stratification (new *vs.* regenerated nutrients) gradient. HABs are found to correspond to conditions of high nutrients and low turbulence. (a) Reprinted from Oceanologica Acta, volume 1, R. Margalef, pages 493–510, Copyright (1978), with permission from Elsevier. (b) Redrawn and simplified from Figueiras and Niell (1987)

under low-nutrient, low-turbulence conditions. Red-tide-forming (*rounded*) dinoflagellates are shown to be associated with high nutrient, low turbulence conditions. In general, given adequate nutrient supply, diatoms tend to have faster growth rates than other phytoplankton groups (Brand and Guillard 1981), but some level of upward advection or turbulent mixing may be required to maintain their populations in sufficiently illuminated conditions (Huisman et al. 2002). In contrast, motility may represent an important asset in a stratified environment by regulating position in the water column thereby determining nutrient uptake and the light environment. The association of red-tide-forming dinoflagellates with low turbulence may result primarily from reduced physical dispersion, thereby permitting the accumulation of algal biomass (see Eq. 10.1), rather than from any direct effects of turbulence on dinoflagellates. The mandala represented a conceptual model stressing the importance of hydrodynamic forcing relative to physiological responses to light or nutrient concentration (Margalef 1978; Margalef et al. 1979) and was never meant to provide a detailed prescription of taxonomic ordination. Its basic application (Fig. 10.1b) has been shown in numerous studies (Bowman et al. 1981; Jones and Gowen 1990; Figueiras and Ríos 1993; Vila and Masó 2005).

The life-form ordination of the *mandala*, which has been expanded by other authors (Cullen and MacIntyre 1998; Smayda and Reynolds 2001), emphasizes the importance of morphotypic properties, such as size and shape, and behavioral features related to the control of buoyancy or motility (see Sect. 10.2). Life-cycle strategies are also important in the context of the physical environment in that cyst or temporary cyst formation may be utilized to avoid dispersion (Garcés et al. 1999) and to reseed the water column (Anderson et al. 2005).

10.4 Algal Communities

The composition of microalgal communities reflects life-form selection modulated by species–specific features. Therefore, species assemblages tend to share similar functional characteristics in response to the ecological history of the body of water in which they are found. Although the dominance of a particular species within an available biodiversity pool may be reduced to a statistical probability, experience shows that phytoplankton communities can be characterized as assemblages of co-occurring species that show relatively persistent temporal and spatial patterns in composition (Blasco et al. 1981; Venrick 1999). Therefore, in spite of their occasional conspicuous dominance, harmful species do not occur in isolation, but rather as a group. Identification of assemblages incorporating HAB species, and their association with particular physical-chemical properties may therefore provide a predictive capability. For example, the toxic dinoflagellate *Gymnodinium catenatum* was observed in the Iberian upwelling system as a regular component of an assemblage of dinoflagellates including *Prorocentrum triestinum*, *P. micans*, *Ceratium fusus* and *Dinophysis acuta*. The detection of this assemblage in the rías could be used to warn of the likely appearance of *G. catenatum* (Figueiras et al. 1994; Fermín et al. 1996).

10.5 Retention and Transport

Ecosystems are often classified according to the relative importance of one or another physical process. These may include systems categorized by retention or reduced exchange, and others categorized by advective processes.

10.5.1 Retention-Reduced Exchange

Pycnoclines (Pingree et al. 1975; Nielsen et al. 1990; Kononen et al. 2003), fronts (Pingree et al. 1975; Pitcher and Boyd 1996; Pitcher et al. 1998) and cyclonic eddies (Trainer et al. 2002) tend to reduce exchange with surrounding waters, thereby favoring the accumulation of phytoplankton, which can be enhanced by decreased sinking speed or increased swimming capacity. Within pycnoclines, harmful algae are frequently found in thin layers (Bjørsen and Nielsen 1991; Gentien et al. 1995; Rines et al. 2002). The mechanisms of formation of these layers, which may vary substantially in thickness, spatial extent and duration (Dekshenieks et al. 2001), remain unclear. They may result primarily from physical processes, such as the interaction between vertical shear gradients and the horizontal advection of phytoplankton patches (Franks 1995), or they may require a particular swimming-sinking behavior (Sullivan et al. 2003; Gallager et al. 2004).

Retention and/or reduced exchange are often considered to favor HAB generation in small reservoirs or embayments. Blooms of *Pyrodinium bahamense* in Oyster Bay, Jamaica (Seliger et al. 1970), and of *Alexandrium ostenfeldii* and *Heterocapsa triquetra* in Salt Pond, Massachusetts, USA (Anderson and Stolzenbach 1985) are formed by avoiding water exchange. These populations are able to perform vertical migrations in synchrony with the tides, thereby avoiding tidal flushing. Reduced dispersion is also a requisite for blooms of the nitrogen-fixing and floating cyanobacteria *Aphanizomenon flos-aquae* and *Nodularia spumigena* in the Baltic Sea, which develop during calm weather conditions after the spring bloom when the water column is stratified and surface waters are almost exhausted of inorganic nitrogen (Sellner 1997). Similar effects can occur when diffusion from a water body is hindered by man-made structures or hydrographical features such as fronts. These considerations, combined with nutrient enrichment or eutrophication owing to land runoff, explain the frequent development of HABs in enclosed or semi-enclosed coastal systems.

The influence of confinement in the development of HAB events was examined in a series of eight harbors distributed along the Catalan coast (NW Mediterranean, Fig. 10.2a, Vila and Masó 2005). These harbors were selected to include anthropogenically impacted areas and a range of harbor sizes. From a total of 171 taxa identified over an annual cycle, 27 were potentially harmful. These included *Alexandrium minutum*, *A. catenella*, several *Dinophysis* species and *Karlodinium* sp. Based on a multivariate analysis, the authors could distinguish four clusters (Fig. 10.2b) that could be assigned to four different functional groups: bloom-forming dinoflagellates, winter diatoms, summer-autumn diatoms and a group including both large dinoflagellates and elongated diatoms. The bloom-forming dinoflagellates included rounded species like *A. minutum*, *Gyrodinium impudicum*, *Prorocentrum triestinum* and *Scrippsiella* spp., which could be considered as *r*-strategists. This group was most abundant in large harbors during the summer. The group comprising large dinoflagellates and elongated diatoms included dinoflagellates such as *Ceratium* spp., with a high surface-to-volume ratio, and the diatom species *Proboscia alata* and *Rhizosolenia* cf. *imbricata*. The diatom groups were named according to their seasonal dominance; namely winter and summer-autumn diatoms. Blooms were typically attributed to the bloom-forming dinoflagellates and to the winter diatoms, while potentially toxic taxa could be found in any one of the four clusters. The study also showed that dinoflagel-

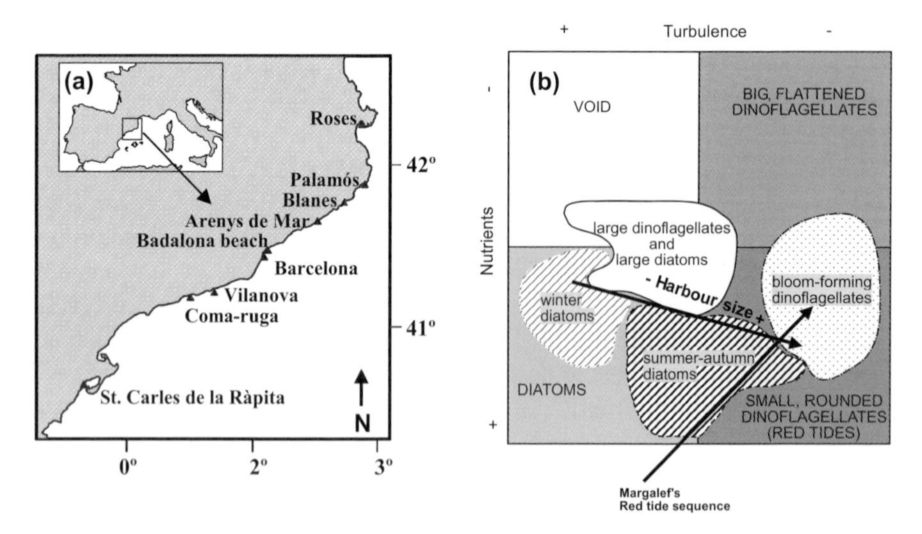

Fig. 10.2. The study of HABs in eight harbors along the Catalan coast (**a**), including a schematic representation of the four functional phytoplankton groups in the framework of Margalef's mandala rotated through 180° (**b**) (from Vila and Masó 2005)

lates were more abundant in large harbors (>15 km^2), and bloom-forming dinoflagellates were less represented in small harbors. The authors considered these variations in the abundance of dinoflagellates to be a function of longer water residence times in larger harbors, which present lower ratios between mouth area and contained water volume than smaller harbors.

10.5.2 Transport

Across-shelf and alongshore currents are known to play a key role in HAB dynamics in coastal waters. Coastal blooms may result from the onshore advection of blooms generated in offshore regions (Pitcher and Boyd 1996; Pitcher et al. 1998; Raine and McMahon 1998; Trainer et al. 2002; Anderson et al. 2005). In other cases, onshore currents interact with coastal morphology and bathymetry to trap dinoflagellate populations in downwelling fronts, where the downward velocity of the water is countered by the ability of dinoflagellates to swim (Fraga et al. 1989; Figueiras et al. 1995). Alongshore currents significantly contribute to the extension of HABs along the coast. Blooms may be transported hundreds of kilometers from their origins by buoyancy currents associated with river outflows (Franks and Anderson 1992) or by surface poleward coastal currents in upwelling systems (Pitcher and Boyd 1996). In other cases, onshore and along-shore currents frequently combine to first accumulate blooms in the coastal region (Pitcher and Boyd 1996; Pitcher et al. 1998; Trainer et al. 2002; Anderson et al. 2005), before transporting them alongshore (Franks and Anderson 1992; Anderson et al. 2005).

Coastal upwelling systems provide an example in which both across-shelf and alongshore currents and retention processes interact to generate HABs. Under upwelling conditions, the general shelf circulation pattern consists of a surface alongshore equatorward current countered by a poleward undercurrent. Ekman transport provides an across-shelf component to the flow with an offshore stream at the surface and compensating onshore flow in deeper layers. The situation is reversed during downwelling, when a surface poleward current is established over the shelf and forces the onshore flow of surface waters. Consequently, the coastal impacts of harmful blooms in upwelling systems coincide with the relaxation phase of upwelling, whereby populations developing offshore during the active phase of upwelling are advected to coastal sites by across-shelf currents and transported alongshore by poleward counter currents. These currents are among the most prominent physical features of coastal upwelling systems and are considered key mechanisms for the initiation of late summer-autumn coastal blooms in several upwelling systems (Fraga et al. 1988; Figueiras et al. 1994; Pitcher and Boyd 1996; Trainer et al. 2002).

Coastline features, such as capes and embayments, particular topographic features of the slope and continental shelf, the presence of buoyant freshwater

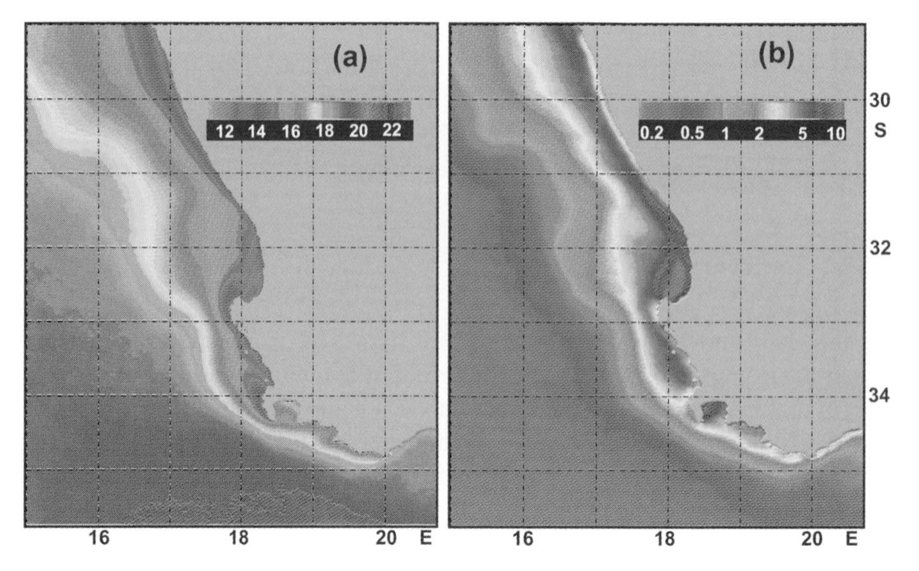

Fig. 10.3. A 5-year (July 1998–June 2003) composite of SST (**a**) and chlorophyll (**b**) derived from daily high-resolution (1 km) NOAA AVHRR and SeaWiFS ocean color data. The role of shelf bathymetry and local topography in influencing the upwelling processes and the distribution of phytoplankton biomass is demonstrated (reprinted from Pitcher and Weeks (2006) with permission from Elsevier)

plumes and the dynamics of oceanic margins interact with wind forcing to generate instabilities in flow that alter the general circulation patterns of upwelling systems at various spatial scales. Alongshore and across-shelf flows are modified, resulting in the amplification and/or reduction of upwelling-downwelling processes (Fig. 10.3a). These coastline discontinuities result in considerable alongshore variability and give rise to areas of convergence or retention, which may favor the development of high biomass blooms (Fig. 10.3b). These areas may also function as sedimentary basins accumulating benthic cysts of HAB species that will later function as bloom inocula.

Interaction between coastline features and across-shelf currents also occur in the Galician Rías Baixas (Fig. 10.4a), comprising four bays located on the northern limit of the Iberian upwelling system. Here HABs dominated by dinoflagellates usually occur during the seasonal transitions in upwelling-downwelling. These bays, which have an almost perpendicular orientation to the coast, act to amplify upwelling and downwelling signals and the exchange between the rías and shelf. Thus, during upwelling, the circulation inside the rías consists of an outflow of surface waters and a compensating inflow of coastal bottom water (Fig. 10.4b). During these upwelling events, phytoplankton composition is characterized by the dominance of diatoms inside the rías and a higher incidence of dinoflagellates on the shelf (Tilstone et al. 1994). However, during downwelling conditions, circulation reverses and coastal

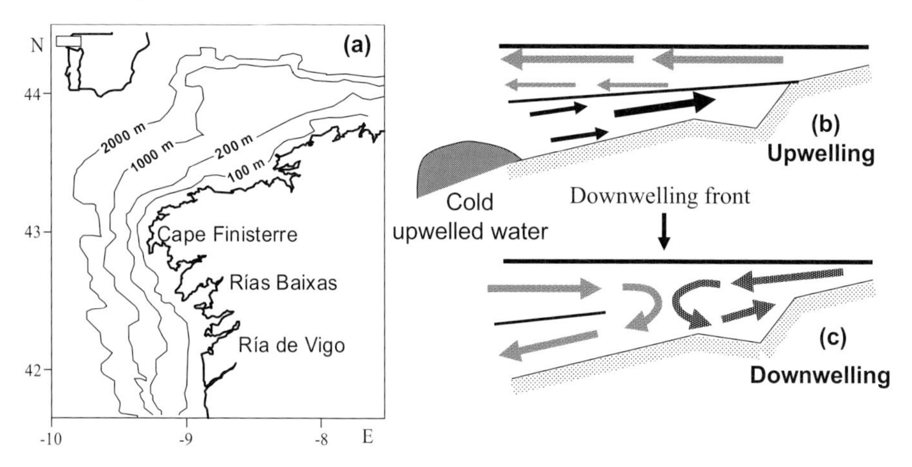

Fig. 10.4. The location of the four Rías Baixas of Galicia on the NW of the Iberian Peninsula (**a**) and a schematic representation of the circulation within the rías during (**b**) upwelling and downwelling (**c**)

surface water enters the rías to form a downwelling front with inner waters of higher continental influence (Fig. 10.4 c). Under these conditions, dinoflagellates are advected from the shelf into the interiors of the rías, where they accumulate in the downwelling front (Fraga et al. 1988; Figueiras et al. 1995; Fermín et al. 1996). This accumulation takes place due to the vertical swimming ability of dinoflagellates which allows them to remain in the water column (Fraga et al. 1989), while diatoms are unable to counteract the downward velocity and are removed from the surface waters.

Bloom transport is also an essential feature of HAB dynamics in the Gulf of Mexico. Blooms of *Karenia brevis*, which are usually initiated offshore, in thermal fronts on the western edge of the Loop Current during upwelling, can be advected towards the coast during periods of wind relaxation. Transport along the Gulf coast of Florida occurs in association with the northward intrusion of the Loop Current. Although *K. brevis* blooms are usually constrained to the west coast of Florida due to the restriction imposed by cyclonic gyres that reduce exchange at both extremes, the blooms are occasionally advected onto the Atlantic side of the Florida coast by the Gulf Stream when the Tortugas gyre debilitates (Tester and Steidinger 1997).

References

Anderson DM, Stolzenbach KD (1985) Selective retention of two dinoflagellates in a well-mixed estuarine embayment: the importance of diel vertical migration and surface avoidance. Mar Ecol Prog Ser 25:39–50

Anderson DM, Keafer BA, Geyer WR, Signell RP, Loder TC (2005) Toxic *Alexandrium* blooms in the western Gulf of Maine: the plume advection hypothesis revisited. Limnol Oceanogr 50:328–345

Berdalet E (1992) Effects of turbulence on the marine dinoflagellate *Gymnodinium nelsonii*. J Phycol 28:267–272

Bjørnsen PK, Nielsen TG (1991) Decimeter scale heterogeneity in the plankton during a pycnocline bloom of *Gyrodinium aureolum*. Mar Ecol Prog Ser 73:263–267

Blasco D, Estrada M, Jones JH (1981) Short-time variability of phytoplankton populations in upwelling regions. The example of northwest Africa. In: Richards FA (ed) Coastal upwelling. Am Geophys Union, Washington, DC, pp 339–347

Bowman MJ, Esaias WE, Schnitzer MB (1981) Tidal stirring and distribution of phytoplankton in Long Island and Block Island Sounds. J Mar Res 39:587–603

Brand LE, Guillard RRL (1981) The effects of continuous light and light intensity on the reproduction rates of twenty-two species of marine phytoplankton. J Exp Mar Biol Ecol 50:119–132

Cullen JJ, MacIntyre JG (1998) Behavior, physiology and the niche of depth-regulating phytoplankton. In: Anderson DM, Cembella AD, Hallegraeff GM (eds) Physiological ecology of harmful algal blooms. NATO ASI Series 41. Springer, Berlin Heidelberg New York, pp 559–579

Dekshenieks MM, Donaghay PL, Sullivan JM, Rines JEB, Osborn TR, Twardowski MS (2001) Temporal and spatial occurrence of thin phytoplankton layers in relation to physical processes. Mar Ecol Prog Ser 223:61–71

Donaghay PL, Osborn TR (1997) Toward a theory of biological-physical control of harmful algal bloom dynamics and impacts. Limnol Oceanogr 42:1283–1296

Fermín EG, Figueiras FG, Arbones B, Villarino ML (1996) Short-time scale development of a *Gymnodinium catenatum* population in the Ría de Vigo (NW Spain). J Phycol 32:212–221

Figueiras FG, Niell FX (1987) Composición del fitoplancton en la ría de Pontevedra, NO de España. Inv Pesq 51:371–409

Figueiras FG, Ríos AF (1993) Phytoplankton succession, red tides, and the hydrographic regime in the Rías Bajas of Galicia. In: Smayda TJ, Shimizu Y (eds) Toxic phytoplankton blooms in the sea. Elsevier, Amsterdam, pp 239–244

Figueiras FG, Jones KJ, Mosquera AM, Álvarez-Salgado XA, Edwards A, MacDougall N (1994) Red tide assemblage formation in an estuarine upwelling ecosystem: Ria de Vigo. J Plankton Res 16:857–878

Figueiras FG, Wyatt T, Álvarez-Salgado XA, Jenkinson I (1995) Advection, diffusion, and patch development of red tide organisms in the Rías Baixas In: Lassus P, Arzul G, Le Denn EE, Gentien P, Marcaillou C (eds) Harmful marine algal blooms. Lavoisier Intercept, Paris, pp 579–584

Fraga S, Anderson DM, Bravo I, Reguera B, Steidinger KA, Yentsch CM (1988) Influence of upwelling relaxation on dinoflagellates and shellfish toxicity in Ria de Vigo, Spain. Est Coast Shelf Sci 27:349–361

Fraga S, Gallager SM, Anderson DM (1989) Chain-forming dinoflagellates: an adaptation to red tides. In: Okaichi T, Anderson DM, Nemoto T (eds) Red tides: biology, environmental science and toxicology. Elsevier, New York, pp 281–284

Franks PJS (1995) Thin layers of phytoplankton: a model of formation by near-inertial wave shear. Deep Sea Res I 42:75–91

Franks PJS (2006) Physics and physical modeling of harmful algal blooms. In: Babin M, Roesler CS, Cullen JJ (eds) Real-time coastal observing systems for ecosystem dynamics and harmful algal blooms. UNESCO, Paris (in press)

Franks PJS, Anderson DM (1992) Alongshore transport of a toxic phytoplankton bloom in a bouyancy current: *Alexandrium tamarense* in the Gulf of Maine. Mar Biol 112:153–164

Gallager SM, Yamazaki H, Davis CS (2004) Contribution of fine-scale vertical structure and swimming behaviour to formation of plankton layers on Georges Bank. Mar Ecol Prog Ser 267:27–43

Garcés E, Masó M, Camp J (1999) A recurrent and localized dinoflagellate bloom in a Mediterranean beach. J Plankton Res 21:2373–2391

Gentien P, Lunven M, Lehaître M, Duvent JL (1995) In-situ depth profiling of particle sizes. Deep Sea Res I 42:1297–1312

Holligan PM (1979) Dinoflagellate blooms associated with tidal fronts around the British Isles. In: Taylor DL, Seliger HH (eds) Toxic dinoflagellate blooms. Elsevier, New York, pp 249–256

Huisman J, Arrayás M, Ebert U, Sommeijer B (2002) How do sinking phytoplankton species manage to persist? Am Nat 159:245–254

Jones KJ, Gowen RJ (1990) Influence of stratification and irradiance regime on summer phytoplankton composition in coastal and shelf seas of the British Isles. Est Coast Shelf Sci 30:557–567

Juhl AR, Latz MI (2002) Mechanisms of fluid shear-induced inhibition of population growth in a red-tide dinoflagellate. J Phycol 38:683–694

Kamykowski D (1981) The simulation of a California red tide using characteristics of a simultaneously measured internal wave field. Ecol Model 12:253–265

Kamykowski D, Yamazaki H, Yamazaki AK, Kirkpatrick GJ (1998) A comparison of how different orientation behaviours influence dinoflagellate trajectories and photoresponses in turbulent water columns. In: Anderson DM, Cembella AD, Hallegraeff GM (eds) Physiological ecology of harmful algal blooms. NATO ASI Series 41. Springer, Berlin Heidelberg New York, pp 581–599

Karp-Boss L, Boss E, Jumars PA (1996) Nutrient fluxes to planktonic osmotrophs in the presence of fluid motion. Oceanogr Mar Biol Annu Rev 34:71–109

Karp-Boss L, Boss E, Jumars PA (2000) Motion of dinoflagellates in a simple shear flow. Limol Oceanogr 45:1594–1602

Kononen K, Huttunen M, Hällfors S, Gentien P, Lunven M, Huttula T, Laanemets J, Lilover M, Pavelson J, Stips A (2003) Development of a deep chlorophyll maximum of *Heterocapsa triquetra* Ehrenb. at the entrance of Gulf of Finland. Limnol Oceanogr 48:594–607

Margalef R (1978) Life-forms of phytoplankton as survival alternatives in an unstable environment. Oceanol Acta 1:493–509

Margalef R, Estrada M, Blasco D (1979) Functional morphology of organisms involved in red tides, as adapted to decaying turbulence. In: Taylor DL, Seliger HH (eds) Toxic dinoflagellate blooms. Elsevier, New York, pp 89–94

Nielsen TG, Kiørboe T, Bjørnsen PK (1990) Effects of a *Chrysochromulina polylepis* subsurface bloom on the planktonic community. Mar Ecol Prog Ser 62:21–35

Pingree RD, Pugh PR, Holligan PM, Forster GR (1975) Summer phytoplankton blooms and red tides along tidal fronts in the approaches to the English Channel. Nature 258:672–677

Pitcher GC, Boyd AJ (1996) Cross-shelf and along-shore dinoflagellate distributions and the mechanism of red tide formation within the southern Benguela upwelling system. In: Yasumoto T, Oshima Y, Fukuyo Y (eds) Harmful and toxic algal blooms. IOC -UNESCO, Paris, pp 243–246

Pitcher GC, Weeks SJ (2006) The variability and potential for prediction of harmful algal blooms in the southern Benguela ecosystem. In: Shannon V et al (eds) Benguela: predicting a large marine ecosystem

Pitcher GC, Boyd AJ, Horstman DA, Mitchell-Innes BA (1998) Subsurface dinoflagellate populations, frontal blooms and the formation of red tide in the southern Benguela upwelling system. Mar Ecol Prog Ser 172:253–264

Raine R, McMahon T (1998) Physical dynamics on the continental shelf off southwestern Ireland and their influence on coastal phytoplankton blooms. Cont Shelf Res 18:883–914

Rines JEB, Donaghay PL, Dekshenieks MM, Sullivan JM, Twardowski MS (2002) Thin layers and camouflage: hidden *Pseudo-nitzschia* spp. (Bacillariophyceae) populations in a fjord in the San Juan Islands, Washington, USA. Mar Ecol Prog Ser 225:123–137

Seliger HH, Carpenter JH, Loftus M, McElroy WD (1970) Mechanisms for the accumulation of high concentration of dinoflagellates in a Bioluminescent Bay. Limnol Oceanogr 15:234–245

Sellner KG (1997) Physiology, ecology, and toxic properties of marine cyanobacteria blooms. Limnol Oceanogr 42:1089–1104

Smayda TJ, Reynolds CS (2001) Community assembly in marine phytoplankton: application of recent models to harmful dinoflagellate blooms. J Plankton Res 23:447–461

Sullivan JM, Swift E, Donaghay PL, Rines JEB (2003) Small-scale turbulence affects the division rate and morphology of two red-tide dinoflagellates. Harmful Algae 2:183–199

Tester PA, Steidinger K (1997) *Gymnodinium breve* red tide blooms: initiation, transport, and consequences of surface circulation. Limnol Oceanogr 42:1039–1051

Tilstone GH, Figueiras FG, Fraga F (1994) Upwelling-downwelling sequences in the generation of red tides in a coastal upwelling system. Mar Ecol Prog Ser 112:241–253

Trainer VL, Hickey BM, Horner RA (2002) Biological and physical dynamics of domoic acid production off the Washington coast. Limnol Oceanogr 47:1438–1446

Tyler MA, Seliger HH (1981) Selection for a red tide organism: physiological responses to the physical environment. Limnol Oceanogr 26:310–324

Venrick EL (1999) Phytoplankton species structure in the central North Pacific 1973–1996: variability and persistence. J Plankton Res 21:1029–1042

Vila M, Masó M (2005) Phytoplankton functional groups and harmful algal species in anthropogenically impacted waters of the NW Mediterranean Sea. Scientia Marina 69:31–45

Yamazaki H, Mackas DL, Denman KL (2002) Coupling small-scale physical processes with biology: towards a Lagrangian approach. In: Robinson AR, McCarthy JJ, Rothschild BJ (eds) The sea: biological-physical interactions in the sea, vol 12. Wiley, New York, pp 51–112

Zingone A, Wyatt T (2005) Harmful algal blooms: keys to the understanding of phytoplankton ecology. In: Robinson AR, Brink KH (eds) The sea: ideas and observations on progress in the study of the seas, vol 13. Harvard University Press, Harvard, pp 867–926

Zirbel MJ, Veron F, Latz MI (2000) The reversible effect of flow on the morphology of *Ceratocorys horrida* (Peridiniales, Dinophyta). J Phycol 36:46–58

11 Ecological Aspects of Harmful Algal In Situ Population Growth Rates

W. Stolte and E. Garcés

11.1 Introduction

The in situ growth of phytoplankton populations, although apparently a simple process, is not consequently dealt with in phytoplankton population ecology. In experimental phytoplankton research, population density is often expressed as cell concentration, and growth is therefore expressed as the cell division rate. In biological oceanography, chemical indicators like chlorophyll-*a* or other pigments are used to describe the phytoplankton population density, and growth rates are consequently expressed as the rate of change of these indicators. Since single phytoplankton cells double in biomass between two subsequent divisions, and may vary in cellular composition, especially pigment content, these measures of growth rate are not necessarily the same. To measure in situ growth rate of individual harmful algal bloom (HAB) species, bulk biomass parameters are usually not suitable, as other phytoplankton may also be present. Moreover, different methods have been applied to assess the in situ growth rate of phytoplankton populations giving different type of results. Therefore, a definition of terms is justified.

Assuming a homogeneous population with density N, the population growth rate dN/dt is proportional to the population density, and the specific or per-capita growth rate μ (in d^{-1}) is used to characterize the population growth.

$$\frac{dN}{dt} = \mu N , \qquad \text{Eq. (11.1)}$$

In case the per-capita growth rate μ is constant over time, the population density will change exponentially according to:

$$N_t = N_0 e^{\mu t} , \qquad \text{Eq. (11.2)}$$

Ecological Studies, Vol. 189
Edna Granéli and Jefferson T. Turner (Eds.)
Ecology of Harmful Algae

where N_t and N_0 are the population densities at time t and time 0, respectively. At optimal availability of resources (light, nutrients) and at given physical circumstances (temperature, salinity) the per-capita growth rate is defined as the maximum per-capita growth rate $\mu_{m(T,S)}$. Typically, this value is determined from the washout rate in chemostat (Pirt 1975).

In this chapter, reported growth rates of HAB and non-HAB species measured in the field are analysed to better understand the ecology of certain HAB groups.

11.2 Ecological Interpretation of In Situ Growth Rate Measurements

Phytoplankton populations need to grow in order to compensate for loss factors such as sedimentation, grazing and lysis. The net per-capita changes in population density (μ_{net}) can be expressed as

$$\mu_{net} = \mu_{gross} - l - g - e + i \qquad \text{Eq. (11.3)}$$

where μ_{gross} is the per-capita population gross growth rate, l is the lysis rate (due to viruses, parasites, or autolysis), g is the per-capita grazing rate, e and i are the per-capita export and import rates.

Phytoplankton show different strategies in order to minimize losses and maximize resource utilization. Populations can be classified as either *r*- or *K*-selected with respect to their growth strategy (MacArthur and Wilson 1967). This theory has its base in the logistic growth equation, which predicts that net growth approaches zero when population density approaches K or the *carrying capacity* of the environment for that particular species

$$\frac{dN}{Ndt} = \mu_{net} = r\left(1 - \frac{N}{K}\right), \qquad \text{Eq. (11.4)}$$

where r is the value of μ at infinitely low population density and can be interpreted as the maximum per-capita growth rate of the species at the prevailing conditions (compare to $\mu_{max(T,S)}$). The cause of this negative correlation (Fig. 11.1) is not specified by the logistic model, but may result from intraspecific competition, grazing, etc. Although simple, the theory has been applied successfully as a framework for better understanding of ecological processes. For phytoplankton populations, succession during a growth season is from *r*-selected phytoplankton species, which have optimized their fitness for conditions with ample resources and low mortality rates, to more *K*-selected species, which have optimized their fitness for conditions with low resource availability and high risk of mortality (Margalef 1958; Sommer 1981). Typi-

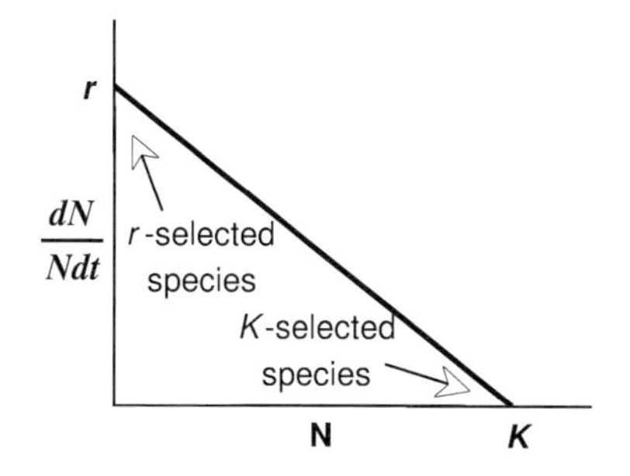

Fig. 11.1. Schematic representation of Eq. (11.4); *r*-selected species have optimized their fitness for conditions that allow for high net per capita growth rates, *K*-selected species for conditions that do not allow for high net per capita growth rates

cally, *r*-selected species have high maximum per-capita growth rates, poor competitive abilities, and/or lack mechanisms to avoid grazing or sedimentation. Typical *K*-selected species have comparatively low maximum per-capita growth rates, are good competitors, and have developed mechanisms to avoid losses due to grazing and sedimentation. Moreover, differences in life-history may play an important role. Unfortunately, data to evaluate the whole suite of characteristics for phytoplankton are scarce or lacking. Besides, there are no objective criteria to judge the "*r*-ness" or "*K*-ness" of a certain characteristic. Alternatively, the observed in situ growth rates (μ_{net}) for individual species should reflect the strategy of a species. "*r*-Strategists" should occur relatively often at conditions that allow for high per-capita growth rates. Contrastingly, *K*-strategists should occur relatively often under conditions that do not allow for high growth rates (Fig. 11.1).

In this chapter, we aim to integrate the current knowledge on phytoplankton in situ growth rates in order to find patterns that may classify different harmful phytoplankton groups as relative *r*- or *K*-strategists. This should lead to simple, testable hypotheses concerning the occurrence and timing of HABs.

Data on field measurements of phytoplankton population growth rates are relatively rare, especially from HAB species. Moreover, different growth and loss-processes are determined depending on the used method (Table 11.1). Traditional methods that require incubation of whole samples and microscope counting of individual species (Methods 1 and 2, Table 11.1) may include the effects of phytoplankton lysis and losses due to grazers present in the incubation bottle. Incubation methods with chemical detection may have the same drawback (Methods 3, 4, 5, 6, Table 11.1). Isolation of individual cells by hand-picking in combination with ^{14}C uptake measurements have been applied to measure specific carbon uptake and growth rate independent of lysis and/or grazing (Method 3, Table 11.1; Granéli et al. 1997). Recently, more advanced methods have been applied to identify the growth rate of individual species in the field, such as cell cycle methods (Method 8, Table 11.1) that do not require incubation at all. This method relies on the fact that representative samples are taken from the same field population under a period of 1–2 days.

Table 11.1. Overview of the ecological interpretation of different in situ growth rate methods. *Numbers* in first column are referred to in Table 11.2

	Method description	Ecological interpretation (see Eq. 11.3)	References
1	Dialysis incubation (by measuring change in cell, particle or chlorophyll concentration or alternative biomass index)	μ_{gross} - l - g^{a}	Prakash et al. 1973; Maestrini and Kossut 1981
2	Diffusion incubation (by measuring change in cell, particle or chlorophyll concentration or alternative biomass index)	μ_{gross} - l - g^{a}	Furnas 1982a; Vargo 1984; Ferrier-Pagès 2001; Tang 2003
3	^{14}C-carbon uptake = rates of carbon and/or chlorophyll accumulation	μ_{gross}(- l - g^{a})	Eppley et al. 1970; Granéli et al. 1997
4	Nutrient assimilation	μ_{gross} - l - g^{a}	Furnas 1982b; Nelson and Smith 1986
5	ATP synthesis	μ_{gross} - l - g^{a}	Sheldon and Sutcliffe 1978
6	Protein synthesis and turnover	μ_{gross} - l - g^{a}	DiTullio and Laws 1983
7	Pigment specific activity = Pigment labelling	μ_{gross} – l	Redalje and Laws 1981; Goericke and Welschmeyer 1993
8	Cell cycle method or cytological index or mitotic index or biochemical cell cycle markers	μ_{gross} or *PGR*[b]	Swift and Durbin 1972; Smayda 1975; Carpenter and Chang 1988; Chang and Carpenter 1991; Garcés et al. 1998a; 1999; Garcés and Masó 2001, and references therein
9	In vivo fluorescence ratios	μ_{gross} or PGR[b]	Heath 1988
10	Thymidine or germanium	μ_{gross} or PGR[b]	Rivkin and Voytek 1986
11	Dilution technique	μ_{gross} – l and g^{a}	Landry et al. 1995; Garcés et al. 2005, and refs therein [a]grazing is taken into account only from grazers that are included in the incubation bottle (often only microzooplankton)

[a] Grazing is taken into account only from grazers that are include in the incubation bottle (ofen only microzooplankton)
[b] PGR = potential growth rate

The application of flow-cytometric DNA-cycle detection for individual species is an especially powerful method to measure gross growth rates in the field. However, in this case it is assumed that populations can be recognized by flow-cytometric detection methods.

Depending on the method, a value between net and gross growth rate is determined (summarized in Table 11.1). Since the effect of cell lysis and grazing can be large, often in the same order of magnitude as growth, gross and net growth rates may differ substantially. However, differences in growth rates between methods in our analysis were not significant (Kruskal-Wallis test).

The measurement of an individual population's growth rate in the field is only possible when individual species can be analysed. Usually, microscopic or flow-cytometric techniques are used for quantification (Methods 1, 2, 8 and 11, Table 11.1) or physical separation of cells before analysis (e.g. Method 3, but in principle applicable for other methods too). Detection of biochemical markers such as pigments (e.g. Method 7, but also applicable with Methods 1, 2 and 11) are taxon-specific rather than species-specific (Mackey et al. 1996), and are influenced by the variability of cellular pigment content (Stolte et al. 2000). In our analysis, we have only used data obtained from individual populations.

11.3 In Situ Growth Rates; Variation Among Taxonomic Groups

In total, 178 entries from six taxonomical classes are listed in Table 11.2, including harmful and non-harmful species. Due to the different methods that were applied to measure the in situ growth rates, values vary between net- and gross-values (compare Table 11.1). Because the different methods represent different concepts (Table 11.1), no statistical significance-tests were applied. Still, there are some interesting trends worth further investigation. When ranking phytoplankton after their observed in situ growth rates, the highest values were recorded for cryptophytes, chrysophytes and diatoms (Fig. 11.2), groups that are rarely involved in harmful events. However, prymnesiophytes, which frequently cause harmful blooms, also show high in situ growth rates. Conversely, cyanobacteria and dinoflagellate in situ growth rates are clearly lower (Fig. 11.2). It may therefore be hypothesized that among HAB-forming classes, prymnesiophytes are more *r*-selected, and dinoflagellates are more *K*-selected species. For cyanobacteria, only non-HAB coccoid species are represented, and no filamentous forms, which are responsible for HABs in brackish and fresh water.

Table 11.2. Reported in situ per-capita growth rates (μ) for different taxonomic groups, and possible harmful effects

Species	μ	Harmful effect*	Reference
Cyanobacteria			
Prochlorococcus sp.[8]	1.21, 0.50, 0.50, 0.67, 0.54, 0.58, 0.72	–	a
Prochlorococcus sp.[2]	0.75, 0.54, 0.80, 0.70, 0.64, 0.56, 0.84	–	a
Synechococcus sp.[2]	0.42	–	b
Bacillariophyceae (centrales)			
Bacteriastrum spp.[2]	1.52	–	b
Cerataulina pelagica[2]	1.32, 0.42	–	b
Cerataulina pelagica[3]	0.20	–	c
Cerataulina pelagica[11]	0.70	–	c
Chaetoceros curvisetum[2]	2.22	–	b
Chaetoceros peruvianum[2]	0.76	–	b
Chaetoceros subtilis[2]	0.28	–	b
Chaetoceros sp. solitary[2]	0.35	–	b
Chaetoceros sp. chained[2]	0.42	–	b
Chaetoceros spp.[2]	2.08	–	b
Chaetoceros sp.[3]	0.30, 0.30	–	c
Chaetoceros sp.[11]	0.30, 0.10	–	c
Cyclotella caspia[3]	2.90	–	c
Cyclotella caspia[11]	2.50	–	c
Cyclotella striata[3]	0.60	–	c
Cyclotella striata[11]	1.70	–	c
Hemiaulus spp.[2]	0.55	–	b
Leptocylindrus danicus[2]	1.11, 0.97	–	b
Leptocylindrus minimus[2]	1.04	–	b
Rhizosolenia alata[2]	0.14	–	b
Rhizosolenia delicatula[2]	0.69	–	b
Rhizosolenia fragilissima[2]	0.83, 0.28	–	b
Rhizosolenia setigera[2]	0.35	–	b
Rhizosolenia stolterfothii[2]	0.35	–	b
Rhizosolenia styliformis[2]	0.14	–	b
Skeletonema costatum[2]	1.94, 0.49	–	b
Skeletonema costatum[3]	1.00, 0.50, 0.30	–	c
Skeletonema costatum[11]	1.60, 1.00, 0.50	–	c
Thalassiosira rotula[3]	0.40, 0.20	–	c
Thalassiosira rotula[11]	0.50, 0.20	–	c
Thalassiosira sp.[2]	1.11	–	b
Thalassiosira sp. „small“[2]	0.97	–	b
Bacillariophyceae (pennates)			
Asterionella glacialis[2]	0.90	–	b
Nitzschia closterium[2]	1.04	–	b
Nitzschia closterium „medium“[2]	1.52	–	b
Nitzschia closterium „medium“[3]	0.35	–	b

Table 11.2. (*Continued*)

Species	μ	Harmful effect*	Reference
Nitzschia closterium „large“[2]	0.90	–	b
Nitzschia curta[3]	0.21	–	b
Nitzschia fraudulenta[2]	0.49	ASP	b
Nitzschia lineola[2]	1.25	–	b
Nitzschia pungens[2]	0.55	ASP	b
Nitzschia pungens[3]	0.20, 0.50	ASP	c
Nitzschia pungens[11]	0.70	ASP	c
Thalassionema (Thalassiosira) nitzschioides[2]	0.90	–	b
Thalassionema (Thalassiosira) frauenfeldii[2]	0.97	–	b
Thalassiothrix spp.[2]	0.62	–	b
Unknown pennate species[2]	0.90	–	b
Cryptophyceae			
Rhodomonas lacustris[3]	3.00, 0.70, 0.40, 0.30	–	c
Rhodomonas lacustris[11]	2.70, 1.60, 0.20, 0.30	–	c
Rhodomonas sp[3]	1.80, 0.60, 0.40	–	c
Rhodomonas sp[11]	1.20, 0.30, 0.60	–	c
Dinophyceae			
Alexandrium catenella[11]	0.22, 0.24	PSP	d
Alexandrium minutum[2]	0.76, 0.97	PSP	e
Alexandrium tamarense[2]	0.23	PSP	f
Alexandrium taylori[11]	0.67, 0.24, 0.30, 0.04, 0.64, 0.17	HBNT	d
Ceratium arietinum[8]	0.09	–	b
Ceratium candelabrum[8]	0.09	–	b
Ceratium contrarium[8]	0.15	–	b
Ceratium declinatum[8]	0.12	–	b
Ceratium furca[3]	0.28, 0.60	HBNT	g
Ceratium horridum[8]	0.08	–	b
Ceratium lineatum[3]	0.48, 0.81	HBNT	g
Ceratium macroceros[8]	0.08	–	b
Ceratium massilience[8]	0.15	–	b
Ceratium pulchellum[8]	0.19	–	b
Ceratium symmetricum (summetricum)[8]	0.14	–	b
Ceratium tripos[3]	0.17	HBNT	g
Dinophysis acuminata[8]	0.22, 0.12	DSP	h
Dinophysis acuminata[3]	0.59, 0.49	DSP	g
Dinophysis acuta[8]	0.11	DSP	h
Dinophysis acuta[3]	0.41, 0.35	DSP	g
Dinophysis caudata[8]	0.19	DSP	h
Dinophysis norvegica[3]	0.63, 0.29, 0.18	DSP	g
Dinophysis norvegica[8]	0.17, 0.13, 0.21, 0.40	DSP	i
Dinophysis tripos[8]	0.21	DSP	h

Table 11.2. (*Continued*)

Species	μ	Harmful effect*	Reference
Gymnodinium spp.[3]	0.30, 0.30	–	c
Gymnodinium spp.[11]	0.10, 0.10	–	c
Gymnodiniacea „medium“[2]	0.07	–	b
Gymnodiniacea „small“[2]	0.21	–	b
Heterocapsa triquetra[8]	0.10, 0.21, 0.17, 0.05, 0.03, 0.02, 0.03, 0.04, 0.05, 0.04, 0.11, 0.06, 0.02, 0.02	–	j
Karlodinium sp.[8]	0.94, 0.60, 0.59, 0.50, 0.39,	FK	k
Katodinium rotundatum[3]	2.60, 1.00, 0.80	–	c
Katodinium rotundatum[11]	2.40, 0.50, 0.10	–	c
Prorocentrum minimum[2]	0.21	UNK	b
Prorocentrum triestinum[2]	0.14	–	b
Chrysophyceae			
Ochromonas minima[3]	1.00, 1.20, 1.10	–	c
Ochromonas minima[11]	0.90, 0.20, 0.60	–	c
Prymnesiophyceae			
Chrysochromulina spp.[3]	0.90, 0.80	FK	c
Chrysochromulina spp.[11]	1.00, 1.00	FK	c
Phaeocystis globosa single[8]	0.68, 0.49	HBNT	l
Phaeocystis globosa colony[8]	0.73, 0.76, 0.93	HBNT	l

Numbers behind species names refer to methods in Table 11.1
[a] (Liu et al. 1998);
[b] (Furnas 1990);
[c] (Fahnenstiel et al. 1995);
[d] (Garcés et al. 2005);
[e] (Garces et al. 1998b);
[f] (Ichimi et al. 2001);
[g] (Granéli et al. 1997);
[h] (Reguera et al. 1996);
[i] (Gisselson et al. 2002);
[j] (Litaker et al. 2002);
[k] (Garcés et al. 1999);
[l] (Veldhuis et al. 2005)
* *HBNT* high biomass but no toxins detected, *FK* fish killing bloom, *ASP* amnesic shellfish poisoning, *PSP* paralytic shellfish poisoning, *DSP* diarrhetic shellfish poisoning, *UNK* unknown toxin, "–" not known to be harmful

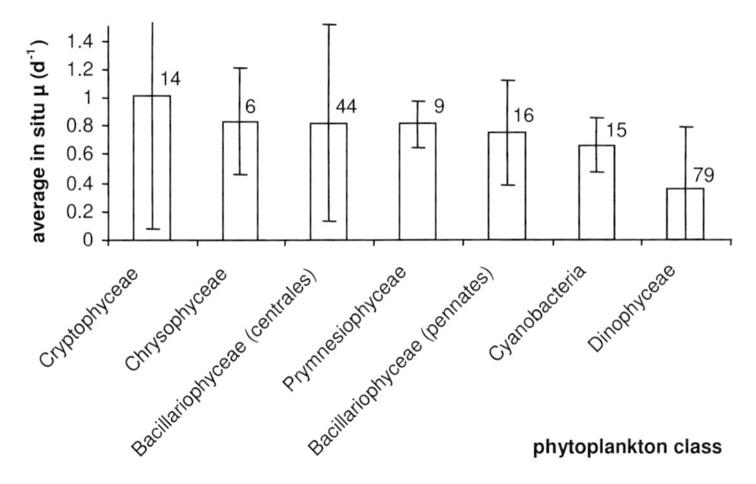

Fig. 11.2. Average in situ per capita growth rates (d^{-1}) with standard deviations for each phytoplankton class in descending order. The number of observations is stated above each column

11.4 Are Harmful Algal Species r- or K-Strategists?

Harmful algal species have on average lower reported in situ growth rates than non-HAB species (Fig. 11.3). This result is in agreement with the hypothesis that harmful algae are relatively *K*-adapted species, capable of reducing their losses due to grazing, either by being toxic or via other mechanisms. However, the diversity of HAB-species is great, hampering the possibility to draw conclusions on a possible general strategy for all HAB species.

It is unlikely that all phytoplankton that are harmful from a human perspective have a general growth strategy. However, it is likely that growth strategies correspond to the type of harmful effect, or more specifically, the mechanism or toxin that is responsible for the harmful effect. The data were therefore analysed according to the type of harmful effect that the particular species may cause.

The highest observed in situ growth rates are reported for high-biomass non-toxic (HBNT) species (Fig. 11.4), and fish-killing (FK) species (both on average 0.59 d^{-1}). Species that are not known for any harmful effect have similar growth rates to HBNT and FK species (Fig. 11.4). Dinoflagellates responsible for diarrhetic shellfish poisoning (DSP) have the lowest in situ growth rates (0.3 d^{-1}). Based on in situ growth rates, DSP-causing species *Dinophysis* spp. could be considered to be relatively *K*-selected species. Indeed, *Dinophysis* spp. are recognized for their relatively low growth rates (Smayda and Reynolds 2001) and poor edibility (Carlsson et al. 1995). Moreover, they often

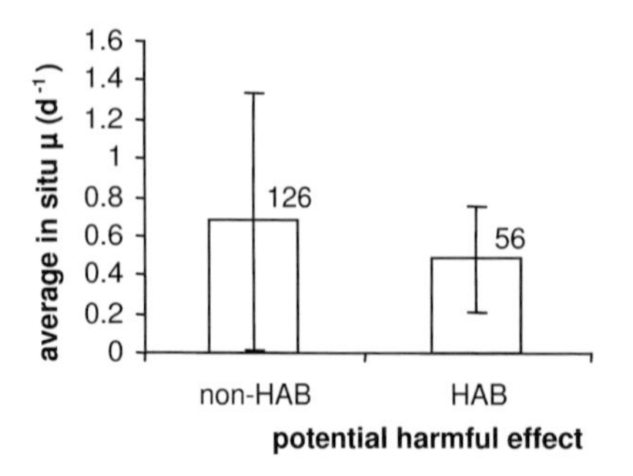

Fig. 11.3. Average in situ per capita growth rates (d^{-1}). Harmful species are any species with a harmful effect according to Table 11.1. The number of cases is reported above each bar

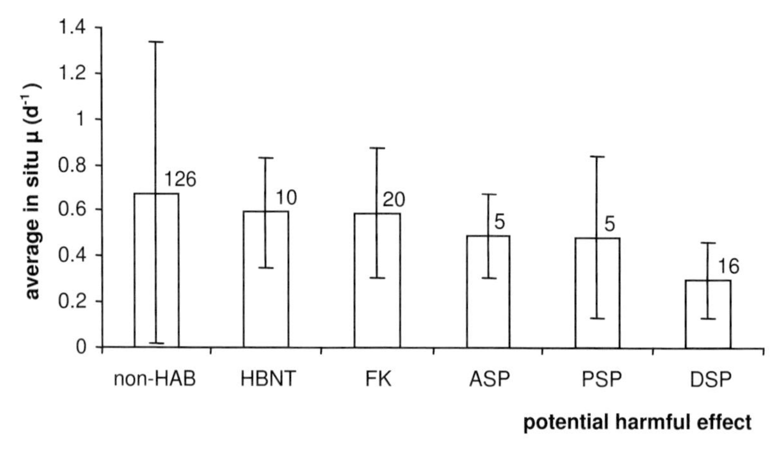

Fig. 11.4. Observed in situ per capita growth rates (d^{-1}) with standard deviations for harmful algae grouped per harmful effect

occur in high numbers at the pycnocline (Gisselson et al. 2002), an environment that is either stable in the case of tropical waters, or at least predictable in seasonally stratified temperate seas. In general, phytoplankton communities in stable and predictable environments would be growing close to their carrying capacity (MacArthur and Wilson 1967), and one would expect *K*-selective species to comprise an important part of those communities.

The dinoflagellates that cause paralytic shellfish poisoning (PSP), displaying intermediate values, do not seem to be clearly *r*- or *K*-selected based on in situ growth rates. This is not in conflict with the earlier classification of these types of dinoflagellates as species of intermediately nutrient-enriched waters (Smayda and Reynolds 2001).

11.5 Conclusions

With the current set of data, we provide support for the hypothesis that HBNT and FK blooms are *r*-selected, and will dominate in areas and during periods of ample nutrients and a low risk of losses due to grazing and sedimentation. These types of blooms can therefore be considered as indicators for eutrophication of the marine environment. Both types could also contribute to significant disturbances of marine coastal environments, either through hypoxia in bottom waters or through the effects of their toxins, contributing to an unstable environment even more suitable for *r*-selected species (MacArthur and Wilson 1967). This positive feedback mechanism might lead to shifts in ecosystem properties upon establishment of such HABs. This may be an additional explanation for recurring high-biomass and fish-killing blooms in eutrophicated areas.

We also provide support that DSP- and to a lesser extent PSP-causing species are more *K*-selected, and probably more resistant to grazing than other species (e.g. Teegarden 1999). Competition for nutrients by these groups is either avoided by using alternative nutrient sources, such as phagotrophy in case of *Dinophysis* spp., or is by choosing habitats with intermediate nutrient concentrations, such as coastal fronts in the case of some PSP-causing dinoflagellates (Smayda and Reynolds 2001). Life-history events that are hard to classify as either *r*- or *K*-selected traits such as cyst formation, swimming behaviour and sexual reproduction most probably contribute to the success of these species.

Compared to earlier efforts to classify phytoplankton, in particular harmful algal species, into a theoretical ecological framework (Margalef 1958; Smayda and Reynolds 2001), the current approach is simple. However, assuming that any growth strategy finally is reflected by the observed growth rate, some preliminary conclusions could be made with respect to different groups of HAB species. We provide additional support for a diversity of strategies within HAB-forming phytoplankton. An increasing problem of HAB occurrence in many coastal waters may therefore have different causes, and a variety of mitigation or prevention measurements must be applied if HAB prevalence is to be reduced.

References

Carlsson P, Granéli E, Finenko G, Maestrini SY (1995) Copepod grazing on a phytoplankton community containing the toxic dinoflagellate *Dinophysis acuminata*. J Plankton Res 17:1925–1938

Carpenter EJ, Chang J (1988) Species specific phytoplankton growth rates via diel DNA synthesis cycles. I. Concept of the method. Mar Ecol Prog Ser 43:105–111

Chang J, Carpenter EJ (1991) Species specific phytoplankton growth rates via diel DNA synthesis cycles. V. Application to natural populations in Long Island Sound. Mar Ecol Prog Ser 78:115–122

DiTullio G, Laws EA (1983) Estimating of phytoplankton N uptake based on ^{14}C incorporation into protein. Limnol Oceanogr 28:177–185

Eppley RW, Reid FMH, Strickland JDH (1970) Estimates of phytoplankton crop size, growth rate and primary production. Bull Scripps Inst Oceanogr 17:33–42

Fahnenstiel GL, McCormick MJ, Lang GA, Redalje DG, Lohrenz SE, Markowitz M, Wagoner B, Carrick HJ (1995) Taxon-specific growth and loss rates for dominant phytoplankton populations from the northern Gulf of Mexico. Mar Ecol Prog Ser 117:229–239

Ferrier-Pagès C, Furla P (2001) Pico- and nanoplankton biomass and production in the two largest atoll lagoons of French Polynesia. Mar Ecol Prog Ser 211:63–76

Furnas MJ (1982a) An evaluation of two diffusion culture techniques for estimating phytoplankton growth rates in situ. Mar Biol 70:63–72

Furnas MJ (1982b) Growth rates of summer nanoplankton populations in lower Narragansett Bay, Rhode Island, USA. Mar Biol 70:105–115

Furnas MJ (1990) In situ growth-rates of marine-phytoplankton – approaches to measurement, community and species growth rates. J Plankton Res 12:1117–1151

Garcés E, Masó M (2001) Phytoplankton potential growth rate versus increase in cell numbers: estimation of cell lysis. Mar Ecol Prog Ser 212:297–300

Garcés E, Delgado M, Maso M, Camp J (1998a) Life history and in situ growth rates of *Alexandrium taylori* (Dinophyceae, Pyrrophyta). J Phycol 34:880–887

Garcés E, Delgado M, Vila M, Camp J (1998b) An *Alexandrium minutum* bloom: in situ growth or accumulation? In: Reguera B, Blanco J, Fernandez ML, Wyatt T (eds) Harmful algae. Xunta de Galicia and IOC- UNESCO, Vigo, pp 167–170

Garcés E, Delgado M, Masó M, Camp J (1999) In situ growth rate and distribution of the ichthyotoxic dinoflagellate *Gyrodinium corsicum* Paulmier in an estuarine embayment (Alfacs Bay, NW Mediterranean Sea). J Plankton Res 21:1977–1991

Garcés E, Vila M, Masó M, Sampedro N, Grazia Giacobbe M, Penna A (2005) Taxon-specific analysis of growth and mortality rates of harmful dinoflagellates during bloom conditions. Mar Ecol Prog Ser 301:67–79

Gisselson L-Å, Carlsson P, Granéli E, Pallon J (2002) *Dinophysis* blooms in the deep euphotic zone of the Baltic Sea: do they grow in the dark? Harmful Algae 1:401–418

Goericke R, Welschmeyer NA (1993) The carotenoid-labeling method: measuring specific rates of carotenoid synthesis in natural phytoplankton communities. Mar Ecol Prog Ser 98:157–171

Granéli E, Anderson DM, Carlsson P, Maestrini SY (1997) Light and dark carbon uptake by *Dinophysis* species in comparison to other photosynthetic and heterotrophic dinoflagellates. Aquat Microb Ecol 13:177–186

Heath MR (1988) Interpretation of in vivo fluorescence and cell division rates of natural phytoplankton using a cell cycle model. J Plankton Res 10:1251–1283

Ichimi K, Yamasaki M, Okumura Y, Suzuki T (2001) The growth and cyst formation of a toxic dinoflagellate, *Alexandrium tamarense*, at low water temperatures in northeastern Japan. J Exp Mar Biol Ecol 261:17–29

Landry MR, Kirshtein J, Constantinou J (1995) A refined dilution technique for measuring the community grazing impact of microzooplankton, with experimental tests in the Central Equatorial Pacific. Mar Ecol Prog Ser 120:53–63

Litaker RW, Warner VE, Rhyne C, Duke CS, Kenney BE, Ramus J, Tester PA (2002) Effect of diel and interday variations in light on the cell division pattern and in situ growth rates of the bloom-forming dinoflagellate *Heterocapsa triquetra*. Mar Ecol Prog Ser 232:63–74

Liu H, Campbell L, Landry MR, Nolla HA, Brown SL, Constantinou J (1998) *Prochlorococcus* and *Synechococcus* growth rates and contributions to production in the Arabian Sea during the 1995 Southwest and Northeast Monsoons. Deep-Sea Res Part II-Topical studies in oceanography 45:2327–2352

MacArthur RH, Wilson EO (1967) The theory of island biogeography. Princeton Univ Press, Princeton, USA, 203 pp

Mackey MD, Mackey DJ, Higgins HW, Wright SW (1996) CHEMTAX - A program for estimating class abundances from chemical markers: application to HPLC measurements of phytoplankton. Mar Ecol Prog Ser 144:265–283

Maestrini SY, Kossut MG (1981) In situ cell depletion of some marine algae enclosed in dialysis sacs and their use for the determination of nutrient-limiting growth in Ligurian coastal waters (Mediterranean Sea). J Exp Mar Biol Ecol 50:1–19

Margalef R (1958) Temporal succession and spatial heterogeneity in phytoplankton. In: Buzzati-Traverso AA (ed) Perspectives of marine biology. Univ Calif Press, Berkeley, pp 323–349

Nelson DM, Smith WO (1986) Phytoplankton bloom dynamics of the western Ross Sea ice edge. II. Mesoscale cycling of nitrogen and silicon. Deep Sea Res 33:1389–1412

Pirt SJ (1975) Principles of microbe and cell cultivation. Blackwell Scientific Publications, Oxford, 274 pp

Prakash A, Skoglund L, Rysttad B, Jensen A (1973) Growth and cell size distribution of marine planktonic algae in batch and dialysis cultures. J Fish Res Board Canada 30:143–155

Redalje DG, Laws EA (1981) A new method for estimating phytoplankton growth rates and carbon biomass. Mar Biol 62:73–79

Reguera B, Bravo I, McCall H, Reyero MI (1996) Phased cell division and other biological observations in field populations of *Dinophysis* spp. during cell cycle studies. In: Yasumoto T, Oshima Y, Fukuyo Y (eds) Harmful and toxic algal blooms. IOC-UNESCO, Paris, pp 257–260

Rivkin R, Voytek M (1986) Cell division rates of eukaryotic algae measured by tritiated thymidine incorporation into DNA: coincident measurements of photosynthesis and cell division of individual species of phytoplankton isolated from natural populations. J Phycol 22:199–205

Sheldon RW, Sutcliffe WH Jr (1978) Generation time of 3 h for Sargasso Sea microplankton determined by ATP analysis. Limnol Oceanogr 23:1051–1055

Smayda TJ (1975) Phased cell division in natural population of the marine diatom *Ditylum brightwelli* and the potential significance of diel phytoplankton behavior in the sea. Deep Sea Res 22:151–165

Smayda TJ, Reynolds CS (2001) Community assembly in marine phytoplankton: application of recent models to harmful dinoflagellate blooms. J Plankton Res 23:447–461

Sommer U (1981) The role of *r*- and *K*-selection in the succession of phytoplankton in Lake Constance. Acta Oecologica 2:327–342

Stolte W, Kraay GW, Noordeloos AAM, Riegman R (2000) Genetic and physiological variation in pigment composition of *Emiliania huxleyi* (Prymnesiophyceae) and the potential use of its pigment ratios as a quantitative physiological marker. J Phycol 36:529–539

Swift E, Durbin EG (1972) The phased division and cytological characteristics of *Pyrocystis* spp. can be used to estimate doubling times of their populations in the sea. Deep Sea Res 19:189–198

Tang KW (2003) Grazing and colony size development in *Phaeocystis globosa* (Prymnesiophyceae): the role of a chemical signal. J Plankton Res 25:831–842

Teegarden GJ (1999) Copepod grazing selection and particle discrimination on the basis of PSP toxin content. Mar Ecol Prog Ser 181:163–176

Vargo G (1984) Growth rates of natural populations of marine diatoms as determined in cage cultures. In: Holm-Hansen O, Bolis L, Gillus R (eds) Lecture notes in coastal and estuarine studies, vol 8. Springer, Berlin Heidelberg New York, pp 113–128
Veldhuis MJW, Brussaard CPD, Noordeloos AAM (2005) Living in a *Phaeocystis* colony: a way to be a successful algal species. Harmful Algae 4:841–858

12 Harmful Algae and Cell Death

M.J.W. Veldhuis and C.P.D. Brussaard

12.1 Introduction

Mortality can be defined as the number of deaths in a given time, place or population, i.e. it is the loss term of population dynamics. For phytoplankton it is the opposite of cell population growth and is responsible with other loss factors for a reduction in the population size of a species (Reynolds 1984). In summary, this can be expressed as:

$$\mu_{gross} = \mu_{net} + \text{grazing} + \text{lysis} + \text{sinking} + \text{physical transport} \qquad \text{(Eq. 12.1)},$$

where μ_{gross} is the specific growth rate; μ_{net} is the net increase in cell abundance; grazing is the loss of algal cells due to herbivore grazing; cell lysis is the rupture of the cell membrane and loss of cytoplasm caused by autocatalytic mortality, pathogenicity (viral, bacterial, fungal) or stress-induced death; sinking is the loss of cells due to sedimentation and burial; and physical transport is dilution by lateral/vertical dispersal.

For a long time, grazing and sedimentation have been considered as the main factors reducing the size of natural phytoplankton populations. In the last decade of the 20th century cell lysis became recognised as an additional and significant loss factor for phytoplankton cells, affecting not only algal population dynamics and species succession, but also the flux of energy and matter to other trophic levels (Van Boekel et al. 1992; Brussaard et al. 1996; Agustí et al. 1998).

The fact that phytoplankton cells die has been noticed by anyone who has ever cultured algal species (Fig. 12.1a). Directly upon transfer into fresh medium, a slight decline in total cell abundance is often noted. More obviously, the cell abundance in batch cultures typically declines after having reached a maximum due to limitation by nutrients or light (shelf-shading of dense cultures). Detailed laboratory studies clearly revealed significant specific cell lysis that was caused by the physiological stress of nutrient limitation (Brussaard et al. 1997; Lee and Rhee 1997) or light (Berges and Falkowski

Ecological Studies, Vol. 189
Edna Granéli and Jefferson T. Turner (Eds.)
Ecology of Harmful Algae

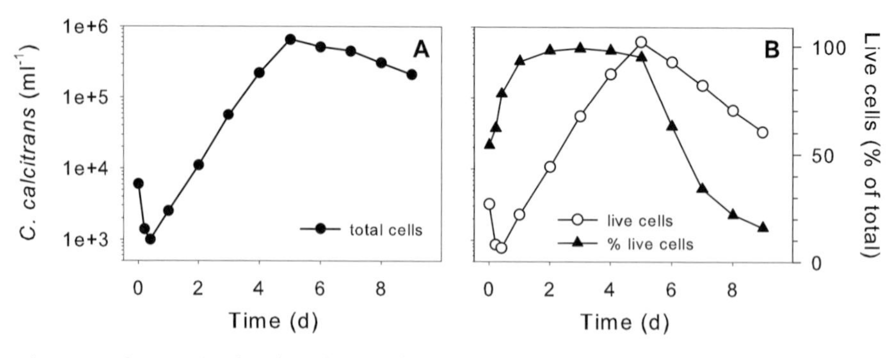

Fig. 12.1. Changes in the abundance of **A** total cells and **B** live cells (abundance and as percentage of total cells) of a standard culture of *Chaetoceros calcitrans* after staining with the live/dead stain SYTOX

1998; Lee and Rhee 1999). Interestingly, a form of intrinsic cell death under *N*-controlled conditions (using continuous cultures) was observed for *Ditylum brightwellii*, which was independent of the growth rate (Brussaard et al. 1997). In addition to environmentally relevant factors such as nutrients and light, infection by pathogens is an important factor inducing algal cell death (Imai et al. 1993; Brussaard 2004). Both growth and mortality are traditionally based on net changes in total population abundance, but this provides no direct information on the viability status of the individual cells (in microbiology, viability refers to an organism being capable of reproducing under appropriate conditions, see Singleton and Sainsbury 1994).

Moreover, the change in photochemical efficiency of the PSII reaction centre (Fv/Fm; e.g. Berges and Falkowski 1998), the reduction in ^{14}C-primary productivity, and the loss of pigmentation (Veldhuis et al. 2001) that are typically associated with declines in cell abundances provide only information on the population as a whole. The recent development of sensitive fluorescent viability or live/dead stains has allowed distinction of the living and dead proportions within a population. Also, the use of dyes in combination with flow cytometry allows the tracing of functional properties at the level of the single cell (Fig. 12.1b and 12.2; Collier 2000; Veldhuis and Kraay 2000). Various vital staining cell assays were specifically developed for phytoplankton, including HABs, and application of these methods, both in the laboratory and the field, showed the co-occurrence of viable and non-viable phytoplankton cells (Crippen and Perrier 1974; Dorsey et al. 1989; Jochem 1999; Brussaard et al. 2001; Veldhuis et al. 2001). The measured viability as a proportion of the total population still needs to be converted into a specific loss rate. This is relatively easy under laboratory conditions using specific formulas (Brussaard et al. 1997), by monitoring batch cultures over time or using continuous cultures in steady state, but measuring such parameters in field populations is more difficult.

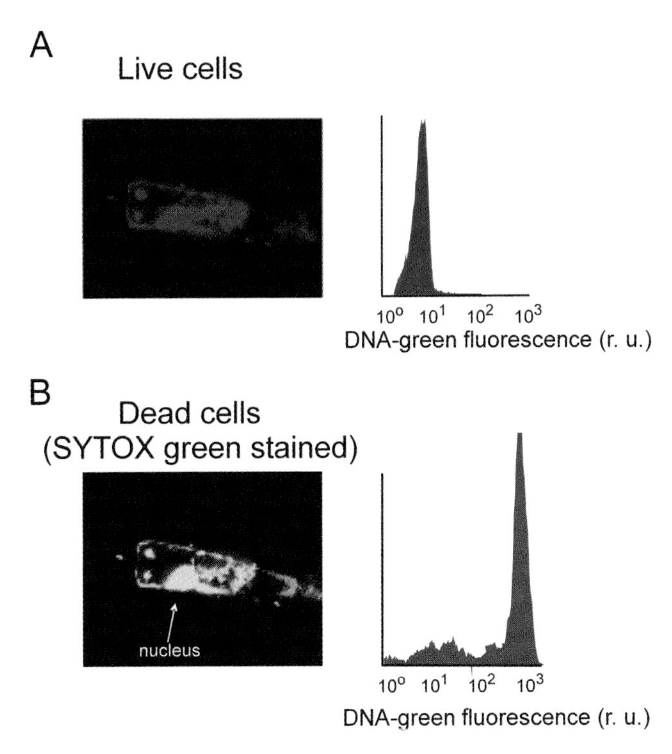

Fig. 12.2. The viability status (live/dead) of phytoplankton cells can be determined using the green nucleic acid-specific live/dead stain SYTOX Green in combination with epifluorescence microscopy or flow cytometry. SYTOX Green can only enter cells with compromised membranes, resulting in non-stained live cells with only the red fluorescence of the chloroplast visible (**A**) and green-stained dead (non-viable) cells (**B**)

Recently, evidence accumulated that the whole process of cell mortality in unicellular algae can be genetically controlled, in a process described as programmed cell death (PCD), often associated with apoptosis (Berges and Falkowski 1998; Segovia et al. 2003; Bidle and Falkowski 2004). Cellular pathways activated by specific conditions (environmental or intercellular) direct a biochemical cascade inside the cell resulting in cell death. A variety of diagnostic markers for apoptosis include: activation of caspases (a family of proteases; Berges and Falkowski 1988; Berman-Frank et al. 2004; Bidle and Falkowski 2004), inversion of an aminophospholipid (phosphatidylserine in particular) from the internal side of the membrane to the external side (Brussaard and Berges unpubl. results), and DNA fragmentation (Berges and Falkowski 1998; Veldhuis et al. 2001; Segovia et al. 2003; Casotti et al. 2005). Interestingly, an alternative morphotype of programmed cell death, paraptosis, was recently suggested for the dinoflagellate *Amphidinium carterae* (Franklin and Berges 2004).

Although research of cell mortality in phytoplankton is still in its infancy, some of the aspects indicated above are very useful in HABs research, since in particular in this group of phytoplankton numerous species produce harmful metabolites, which can induce cell mortality in a large variety of aquatic organisms.

12.2 Mortality of HABs

As it is commonly accepted that HAB species can reduce growth and survival of a variety of co-occurring organisms, there might be a general perception that HABs have a clear competitive advantage over other phytoplankton species, and therefore would generally dominate the phytoplankton community. Fortunately, this differs from reality. A recent observation showed that HAB exochemicals also induce autoinhibition of growth (e.g. *Karina brevis*; Kubanek et al. 2005), which may explain why numerous HABs are categorised as 'low biomass' species. HABs themselves are also susceptible to different forms of mortality. Comparing the DNA-cell-cycle analysis (μ_{gross} based on the genomic DNA staining assay; Carpenter and Chang 1988) with the actual changes in cell abundance, Garcés and Masó (2001) calculated the mortality rates of 6 HAB species to be between 0 and 0.4 d^{-1} during natural blooms in the Mediterranean.

While studies on how environmental variables (e.g. nutrients, irradiance) influence the production of the toxins that affect humans are readily available (see also Chap. 17), information on the direct effect of environmental variables on the process of cell mortality of HABs is virtually nonexistent. We do, however, know that pathogens can be a significant cause of cell death for HABs (Brussaard 2004; Park et al. 2004; Chap. 21). Specifically viruses have been shown to be ecologically important loss agents for bloom-forming HAB-species such as *Aureococcus anophagefferens* (Milligan and Cosper 1994), *Chrysochromulina ericina* (Sandaa et al. 2001), *Heterosigma akashiwo* (Nagasaki and Yamaguchi 1997; Tai et al. 2003), *Phaeocystis pouchetii* (Jacobsen et al. 1996) and *Phaeocystis globosa* (Brussaard et al. 2004). Indeed, viruses affecting bloom-forming HABs have been among the most common phytoplankton viruses isolated, although this bias might be due to the high biomass

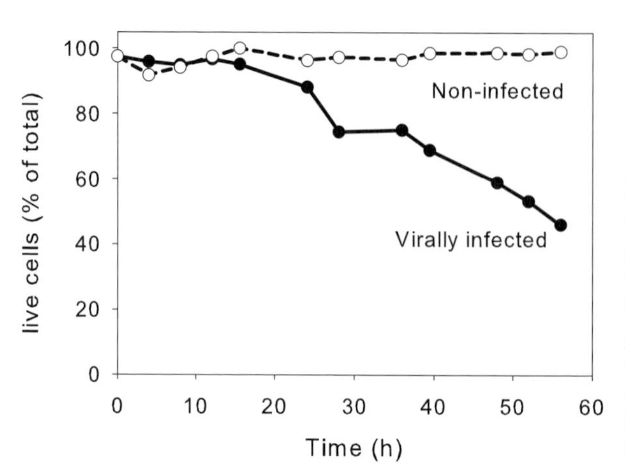

Fig. 12.3. Percentage live cells of *Phaeocystis pouchetii* after staining with SYTOX Green upon infection with *P. pouchetii*-specific viruses at the start of the experiment (t=0 h). A non-infected control culture is added for reference

of these bloom-formers (and thus enhanced virus-phytoplankton contact rates). Detailed studies of the lytic cycle of viruses infecting the HAB-species *Phaeocystis pouchetii*, *P. globosa*, and *Heterosigma akashiwo* (Brussaard 2004; Brussaard and Lawrence unpubl. data) showed a loss of photosynthetic efficiency upon infection. Live/dead stains clearly revealed the devastating impact that viral infection can have on the algal population (Brussaard et al. 2005a), resulting in specific cell lysis rates up to 0.8 d^{-} (Fig. 12.3). It has been suggested that viruses might be used as agents to mitigate bloom forming HABs (Nagasaki et al. 1999), but despite their potential effectiveness, problems facing the use of viruses as a biocontrolling agent to mitigate HABs include the high degree of host specificity and the strong dilution factor upon release into natural waters.

12.3 Death Due to HABs

By definition, harmful algae are characterised by their ability to be toxic, produce harmful secondary metabolites, or have properties that are deemed as harmful for humans and/or other life. HABs can kill marine life, cause losses to aquaculture operations, result in human health problems, and disrupt aquatic ecosystems as a result of unpalatability, physical damage, or oxygen depletion (resulting from the decay of high-biomass blooms). The secondary metabolites excreted by HABs can act as growth-inhibiting agents for other phytoplankton species, influencing species competition and succession, and hence affecting the structure of the plankton community. This type of relationship is often referred to as allelopathy (see also Chap. 15). There is a great range in the susceptibility to these compounds (Tang et al. 2001; Kubanek et al. 2005), sometimes also with a direct negative effect for zooplankton (Fistarol et al. 2004; Ianora et al. 2004). On rare occasions, such as the outbreak of *Chrysochromulina polylepis* in the Kattegat/Skagerrak area in 1988, the toxin produced can be lethal to virtually all pelagic and benthic organisms (Nielsen et al. 1990).

Due to the excretion of allelopathic substances by certain HAB species, co-occurring phytoplankton species may undergo haemolysis (Arzul et al. 1999), resulting in damage to the cytoplasmic membrane and subsequent cell lysis. In general, growth-inhibiting dose-response tests are based on microscopic observations, and may last several hours to days before first signs of adverse effects are visible. Using newly developed fluorescent dyes, a much faster response of the target cells to allelochemicals can be detected (Veldhuis unpubl. data; Brussaard and Legrand unpubl. data). For example, cell-free extracts of *Alexandrium catenella* cause rapid disintegration of the cell membrane of *Tetraselmis* sp. and *Rhodomonas baltica* within a time span of 15 to 30 min when applying a membrane permeability test (SYTOX Green in com-

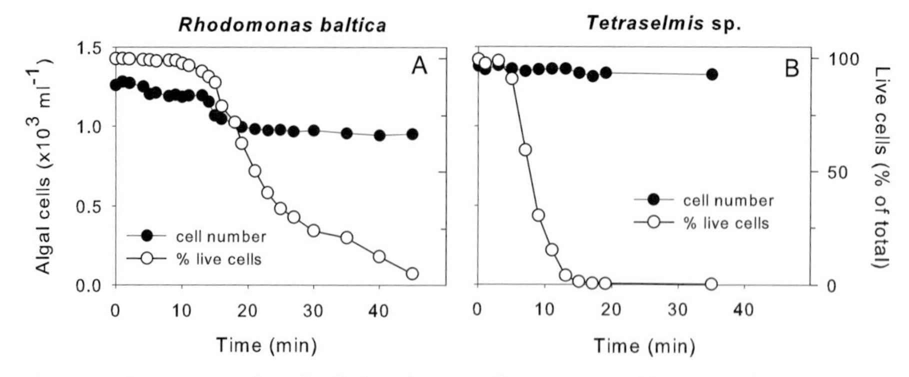

Fig. 12.4. Time course of total cell abundance and percentage of live cells of *Rhodomonas baltica* (**A**) and *Tetraselmis* sp. (**B**) upon addition of a cell free extract of an *Alexandrium catenella* culture

bination with flow cytometry, (Fig. 12.4; Veldhuis unpubl. data). The effectiveness of these compounds was comparable to a 0.4 % solution of paraformaldehyde for rate of cell membrane disintegration. Once the cell membranes had lost their function in *R. baltica* (in about 25 % of the population) the cells disintegrated quickly, which subsequently resulted in a decline of total cell numbers. In contrast, the use of live/dead stains indicated that cell death of *Tetraselmis* sp. occurred faster, but no cell loss was observed, based on the cellular red chlorophyll autofluorescence of the algal cells. A more detailed study executed using allelochemicals of the HAB species *Prymnesium parvum* showed loss of viability of *R. baltica* after only a few minutes (Brussaard and Legrand unpubl. data). However, loss of cells was not detected until hours later. Certainly, the physiological response of the targeted cells to these substances is much faster than previously thought.

12.4 Mechanisms to Avoid Cell Mortality

Although it may not appear so at first (and especially from a socio-economic perspective), the mechanisms that HAB cells have to avoid cell death are at least as important as the factors that cause cell death, because such mechanisms affect the magnitude of HAB blooms and consequently their harmful effects. In general, the effectiveness of allelochemicals can be judged with a dose-response relationship, ranging from no immediate effects via sublethal effects to finally inducing cell mortality, as concentration of the metabolite increases. To minimise the devastating effect, target cells may transfer from vegetative to more resilient cell types (resting stages or cysts, which possess a thicker cell wall, and may therefore not be so easily affected by cell membrane disrupting substances).

A different mechanism to avoid cell mortality, also involving a novel life-cycle stage, has been observed for the genus *Phaeocystis*. Although *P. globosa* is also a HAB species, its harmful aspect is mainly due its formation of massive blooms that consist mainly of gelatinous colonies (Schoemann et al. 2005). The colonial mucilage disturbs the food-web structure in the coastal zone of the southern bight of the North Sea, and causes severe economical damage to shellfish aquaculture and tourism (Lancelot and Mathot 1987). In spring, with ample nutrients present, colonial cells growing much faster than the single cells (Veldhuis et al. 2005), and cell mortality by grazing of colonies much reduced (Tang et al. 2001; Koski et al. 2005).

A mechanistic study to examine the relevance of virally mediated cell lysis of *P. globosa* in mesocosms indicated that mainly the single cell morphotype was affected by viruses specific for *P. globosa* (Brussaard et al. 2005a). A newly developed ecosystem model, including a detailed virus module, also suggested that the colonial form of *P. globosa* largely escaped viral infection due to their size (Ruardy et al. 2005), When nutrients are depleted cells are shedded from the colonies (Veldhuis et al. 1991) and subsequently massive viral lysis of the released cells occurred (Brussaard et al. 2005a).

12.5 Ecological Implications

HABs are increasing on a global scale, in geographical extent, species diversity, and economic impact (Smayda 1990; Anderson et al. 2002). Although human-induced activities (eutrophication and pollution) have been implicated as the main causes for this increase, climate change has also been suggested because of its associated effects (changes in upwelling patterns and other hydrological features, and increases in water temperature). In addition, the accidental introduction of alien HABs into new environments, such as through ballast water discharge will also enhance the global expansion of HABs (Hallegraeff and Bolch 1992). Cell mortality clearly affects population dynamics as well as the structure of the food web, particularly for high-biomass blooms such as those of *P. globosa* (Brussaard et al. 1996, 2005b). However, even low-biomass HABs can alter the phytoplankton diversity as well as the composition of other trophic levels in an undesirable manner. The responses of allelochemicals acting as bioactive metabolites in other organisms is usually rather fast, on a scale of minutes rather than hours or days (Fig. 12.4), and is often not very species selective (Fistarol et al. 2004). Resistance to these substances will therefore depend on a cell's ability to quickly detect (extra- or intracellular), and respond to the allelochemical by a physiological alteration or activation of biochemical pathways to neutralise the reactive components. It is particularly interesting to understand the intracellular pathways leading to growth inhibition and cell mortality, and to determine if these are linked to the genetically controlled PCD and apoptosis. On the other hand, these types of secondary metabolites are probably more universally present and not entirely restricted to HABs. In diatoms, thus far these metabolites have been found in only one toxin-producing species, namely *Pseudo-nitzschia pseudodelicatissima* (Pan et al. 2001), but numerous others produce aldehydes in concentrations sufficiently high to inhibit population growth of other phytoplankton and even copepods (Ianora et al. 2004; Casotti et al. 2005). Chemical warfare could therefore be more common in phytoplankton than has generally been assumed, and not just restricted to HAB species.

Acknowledgements. We would like to express special thanks to one of the anonymous reviewers for constructive comments.

References

Agustí S, Satta MP, Mura MP, Benavent E (1998) Dissolved esterase activity as a tracer of phytoplankton lysis: evidence of high phytoplankton lysis rates in the northwestern Mediterranean. Limnol Oceanogr 43:1836–1849

Anderson DM, Glibert MP, Burkholder J (2002) Harmful algal blooms and eutrophication: nutrient sources, composition, and consequences. Estuaries 25:704–726

Arzul G, Sequel M, Guzman L, Erard-Le Denn E (1999) Comparison of allelopathic properties in three toxic *Alexandrium* species. J Exp Mar Biol Ecol 232:285–295

Berges JA, Falkowski PG (1998) Physiological stress and cell death in marine phytoplankton: induction of proteases in response to nitrogen or light limitation. Limnol Oceanogr 43:129–135

Berman-Frank I, Bidle KD, Haramaty L, Falkowski PG (2004) The demise of the marine cyanobacterium *Trichodesmium* spp., via an autocatalyzed cell death pathway. Limnol Oceanogr 49:997–1005

Bidle KD, Falkowski PG (2004) Cell death in planktonic, photosynthetic microorganisms. Nature Rev Microbiol 2:643–655

Brussaard CPD (2004) Viral control of phytoplankton populations – a review. J Eukaryot Microbiol 51:125–138

Brussaard CPD, Gast GJ, Van Duyl FC, Riegman R (1996) Impact of phytoplankton bloom magnitude on a pelagic microbial food web. Mar Ecol Prog Ser 144:211–221

Brussaard CPD, Noordeloos AAM, Riegman R (1997) Autolysis kinetics of the marine diatom *Ditylum brightwellii* (Bacillariophyceae) under nitrogen and phosphorus limitation and starvation. J Phycol 33:980–987

Brussaard CPD, Marie D, Thyrhaug R, Bratbak G (2001) Flow cytometric analysis of phytoplankton viability following viral infection. Aquat Microb Ecol 26:157–166

Brussaard CPD, Short SM, Frederickson CM, Suttle CA (2004) Isolation and phylogenetic analysis of novel viruses infecting the phytoplankter *Phaeocystis globosa* (Prymnesiophyceae). Appl Environ Microbiol 70:3700–3705

Brussaard CPD, Kuipers B, Veldhuis MJW (2005a) A mesocosms study of *Phaeocystis globosa* population dynamics. I. Regulatory role of viruses in bloom control. Harmful Algae 4:859–874

Brussaard CPD, Mari X, Van Bleijswijk JDL, Veldhuis MJW (2005b) A mesocosm study of *Phaeocystis globosa* population dynamics. II. Significance for the microbial community. Harmful Algae 4:875–893
Carpenter EJ, Chang J (1988) Species-specific phytoplankton growth rates via diel DNA synthesis cycles. I. Concept of the method. Mar Ecol Prog Ser 43:105–111
Casotti R, Mazza C, Brunet C, Vantrepotte V, Ianora A, Miralto A (2005) Growth inhibition and toxicity of the diatom aldehyde 2-trans-4-trans decadienal on *Thalassiosira weissflogii* (Bacillariophyceae). J Phycol 41:7–20
Collier JL (2000) Flow cytometry and the single cell in phycology. J Phycol 36:628–644
Crippen RW, Perrier JL (1974) The use of neutral red and Evans blue for live-dead determination of marine plankton. Stain Technol 49:97–104
Dorsey J, Yentsch CM, Mayo S, McKenna C (1989) Rapid analytical technique for the assessment of cell metabolic activity in marine microalgae. Cytometry 10:622–628
Fistarol GO, Legrand C, Selander E, Hummert C, Stolte W, Granéli E (2004) Allelopathy in *Alexandrium* spp.: effect on a natural plankton community and on algal monocultures. Aquat Microb Ecol 35:45–56
Franklin DJ, Berges JA (2004) Mortality in cultures of the dinoflagellate *Amphidinium carterae* during culture senescence and darkness. Proc R Soc Lond B 271:2099–2107
Garcés E, Masó M (2001) Phytoplankton potential growth rate versus increase in cell numbers: estimation of cell lysis. Mar Ecol Prog Ser 212:297–300
Hallegraeff GM, Bolch CJ (1992) Transport of diatoms and dinoflagellate resting spores via ship's ballast water: implication for plankton biogeography and aquaculture. J Plankton Res 14:1067–1084
Ianora A, Miralto A, Poulet SA, Carotenuto Y, Buttino I, Romano G, Casotti R, Pohnert G, Wichard T, Colucci-D'Amato L, Terrazzano G, Smetacek V (2004) Aldehyde suppression of copepod recruitment in blooms of a ubiquitous planktonic diatom. Nature 429:403–407
Imai I, Ishida Y, Hata Y (1993) Killing of marine phytoplankton by a gliding bacterium *Cytophaga* sp., isolated from the coastal Sea of Japan. Mar Biol 116:527–532
Jacobsen A, Bratbak G, Heldal M (1996) Isolation and characterization of a virus infecting *Phaeocystis pouchetii* (Prymnesiophyceae). J Phycol 32:923–927
Jochem FJ (1999) Dark survival strategies in marine phytoplankton assessed by cytometric measurement of metabolic activity with fluorescein diacetate. Mar Biol 135:721–728
Koski M, Dutz J, Klein-Breteler WCM (2005) Selective grazing of *Temora longicornis* in different stages of a *Phaeocystis globosa* bloom – a mesocosm study. Harmful Algae 4:915–927
Kubanek J, Hicks MK, Naar J, Villareal TA (2005) Does the red tide dinoflagellate *Karinia brevis* use allelopathy to outcompete other phytoplankton? Limnol Oceanogr 50:883–895
Lancelot C, Mathot S (1987) Dynamics of a *Phaeocystis*-dominated spring bloom in Belgian coastal waters. I. Phytoplanktonic activities and related parameters. Mar Ecol Prog Ser 37:239–248
Lee D-Y, Rhee G-Y (1997) Kinetics of cell death in the cyanobacterium *Anabaena flos-aquae* and the production of dissolved organic carbon. J Phycol 33:991–998
Lee D-Y, Rhee G-Y (1999) Kinetics of growth and death in *Anabaena flos-aquae* (Cyanobacteria) under light limitation and supersaturation. J Phycol 35:700–709
Milligan KDL, Cosper EM (1994) *Aureococcus anophagefferens* (Pelagophyceae). Science 266:805–807
Nagasaki K, Yamaguchi M (1997) Isolation of a virus infectious to the harmful bloom causing microalga *Heterosigma akashiwo* (Raphidophyceae). Aquat Microb Ecol 13:135–140

Nagasaki K, Tarutani K, Yamaguchi M (1999) Growth characteristics of *Heterosigma akashiwo* virus and its possible use as a microbiological agent for red tide control. Appl Environ Microbiol 65:898–902

Nielsen TG, Kiørboe T, Bjørnsen PK (1990) Effects of a *Chrysochromulina polylepis* subsurface bloom on the planktonic community. Mar Ecol Prog Ser 62:21–35

Pan Y, Parsons ML, Busman M, Moller P, Dortch Q, Powell CL, Fryxell GA, Doucette GJ (2001) *Pseudo-nitzschia pseudodelicatissima*, a confirmed producer of domoic acid from the northern Gulf of Mexico. Mar Ecol Prog Ser 220:83–92

Park MG, Yih W, Coats DW (2004) Parasites and phytoplankton, with a special emphasis on dinoflagellate infections. J Eukaryot Microbiol 51:145–155

Reynolds CS (1984) The ecology of freshwater phytoplankton. Cambridge University Press, Cambridge, 384 pp

Ruardy P, Veldhuis MJW, Brussaard CPD (2005) Modelling bloom dynamics of the polymorphic phytoplankter *Phaeocystis globosa*: impact of grazers and viruses. Harmful Algae 4:941–963

Sandaa R-A, Heldal M, Castberg T, Thyrhaug R, Bratbak G (2001) Isolation and characterization of two viruses with large genome size infecting *Chrysochromulina ericina* (Prymnesiophyceae) and *Pyramimonas orientalis* (Prasinophyceae). Virology 290:272- 280

Schoemann V, Becquevort S, Stefels J, Rousseau V, Lancelot C (2005) *Phaeocystis* blooms in the global ocean and their controlling mechanisms: a review. J Sea Res 53:43–66

Segovia M, Haramaty L, Berges JA, Falkowski PG (2003) Cell death in the unicellular chlorophyte *Dunaliella tertiolecta.* A hypothesis on the evolution of apoptosis in higher plants and metazoans. Plant Physiol 132:99–105

Singleton P, Sainsbury D (1994) Dictionary of microbiology and molecular biology. Wiley Interscience, Chichester, UK 1,019 pp

Smayda TJ (1990) Novel and nuisance phytoplankton blooms in the sea: evidence for a global epidemic. In: Granéli E, Sundstrom B, Edler L, Anderson DM (eds) Toxic marine phytoplankton. Elsevier, New York, pp 29–40

Tai V, Lawrence JE, Lang AS, Chan AM, Culley AI, Suttle CA (2003) Characterization of HaRNAV, a single-stranded RNA virus causing lysis of *Heterosigma akashiwo* (Raphidophyceae). J Phycol 39:343–352

Tang KW, Jakobsen HH, Visser AW (2001) *Phaeocystis globosa* (Prymnesiophyceae) and the planktonic food web: feeding, growth, and trophic interactions among grazers. Limnol Oceanogr 46:1860–1870

Van Boekel WHM, Hansen FC, Riegman R, Bak RPM (1992) Lysis-induced decline of a *Phaeocystis* spring bloom and coupling with the microbial food web. Mar Ecol Prog Ser 81:269–276

Veldhuis MJW, Kraay GW (2000) Application of flow cytometry in marine phytoplankton research: current applications and future perspectives. Sci Mar 64:121–134

Veldhuis MJW, Colijn F, Admiraal W (1991) Phosphate utilization in *Phaeocystis pouchetii* (Haptophyceae). Mar Ecol 12:53–62

Veldhuis MJW, Kraay GW, Timmermans KR (2001) Cell death in phytoplankton: correlation between changes in membrane permeability, photosynthetic activity, pigmentation and growth. Eur J Phycol 36:167–177

Veldhuis MJW, Brussaard CPD, Noordeloos AAM (2005) Living in a *Phaeocystis* colony; a way to be a successful algal species. Harmful Algae 4:841–858

13 The Diverse Nutrient Strategies of Harmful Algae: Focus on Osmotrophy

P. M. Glibert and C. Legrand

13.1 Introduction and Terminology

Ever since the discovery of microscopic flagellates, the manner in which they sustain themselves has been a topic of interest. Indeed, in 1677 Anthoni van Leeuwenhoek proposed to call the flagellates that he observed 'animalcules' after observing their motion, "When these animalcules bestirred 'emselves, they sometimes stuck out two little horns, which were continually moved, after the fashion of a horse's ears" (re-quoted by Milius 1999). Today, there is much interest in identifying the mode of nutrition of plankton, especially harmful algae, for a variety of reasons. As many harmful algal blooms (HABs) have been associated with eutrophication and/or alterations in the amount and form of nutrients supplied (reviewed by Glibert et al. 2005), linking HABs nutrition with specific nutrient forms is a key management issue. Additionally, as models are developed for the prediction of HABs, accurately relating growth to the nutrient supply is essential. Lastly, there is much to be learned about adaptive physiology from cells that have a diversity of nutritional mechanisms.

Many HABs rely strictly upon photosynthesis for their carbon, and use inorganic nutrients for their nutrition: these are the *autotrophs*. Some phototrophic species, however, have alternative pathways for acquiring carbon or nutrients: these are the *mixotrophs*. They use dissolved or particulate organic substances to renew their cellular reserves of carbon, macronutrients, amino acids, trace elements or phospholipids (Raven 1997; Stoecker 1999; Granéli et al. 1999). In applying this terminology to nutrition, mixotrophy encompasses several processes, including *osmotrophy*, nutrition by direct absorption and uptake of organic molecules, and *phagotrophy*, ingestion of prey or other food particles. When carbon is incorporated, the process is *heterotrophy*.

Many mixotrophic species display a continuum of dependence on these alternative pathways. Some species only display osmotrophy or phagotrophy

Ecological Studies, Vol. 189
Edna Granéli and Jefferson T. Turner (Eds.)
Ecology of Harmful Algae

when particular cellular requirements can apparently not be met by autotrophy. Other species have lost the ability for autotrophy and rely exclusively on phagotrophy and osmotrophy (Jones 1994). Heterotrophic dinoflagellates are among the species that have lost the ability for autotrophy. In fact, roughly half of all dinoflagellates are obligate heterotrophs (Smayda 1997). *Noctiluca* is a well-recognized example of a phagotrophic, heterotrophic dinoflagellate (Kimor 1979). Some heterotrophic dinoflagellates (e.g., *Pfiesteria*) can 'borrow' or 'steal' functional chloroplasts from its prey; this strategy is termed *kleptochloroplastidy* (Skovgaard 1998). The continuum of processes also includes such strategies as *dasmotrophy*, as found for the genus *Chrysochromulina*, in which extracellular toxins perforate the cell membrane of prey, inducing osmosis and leakage of organic compounds available for uptake or incorporation (Estep and MacIntyre 1989).

This chapter focuses on our current understanding of osmotrophy in HABs. This chapter is not intended to be a review of the uptake of organic compounds by algae (see recent review by Berman and Bronk 2003), but rather an introduction to the importance and diversity of this nutritional strategy for HABs.

13.2 Osmotrophy Pathways and Methods to Explore Them

One of the factors complicating the ability to trace the pathways of uptake of organic nutrients is the fact that in nature, the range of organic compounds is large, including dissolved amino acids, nucleic acids, polypeptides, polyphenolic substances and polysaccharides. Many are unknown or not well characterized, many are refractory, and others are highly variable in time and space (Antia et al. 1991; Bronk 2002). Ultra-filtration techniques have been used to characterize DOM by broad molecular weight fraction (e.g., Benner et al. 1997), but such approaches do not allow characterization or identification of individual compounds. Using ^{15}N-NMR spectroscopy, it has recently been found that there are different pools of high molecular weight DON in the oceans and they may have different chemical and biological reactivity (Aluwihare et al. 2005). This approach has yet to be applied to eutrophic coastal regions and may be useful in characterizing which DOM compounds are bioavailable during HAB outbreaks.

For the determination of the rates of uptake of specific organic compounds, stable isotopic or radiotracer techniques can be used. These are analogous to the techniques used for inorganic nutrient uptake. These techniques are limited, however, because only selected compounds can be enriched with a stable or radioactive tracer. Studies in which radioisotopes have been used to trace the uptake of specific organic compounds have a long history in the literature (e.g., Wheeler et al. 1974). Some natural dissolved organic nitrogen

(DON) compounds have been prepared with ^{15}N labels, but this involves growing natural assemblages or cultures with a ^{15}N substrate, isolating and concentrating the released DON and using this concentrate as a tracer in subsequent experiments (e.g., Bronk and Glibert 1993; Veuger et al. 2004). These experiments have not been widely conducted because of the difficulty and variability in the DON isolates. Other investigators have used organic compounds that are labeled with two tracers, such as ^{15}N and ^{13}C to assess the extent to which an organic compound may be used for nutrient or carbon (e.g., Mulholland et al. 2004).

Osmotrophy pathways are numerous and include direct uptake as well as extracellular oxidation and hydrolysis, and pinocytosis (Fig. 13.1). Enzymatic measurements have been used to determine some of the pathways involved in the incorporation and degradation of organic compounds (Chrost 1991). Several enzymes illustrate these pathways. Urease is an intracellular enzyme that has been characterized in relation to the uptake of urea. Urease activity appears to be constitutive for many algal species but may be higher in many

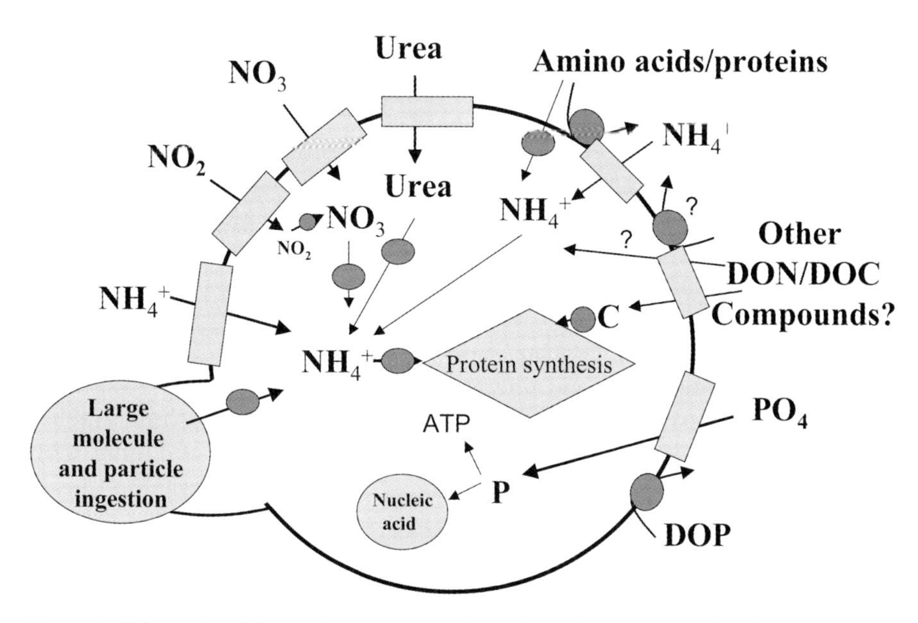

Fig. 13.1. Schematic of the various pathways by which HABs may acquire their nutrients. All phytoplankton can transport nitrate, nitrite, ammonium, urea and phosphate across cell membranes via passive diffusion or active transport. Some species can transport large organic molecules across their cell membrane, and some have cell surface enzymes for breaking down larger organic molecules before transporting the nutrients. Some species have the ability to phagocytize other cells, particles and/or large molecules. Although many pathways are shown, this is not intended to imply that all HABs have all capabilities. Pathways that involve enzymatic reactions are indicated with a *circle*. Pathways for which there is much uncertainty are indicated with a *question mark*

HAB species compared to non-HABs. For example, urease activity has been found to be sufficiently high in *Aureococcus anophagefferens* and *Prorocentrum minimum* to meet the cellular nitrogen demand for growth but insufficient to meet this demand in the diatom *Thalassiosira weissflogii* (Fan et al. 2003). In *Alexandrium fundyense* urease activity was shown to be seasonally variable and related to the toxin content of the cells (Dyhrman and Anderson 2003).

Another ecologically important enzyme is alkaline phosphatase (AP). This enzyme is responsible for the hydrolysis of organic phosphorus sources, such as sugar phosphates, phospholipids, and nucleotide phosphates, among other compounds. This enzyme is located on the cell surface, membrane-bound. In general, activity of AP increases upon phosphate stress and thus is an excellent ecological indicator of phosphorus limitation (Dyhrman 2005). Polyclonal antibodies have been developed to AP in selected species and whole-cell immunolabeling approaches have been applied in the field to determine the AP activity of *A. fundyense* in mixed assemblages (Dyhrman and Palenik 1999). Such approaches allow for species-specific activities to be determined, and advance our understanding of the nutrient status of individual species within mixed assemblages.

Some species have the capability for the breakdown of other organic compounds at the cell surface. These species have enzymes that can oxidize or hydrolyze amino acids or proteins. Extracellular amino acid oxidation has been shown to be expressed to a greater degree by dinoflagellates compared to some diatoms, and furthermore, to be expressed to a greater degree when inorganic nutrients are at or near depletion (Palenik and Morel 1990; Mulholland et al. 1998). Also occurring at the cell surface of many flagellates is peptide hydrolysis. Both peptide hydrolysis and amino acid oxidation may be important in some HABs, e.g., *A. anophagefferens* in natural communities (Mulholland et al. 2002). Leucine amino peptidase is another protease that hydrolyses peptide bonds and liberates amino acids (Dyhrman 2005). It is measured by assessing the rate of hydrolysis of an artificial substrate, and has been shown to be of potential significance in dinoflagellates and other HAB classes. Stoecker and Gustafson (2003), for example, demonstrated in Chesapeake Bay that Leu-amino-peptidase activity was associated with a dinoflagellate bloom, and that in non-axenic cultures of *Akashiwo sanguinea, Gonyaulax grindleyi, Gyrodinium uncatenum, Karlodinium micrum* and *P. minimum* the activity was associated with the dinoflagellates and not the bacteria.

The direct uptake of larger organic molecules by osmotrophic dinoflagellates is largely unknown, but there is evidence that the mechanism involves pinocytosis (endocytotic uptake), at least in some species. For example, *Prorocentrum micans* can accumulate fluorescently labeled proteins in the pusule, an organelle that functions in osmoregulation, waste disposal and/or nutrition (Klut et al. 1987). Fluorescently labeled high molecular weight dextrans were also found to be taken up by *Alexandrium catenella* and to accumulate in small vesicles in the cells (Legrand and Carlsson 1998). In the same species, Doblin et al. (2001), using autoradiography, found that radioactively labeled humic substances were also taken up by the cells. The process of pinocytosis begins to blur the distinction between osmotrophy and phagotrophy and many species, such as *Alexandrium* spp., can use the same pathway to capture large organic molecules or prey.

Several new methods are being applied to the study of the large suite of individual compounds in natural mixes of dissolved organic matter (DOM) and their uptake by HABs (Glibert et al. 2005). Electrospray mass spectrometry is a tool that can be used to identify and quantify the presence of a broad suite of compounds in a natural sample and to track their rate of change under experimental conditions (Seitzinger et al. 2005). Other advances in flow cytometry, epifluorescence microscopy and the detection of gene expression, using quantitative reverse transcription PCR, show promise in detecting expression of enzymes and can help to isolate the source of the signal in a natural sample (Dyhrman 2005). New in situ nutrient monitoring systems provide data on both inorganic and organic nutrient fluxes on the time and space resolution necessary to define the ephemeral changes that may be significant to phytoplankton cells. There are thus many new tools to address the composition and fluxes of DOM and the physiological capability of cells to use distinct fractions of available DOM. Although many approaches are now available, there is a need to integrate both qualitative and quantitative approaches to estimate the importance of osmotrophy to HABs in nature.

13.3 Cellular Costs and Benefits of Osmotrophy

The photosynthetic apparatus constitutes up to 50 % of the cell biomass and uses a high proportion of the cellular budget of energy and nutrients. All autotrophs have the ability to assimilate organic molecules since they possess enzymes in the Krebs cycle. Thus, the costs and maintenance of transplasma membranes, lysosomes and specific enzymes used in the uptake and digestion of organic molecules is similar in osmotrophs and strict autotrophs. In contrast, strict heterotrophs have higher growth efficiencies than strict autotrophs and mixotrophs because they do not have the cellular maintenance costs of the photosynthetic apparatus (Raven 1997; Stoecker 1999). There is some evidence that the energetic costs for mixotrophy may be higher in brackish waters than in marine environments, due to the physiological constraints of osmotic stress in a brackish system (Moorthi 2004). Although there may be higher cellular costs to maintaining mixotrophy, there are obvious advantages. Facultative mixotrophs can switch from autotrophic to heterotrophic nutrition under unfavorable environmental conditions (light, nutrients), resulting in survival and potentially increased growth compared to those species unable to accommodate to unfavorable conditions. It has thus been hypothesized that osmotrophy in HABs is an advantage that may help them thrive during periods when inorganic nutrients may not be sufficient.

13.4 Ecological Significance of Osmotrophy

For osmotrophy to be of significance ecologically, the organisms present must have the capability for uptake of complex organic nutrients and/or carbon, and such compounds must be available in the environment. To illustrate the ecological importance of osmotrophy, several examples are described, which show both the diversity of the sources supporting HAB osmotrophy and the significance of these sources in their proliferation.

Blooms of the dinoflagellate *Karenia brevis* generally originate offshore, in oligotrophic waters, and may be stimulated by sources of nutrients that are not of anthropogenic origin. Indeed, it is thought that these blooms are stimulated by organic matter produced by blooms of yet another HAB species of the cyanobacterium genus *Trichodesmium* (Walsh and Steidinger 2001). *Trichodesmium* spp. are diazotrophs that can form large blooms in tropical and subtropical regions of the world's oceans (LaRoche and Breitbarth 2005). When these cells are actively fixing N_2 they release ammonium and DON, mostly in the form of low molecular weight compounds (Capone et al. 1994; Glibert and Bronk 1994). The release of DON during blooms can enrich the ambient DON within blooms by 50–500 % (Karl et al. 1992; Glibert and O'Neil 1999). Using isotopic techniques in which the *Trichodesmium* were first enriched with $^{15}N_2$, and the released DON was subsequently collected and provided to natural *K. brevis*, the direct uptake of this source of nitrogen was documented (Bronk et al. 2004).

DON stimulates another dinoflagellate, *P. minimum*, during its early stages of development, but in this case, DON is typically of terrestrial or anthropogenic origin. The occurrence of large *P. minimum* blooms has often been linked to eutrophication in coastal environments (Heil et al. 2005). A large proportion of river nutrient loading is in the form of humic substances, high molecular weight organic compounds which, while generally considered refractory, may stimulate *P. minimum*. In mesocosm experiments, Granéli et al. (1985) showed that dinoflagellate populations, including *P. minimum*, were stimulated by inorganic nitrogen only when added in combination with humic acids. Carlsson et al. (1999) found that humic additions stimulated the growth of this dinoflagellate, and that *P. minimum* used 35 % of the DON from the added humics. Heil (2005) also found that organic additions stimulated growth rates and cell yields of *P. minimum*, but the extent of stimulation varied with the organic fraction examined and its molecular weight: stimulation was greatest with humic and fulvic acids and lowest with hydrophilic acids.

Aureococcus anophagefferens provides yet another model of the role of DON in stimulating brown-tides. Development of these blooms has previously been correlated with years of low rainfall (LaRoche et al. 1997), in turn leading to higher ratios of organic: inorganic nitrogen. Studies in natural blooms have substantiated high uptake rates of organic nitrogen (e.g., Berg et al. 1997). Lomas et al. (2004) identified the benthos as an important source of both organic nitrogen and phosphorus in the Long Island, New York embayments where these blooms are most common. Recent reports underscore the importance of DON, specifically low molecular weight DON (urea and amino acids), in promoting growth of these cells and therefore the development of blooms, particularly under low light (Pustizzi et al. 2004; Mulholland et al. 2004). Peptides, acetamide and urea contribute significantly to the growth of *A. anophagefferens* under axenic conditions (Berg et al. 2002).

Urea has been related to several HAB outbreaks. Urea is a simple organic molecule with significant ecological implications because of the growing use of urea as an agricultural fertilizer (Glibert et al. 2006). Blooms of cyanobacteria have been found to be directly related to urea availability or rates of urea uptake (e.g., Berman 1997; Glibert et al. 2004). In Florida Bay, the fraction of the algal community composed of cyanobacteria was positively correlated with the fraction of urea-N uptake and negatively correlated with the fraction of nitrate-N uptake (Glibert et al. 2004). In the Baltic Sea, urea was among the nitrogen compounds shown to contribute significantly to the growth of the filamentous cyanobacteria *Aphanizomenon* sp. and *Nodularia spumigena* (Panosso and Granéli 2000; Berg et al. 2003). Urea has also been found to promote faster growth than other nitrogen sources in *A. ovalisporum* and *T. theibautii* (Berman and Chava 1999; Mulholland et al. 1999). Urea has also been found to be significant for several HAB dinoflagellate outbreaks, such as in blooms of *Lingulodinium polyedrum* off the coast of California (Kudela and Cochlan 2000) and *Alexandrium catenella* from the Thau lagoon, Mediterranean Sea (Collos et al. 2004).

Several HAB events are related to an elevation in the ratio of dissolved organic carbon (DOC):DON. Conditions where dissolved organic nutrients and DOC concentrations are high will likely favor osmotrophs over strictly photosynthetic protists. Three separate blooms of *Pfiesteria piscicida, P. minimum* and *A. anophagefferens* in Chesapeake Bay were all correlated with elevated DOC:DON ratios relative to the long-term mean (Glibert et al. 2001). The elevation in this ratio for these particular blooms was a reflection of both elevated levels of DOC as well as a depletion of DON. Lomas et al. (2001) showed this relationship to be robust for numerous brown tide blooms in Long Island, New York. Furthermore, during a bloom of *Gymnodinium* spp. in Kuwait Bay, the ratios of DOC:DON for stations collected within the bloom were approximately twice those determined for non-bloom stations with a mixed phytoplankton assemblage (Heil et al. 2001).

The extent to which carbon from DOC is used by osmotrophs is highly variable. Amino acid carbon may be used more readily than urea-carbon in *A. anophagefferens* (Mulholland et al. 2004). It is thought that DOC is important in the growth and maintenance of blooms of many *Dinophysis* species, including those responsible for diarrhetic shellfish poisoning (DSP). These blooms are often found in light limited environments, such as off the French coast or in the Baltic Sea, where they proliferate in thin layers in the region of the thermocline where light is less than 1 % (Gisselson et al. 2002; Gentien et al. 2005). Utilization of DOC in these species has been documented by the uptake of ^{14}C in organic compounds in the dark (Granéli et al. 1997) and enrichment experiments with natural dinoflagellate communities (Purina et al. 2004). These results suggest that *Dinophysis* can compensate for reduced irradiance by feeding heterotrophically.

13.5 A Comment on Evolutionary Aspects of Osmotrophy

The endosymbiotic theory of eukaryote evolution was introduced by Margulis (1975) and more formally described by Margulis (1981). The concept that symbiotic relationships gave rise to many organelles, including chloroplasts, has now been vastly extended. Whereas the primary plastids all likely trace back to a cyanobacterial endosymbiotic event, the phylogenetic evidence now suggests that early branching of the eukaryotic phototrophs gave rise to two major clades, the “green algae”, which contain chlorophyll-*b* (chl-*b*) and the “red algae” which contain chlorophyll-*c* (chl-*c*) (Delwiche 1999; Grzebyk et al. 2003). With respect to the dinoflagellates, there is evidence for secondary plastid endosymbiotic events, leading to dinoflagellates with either chl-*b* or *c* (Delwiche 1999). Future genomic analyses will be of value in resolving the number and nature of endosymbiotic plastid events (Palmer 2003).

Even with the uncertainties in understanding plastid endosymbiotic evolution, the evolution of mixotrophy is more complex than autotrophy. Raven (1997) suggested that phagotrophy is a primitive character and its absence in photosynthetic protists a derived trait. Osmotrophs take up organic substances via importers localized in the plasma membrane, just like phagotrophs, but they do not have proper food vacuoles and digestion enzymes are likely excreted into the environment. This suggests that in an evolutionary sense, osmotrophy may have preceded phagotrophy because without membrane importers food vacuoles would have no function (Martin et al. 2003). However, this idea is not without controversy, given the traditional view that phagocytosis was necessary for eukaryotic evolution. Mixotrophy is found in a large number of planktonic groups, suggesting independent evolution (Jones 1994). How eukaryotes became osmotrophs in the first place remains unclear, and genomic analyses are needed to see if eukaryotic genes involved in osmotrophy (e.g., cell membrane transport proteins involved in the uptake of specific organic compounds) share similarities with those found among bacteria.

13.6 Conclusions

Organic forms of nutrients originate from various sources, natural and anthropogenic. Organic nutrients are, in turn, used by many HAB species that have multiple acquisition mechanisms. The strategies for nutrient and carbon acquisition by HABs are thus far more complex than were thought a decade or two ago. With the application of the host of methods now available to characterize and quantify organic matter and to measure nutrient and carbon fluxes, the role of osmotrophy in HAB dynamics will increasingly be recognized. However complex, these processes must be better understood, quantified and incorporated into models in order to advance our ability to understand population and food web dynamics and to predict the occurrences of HABs.

Acknowledgements. The authors thank E. Granéli for the invitation to prepare this chapter. This is contribution number 3895 from the University of Maryland Center for Environmental Science. P. Glibert acknowledges financial support from Maryland Sea Grant and NOAA MERHAB. C. Legrand acknowledges financial support from the Swedish Research Council and The European Commission through the FP 6-THRESHOLDS project (GOCE 003933).

References

Antia NJ, Harrison PJ, Oliveira L (1991) The role of dissolved organic nitrogen in phytoplankton nutrition, cell biology and ecology. Phycologia 30:1–89

Aluwihare LI, Repeta DJ, Pantoja S, Johnson CG (2005) Two chemically distinct pools of organic nitrogen accumulate in the ocean. Science 308:1007–1010

Benner RB, Biddanda B, Black B, McCarthy M (1997) Abundance, size distribution, and stable carbon and nitrogen isotopic compositions of marine organic matter isolated by tangential-flow ultrafiltration. Mar Chem 57:243–263

Berg GM, Repeta DJ, LaRoche J (2002) Dissolved organic nitrogen hydrolysis rates in axenic cultures of *Aureococcus anophagefferens* (Pelagophyceae): comparison with heterotrophic bacteria. Appl Environ Microbiol 68:401–404

Berg GM, Balode M, Purina I, Bekere S, Béchemin C, Maestrini SY (2003) Plankton community composition in relation to availability and uptake of oxidized and reduced nitrogen. Aquat Microb Ecol 30:263–274

Berg GM, Glibert PM, Lomas MW, Burford MA (1997) Organic nitrogen uptake and growth by the chrysophyte *Aureococcus anophagefferens* during a brown tide event. Mar Biol 129:377–387

Berman T (1997) Dissolved organic nitrogen utilization by an *Aphanizomenon* bloom in Lake Kinneret. J Plankton Res 19:577–586

Berman T, Bronk DA (2003) Dissolved organic nitrogen: a dynamic participant in aquatic ecosystem. Aquat Microb Ecol 31:279–305

Berman T, Chava S (1999) Algal growth on organic compounds as nitrogen sources. J Plankton Res 21:1423–1437

Bronk DA (2002) Dynamics of DON. In: Hansell DA, Carlson CA (eds) Biogeochemistry of marine dissolved organic matter. Academic Press, San Diego, pp 153–247

Bronk DA, Glibert PM (1993) Application of a new ^{15}N tracer method to the study of dissolved organic nitrogen uptake during spring and summer in Chesapeake Bay. Mar Biol 115:501–508

Bronk DA, Sanderson MP, Mulholland MR, Heil CA, O'Neil JM (2004) Organic and inorganic nitrogen uptake kinetics in field populations dominated by *Karenia brevis*. In: Steidinger KA, Landsberg JH, Tomas CR, Vargo GA (eds) Harmful algae 2002. Florida Fish and Wildlife Conservation Comm, Fl Inst Oceanogr, IOC-UNESCO, St. Petersburg, USA, pp 80–82

Capone DG, Ferrier MD, Carpenter EJ (1994) Amino acid cycling in colonies of the planktonic marine cyanobacterium *Trichodesmium thiebautii*. Appl Environ Microbiol 60:3989–3995

Carlsson P, Granéli E, Segatto AZ (1999) Cycling of biological available nitrogen in riverine humic substances between marine bacteria, a heterotrophic nanoflagellate and a photosynthetic dinoflagellate. Aquat Microb Ecol 18:23–36

Chrost RJ (1991) Microbial enzymes in aquatic environments. Springer, Berlin Heidelberg New York, 317 pp

Collos Y, Gagné C, Laabir M, Vaquer A, Cecchi P, Souchu P (2004) Nitrogenous nutrition of *Alexandrium catenella* (Dinophyceae) in cultures and in Thau lagoon, southern France. J Phycol 40:96–103

Delwiche CF (1999) Tracing the web of plastid diversity through the tapestry of life. Am Nat 154:S164–S177

Doblin M, Legrand C, Carlsson P, Hummert C, Granéli E, Hallegraeff G (2001) Uptake of radioactively labeled humic substances by the toxic dinoflagellate *Alexandrium catenella*. In: Hallegraeff GM, Blackburn SI, Bolch CJ, Lewis RJ (eds) Harmful algal blooms 2000. IOC-UNESCO, Paris, pp 336–339

Dyhrman ST (2005) Ectoenzymes in *Prorocentrum minimum*. Harmful Algae 4:619–627

Dyhrman ST, Anderson DM (2003) Urease activity in cultures and field populations of the toxic dinoflagellate *Alexandrium*. Limnol Oceanogr 48:647–655

Dyhrman ST, Palenik BP (1999) Phosphate stress in cultures and field populations of the dinoflagellate *Prorocentrum minimum* detected using a single cell alkaline phosphatase activity assay. Appl Environ Microbiol 65:3205–3212

Estep K, MacIntyre F (1989) Taxonomy, life cycle, distribution and dasmotrophy of *Chrysochromulina*: a theory accounting for scales, haptonema, muciferous bodies and toxicity. Mar Ecol Prog Ser 57:11–21

Fan C, Glibert PM, Alexander J, Lomas MW (2003) Characterization of urease activity in three marine phytoplankton species. Mar Biol 142:949–958

Gentien P, Donoghay P, Yamazaki H, Raine R, Reguera B, Osborn T (2005) Harmful algal blooms in stratified environments. Oceanography 18:172–183

Gisselson L-Å, Carlsson P, Granéli E, Pallon J (2002) *Dinophysis* blooms in the deep euphotic zone of the Baltic Sea: do they grow in the dark? Harmful Algae 1:401–418

Glibert PM, Bronk DA (1994) Release of dissolved organic nitrogen by the marine diazotrophic cyanobacterium *Trichodesmium* spp. Appl Environ Microbiol 60:3996–4000

Glibert PM, O'Neil JM (1999) Dissolved organic nitrogen release and amino acid oxidase activity by *Trichodesmium* spp. In: Charpy L, Larkum T (eds), Marine cyanobacteria. Bull Inst Océanogr, special no. 19:265–271

Glibert PM, Harrison J, Heil C, Seitzinger S (2006) Escalating worldwide use of urea – a global change contributing to coastal eutrophication. Biogeochemistry (in press)

Glibert PM, Heil CA, Hollander D, Revilla M, Hoare A, Alexander J, Murasko S (2004) Evidence for dissolved organic nitrogen and phosphorus uptake during a cyanobacterial bloom in Florida Bay. Mar Ecol Prog Ser 280:73–83

Glibert PM, Magnien R, Lomas MW, Alexander J, Fan C, Haramoto E, Trice TM, Kana TM (2001) Harmful algal blooms in the Chesapeake and Coastal Bays of Maryland, USA: Comparison of 1997, 1998, and 1999 events. Estuaries 24:875–883

Glibert PM, Seitzinger S, Heil CA, Burkholder JM, Parrow MW, Codispoti LA, Kelly V (2005) The role of eutrophication in the global proliferation of harmful algal blooms. Oceanography 18:198–209

Granéli E, Anderson DM, Carlsson P, Maestrini SY (1997) Light and dark carbon uptake by *Dinophysis* species in comparison to other photosynthetic and heterotrophic dinoflagellates. Aquat Microb Ecol 13:177–186

Granéli E, Carlsson P, Legrand C (1999) The role of C, N and P in dissolved and particulate matter as a nutritional source for phytoplankton growth, including toxic species. Aquat Ecol 33:17–27

Granéli E, Edler L, Gedziorowska D, Nyman U (1985) Influence of humic and fulvic acids on *Prorocentrum minimum* (Pav) J Schiller. In: Anderson DM, White AW, Baden DG (eds) Toxic dinoflagellates. Elsevier, New York, pp 201–206

Grzebyk D, Schofield O, Vetriani C, Falkowski PG (2003) The Mesozoic radiation of eukaryotic algae: The portable plastid hypothesis. J Phycol 39:259–267

Heil CA (2005) Influence of humic, fulvic and hydrophilic acids on the growth, photosynthesis and respiration of the dinoflagellate *Prorocentrum minimum* (Pavillard) Schiller. Harmful Algae 4:603–618

Heil CA, Glibert PM, Al-Sarawi MA, Faraj M, Behbehani M, Husain M (2001) First record of a fish-killing *Gymnodinium* sp. bloom in Kuwait Bay, Arabian Sea: Chronology and potential causes. Mar Ecol Prog Ser 214:15–23

Heil CA, Glibert PM, Fan C (2005) *Prorocentrum minimum* (Pavillard) Schiller –A review of a harmful algal bloom species of growing worldwide importance. Harmful Algae 4:449–470

Jones RI (1994) Mixotrophy in planktonic protists as a spectrum of nutritional strategies. Mar Microb Food Webs 8:87–96

Karl DM, Letelier R, Hebel RV, Bird DF, Winn CD (1992) *Trichodesmium* blooms and new nitrogen in the North Pacific Gyre. In: Carpenter EJ, Capone DG, Reuter JG (eds) Marine pelagic cyanobacteria: *Trichodesmium* and other diazotrophs. NATO ASI Series 362. Kluwer Academic Publishers, Dordrecht, The Netherlands, pp 219–238

Kimor B (1979) Predation by *Noctiluca miliaris* Souriray on *Acartia tonsa* Dana eggs in the inshore waters of southern California. Limnol Oceanogr 24:568–572

Klut ME, Bisalputra T, Antia NJ (1987) Some observations on the structure and function of the dinoflagellate pusule. Can J Bot 65:736–744

Kudela RM, Cochlan WP (2000) Nitrogen and carbon uptake kinetics and the influence of irradiance for a red tide bloom off southern California. Aquat Microb Ecol 21:31–47

Legrand C, Carlsson P (1998) Uptake of high molecular weight dextran by the dinoflagellate *Alexandrium catenella*. Aquat Microb Ecol 16:81–86

LaRoche J, Breitbarth E (2005) Importance of the diazotrophs as a source of new nitrogen in the ocean. J Sea Res 53:67–91

LaRoche J, Nuzzi R, Waters R, Wyman K, Falkowski PG, Wallace DWR (1997) Brown tide blooms in Long Island's coastal waters linked to variability in groundwater flow. Global Change Biol 3:397–410

Lomas MW, Glibert PM, Clougherty DA, Huber DR, Jones J, Alexander J, Haramoto E (2001) Elevated organic nutrient ratios associated with brown tide algal blooms of *Aureococcus anophagefferens* (Pelagophyceae). J Plankton Res 23:1339–1344

Lomas MW, Kana TM, MacIntyre HL, Cornwell JC, Nuzzi R, Waters R (2004) Interannual variability of *Aureococcus anophagefferens* in Quantuck Bay, Long Island: natural test of the DON hypothesis. Harmful Algae 3:389–402

Margulis L (1975) Symbiotic theory of the origin of eukaryotic organelles; criteria for proof. Symp Soc Exp Biol. 29:21–38

Margulis L (1981) Symbiosis in cell evolution. W. H. Freeman and Company, San Francisco, pp 206–227

Martin W, Rotte C, Hoffmeister M, Theissen U, Gelius-Dietrich G, Ahr S, Henze K (2003) Early cell evolution, eukaryotes, anoxia, sulfide, oxygen, fungi first (?), and a tree of genomes revisited. IUBMB Life 55:193–204

Milius S (1999) Myriad monsters confirmed in water droplets. Science News Online, vol 156, www.sciencenews.org/pages/sn_arc99/12_18_99b/fob1.htm

Moorthi S (2004) Mixotrophic flagellates in coastal marine sediments: quantitative role and ecological significance. PhD Thesis, Leibniz Institute of Marine Sciences, Kiel University, Germany, 128 pp

Mulholland MR, Boneillo G, Minor EC (2004) A comparison of N and C uptake during brown tide (*Aureococcus anophagefferens*) blooms from two coastal bays on the east coast of the USA. Harmful Algae 3:361–376

Mulholland MR, Gobler CJ, Lee C (2002) Peptide hydrolysis, amino acid oxidation and N uptake in communities seasonally dominated by *Aureococcus anophagefferens*. Limnol Oceanogr 47:1094–1108

Mulholland MR, Glibert PM, Berg GM, Van Heukelem L, Pantoja S, Lee C (1998) Extracellular amino acid oxidation by microplankton: a cross-ecosystem comparison. Aquat Microb Ecol 15:141–152

Mulholland MR, Ohki K, Capone DG (1999) Nitrogen utilization and metabolism relative to patterns of N_2 fixation in cultures of *Trichodesmium* NIBB1067. J Phycol 35:977–988

Palenik B, Morel FMM (1990) Comparison of cell-surface L-amino acid oxidases from several marine phytoplankton. Mar Ecol Prog Ser 59:195–201

Palmer JD (2003) The symbiotic birth and spread of plastids: how many times and whodunit? J Phycol 39:4–11

Panosso R, Granéli E (2000) Effect of dissolved organic matter on the growth of *Nodularia spumigena* (Cyanophyceae) cultivated under N and P deficiency. Mar Biol 136:331–336

Purina I, Balode M, Béchemin C, Pöder T, Maestrini S (2004) Influence of dissolved organic matter from terrestrial origin on the changes of dinoflagellates species composition in the Gulf of Riga, Baltic Sea. Hydrobiologia 514:127–137

Pustizzi F, MacIntyre H, Warner ME, Hutchins DA (2004) Interactions of nitrogen source and light intensity on the growth and photosynthesis of the brown tide alga *Aureococcus anophagefferens*. Harmful Algae 3:343–360

Raven JA (1997) Phagotrophy in phototrophs. Limnol Oceanogr 42:198–205

Seitzinger SP, Hartnett HE, Lauck R, Mazurek M, Minegishi T, Spyres G, Styles RM (2005) Molecular level chemical characterization and bioavailability of dissolved organic matter in stream water using ESI mass spectrometry. Limnol Oceanogr 50:1–12

Skovgaard A (1998) Role of chloroplast retention in a marine dinoflagellate. Aquat Microb Ecol 15:293–301
Smayda TJ (1997) Harmful algal blooms: their ecophysiology and general relevance to phytoplankton blooms in the sea. Limnol Oceanogr 42:1137–1153
Stoecker DK (1999) Mixotrophy among dinoflagellates. J Eukaryot Microbiol 46:397–401
Stoecker DK, Gustafson D (2003) Cell-surface proteolytic activity of photosynthetic dinoflagellates. Aquat Microb Ecol 30:175–183
Veuger B, Middelburg JJ, Boschker HTS, Nieuwenhuize J, van Rijswijk P, Rochelle-Newall EJ, Navarro N (2004) Microbial uptake of dissolved organic and inorganic nitrogen in Randers Fjord. Est Coast Shelf Sci 61:507–515
Walsh JJ, Steidinger KA (2001) Saharan dust and Florida red tides: the cyanophyte connection. J Geophys Res 106:11597–11612
Wheeler PA, North BB, Stephens GC (1974) Amino acid uptake by marine phytoplankters. Limnol Oceanogr 19:249–259

14 Phagotrophy in Harmful Algae

D. Stoecker, U. Tillmann, and E. Granéli

14.1 Introduction

The contributions of phagotrophy to bloom formation and toxicity of harmful algae have not received much attention, although phagotrophy is common among toxic and/or red-tide dinoflagellates and haptophytes (Figs. 14.1 and 14.2). Ingestion of prey can dramatically increase the growth rate of some harmful algae (Li et al. 1999; Jeong et al. 2004, 2005a, 2005b). Predation on competitors or potential grazers may contribute to the ability of some HAB species to form dense, mono-specific blooms (Legrand et al. 1998; Granéli and Johansson 2003b). In some species, grazing may be related to toxicity (Tillmann 2003; Skovgaard and Hansen 2003).

Some toxic dinoflagellates are strict heterotrophs (*Protoperidinium crassipes*) or heterotrophs with a limited capacity for mixotrophy (*Pfiesteria piscicida, P. shumwayae*) (Jeong and Latz 1994; Burkholder et al. 2001; Glasgow et al. 2001). Most harmful dinoflagellates are plastidic, bloom-forming species that traditionally have been regarded as photoautotrophs, but evidence is accumulating that many can ingest other cells. Some are harmful because they contain toxins and others because they form high-biomass blooms that disrupt food webs. There is evidence for phagotrophy in many toxic and/or red-tide planktonic dinoflagellates including *Akashiwo sanguinea* (=*Gymnodinium sanguineum*) (Bockstahler et al. 1993; Jeong et al. 2005a), *Alexandrium ostenfeldii* and *A. tamarense* (Jacobson and Anderson 1996; Jeong et al. 2005a), *Ceratium furca* (Smalley et al. 2003), *Cochlodinium polykrikoides* (Jeong et al. 2004), *Dinophysis acuminata, D. norvegica* (Jacobson and Andersen 1994; Gisselson et al. 2002), *Gonyaulax polygramma* (Jeong et al. 2005b), *Gymnodinium catenatum, G. impudicum*, (Jeong et al. 2005a), *Heterocapsa triquetra* (Legrand et al. 1998; Jeong et al. 2005a), *Karlodinium micrum* (=*Gyrodinium galatheanum*) (Li et al. 2000), *Lingulodinium polyedra* (Jeong et al. 2005a), *Prorocentrum donghaiense, P. triestinum, P. micans* and *P. minimum* (Jeong et al. 2005a), *Protoceratium reticulatum* (Jacobson and Anderson 1996) and *Scrippsiella trochoidea* (Jeong et al. 2005a). There is also evidence for

Ecological Studies, Vol. 189
Edna Granéli and Jefferson T. Turner (Eds.)
Ecology of Harmful Algae

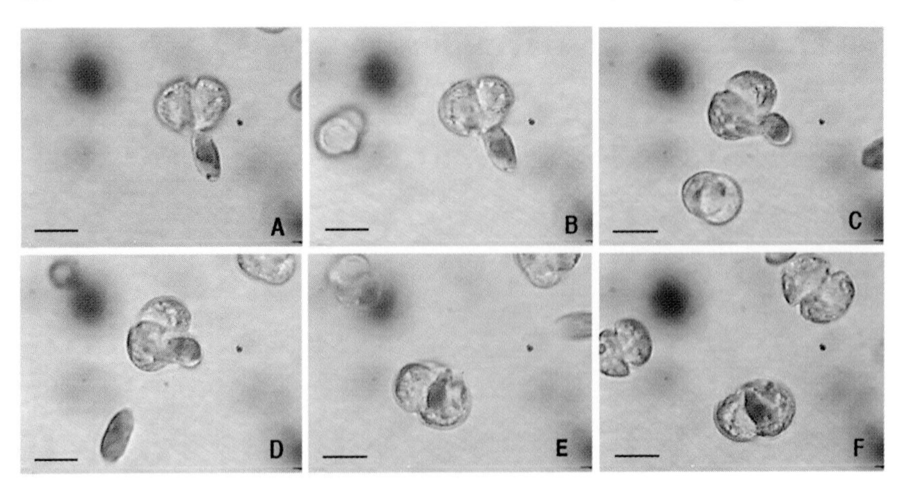

Fig. 14.1. Sequence showing a cell of *Karlodinium micrum* feeding on a cryptophyte alga

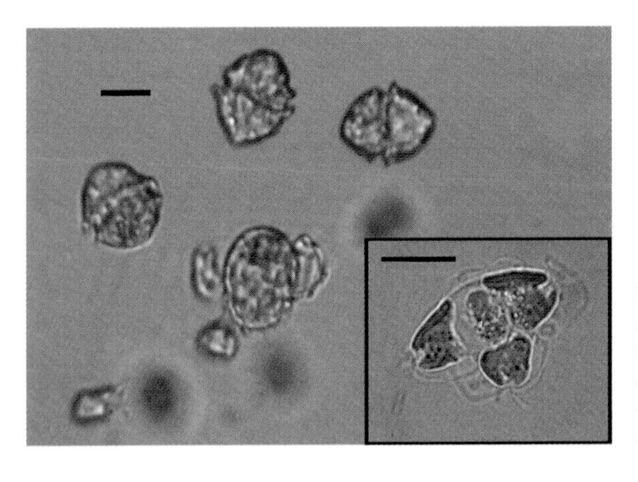

Fig. 14.2. Three *Prymnesium parvum* cells catching and preying on a cell of the larger species *Heterocapsa triquetra. Scale bars* 10 μm

phagotrophy in the toxic benthic dinoflagellates *Gambierdiscus toxicus*, *Ostreopsis lenticularis*, *O. ovata*, *O. siamensis*, and *Prorocentrum arenarium* and *P. belizeanum* (Faust 1998) (Table 14.1).

Among the toxic haptophytes, phagotrophy is known in *Chrysochromulina leadbeateri*, *C. polylepis and Prymnesium parvum* (including *P. parvum* f. *patelliferum*) (Johnsen et al. 1999; Jones et al. 1993; Tillmann 1998, 2003) (Table 14.2). In addition to *Chrysochromulina leadbeateri* and *C. polylepis*, whose toxic potential seems to be well established, three other species of *Chrysochromulina* (*C. brevifilum*, *C. kappa*, *C. strobilus*) were found to be toxic to the bryozoan *Electra pilosa* when given as the only food source (Jebram 1980). All these, at least potentially toxic species, have been shown to be phagotrophic as well (Parke et al. 1955, 1959; Jones et al. 1993). Other families

Table 14.1. Evidence for phagotrophy in harmful dinoflagellate species

Species	Evidence for Phagotrophy[a]	References
Alexandrium ostenfeldii	Obs. of food vacuoles with electron microscopy	Jacobsen and Anderson 1996
Alexandrium tamarense	Uptake of radio-labeled bacteria; Obs. of bacteria and flagellates inside cell; Observations of food vacuoles containing algal cells	Nygaard and Tobiesen 1993; Sorokin et al. 1996; Jeong et al. 2005
Cochlodinium polykrikoides	Obs. of feeding and measurement of ingestion of small cryptophytes and other small phytoflagellates	Jeong et al. 2004
Dinophysis acuminata	Obs. of food vacuoles with light and electron microscopy	Jacobsen and Andersen 1994
Dinophysis norvegica	Obs. of food vacuoles with light and electron microscopy	Jacobsen and Andersen 1994; Gisselson et al. 2002
Gambierdiscus toxicus	Obs. of food vacuoles with light microscopy	Faust 1998
Gymnodinium catenatum	Observations of food vacuoles containing algal cells	Jeong et al. 2005
Karlodinium micrum (syn.=*Gyrodinium galatheanum*, *Gymnodinium galatheanum*)	Obs. of feeding on cryptophytes and feeding and growth experiments	Li et al. 1999
Lingulodinium polyedra[b]	Obs. of feeding and of food vacuoles containing algal cells, feeding and growth experiments	Jeong et al. 2005
Ostreopsis lenticularis, *O. ovata*, *O. siamensis*	Obs. of food vacuoles with light microscopy	Faust 1998
Pfiesteria piscicida. *P. shumwayae*	Primarily heterotrophic; obs. of feeding on cryptophytes, other protists and fish tissues; grazing and growth experiments	Burkholder et al. 2001, Lin et al. 2004
Prorocentrum arenarium, *P. belizeanum*, *P. hoffmannianum*	Obs. of food vacuoles with light microscopy	Faust 1998
Prorocentrum minimum (syn.=*P. cordatum*)	Obs. of food vacuoles with ingested cryptophyte material; feeding experiments; obs. of food vacuoles containing algal cells	Stoecker et al. 1997, Jeong et al. 2005
Protoceratium reticulatum (syn.=*Peridinium reticulatum*, *Gonyaulax grindleyi*)	Obs. of food vacuoles	Jacobson and Anderson 1996
Protoperidinium crassipes	Heterotroph; preys on other dinoflagellates	Jeong and Latz (1994)

[a] Obs. = observation(s)

[b] Not on IOC Taxonomic Reference List of Toxic Algae (Moestrup 2004) but toxic (Steidinger and Tangen 1997)

Table 14.2. Evidence for phagotrophy in harmful haptophyte species

Species	Evidence for phagotrophy	Reference
Chrysochromulina leadbeateri	Ingestion of FLA	Johnsen et al. 1999; Legrand pers. com
Chrysochromulina polylepis	Ingestion of bacteria, small particles and small protists	Manton and Parke 1962; Jones et al. 1993; Nygaard and Tobiesen 1993
Prymnesium parvum f. *parvum* f. *patelliferum*	Feeding measurement of bacteria, observation and measurements of ingestion of various protists	Nygaard and Tobiesen 1993; Tillmann 1998; Legrand et al. 2001; Skovgaard et al. 2003

containing harmful eukaryotic phytoplankton are the diatoms, raphidophytes, and pelagophytes. The ability to ingest other cells is probably lacking in vegetative cells of diatoms due to their valve structure. Nygaard and Tobiesen (1993) measured uptake of radio-labeled bacteria by the toxic raphidophyte, *Heterosigma akashiwo*, but this is the only report of phagotrophy in this family. Phagotrophy has not been reported in the pelagophytes responsible for brown tides. The prokaryotic cyanobacteria also cause harmful blooms, but are not phagotrophic.

Feeding may go undetected in many species because it is sporadic and because the presence of plastids can make food vacuoles difficult to observe. Most, but not all, mixotrophic flagellates can be grown in the absence of prey, and thus their capacity to ingest other cells is often overlooked. More harmful algal species will probably be added to the list of phagotrophs.

14.2 Phagotrophy and its Advantages

Ingestion of prey supports the growth of harmful heterotrophic dinoflagellates, including *Protoperidinium crassipes* (Jeong and Latz 1994) and *Pfiesteria* spp. (Burkholder et al. 2001; Glasgow et al. 2001). *P. crassipes* feeds on other dinoflagellates, including toxic *Lingulodinium polyedra* (Jeong and Latz 1994). Although *Pfiesteria* spp. can retain cryptophyte plastids for short periods, they are mostly heterotrophic and use a peduncle to feed on prey ranging in size from bacteria to fish tissues (Burkholder et al. 2001). The maintenance cost for such a feeding apparatus is calculated to be low compared to the costs for maintaining photosynthetic apparatus (Raven 1997), which may explain the relatively high growth rates of small heterotrophic harmful algae such as *Pfiesteria* compared to their more autotrophic relatives.

The advantages of phagotrophy for photosynthetic algae are diverse (Stoecker 1998, 1999). Most dinoflagellates and haptophytes can grow strictly autotrophically, but some species of dinoflagellates (for example, *Karlodinium micrum, Cochlodinium polykrikoides, Gonyaulax polygramma, Lingulodinium polyedra*) grow much faster with prey than strictly autotrophically (Li et al. 1999; Jeong et al. 2004, 2005a, 2005b). Phagotrophy can make an important contribution to C, N and P budgets in these species. In contrast, although some haptophytes ingest prey when light and inorganic nutrients are sufficient (Skovgaard et al. 2003), growth rates with and without prey are usually similar (Larsen et al. 1993; Pintner and Provasoli 1968). An exception to this generalization is that the simultaneous addition of small diatoms and bacteria can enhance the growth of *Prymnesium parvum* even in nutrient-replete media (Martin-Cereceda et al. 2003).

Some phagotrophic dinoflagellates only feed (or feed at higher rates) when they are P or N limited, but not when they are C or light limited (Li et al. 1999, 2000; Smalley et al. 2003). In the bloom-forming dinoflagellate *Ceratium furca*, internal nutrient ratios (C:P and N:P) regulate feeding, which is induced when cellular ratios deviate from optimum due to N or P limitation (Smalley et al. 2003). In many haptophytes, phagotrophy is also stimulated by nutrient limitation. Both *Prymnesium parvum* and *Chrysochromulina polylepis* have been shown to ingest more bacteria under phosphate deficiency than nutrient-replete conditions (Nygaard and Tobiesen 1993; Legrand et al. 2001). However, ingestion of algal prey (*Rhodomonas baltica*) by *P. parvum* was equally high when the haptophyte was grown under nutrient-replete or N- or P-limiting conditions (Skovgaard et al. 2003).

In mixotrophic dinoflagellates, feeding does not appear to be a response to light limitation and most, but not all, mixotrophic dinoflagellates will not survive or feed in prolonged darkness (reviewed in Stoecker 1999). For example, feeding by *Karlodinium micrum* decreases at low irradiance and is highest at irradiances optimal for autotrophic growth (Li et al. 2000). In contrast, in some haptophytes ingestion is inversely proportional to light intensity, indicating that mixotrophy can be stimulated by both low nutrient and low light conditions (Jones et al. 1995). However, *Chrysochromulina* and *Prymnesium parvum* are not able to survive or grow in the dark, even when bacteria are added as prey (Pintner and Provasoli 1968; Jochem 1999). In contrast, ingestion of fluorescent-labeled bacteria (FLB) by *C. polylepis* has been shown to be higher under high light conditions and with the addition of humic substances, and these differences were explained by greater quantities of bacteria being found in these treatments (Granéli and Carlsson 1998). This indicates that in the presence of abundant bacterial prey, *C. polylepis* ingests prey independent of the light regime.

It is also possible that some harmful algae acquire organic growth factors from ingestion of prey. The photosynthetic dinoflagellate *Gyrodinium resplendens* (not known to be toxic) feeds on other dinoflagellates and can only be cultivated when provided with prey (Skovgaard 2000). Investigators have not been able to successfully cultivate toxic *Dinophysis* spp. (Nishitani et al. 2003). Food vacuoles containing ciliates and other protists have been observed (Jacobson and Andersen 1994; Gisselson et al. 2002) and it is possible that toxic *Dinophysis* spp., although they are photosynthetic, are obligate phagotrophs requiring specific prey. For toxic haptophytes there is no evidence of obligate phagotrophy; *Prymnesium* and toxic *Chrysochromulina* species have been successfully grown in axenic culture (Edvardsen and Paasche 1998).

In addition to the physiological advantages of prey ingestion, phagotrophy can also be an ecological strategy for eliminating or reducing predators and competitors. The dinoflagellate *Pfiesteria piscicida*, when actively toxic, can kill and often consume many of its protistan grazers (Burkholder and Glasgow 1995; Stoecker et al. 2002). Many of the non-thecate mixotrophic dinoflagellates consume cryptophytes and other phytoflagellates, their potential competitors for light and nutrients (Li et al. 1999; Jeong et al. 2005a, 2005b). For example, the estimated grazing impact of dense *Cochlodinium polykrikoides* blooms on cryptophyte populations is high, with the dinoflagellate populations having the potential to remove the entire cryptophyte standing stock within a day (Jeong et al. 2004).

Toxic haptophytes cause negative effects on a range of other planktonic organisms. Field observations during the *Chrysochromulina polylepis* bloom in 1988 in Scandinavian waters led to the hypothesis that *C. polylepis* toxins severely affect other protists (Nielsen et al. 1990) and subsequent laboratory experiments confirmed both allelopathy of *C. polylepis* (Schmidt and Hansen 2001) as well as negative effects on protistan grazers (John et al. 2002). Unfortunately, in these studies on *C. polylepis,* phagotrophy was not addressed. Like *C. polylepis*, *Prymnesium parvum* is known to severely affect other plankton organisms, as has been shown using algal cultures (Granéli and Johansson 2003b), heterotrophic protists (Granéli and Johansson 2003a; Rosetta and McManus 2003; Tillmann 2003) and natural communities (Fistarol et al. 2003).

14.3 Relationship of Phagotrophy to Toxicity

It is interesting that many toxic algae are phagotrophic or closely related to known phagotrophs. Is there a relationship between phagotrophy and toxicity? In the case of the heterotrophic *Protoperidinium crassipes*, it is possible that the toxin or its precursor is derived from prey. *P. crassipes* feeds on other dinoflagellates including toxic red-tide species (Jeong and Latz 1994). However, most mixotrophic toxic algae can be grown strictly autotrophically and retain their toxicity, thus it is unlikely that prey are a source of toxin for these species.

It is also possible that some toxins are involved in prey capture. Dinoflagellates have been observed to "trap" or "paralyze" motile prey before they are ingested (reviewed in Hansen and Calado 1999). Mechanical contact between the dinoflagellate and the prey appears to usually be necessary for the prey to be subdued. It is possible that trichocysts involved in prey capture contain toxins as well as "sticky" material. Interactions between *Pfiesteria piscicida* zoospores and their predators are interesting. Many ciliate species readily consume non-toxic zoospores and grow well, but many types of ciliates die if they are fed toxic zoospores (Stoecker et al. 2002). In some cases, the zoospores then consume the ciliates (Burkholder and Glasgow 1995).

Among the haptophytes, there is good evidence that toxins are involved in prey capture and defense against predation. The feeding frequency of *Prymnesium parvum* on motile prey is positively correlated with toxicity (Skovgaard and Hansen 2003). Non-motile prey is ingested at high rates irrespective of toxin concentration. Moreover, the addition of filtrate of a toxic *Prymnesium* culture to a dilute (and low toxic) *P. parvum* culture with an initial low feeding frequency also results in an increase of feeding frequency on motile prey. Toxicity has also been shown to be a key factor in determining the interaction of *P. parvum* with protistan grazers (Granéli and Johansson 2003b; Tillmann 2003). Whereas low toxicity *P. parvum* is a suitable prey for the heterotrophic dinoflagellate *Oxyrrhis marina*, the dinoflagellate is rapidly killed by *P. parvum* at high toxicity levels. Moreover, under these conditions the former predator (*O. marina*) is ingested by *P. parvum*, thus reversing the normal direction of grazing interactions (Tillmann 2003).

In contrast to *Prymnesium* with its short and stiff haptonema, species of *Chrysochromulina* have a longer and mostly flexible haptonema that may be involved in the feeding process by capturing and transporting prey to the cytostome positioned in the posterior end of the cell (Kawachi et al. 1991; Kawachi and Inouye 1995). In addition, in some species of *Chrysochromulina*, spine scales may have a functional role in food capture (Kawachi and Inouye 1995). The potential role of *Chrysochromulina* toxins in phagotrophy, however, is unknown. Both *Chrysochromulina* and *Prymnesium* species have organelles that can quickly discharge mucus, but the chemical composition of the mucus is not known (Green et al. 1982; Estep and MacIntyre 1989). It is tempting to speculate that muciferous bodies may be involved in phagotrophy as discharge was observed when certain other protist species were added to cultures of the phagotrophic *Chrysochromulina ericina* (Parke et al. 1956). It is not known if toxicity is related to mucus discharge in some species.

It seems likely that phagotrophy and the need to "trap" or "paralyze" prey is related to the evolution of toxicity in at least some toxic phytoplankton. Production of toxin by *Pfiesteria* spp. and immobilization of fish may be an extreme example of this phenomenon (Hansen and Calado 1999), but it may also occur with more subtlety and routinely during predation on protistan prey. A possibly related observation is that toxicity of some, but not all, dinoflagellates and haptophytes increases when they are nutrient limited (see Chap. 18 Granéli and Flynn, this book). In many mixotrophic dinoflagellates and haptophytes, nutrient limitation stimulates feeding (Li et al. 2000; Smalley et al. 2003; Legrand et al. 2001). A connection between toxicity and feeding might explain the link between toxicity and nutrient limitation. However, in some cases, toxicity and feeding are not linked; for example, *Karlodinium micrum* grown as an autotroph in the absence of prey or as a mixotroph with prey do not differ in toxicity (Deeds et al. 2002).

14.4 Significance of Phagotrophy

The recognition that many toxic phytoplankton are phagotrophic challenges traditional approaches to investigating the physiological ecology and community ecology of harmful algal blooms. Most of our estimates of the maximum growth rate of harmful algae are based on laboratory studies using monocultures. Phagotrophy can more than double the growth rates of some harmful algae. For many taxa we have probably underestimated their potential growth rates in nature. Phagotrophy is also important to consider in regard to inorganic nutrient acquisition—phagotrophs can acquire elements from prey under conditions that would appear limiting based on uptake kinetics for dissolved N, P and Fe. Some toxic algae may be obligate phagotrophs, and perhaps by providing suitable prey we may be able to cultivate them, and thus be able to investigate toxin production and other aspects of their physiological ecology.

It is also important to take phagotrophy into account in order to understand toxicity and its variability. Toxicity often declines in culture. If toxins are involved in prey capture or defense against predators, then it is likely that growth in monoculture should select against toxicity. In some cases, it is possible that toxic algae acquire their toxins (or the precursors) from prey and this needs further investigation, particularly for heterotrophic species.

The phagotrophic tendencies of dinoflagellates and haptophytes also may partially explain some of the puzzling aspects of their bloom dynamics and population ecology. Many different types of phytoplankton grow rapidly in response to a combination of light and nutrients, but most do not form monospecific blooms. The ability of some phagotrophic mixotrophs to eat their competitors must contribute to their ability to monopolize resources. The dynamics of some harmful algal blooms are unusual in that high cell densities tend to persist after nutrients are exhausted. Phagotrophic mixotrophs may survive because they are able to eat to obtain limiting elements.

Phagotrophy is not an isolated aspect of the physiological ecology of phytoplankton; in many taxa it may have coevolved with other capabilities such as the ability to use dissolved organic material and allelopathic tendencies.

Several authors, perhaps beginning with Estep and MacIntyre (1989) have suggested that some algal toxins may immobilize, make "leaky" or kill other types of cells in the plankton as part of a nutritional strategy. Allelochemicals may reduce competition or predation, but they can also make resources available. Phagotrophy is one of the elements in the interconnected bag of tricks that lets certain species of phytoplankton dominate resources in the plankton and form toxic or high biomass harmful blooms.

References

Burkholder JM, Glasgow HB (1995) Interactions of a toxic estuarine dinoflagellate with microbial predators and prey. Arch Protistenkd 145:177–188

Burkholder JM, Glasgow HB, Deamer-Melia N (2001) Overview and present status of the toxic *Pfiesteria* complex (Dinophyceae). Phycologia 40:186–214

Bockstahler KR, Coats DW (1993) Spatial and temporal aspects of mixotrophy in Chesapeake Bay dinoflagellates. J Eukaryot Microbiol 40:49–60

Deeds JR, Terlizzi DE, Adolf JE, Stoecker DK, Place AR (2002) Toxic activity from cultures of *Karlodinium micrum* (=*Gyrodinium galatheanum*) (Dinophyceae) – a dinoflagellate associated with fish mortalities in an estuarine aquaculture facility. Harmful Algae 1:169–189

Edvardsen B, Paasche E (1998) Bloom dynamics and physiology of *Prymnesium* and *Chrysochromulina*. In: Anderson DM, Cembella AD, Hallegraeff GM (eds) Physiological ecology of harmful algal blooms. NATO ASI Series 41. Springer, Berlin Heidelberg New York, pp 193–208

Estep KW, MacIntyre F (1989) Taxonomy, life cycle, distribution and dasmotrophy of *Chrysochromulina*: a theory accounting for scales, haptonema, muciferous bodies and toxicity. Mar Ecol Prog Ser 57:11–21

Faust MA (1998) Mixotrophy in tropical benthic dinoflagellates. In: Reguera B, Blanco J, Fernandez ML, Wyatt T (eds) Harmful algae. Xunta de Galicia and IOC-UNESCO, Paris, pp 390–393

Fistarol GO, Legrand C, Granéli E (2003) Allelopathic effect of *Prymnesium parvum* on a natural plankton community. Mar Ecol Prog Ser 255:115–125

Gisselson L-A, Carlsson P, Granéli E, Pallon J (2002) Dinophysis blooms in the deep euphotic zone of the Baltic Sea: do they grow in the dark? Harmful Algae 1:401–418

Glasgow HB, Burkholder JM, Morton SL, Springer J (2001) A second species of ichthyotoxic *Pfiesteria* (Dinamoebales, Dinophyceae). Phycologia 40:234–245

Granéli E, Carlsson P (1998) The ecological significance of phagotrophy in photosynthetic flagellates. In: Anderson DM, Cembella AD, Hallegraeff GM (eds) Physiological ecology of harmful algal blooms. NATO ASI Series 41. Springer, Berlin Heidelberg New York, pp 539–557

Granéli E, Johansson N (2003a) Effects of the toxic haptophyte *Prymnesium parvum* on the survival and feeding of a ciliate: the influence of different nutrient conditions. Mar Ecol Prog Ser 254:49–56

Granéli E, Johansson N (2003b) Increase in the production of allelopathic substances by *Prymnesium parvum* cells grown under N- or P-deficient conditions. Harmful Algae 2:135–145

Green JC, Hibberd DJ, Pienaar RN (1982) The taxonomy of *Prymnesium* (Prymnesiophyceae) including a description on a new cosmopolitan species, *P. patellifera* sp. nov., and further observations on *P. parvum* N. Carter. Brit Phycol J 17:363–382

Hansen PJ, Calado AJ (1999) Phagotrophic mechanisms and prey selection in free-living dinoflagellates. J Eukaryot Microbiol 46:382–389

Jacobson DM, Andersen RA (1994) The discovery of mixotrophy in photosynthetic species of Dinophysis (Dinophyceae): light and electron microscopical observations of food vacuoles in *Dinophysis acuminata*, *D. norvegica* and two heterotrophic dinophysoid dinoflagellates. Phycologia 33:97–110

Jacobson DM, Anderson DM (1996) Widespread phagocytosis of ciliates and other protists by marine mixotrophic and heterotrophic thecate dinoflagellates. J Phycol 32:279–285

Jebram D (1980) Prospection for sufficient nutrition for the cosmopolitic marine bryozoan *Electra pilosa* (Linnaeus). Zoologische Jahrbücher, Abteilung für Systematik, Ökologie und Geographie der Tiere 107:368–390

Jeong HJ, Latz MI (1994) Growth and grazing rates of the heterotrophic dinoflagellates *Protoperidinium* spp. on red tide dinoflagellates. Mar Ecol Prog Ser 106:173–185

Jeong HJ, Yoo YD, Kim JS, Kim TH, Kim JH, Kang NS, Yih W (2004) Mixotrophy in the phototrophic harmful alga *Cochlodinium polykrikoides* (Dinophycean): Prey species, the effects of prey concentration, and grazing impact. J Eukaryot Microbiol 51:563–569

Jeong HJ, Yoo YD, Park JY, Song JY, Kim ST, Lee SH, Kim KY, Yih WH (2005a) Feeding by the phototrophic red-tide dinoflagellates: 5 species newly revealed and 6 species previously known to be mixotrophic. Aquat Microb Ecol 40:133–150

Jeong HJ, Yoo YD, Seong KA, Kim JH, Park JY, Kim S, Lee SH, Ha JH, Yih WH (2005b) Feeding by the mixotrophic red-tide dinoflagellate *Gonyaulax polygramma*: mechanisms, prey species, effects of prey concentration, and grazing impact. Aquat Microb Ecol 38:249–257

Jochem FJ (1999) Dark survival strategies in marine phytoplankton assessed by cytometric measurement of metabolic activity with fluorescein diacetate. Mar Biol 135:721–728

John U, Tillmann U, Medlin L (2002) A comparative approach to study inhibition of grazing and lipid composition of a toxic and non-toxic clone of *Chrysochromulina polylepis* (Prymnesiophyceae). Harmful Algae 1:45–57

Johnsen G, Dallokken R, Eikrem W, Legrand C, Aure J, Skoldal HR (1999) Eco-physiology, bio-optics and toxicity of the ichthyotoxic *Chrysochromulina leadbeateri* (Prymnesiophyceae). J Phycol 35:1465–1474

Jones HLJ, Leadbeater BSC, Green JC (1993) Mixotrophy in marine species of *Chrysochromulina* (Prymnesiophyceae): Ingestion and digestion of a small green flagellate. J Mar Biol Assoc UK 73:283–296

Jones HLJ, Durjun P, Leadbeater BSC, Green JC (1995) The relationship between photoacclimation and phagotrophy with respect to chlorophyll a, carbon and nitrogen content, and cell size of *Chrysochromulina brevifilum* (Prymnesiophyceae). Phycologia 34:128–134

Kawachi M, Inouye I (1995) Functional roles of the haptonema and the spine scales in the feeding process of *Chrysochromulina spinifera* (Fournier) Pienaar et Norris (Haptophyta = Prymnesiophyta). Phycologia 34:193–200

Kawachi M, Inouye I, Maeda O, Chihara M (1991) The haptonema as a food-capturing device: observations on *Chrysochromulina hirta* (Prymnesiophyceae). Phycologia 30:563–573

Larsen A, Eikrem W, Paasche E (1993) Growth and toxicity in *Prymnesium patelliferum* (Prymnesiophyceae) isolated from Norwegian waters. Can J Bot 71:1357–1362

Legrand C, Granéli E, Carlsson P (1998) Induced phagotrophy in the photosynthetic dinoflagellate *Heterocapsa triquetra*. Aquat Microb Ecol 15:65–75

Legrand C, Johansson N, Johnsen G, Borsheim KY, Granéli E (2001) Phagotrophy and toxicity variation in the mixotrophic *Prymnesium patelliferum* (Haptophyceae). Limnol Oceanogr 46:1208–1214
Li A, Stoecker DK, Adolf JE (1999) Feeding, pigmentation, photosynthesis and growth of the mixotrophic dinoflagellate *Gyrodinium galatheanum*. Aquat Microb Ecol 19:163–176
Li A, Stoecker DK, Coats DW (2000) Mixotrophy in *Gyrodinium galatheanum* (Dinophyceae): grazing responses to light intensity and inorganic nutrients. J Phycol 36:33–45
Martin-Cereceda M, Novarino G, Young JR (2003) Grazing by *Prymnesium parvum* on small planktonic diatoms. Aquat Microb Ecol 33:191–199
Nielsen TG, Kiorboe T, Bjornsen PK (1990) Effects of a *Chrysochromulina polylepis* subsurface bloom on the planktonic community. Mar Ecol Prog Ser 62:21–35
Nishitani G, Miyamura K, Imai I (2003) Trying to cultivation of *Dinophysis caudata* (Dinophyceae) and the appearance of small cells. Plankton Biol Ecol 50:31–36
Nygaard K, Tobiesen A (1993) Bacterivory in algae; a survival strategy during nutrient limitation. Limnol Oceanogr 38:273–279
Parke M, Manton I, Clarke B (1955) Studies on marine flagellates. II. Three new species of *Chrysochromulina*. J Mar Biol Assoc UK 34:579–609
Parke M, Manton I, Clarke B (1956) Studies on marine flagellates. III. Three further species of *Chrysochromulina*. J Mar Biol Assoc UK 35:387–414
Parke M, Manton I, Clarke B (1959) Studies on marine flagellates. V. Morphology and microanatomy of *Chrysochromulina strobilus*. J Mar Biol Assoc UK 38:169–188
Pintner IJ, Provasoli L (1968) Heterotrophy in subdued light of 3 *Chrysochromulina* species. Bull Misaki Mar Biol Inst, Kyoto Univ, 12:25–31
Raven JA (1997) Phagotrophy in phototrophs. Limnol Oceanogr 42:198–205
Rosetta CH, McManus GB (2003) Feeding by ciliates on two harmful algal bloom species, *Prymnesium parvum* and *Prorocentrum minimum*. Harmful Algae 2:109–126
Schmidt LE, Hansen PJ (2001) Allelopathy in the prymnesiophyte *Chrysochromulina polylepis*: effect of cell concentration, growth phase and pH. Mar Ecol Prog Ser 216:67–81
Skovgaard A (2000) A phagotrophically derivable growth factor in the plastidic dinoflagellate *Gyrodinium resplendens* (Dinophyceae). J Phycol 36:1069–1078
Skovgaard A, Hansen PJ (2003) Food uptake in the harmful alga *Prymnesium parvum* mediated by excreted toxins. Limnol Oceanogr 48:1161–1166
Skovgaard A, Legrand C, Hansen PJ, Granéli E (2003) Effects of nutrient limitation on food uptake in the toxic Haptophyte *Prymnesium parvum*. Aquat Microb Ecol 31:259–265
Smalley GW, Coats DW, Stoecker DK (2003) Feeding in the mixotrophic dinoflagellate *Ceratium furca* is influenced by intracellular nutrient concentrations. Mar Ecol Prog Ser 262:137–151
Stoecker DK (1998) Conceptual models of mixotrophy in planktonic protists and some ecological and evolutionary implications. Eur J Protistol 34:281–290
Stoecker DK (1999) Mixotrophy among dinoflagellates. J Eukaryot Microbiol 46:397–401
Stoecker DK, Parrow MW, Burkholder JM, Glasgow HB (2002) Grazing by microzooplankton on *Pfiesteria piscicida* cultures with different histories of toxicity. Aquat Microb Ecol 28:79–85
Tillmann U (1998) Phagotrophy by a plastidic haptophyte, *Prymnesium patelliferum*. Aquat Microb Ecol 14:155–160
Tillmann U (2003) Kill and eat your predator: a winning strategy of the planktonic flagellate *Prymnesium parvum*. Aquat Microb Ecol 32:73–84

15 Allelopathy in Harmful Algae: A Mechanism to Compete for Resources?

E. GRANÉLI and P.J. HANSEN

15.1 Harmful Algal Species Known of Allelopathy

Some phytoplankton species produce and release secondary metabolites that negatively affect the growth of other organisms; i.e., they are allelopathic (e.g., Rizvi and Rizvi 1992). The production of such allelopathic chemicals by phytoplankton is known among several different algal groups: cyanobacteria, dinoflagellates, prymnesiophytes, and raphidophytes (Table 15.1). Some reports suggest that allelochemicals are also produced in diatoms and green algae (e.g., Subba Rao and Smith 1995; Chiang et al. 2004). A comprehensive list of freshwater and marine algae suspected of production of allelopathic substances was compiled by Legrand et al. (2003). In this chapter, we focus on unicellular organisms (protists and cyanobacteria) as target cells, because many so-called phytoplankton cells actually are mixotrophic or heterotrophic, and because many so-called protozoa functionally are mixotrophic. This requires that we address the negative effects of allopathic marine phytoplankton on both competitors and grazers.

15.2 Approaches to Demonstrate/Study Allelopathy – Pitfalls and Strength/Weaknesses of Experimental Approaches

Judging from the long list of reports of allelopathic effects of toxic algae on competitors and protistan grazers, allelopathy seems to be common among planktonic algae (Table 15.1, see also review by Legrand et al. 2003). Several approaches have been used to study and document allelopathic interactions, all of which have their strengths and weaknesses. In some cases, the proof of the production and release of allelochemicals in algae is an easy task. This is the case for some dinoflagellates (*Alexandrium* spp. *Heterocapsa circular-*

Ecological Studies, Vol. 189
Edna Granéli and Jefferson T. Turner (Eds.)
Ecology of Harmful Algae

Table 15.1. Algae species, their substances and allelopathic effect

Allelopathic species	Allelopathic substance[a]	Effect[b]	Reference
Anabaena spp.	U	GI	Suikkanen et al. (2004, 2005)
Anabaena cylindrical	EP	GI	Legrand et al. (2003)
Anabaena flos-aquae	U, A, M, HX	GI, D	Legrand et al. (2003)
Aphanizomenon spp.	U	GI, D	Legrand et al. (2003), Suikkanen et al. (2004, 2005)
Gomphosphaeria aponina	Aponin	GI, D	Legrand et al. (2003)
Hapalosiphon fontinalis	Hapalindole A	D	Legrand et al. (2003)
Fischerella muscicola	Fischerellin	GI, PI, D	Legrand et al. (2003)
Microcystis spp.	Microcystin	GI, PI	Pflugmacher (2002)
Nodularia spumigena	U	GI	Suikkanen et al. (2004, 2005)
Oscillatoria spp.	U	GI, PI, D	Legrand et al. (2003)
Alexandrium spp.	U	GI, D, CP, HC	Tillmann and John (2002), Fistarol et al. (2004a), Arzul et al. (1999)
A. minutum	U	CP	Fistarol et al. (2004a), Arzul et al. (1999)
A. ostenfeldii	U	GI, D	Hansen et al. (1992), Tillmann and John (2002)
A. tamarense	U	GI, D	Arzul et al. (1999), Tillmann and John (2002), Fistarol et al. (2004a, 2004b)
Amphidinium klebsii	U	GI	Sugg and VanDolah (1999)
Ceratium sp.	U	GI	Legrand et al. (2003)
Coolia monotis	Cooliatoxin, U	PI, GI, D	Legrand et al. (2003), Sugg and VanDolah (1999)
Dinophysis spp.	OA, DTX, U	GI	Kubanek et al. (2005)
Gamberidiscus toxicus	U	HC	Sugg and VanDolah (1999)
Gymnodinium spp.	Hemagglutenin	GI	Ahmed et al. (1995)
Gyrodinium aureolum	H	HC	Yasumoto et al. (1990)
Heterocapsa circularisquama	U, cell contact	Paralysis, D	Uchida (2001)
Karenia brevis	U	GI	Kubanek et al. (2005)
Karenia mikimotoi	H	HC, CP, GI	Legrand et al. (2003), Yasumoto et al. (1990), Fistarol et al. (2004b), Uchida (2001)
Karlodinium micrum	Karlotoxin	GI	Adolf et al. (2004)
Ostreopsis lenticularis	U	GI	Sugg and VanDolah (1999)
Peridinium aciculiferum	U	GI, D	Legrand et al. (2003)
Peridinium bipes	U	D	Legrand et al. (2003)
Prorocentrum lima	U	GI	Windust et al. (1996), Sugg and VanDolah (1999)
Chattonella marina	Hemagglutinin	GI	Ahmed et al. (1995)
Chaetoceros didymus	U	CP	Uchida (2001)
Pseudo-nitzschia pungens	U	GI	Legrand et al. (2003)
Rhizosolenia alata	U	GI	Legrand et al. (2003)

Table 15.1. (*Continued*)

Allelopathic species	Allelopathic substance[a]	Effect[b]	Reference
Chrysochromulina polylepis	H, fatty acids	D, GR, CP	Legrand et al. (2003), Fistarol et al. (2004b), Johansson and Granéli (1999b)
Prymnesium parvum	Prymnesin	D	Granéli and Johansson (2003), Fistarol et al. (2003), Barreiro et al. (2005), Johansson and Granéli (1999a)
Bothryococcus braunii	Free fatty acids	GI	Chiang et al. (2004)

[a] *U* unknown, *A* anatoxin, *M* microcystin, *OA* okadaic acid, *DTX* dinophysistoxin, *H* haemolysin, *EP* extracellular peptides, *PC* proteinaceous compounds, *HX* hydroxamates chelators

[b] *AI* autoinhibition, *GI* growth inhibition, *PI* Photosynthesis inhibition, *CP* cyst promotion, *GR* grazing inhibition, *HC* haemolytic/cytotoxic, *D* Death

isquama) and prymnesiophytes (*Prymnesium parvum*, *Chrysochromulina polylepis*). These species quickly cause lysis of the cells of most of their competitors (within minutes), when the latter are exposed to either certain amounts of the allelochemicals or to certain cell densities of the allelopathic algae (Fig. 15.1). Such allelopathy is thought to avoid competition for nutrients, vitamins, etc. (e.g., Hansen 1989; Uchida et al. 1995, cited in Uchida 2001; Tillmann and John 2002; Fistarol et al. 2004a). In such cases, the allelopathic effects of a specific alga on other algae can be documented using mixed growth experiments. The potentially toxic species is typically mixed with another species in batch cultures using a rich growth medium, and the culture is allowed to grow. Growth media are typically so rich in macronutrients that often less than 10–15 % of the nutrients have been taken up by the algae when they reach stationary growth phase. Thus, factors other than macronutrients, such as pH and light, limit the growth of algae in such batch cultures (Schmidt and Hansen 2001; Hansen 2002; Lundholm et al. 2005).

Algae have different tolerances to pH, such that one marine species can grow even at pH exceeding 10, while the growth of other species is reduced at pH 8.4 (Hansen 2002). Thus, in order to separate the negative effects of high pH from allelopathic effects, bubbling is by far the best way to avoid elevated pH in the growth medium. The use of non-carbon buffers should be avoided because algal cultures may then experience dissolved inorganic carbon (DIC) limitation, when the cultures grow dense. Thus, for many reports of allelopathy in the literature, which have used mixed batch cultures to demonstrate the negative effects of one alga on another but have not taken the pH into account, results should be interpreted cautiously.

Potential competition for nutrients between the donator and the target, as well as other interspecific interactions like mixotrophy or predation can be avoided by using filtrates of allelopathic compounds. Filtrates are particularly useful in studies of potentially toxic effects on natural planktonic communities (see Fistarol et al. 2004a). The drawback of using filtrates is that some toxins are labile, so toxic effects may cease after some time (hours to days), because the toxins are not exuded continuously as is the case in mixed cultures (Fistarol et al. 2004a). However, daily additions of the filtrates can overcome such problem (Fistarol et al. 2005).

There are clear advantages of using the mixed algae culture approach, despite the potential problems described above. Mixed cultures allow the continuing production of allelochemicals by the toxic algae, and therefore the opportunity of studying the toxic effects at "natural cell concentrations". This approach also allows for studies of the growth dynamics of both the toxic species and the target species. It is, after all, not always the toxic species that actually wins the competition. In many cases, the outcome of the experiment will depend on the initial concentrations, the concentration ratio of the toxic and target cells, their growth rates, and the pH tolerance.

15.3 Which Toxins are Involved in the Allelopathic Effects?

Many allelochemicals are believed to be secondary metabolites (toxins) with hemolytic capacity, i.e., they perforate cell membranes of other phytoplankton species, their grazers or even fish gills. For the ichthyotoxic species, *P. parvum*, some of the toxins (prymnesins) have been identified as hemolytic compounds. Igarashi et al. (1995, cited in Igarashi et al. 1998) characterized two of them, prymnesin 1 and 2, as polyoxy-polyene-polyethers and Kozakai et al. (1982) have described two other hemolytic toxins formed by a mixture of galactolipids. Prymnesins 1 and 2 are known to be among the most potent toxins with hemolytic activity. They have a toxic potency, which is about 50,000 times greater than plant saponin (on a molar basis, Igarashi et al. 1998; Sasaki et al. 2001).

For the fish-killing algae, *C. polylepis* and *Karenia mikimotoi* (=*Gyrodinium aureolum*), glycosylglycerolipids are among the identified toxins and they have ichthyotoxic and hemolytic properties (Yasumoto et al. 1990). Gentien and Arzul (1990) have found three hemolytic fractions in the extracts of *K. mikimotoi* filtrates. Other ichthyotoxic HA with toxins which display hemolytic properties (and are suspected of allelopathy) are the hemolysins from *Amphidinium carterae* (Echigoya 2005).

Adolf et al. (2004) tested the additions of karlotoxins (KmTx2) isolated from *Karlodinium micrum* (a fish-killer) on three raphidophytes (*Heterocapsa akashiwo*, *Fibrocapsa japonica* and *Chattonella subsalsa*), two species of the dinoflagellates *Pfiesteria* spp., a cryptophyte (*Storeatula major*) and on *K. micrum* itself. Karlotoxins in concentrations between 500 and 1,000 mg L^{-1} inhibited the growth of all species, although moderately, while *K. micrum* growth was not affected even at 1,000 mg L^{-1} additions. These are very high concentrations and it is unclear if the karlotoxins are the same substances as the released allelochemicals.

For the Florida dinoflagellate *Karenia brevis*, which is a well-known fish-killer, it seems that the identified toxins, brevetoxins, are not connected to a great extent to their allelopathic activity. Although they produce allelopathic compounds with negative effect on several phytoplankton species, additions of their internally produced toxin, brevetoxins in quite high concentrations could only slightly inhibit the growth of the diatom *Skeletonema costatum* (Kubanek et al. 2005).

Species of the dinoflagellate genus *Alexandrium* are well known for their production of paralytic shellfish poisoning (PSP) toxins, which accumulate in mussels (*Mytilus edulis*). However, they also form red tides that kill fish (e.g., Mortensen 1985). Allelopathy in *Alexandrium* spp. was first described by Hansen (1989) and Hansen et al. (1992), who found that *A. tamarense* and *A. ostenfeldii* exudates were toxic to the ciliate *Favella ehrenbergii*. Tillmann and John (2002) investigated the effects of many strains of *A. tamarense* on the mobility of the heterotrophic dinoflagellates, *Oblea rotundata* and *Oxyrrhis marina*. Their results reveal that the amounts of PSP in the cells do not correlate with immobilization of the target species. Some of their *Alexandrium* strains did not even contain PSP toxins, but were still allelopathic. Thus, the allelopathy displayed in *Alexandrium* spp. must be due to some yet-unknown toxins. Recent evidence suggests that the hemolytic compound may be a protein-like compound with a molecular weight of >10 kDa (Emura et al. 2004)

Species of the diatom genus *Pseudo-nitzschia*, which produce the amnesic shellfish poisoning (ASP) toxin domoic acid (DA), apparently have no allelopathic effects on other algae (Lundholm et al. 2005). Tests of pure DA revealed that this toxin had no effect on a range of phytoplankton species, even at very high concentrations, ten times the concentrations maximally reported to be found in the medium of a dense *Pseudo-nitzschia* culture.

Windust et al. (1996, 1997) documented that micromolar concentrations of the diarrheic shellfish poisoning (DSP) toxins, okadaic acid (OA) and dinophysistoxin-1 (DTX-1), effectively inhibited the growth of several microalgae. However, although Sugg and Van Dolah (1999) confirmed that OA can inhibit the growth of some phytoplankton species, they concluded that there must be other compounds involved in the toxic effects of the filtrates from the OA-producing species *Prorocentrum lima* in order to explain the observed results.

Some reports have suggested allelopathic effects of marine and brackish water cyanobacteria, like *Nodularia spumigena*, *Aphanizomenon flos-aquae* and *Anabaena* spp. on cryptophytes, while other tested species were unaffected (Suikkanen et al. 2004, 2005).

Thus, it is only rarely that the main known toxins, known from a specific harmful algal species, are the most likely allelopathic compounds (e.g., the prymnesins), while in most cases the known toxins and the allelopathic substances are different chemical compounds. It is particularly interesting that the shellfish toxins (PSP, DSP, and ASP) do not seem to be involved in any allelopathic effect on other protists, at least not at concentrations found in nature.

15.4 Influence of Abiotic and Biotic Factors on Allelopathy

15.4.1 Abiotic Factors

The influence of abiotic factors on allelopathic effects of HA on other planktonic organisms are in many cases unknown. Most of our knowledge comes from the studies of the prymnesiophytes, *P. parvum* and *C. polylepis*, which produce toxins (hemolysins) that kill other protists. Light, pH and nutrient deficiency have been reported to affect the production and fate of allelopathic substances from these algae (Fig. 15.2).

Light is essential for the production of hemolysins in *P. parvum*. Both intracellular and extracellular concentrations decrease over time when the culture is transferred to the dark, even in experiments using a synthetic organic medium which ensures fast growth of *P. parvum* (Shilo 1967). The extracellular hemolysins of prymnesiophytes are nevertheless degraded via photo-oxidation in the light, as shown by Simonsen and Moestrup (1997) using filtrates from *C. polylepis*. Recent experiments have shown that the hemolysins of *P. parvum* incubated in the dark for 72 h did not lose toxicity (Hagström and Granéli 2005).

The first reports of pH-dependency of toxicity of *P. parvum* (to fish) were provided by Shilo and Aschner (1953) and Ulitzur and Shilo (1964). They observed that a change of pH from 8 to 9 increased the toxicity of *P. parvum* to fish. A similar observation was made by Schmidt and Hansen (2001) studying the allelopathic effects of the prymnesiophyte, *C. polylepis*, on the dinoflagellate *Heterocapsa triquetra*. Schmidt and Hansen found that low-cell-density cultures increased their toxicity when pH was elevated from 8 to 9, while toxicity had disappeared at pH 6.5–7. They also found a quantitative shift in the toxicity response as a function of pH for three different cell concentra-

Fig. 15.1. Allelopathic effect of *Prymnesium parvum* filtrate on A *Thalassiosira weissflogii* and B *Rhodomonas baltica*. *Left panels*; sea water control, *right panels*; after exposure to the filtrate

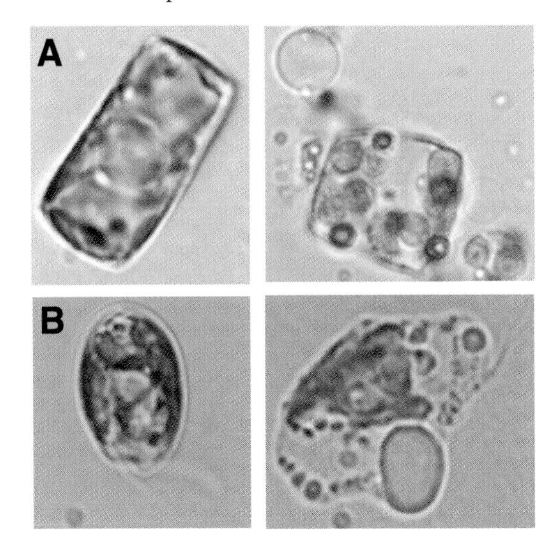

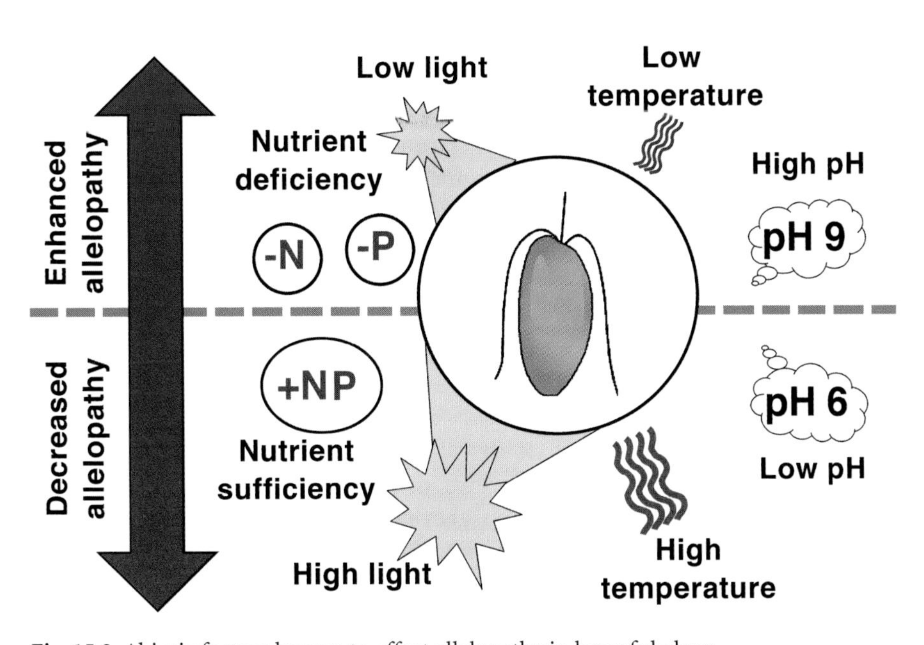

Fig. 15.2. Abiotic factors known to affect allelopathy in harmful algae

tions of the toxic alga. For example, a very dense and highly toxic culture of *C. polylepis* was equally toxic at pH 8 and 9.

The surrounding nutrient concentrations can exert a major influence on the amount of allelopathic substances being produced by phytoplankton. For the prymnesiophytes *C. polylepis* and *P. parvum* Johansson and Granéli (1999a, 1999b) have shown that the concentration of hemolytic substances in the cells increased when the cells were under nutrient limitation, so cells that were phosphorus and nitrogen limited contained more toxin than nutrient-replete cells. The addition of filtrates from *P. parvum* cultures grown under nutrient-deficient conditions (N or P) had a strong negative effect on other protists compared to a positive growth when exposed to cell-free filtrates from non-deficient cultures (Granéli and Johansson 2003; Fistarol et al. 2003). Analyses of *P. parvum* or *C. polylepis* filtrates showed higher toxicity in the filtrates of nutrient-deficient cultures than in the filtrates from N- and P-sufficient cultures (Johansson and Granéli 1999a, 1999b; Uronen et al. 2005). Many parameters may complicate interpretations of studies performed with filtrates. For example, degree of bacterial degradation can affect accumulation of allelochemicals. Light degradation of the allelochemicals will also diminish the toxic effects on the target organisms, but this can be avoided if both the target and the producer are growing together as the allelochemicals will be continued released into the media. Another advantage of co-growth is that both the target and the producers will be suffering from the same chemical condition of the medium. As shown by Granéli and Johansson (2003), the target was much more sensitive to the allelochemicals when grown under the same N and P limitation as the producer and Fistarol et al. (2005) showed higher sensitivity for *P. parvum* filtrate in target cells grown under nutrient limitation than cells grown under nutrient sufficiency.

15.4.2 Biotic Factors

The cell densities of both the toxic alga and its target seem to be important for the outcome of the interaction. However, this is not a well-studied area. Also, the taxonomic position and size of target cells seems to matter. In most cases, we assume that the toxins are exuded into the medium and thereby come into contact with cell membranes of target cells. This is probably the most common way that allelopathy works. However, this may not always be the case. Uchida et al. (1995, 1999, both cited in Uchida 2001) found that the dinoflagellate *Heterocapsa circularisquama* kills at least some other algae by cell contact. Similarly, Kamiyama (1997) and Kamiyama and Arima (1997) found that cell contact was required for *H. circularisquama* to kill ciliates as filtrates were non-toxic. However, *H. circularisquama* only killed ciliates at high cell concentrations, while the ciliates were able to ingest and grow well on less-dense cultures of *H. circularisquama*. The exact mechanism behind toxic effects of

H. circularisquama is still unresolved, and it is still not documented that a toxin is actually involved. Other allopathic algae (*Heterosigma akashiwo*, *Alexandrium* spp and *P. parvum*) are toxic to their protistan grazers through exudates (Hansen 1989; Tillmann and John 2002; Tillmann 2004).

Higher cell numbers of the HA species will result in the release of larger amounts of exudates into the water, and thus a greater fraction of target cells will be affected by the toxins. This has shown to be the case for *Alexandrium* spp. *C. polylepis* and *P. parvum* (Hansen 1989; Schmidt and Hansen 2001; Skovgaard and Hansen 2003; Tillman 2003). However, some species (e.g., *C. polylepis*) may lose their toxicity after some days in stationary growth phase, even though the cell density is very high (Schmidt and Hansen 2001). Also, the growth condition of the toxic algae is important for allelopathy. It has also recently been shown that the allelopathic effects of algae like *P. parvum* and *Alexandrium* spp. depend on the target cell concentrations, such that toxic effects decreased with increasing target cell concentrations (Tillmann 2003). This indicates that the toxin is removed from the system, presumably by binding to the targets membranes. Photo-oxidation is a process that also decreases toxicity by degrading *Prymnesium*-toxins (Fistarol et al. 2003)

The effects of HA on other algae may be size-, species-, and in some cases, group-specific. Thus, their impacts on the species composition of natural planktonic communities may be "selective". *Prymnesium parvum* is a species that produces allelochemicals capable of negatively affecting the entire plankton community. Incubation of natural plankton communities with toxic filtrates showed however, that some plankton groups (like diatoms, small flagellates and ciliates) were affected more quickly than others (e.g., dinoflagellates and cyanobacteria) (Fistarol et al. 2003). *Chrysochromulina polylepis* kills most other algae belonging to different algal taxa, but some species are not affected by its toxin. Size of target algae does not play a role in the sensitivity (= the concentration required for an effect) of the target cells to the toxins of *C. polylepis*, but small target organisms are affected more rapidly. Thus, longer incubations times are required for larger target cells. Some targets may escape the inhibitory effects of allelochemicals by producing temporary cysts, such as been found for *Heterocapsa circularisquama* (Uchida et al. 1999, cited in Uchida 2001) and *Scrippsiella trochoidea* (Fistarol et al. 2004b). Cysts were produced when *H. circularisquama* were exposed to higher *Karenia mikimotoi* cell densities than their own, and when *S. trochoidea* were exposed to filtrates from *A. tamarense*, *K. mikimotoi* and *C. polylepis*.

15.5 Ecological Significance of Allelopathy in Marine Ecosystems

Allelopathic phytoplankton such as *P. parvum* and C. *polylepis* have been responsible for toxic incidents with severe ecological impacts in many parts of the world (Moestrup 1994; Edvardsen and Paasche 1998). These highly potent exotoxins have been shown to have several biological effects on a variety of marine organisms, including ichthyotoxicity, neurotoxicity, cytotoxicity, hepatotoxicity and hemolytic activity (Shilo 1981; Igarashi et al. 1996, cited in Igarashi et al. 1996). Blooms of *P. parvum* develop in inshore localities and brackish waters and are usually monospecific, suggesting a mechanism to outcompete other phytoplankton species. The complete dominance of *P. parvum* cannot be explained by virtue of its own growth rate, since *P. parvum* is known to have moderate growth rates under a range of environmental conditions (Holdway et al. 1978; Brand 1984; Larsen and Bryant 1998). Nevertheless, the total absence of coexisting species during certain blooms is indicative of the high competitive ability of *P. parvum*.

As shown for terrestrial plants, the capability of allelopathy in phytoplankton is closely associated with competition for resources. For marine environments, limiting amounts of nutrients may not only increase the production of allelochemicals, but also accentuate their action (Einhellig 1995). This is a very similar mode of action to the elevated extracellular release of organic compounds that occurs under conditions of nutrient limitation (Myklestad 1977; Soeder and Bolze 1981; Myklestad et al. 1995; von Elert and Jüttner 1997). In marine and coastal waters, nitrogen and phosphorus are seldom in high enough concentrations to sustain the growth of the full array of phytoplankton species co-existing in time and space. The nutrient that is found at the lowest concentration in relation to the algal need (limiting nutrient) is the nutrient that will limit algal growth. Thus, the ability to compete for the limiting nutrient is crucial for the proliferation of a specific species. The species that is able to compete successfully for the available growth-limiting nutrient has the potential to become dominant by increasing its biomass, ultimately forming blooms. Thus, the ability to produce allelopathic compounds can give an advantage to a HA species to outcompete other phytoplankton species for limiting resources such as nitrogen and phosphorus.

Acknowledgements. The authors thank Christina Esplund for help with the table and figure. E. Granéli acknowledges financial support from the Swedish Research Council and The European Commission through the FP 6-THRESHOLDS project (GOCE 003933).

References

Adolf JE, Krupatkina DN, Place AR, Brown PJP, Lewitus AJ (2004) The allelopathic specificity of *Karlodinium micrum* toxins (Karlotoxins). In: 11th Int Conf Harmful Algal Blooms, Cape Town, S Africa, Abstracts, p 52
Ahmed MS, Khan S, Arakawa O, Onoue Y (1995) Properties of hemagglutinins newly separated from toxic phytoplankton. Biochem Biophys Acta 1243:509–512
Arzul G, Seguel M, Guzman L, Erard-Le Denn E (1999) Comparison of allelopathic properties in three toxic *Alexandrium* species. J Exp Mar Biol Ecol 232:285–295
Barreiro A, Guisande C, Maneiro I, Lien TP, Legrand C, Tamminen T, Lehtinen S, Uronen P, Granéli E (2005) Relative importance of the different negative effects of the toxic haptophyte *Prymnesium parvum* on *Rhodomonas salina* and *Brachionus plicatilis*. Aquat Microb Ecol 38:259–267
Brand LE (1984) The salinity tolerance of forty-six marine phytoplankton isolates. Est Coast Shelf Sci 18:543–556
Chiang I-Z, Huang W-Y, Wu J-T (2004) Allelochemicals of *Botryococcus braunii* (Chlorophyceae). J Phycol 40:474–480
Echigoya R, Rhodes L, Oshima Y, Satake M (2005) The structures of five new antifungal and hemolytic amphidinol analogs from *Amphidinium carterae* collected in New Zealand. Harmful Algae 4:383–389
Edvardsen B, Paasche E (1998) Bloom dynamics and physiology of *Prymnesium* and *Chrysochromulina* In: Anderson DM, Cembella AD, Hallegraeff GM (eds) Physiological ecology of harmful algal blooms. NATO ASI Series 41. Springer, Berlin Heidelberg New York, pp 193–208
Einhellig FA (1995) Allelopathy: current status and future goals. In: Derjik EV, Dakshini KMM, Einhellig FA (eds) Allelopathy: organisms, processes, and applications. ACS Symp Ser 582, pp 1–24
Emura A, Matsuyama Y, Oda T (2004) Evidence for the production of a novel proteinaceous hemolytic exotoxin by dinoflagellate *Alexandrium taylori*. Harmful Algae 3:29–37
Fistarol GO, Legrand C, Granéli E (2003) Allelopathic effect of *Prymnesium parvum* on a natural plankton community. Mar Ecol Prog Ser 255:115–125
Fistarol GO, Legrand C, Selander E, Hummert C, Stolte W, Granéli E (2004a) Allelopathy in *Alexandrium* spp.: effect on a natural plankton community and on algal monocultures. Aquat Microb Ecol 35:45–56
Fistarol GO, Legrand C, Rengefors K, Granéli E (2004b) Temporary cyst formation in phytoplankton: a response to allelopathic competitors? Env Microbiol 6:791–798
Fistarol GO, Legrand C, Granéli E (2005) Allelopathic effect on a nutrient-limited phytoplankton species. Aquat Microb Ecol 41:153–161
Gentien P, Arzul G (1990) Exotoxin production by Gyrodinium cf. aureolum (Dinophyceae). J Mar Biol Ass UK 70:571–581
Granéli E, Johansson N (2003) Increase in the production of allelopathic substances by *Prymnesium parvum* cells grown under N- or P- deficient conditions. Harmful Algae 2:135–145
Hagström JA, Granéli E (2005) Removal of *Prymnesium parvum* (Haptophyceae) cells under different nutrient conditions by clay. Harmful Algae 4:249–260
Hansen PJ (1989) The red tide dinoflagellate *Alexandrium tamarense*: effects on behaviour and growth of a tintinnid ciliate. Mar Ecol Prog Ser 53:105–116
Hansen PJ (2002) The role of pH and CO_2 limitation in marine plankton: implications for species succession. Aquat Microb Ecol 28:279–288

Hansen PJ, Cembella AD, Moestrup Ø (1992) The marine dinoflagellate *Alexandrium ostenfeldii*: paralytic shellfish toxin concentration, composition, and toxicity to a tintinnid ciliate. J Phycol 28:597–603

Holdway PA, Watson RA, Moss B (1978) Aspects of the ecology of *Prymnesium parvum* (Haptophyta) and water chemistry in the Norfolk Broads England. Freshwater Biol 8:295–311

Igarashi T, Aritake S, Yasumoto T (1998) Biological activities of prymnesin-2 isolated from a red tide alga *Prymnesium parvum*. Nat Toxins 6:35–41

Johansson N, Granéli E (1999a) Influence of different nutrient conditions on cell density, chemical composition and toxicity of *Prymnesium parvum* (Haptophyta) in semi-continuous cultures. J Exp Mar Biol Ecol 239:243–258

Johansson N, Granéli E (1999b) Cell density, chemical composition and toxicity of *Chrysochromulina polylepis* (Haptophyta) in relation to different N:P supply ratios. Mar Biol 135:209–217

Kamiyama T (1997) Growth and grazing responses of tintinnid ciliates feeding on the toxic dinoflagellate *Heterocapsa circularisquama*. Mar Biol 128:509–515

Kamiyama T, Arima S (1997) Lethal effects of the dinoflagellate *Heterocapsa circularisquama* upon tintinnids ciliates. Mar Ecol Prog Ser 160:27–33

Kozakai H, Oshima Y, Yasumoto T (1982) Isolation and structural elucidation of hemolysin from the phytoflagellate *Prymnesium parvum*. Agric Biol Chem 46:233–236

Kubanek J, Prince E, Hicks MK, Naar J, Villareal T (2005) Does the Florida red tide dinoflagellate use allelopathy to outcompete other phytoplankton? Limnol Oceanogr 50:883–895

Larsen A, Bryant S (1998) Growth rate and toxicity of *Prymnesium parvum* and *Prymnesium patelliferum* (Haptophyta) in response to changes in salinity, light and temperature. Sarsia 83:409–418

Legrand C, Rengefors K, Granéli E, Fistarol GO (2003) Allelopathy in phytoplankton – biochemical, ecological and evolutionary aspects. Phycologia 42:406–419

Lundholm N, Hansen PJ, Kotaki Y (2005) Lack of allelopathy effects of the domoic acid producing marine diatom *Pseudo-nitzschia multiseries*. Mar Ecol Prog Ser 288:21–33

Moestrup Ø (1994) Economic aspects: ′blooms′, nuisance species, and toxins. In: Green JC, Leadbeater BSC (eds) The haptophyte algae. Clarendon Press, Oxford, pp 265–285

Mortensen AM (1985) Massive fish mortalities in the Faroe Islands caused by a bloom of *Gonyaulax excavata*. In: Anderson DM, White AW, Baden DG (eds) Toxic dinoflagellates. Elsevier, New York, pp 165–170

Myklestad S (1977) Production of carbohydrates by marine phytoplanktonic diatoms. II Influence of the N/P ratio in the growth medium on the assimilation ratio, growth rate, and production of cellular and extracellular carbohydrates by *Chaetoceros affinis* var. Willei (Gran) Hustedt and Skeletonema costatum (Grev.) Cleve. J Exp Mar Biol Ecol 29:161–179

Myklestad S, Ramlo B, Hestman S (1995) Demonstration of strong interaction between the flagellate *Chrysochromulina polylepis* (Prymnesiophyceae) and a marine diatom. In: Lassus P, Arzul G, Le Denn EE, Gentien P, Marcaillou C (eds) Harmful marine algal blooms. Lavoisier Intercept, Paris, pp 633–638

Pflugmacher S (2002) Possible allelopathic effects of cyanotoxins, with reference to Microcystin-LR, in aquatic ecosystems. Environ Toxicol 17:407–413

Rizvi SJH, Rizvi V (eds) (1992) Allelopathy: basic and applied aspect. Chapman & Hall, London, 480 pp

Sasaki M, Shida T, Tachibana K (2001) Synthesis and stereochemical confirmation of the HI/JK ring system of prymnesins, potent hemolytic and ichthyotoxic glycoside toxins isolated from the red tide alga. Tetrahedron Lett 42:5725–5728

Schmidt LE, Hansen PJ (2001) Allelopathy in the prymnesiophyte *Chrysochromulina polylepis*: effect of cell concentration, growth phase and ph. Mar Ecol Prog Ser 216:67–81
Shilo M (1967) Formation and mode of action of algal toxin. Bacteriol Rev 31:180–193
Shilo M (1981) The toxic principles of *Prymnesium parvum*. In: Carmichael WW (ed) The water environment. Algal toxins and health. Plenum Press, New York, pp 37–47
Shilo M, Aschner M (1953) Factors governing the toxicity of cultures containing the phytoflagellate *Prymnesium parvum* Carter. J Gen Microbiol 36:333–343
Simonsen S, Moestup Ø (1997) Toxicity tests in eight species of *Chrysochromulina* (Haptophyta). Can J Bot 75:129–136
Skovgaard A, Hansen PJ (2003) Food uptake in the harmful alga *Prymnesium parvum* mediated by excreted toxins. Limnol Oceanogr 48:1161–1166
Soeder CJ, Bolze A (1981) Sulphate deficiency stimulates release of dissolved organic matter in synchronous cultures of *Scenedesmus obliquus*. Physiol Plant 52:233–238
Subba Rao DV, Pan Y, Smith SJ (1995) Allelopathy between *Rhizosolenia alata* (Brightwell) and the toxigenic *Pseudo-nitzschia pungens* f. *multiseries* (Hasle). In: Lassus P, Arzul G, Le Denn EE, Gentien P, Marcaillou C (eds) Harmful marine algal blooms. Lavoisier Intercept Ltd, Paris, pp 681–686
Sugg LM, VanDolah FM (1999) No evidence for an allelopathic role of okadaic acid among ciguatera-associated dinoflagellates. J Phycol 35:93–103
Suikkanen S, Fistarol GO, Granéli E (2004) Allelopathic effects of the Baltic cyanobacteria *Nodularia spumigena*, *Aphanizomenon flos-aquae* and *Anabaena lemmermannii* on algal monocultures. J Exp Mar Biol Ecol 308:85–101
Suikkanen S, Fistarol GO, Granéli E (2005) Effects of cyanobacterial allelochemicals on a natural plankton community. Mar Ecol Prog Ser 287:1–9
Tillmann U (2003) Kill and eat your predator: a winning strategy of the planktonic flagellate *Prymnesium parvum*. Aquat Microb Ecol 32:73–84
Tillmann U (2004) Interactions between planktonic microalgae and protozoan grazers. J Eukaryot Microbiol 51:156–168
Tillmann U, John U (2002) Toxic effects of *Alexandrium* spp. on heterotrophic dinoflagellates: an allelochemical defence mechanism independent of PSP-toxin content. Mar Ecol Prog Ser 230:47–58
Uchida T (2001) The role of cell contact in the life cycle of some dinoflagellate species. J Plankton Res 23:889–891
Ulitzur S, Shilo M (1964) A sensitive assay system for the determination of the ichthyotoxicity of *Prymnesium parvum*. J Gen Microbiol 36:161–169
Uronen P, Lehtinen S, Legrand C, Kuuoppo P, Tamminen T (2005) Haemolytic activity and allelopathy of the haptophyte *Prymnesium parvum* in nutrient-limited and balanced growth conditions. Mar Ecol Prog Ser 299:137–148
von Elert E, Jüttner F (1997) Phosphorus limitation and not light controls the extracellular release of allelopathic compounds by *Trichormus doliolum* (Cyanobacteria). Limnol Oceanogr 42:1976–1802
Windust AJ, Wright JLC, McLachlan JL (1996) The effects of the diarrhetic shellfish poisoning toxins, okadaic acid and dinophysistoxin-1, on growth of microalgae. Mar Biol 126:19–25
Windust AJ, Quilliam MA, Wright JLC, McLachlan JL (1997) Comparative toxicity of the diarrhetic shellfish poisons, okadaic acid, okadaic acid diol-ester and dinophysistoxin-4, to the diatom *Thalassiosira weissflogii*. Toxicon 35:1591–1603
Yasumoto T, Underdahl B, Aune T, Hormazabal V, Skulberg OM, Oshima Y (1990) Screening for hemolytic and ichthyotoxic components of *Chrysochromulina polylepis* and *Gyrodinium aureolum* from Norwegian coastal waters. In: Granéli E, Sundström B, Edler L, Anderson DM (eds) Toxic marine phytoplankton. Elsevier, New York, pp 436–440

16 Trace Metals and Harmful Algal Blooms

W.G. SUNDA

16.1 Introduction

Trace metals influence phytoplankton both as limiting nutrients and as toxicants. Until recently, phytoplankton productivity in the ocean was thought to be primarily limited by major nutrients (N, P, and Si). However, enrichment experiments in bottles and in mesoscale patches of surface water have shown that iron limits algal growth in major regions of the ocean (Coale et al. 1996; Tsuda et al. 2003; Boyd et al. 2000; Coale et al. 2004), and some coastal upwelling systems (Hutchins et al. 1998, 2002). In addition, it now appears that iron limits N_2-fixation by cyanobacteria in the ocean, and thus may control oceanic nitrogen inventories (Rueter 1983; Falkowski 1997; Kustka et al. 2003; Mills et al. 2004). Several other micronutrient metals (Zn, Co, Mn, and Cu) have also been shown to stimulate phytoplankton growth in ocean waters, but their effect is usually less than that of iron (Coale 1991; Crawford et al. 2003; Franck et al. 2003). However, these metals may play an important role in regulating the composition of phytoplankton communities because of large differences in trace metal requirements among species (Brand et al. 1983; Coale 1991; Sunda and Huntsman 1995a). For example, the addition of zinc to surface waters of the subarctic Pacific stimulated the growth of coccolithophores over that of diatoms, while the addition of iron preferentially favored the growth of large diatoms (Crawford et al. 2003). In general, iron can have a critical influence on the composition and structure of algal communities because of differences in requirements among species, particularly coastal and oceanic ones (Brand et al. 1983; Sunda and Huntsman 1995b) and large-celled and small-celled species (Price et al. 1994; Sunda and Huntsman 1997). Trace metals are believed to be less important in limiting algal growth in most coastal and freshwater systems. An important issue in the influence of trace metals on harmful blooms is the relative sensitivity of HAB species to trace metal limitation or toxicity relative to that of competing non-HAB algae. Although such differences are likely, they are as yet largely unknown.

Ecological Studies, Vol. 189
Edna Granéli and Jefferson T. Turner (Eds.)
Ecology of Harmful Algae

16.2 Chemistry and Availability of Metals

The availability of trace metals varies widely in natural waters due to large variations in total metal concentrations and chemical speciation. Like major nutrients, surface concentrations of many bioactive trace metals (Fe, Zn, Co, Cu, Cd, and Ni) in the ocean and in lakes are often depleted substantially relative to concentrations in deeper waters because of biological uptake in sunlit surface waters and regeneration at depth, as occurs for major nutrients (Bruland 1980; Martin et al. 1989). This efficient removal by phytoplankton results in extremely low dissolved (i.e., filterable) concentrations of iron (0.02–0.5 nM), Zn (0.06–0.2), and Cu (0.4–1.4 nM) in surface ocean waters. Concentrations of these metals increase substantially in transects from oceanic to coastal and estuarine waters because of inputs from continental sources, including rivers, ground water, aeolian dust, and coastal sediments. Filterable iron concentrations can approach 10–20 μM in rivers, orders-of-magnitude higher than surface ocean values (Boyle et al. 1977). This filterable iron occurs largely as hydrous iron oxide-organic colloids, which are rapidly lost from estuarine and coastal waters via salt-induced coagulation and particulate settling (Boyle et al. 1977). Because of this efficient removal, very little of the iron in rivers reaches the sea, and most of the iron in ocean waters is derived from the deposition of mineral dust blown on the wind from arid regions of the continents (Duce and Tindale 1991). These aeolian inputs change seasonally with variations in prevailing winds and are highest in waters downwind of large arid regions such as North Africa and Central Asia.

Trace metals exist as a variety of chemical species in natural waters, which strongly influences their chemical behavior, cycling, and availability to phytoplankton. Most occur as cations that are complexed to varying degrees by inorganic and organic ligands or are adsorbed onto, or bound within particles. Many biologically active metals (Fe, Cu, Mn, and Co) cycle between different oxidation states, which have quite different kinetic lability, solubility, and reactivity toward complex formation.

Iron is the most biologically important metal, and its chemical behavior is also the most complex. Its stable oxidation state in oxygen-containing waters, Fe(III), undergoes substantial hydrolysis and polymerization, which results in the formation of sparingly soluble hydrous iron oxides (Lui and Millero 2002). This oxide formation and the tendency of ferric ions to adsorb onto particle surfaces results in the particulate removal of iron from seawater and consequent low iron concentrations in ocean waters (Johnson et al. 1997). Most (>99 %) of the dissolved ferric iron in seawater is bound to organic ligands, which minimizes iron adsorption or precipitation, and thus minimizes the particulate scavenging of iron from seawater (Rue and Bruland 1995; Wu and Luther III 1995). Some of these organic complexes may be siderophores produced by cyanobacteria and heterotrophic bacteria to facilitate intracellular iron uptake. These ferric chelates do not appear to be taken up by eukaryotic algae; however, the iron in these chelates is released for cellular uptake by cell-surface reductase systems (Maldonado and Price 2001) and by photolytic reactions, which reductively dissociate ferric chelates (Barbeau et al. 2001). The released ferrous ions are unstable in seawater, and are re-oxidized to soluble ferric hydrolysis species on time scales of minutes (Sunda 2001).

Other bioactive metals such as copper and zinc are also heavily chelated (generally >99 %) in marine waters by unidentified organic ligands of likely biological origin (Sunda and Hanson 1987; Bruland 1989; Coale and Bruland 1990; Moffett et al. 1997). Available evidence indicates that organic complexes of zinc and copper are not directly available to phytoplankton, and that uptake of these metals is instead controlled by the concentration of free aquated metal ions or that of kinetically labile dissolved inorganic species (Sunda and Guillard 1976; Anderson et al. 1978). By lowering free metal ion concentrations, these organic ligands act to buffer free metal ion concentrations, and to minimize algal removal of these metals with algal growth in natural waters.

16.3 Trace Metals as Limiting Nutrients

Trace metal micronutrients (Fe, Mn, Zn, Co, Cu, Ni, and Mo) are essential for the growth and metabolism of all aquatic algae. They play critical roles in photosynthesis and the assimilation of essential macronutrients (N, P, and Si); thus trace metal requirements can be influenced by the availability of light, CO_2, and major nutrients. Of the micronutrient metals, iron is needed in the greatest amount and is the metal that most frequently limits algal growth. Iron serves essential metabolic functions in photosynthetic electron transport, respiration, nitrate assimilation, N_2-fixation, and detoxification of reactive oxygen species (superoxide radicals and hydrogen peroxide). Because of its heavy involvement in photosynthetic electron transport, cellular iron requirements increase with decreasing light intensity (Raven 1990; Sunda and Huntsman 1997). Such effects could increase iron requirements during dense algal blooms where light is attenuated to low levels. Because of the high metabolic cost for iron in nitrate reduction, iron requirements are higher for cells grown on nitrate than on ammonium; consequently, iron can limit nitrate assimilation during blooms (Raven 1988; Price et al. 1994; Maldonado and Price 1996). Because of the even higher iron requirement for N_2-fixation, iron is needed in much greater amounts by cyanobacteria growing diazotrophically (i.e., growing on N_2) than those growing on ammonia (Berman-Frank et al. 2001; Kustka et al. 2003). Iron can thus limit the development of blooms of diazotrophic cyanobacteria in both the ocean (Kustka et al. 2003) and some lakes (Wurtsbaugh and Horne 1983; Evans and Prepas 1997).

Manganese, like iron, is essential for photosynthesis and thus is needed in higher amounts for growth at low light intensities (Sunda and Huntsman 1998). Mn also occurs in superoxide dismutase, an enzyme that removes toxic superoxide radicals (Peers and Price 2004). Because it has fewer metabolic functions, its requirement for growth is much less than that of iron (Raven 1990; Sunda 2000).

Zinc has a variety of metabolic functions and has a cellular requirement similar to that for Mn (Sunda 2000). It occurs in carbonic anhydrase (CA), an enzyme critical to CO_2 transport and fixation (Badger and Price 1994). Higher amounts of this enzyme are needed under CO_2-limiting conditions, and the cellular requirement for zinc increases at low CO_2 (Morel et al. 1994; Lane and Morel 2000; Sunda and Hunstman 2005). Thus, cells may become co-limited by CO_2 and zinc during blooms when CO_2 is consumed to low levels by high levels of C-fixation (Sunda and Hunstman 2005). In Lake Kinneret, Israel spring blooms of the dinoflagellate *Peridinium gatunense* caused a ten-fold decrease in CO_2, and an accompanying up to 100-fold increase in CA activity per dinoflagellate cell (Berman-Frank et al. 1994). Whether this large increase in CA resulted in a Zn-CO_2 co-limitation, however, is not known.

Zinc also occurs in zinc finger proteins, involved in DNA transcription, and in alkaline phosphatase, needed to acquire phosphorus from organic phosphate esters, which dominate phosphate pools under P-limitation. Consequently, Zn and P may co-limit algal growth in regions where both nutrients occur at low concentrations such as the Sargasso Sea (Wu et al. 2000). Cobalt, and sometimes cadmium, can substitute for zinc in many zinc enzymes such as CA, leading to complex interactions among the three metals in marine algae (Morel et al. 1994; Sunda and Huntsman 1995a; Lane and Morel 2000). Cobalt has a unique requirement in vitamin B_{12}, but the need for this cofactor is usually quite low in phytoplankton (Sunda and Huntsman 1995a). A specific requirement for cobalt is seen in marine cyanobacteria and bloom-forming prymnesiophytes (*Chrysochromulina polylepis* and *Emiliania huxleyi*), but the biochemical basis for this is not known (Granéli and Risinger 1994; Sunda and Huntsman 1995a; Saito et al. 2002). Leaching of Co from acidified soils has been suggested to have played a role in the 1988 *C. polylepsis* bloom in Scandinavian coastal waters (Granéli and Haraldsson 1993).

Copper occurs in cytochrome oxidase, a key protein in respiratory electron transport, and in plastocyanin, which can substitute for the iron protein cytochrome c_6 in photosynthetic electron transport (Raven et al. 1999). It is also an essential component of the high-affinity iron transport system of many eukaryotic algae (LaFontaine et al. 2002; Peers et al. 2005). Because copper is needed for iron uptake and can metabolically substitute for iron, co-limitations can occur for Cu and Fe, as observed in centric diatoms (Maldonado unpubl. data) and toxic species of *Pseudo-nitzschia* (Wells et al. 2005).

Nickel and molybdenum, like iron, play important roles in nitrogen assimilation. Ni occurs in the enzyme urease, and thus is required by phytoplankton grown on urea (Price and Morel 1991). Mo occurs with iron in nitrate reductase and nitrogenase, and thus is needed for nitrate assimilation and N_2 fixation (Raven 1988). It occurs in natural waters primarily as the oxyanion molybdate (MoO_4^{2-}). Because of its high concentration in seawater (~100 nM), Mo should not limit algal growth in marine and estuarine environments.

16.4 Trace Metal Toxicity

Many reactive nutrient metals, such as Cu, Zn, Ni, and Cd, are toxic at high concentrations (Brand et al. 1986). For these metals, optimal growth occurs at intermediate concentrations, below which growth rate is limited, and above which growth is inhibited.

Toxic metals often are taken up into cells via the transport systems of nutrient metals such as Mn and Zn (Sunda 2000). In addition, a common mode of toxic action is interference with nutrient metal uptake and metabolism (Harrison and Morel 1983; Sunda and Huntsman 1996). As a consequence, antagonistic interactions occur between toxic and nutrient metals, which can influence algal growth. A good example is the interaction between toxic levels of Cu, Cd, or Zn and the nutrient metal Mn (Sunda and Huntsman 1996; Sunda and Huntsman 1998). In these interactions, growth rate inhibition by these metals is related to an inhibition of cellular Mn uptake and the induction of Mn-deficiency. The deficiency is alleviated either by increasing the concentration of Mn ions or by decreasing that of the toxic metal.

Large differences exist in algal sensitivity to toxic metals. In a survey of 38 clones, wide variations were observed in growth inhibition by Cu both within and between algal groups (Brand et al. 1986). Isolates of the cyanobacterium *Synechococcus* were the most sensitive to Cu, and showed a 50 % growth rate inhibition at an average free cupric ion concentration of 50 pM, a level that had little effect on the growth of diatoms, the least-sensitive algal group. Thus Cu pollution in coastal estuaries could influence the species composition of algal communities by selectively inhibiting the growth of more sensitive species. This prediction was confirmed in a study of coastal bays in Cape Cod, Massachusetts. In this study, very low *Synechococcus* cell abundances were found in bays contaminated with copper (Eel Pond and Falmouth Harbor) where free cupric ion concentrations (~100 pM) exceeded toxic levels for this genus (Moffett et al. 1997). In contrast, much higher *Synechococcus* cell concentrations were seen in adjacent bays where there was no measurable Cu contamination and free Cu ion concentrations were <0.1 pM, well below the toxic range.

Although trace metals can inhibit algal growth, algal grazers such as copepods (Sunda et al. 1987) and ciliates (Stoecker et al. 1986) often are more sensitive to toxic metals (e.g., Cu and Zn) than most phytoplankton. Consequently, heavy metal pollution may promote algal blooms by inhibiting algal grazers and thereby decreasing grazing pressure (Sunda et al. 1987). This effect was observed in a mesocosm study in Saanich Inlet, British Columbia, where the addition of 79 nM Cu resulted in a ten-fold increase in chl-*a* relative to a no-addition control owing to a preferential poisoning of algal grazers (Thomas et al. 1977).

16.5 Trace Metal Effects on HABs: Domoic Acid Production in *Pseudo-nitzschia*

Because trace metal concentrations are much higher in rivers and estuaries than in the ocean, they are much less likely to limit algal growth and influence algal species composition in these waters than in the ocean. This should also be true of HAB species whose emergence in blooms is determined by their ability to outcompete non-HAB species under prevailing conditions. Biogeochemical models indicate that carbon fixation in ~40 % of the ocean's surface is limited by an insufficient supply of iron (Moore et al. 2002). Most of these regions are oceanic upwelling systems, in which deep waters rich in major nutrients (N, P, and Si), but with insufficient iron to support maximal algal growth, are advected to the surface. These low-iron upwelling regions occur in the subarctic and equatorial Pacific and most of the Southern Ocean. They also occur in many coastal upwelling systems, as found off the west coasts of North and South America (Hutchins et al. 1998, 2002). In upwelling waters off California, high nitrate concentrations (15–35 μM) can support diatom blooms with chl-*a* values of up to 10–35 μg L^{-1}, but for this to occur the upwelling ocean water must be supplemented with iron from external sources, such as rivers or continental margin sediments (Bruland et al. 2001). Off California there is little river input during the upwelling season, and the major external iron source is from shelf sediments. Regions with broad continental shelves generally receive sufficient iron to support substantial diatom blooms, while upwelling waters in areas with narrow shelves, such as the Big Sur region, receive little additional iron, which curtails bloom development. Because of these differences in iron inputs, iron limitation in these waters is spatially and temporally quite variable (Bruland et al. 2001). A similar situation is found in other upwelling regions along the western margins of North and South America and in New Zealand where there are coastal mountain ranges and associated narrow continental shelves.

The diatom blooms in upwelling waters off the west coast of North America frequently contain toxic species of the diatom genus *Pseudo-nitzschia* (see Bates and Trainer, Chap. 7). These algae produce the neurotoxin domoic acid, which has caused poisoning in marine mammals, seabirds, and humans

(Bates et al. 1998; Scholin et al. 2000; Trainer et al. 2002). Not all species of this genus produce domoic acid, and at present 11 are confirmed to produce the toxin (Bates and Trainer, Chap. 7). But even within toxic species, domoic acid production can vary substantially with a number of environmental variables, including limitation by Si, P, and trace metals (Fe and Cu) (Bates 1998; Maldonado et al. 2002; Wells et al. 2005). The genus appears to be widespread in iron-limited oceanic upwelling systems and its growth was stimulated by iron-addition in mesoscale iron fertilization experiments in the eastern equatorial Pacific (Coale et al. 1996), the western subarctic Pacific (Tsuda et al. 2003) and the subantarctic ocean (Coale et al. 2004). Stimulation of *Pseudo-nitzschia* growth by iron addition has also been observed in shipboard bottle-incubation experiments conducted in upwelling systems off Big Sur, California (Hutchins et al. 1998) and the Juan de Fuca Eddy off Washington (Wells et al. 2005). In the latter experiment, the addition of a non-marine siderophore, ferrichrome (which strongly chelates iron and reduces its uptake by eukaryotic algae), reduced cellular iron uptake rates by >80 % and reduced *Pseudo-nitzschia* cell concentrations by 33 % after 2 days of incubation. Interestingly, the ferrichrome addition also increased the average domoic acid content of *Pseudo-nitzschia* cells by 3.4-fold and increased the total cellular domoic acid concentration in the seawater by over two-fold (Wells et al. 2005). Thus, although increased iron concentrations may stimulate blooms of *Pseudo-nitzschia*, the effect of iron on cellular domoic acid concentrations, and thus toxin transfer to higher trophic levels (including humans) is not so easily predicted.

It now appears that at least part of the variation in domoic acid production with changes in iron availability may be related to a role of this molecule in cellular iron uptake. Rue and Bruland (2001) noted that domoic acid has a chemical structure similar to that of the siderophore mugineic acid, which is produced by some terrestrial plants to complex iron and facilitate its uptake from soils. They demonstrated that domoic acid chelates both iron and copper and suggested that it may be released by *Pseudo-nitzschia* cells to either facilitate iron uptake or detoxify copper. This hypothesis has since been confirmed in both laboratory and field experiments. In culture experiments with *P. multiseries* and *P. australis*, isolated from Monterey Bay, California, iron limitation of growth increased the rate of extracellular release of domoic acid by five- to ten-fold and the addition of domoic acid increased iron uptake rates in *P. multiseries* by three-fold (Maldonado et al. 2002). Moreover, in the deck-incubation experiments described above in the Juan de Fuca Eddy, the addition of domoic acid increased iron uptake rates by the algal community and had a similar stimulatory effect on the growth of the *Pseudo-nitzschia*-dominated algal community as the addition of iron (Wells et al. 2005). Thus, by promoting iron uptake, domoic acid production could convey a competitive advantage to toxigenic *Pseudo-nitzschia* species in iron-limited oceanic and coastal upwelling systems.

Copper toxicity also increased the rate of extracellular production of domoic acid by over ten-fold in *P. multiseries* cultures, and the addition of an excess of domoic acid alleviated copper toxicity (Maldonado et al. 2002). Thus another physiological function of domoic acid may be copper detoxification. Copper toxicity could be particularly problematic in freshly upwelled seawater due to very high ratios of free Cu to Mn ions in some upwelling source waters, and to a physiological antagonism between Cu toxicity and Mn limitation in diatoms (Sunda 1988/89).

Paradoxically, while Cu toxicity promoted domoic acid production, Cu limitation of *P. australis* caused an equally large increase in intracellular and extracellular domoic acid concentrations (Wells et al. 2005). As noted earlier, copper is needed for intracellular transport of iron by a high-affinity transport system; thus copper limitation may give rise to iron limitation, and thereby trigger domoic acid production. Alternatively, as suggested by Wells et al. (2005), domoic acid may be involved in facilitating copper uptake by *Pseudo-nitzschia* by a yet-unidentified mechanism. Such an effect could be important given the high level of copper complexation by natural chelators, and the resultant low free cupric ion concentrations in surface seawater (Coale and Bruland 1990).

It is clear that iron and copper can have important direct effects on domoic acid production. However, iron may also indirectly influence cellular domoic acid concentrations through its effect on silicification in diatoms. In both laboratory and field experiments, iron limitation was shown to increase Si:N ratios in diatoms by up to three-fold (Hutchins and Bruland 1998; Takeda 1998). As a result, there is a higher depletion of silica relative to nitrate in iron-limited ocean waters, which drives these systems toward Si-limitation (Hutchins and Bruland 1998). Because domoic acid content per cell increases substantially under silica limitation in *Pseudo-nitzschia* (Bates 1998), iron limitation could indirectly increase cellular domoic acid levels by promoting Si-limitation. This effect will act in concert with the direct effects of iron limitation on domoic acid production discussed above.

16.6 Trace Metal Effects on Other HAB Species

Although iron and other trace metals are likely to influence the growth and toxin production in other HAB species, there is little firm evidence for such effects. One of the more recently proposed linkages between HABs and trace metals is that among atmospheric iron inputs to the ocean, blooms of the N_2-fixing cyanobacterium *Trichodesmium*, and toxic blooms of the dinoflagellate *Karenia brevis* along the west coast of Florida (Lenes et al. 2001; Walsh and Steidinger 2001). These authors noted that *Karenia brevis* blooms were often preceded by or co-occurred with blooms of *Trichodesmium* in stratified

outer-shelf waters during summer and fall. They argued that fixed nitrogen produced by *Trichodesmium* was subsequently used to fuel *Karenia brevis* blooms. As noted earlier, N_2-fixation requires high levels of cellular iron, which is largely supplied from aeolian deposition. During the summer, the west Florida shelf receives large inputs of Saharan dust transported by the Trade Winds from North Africa, which have been argued to fuel *Trichodesmium* blooms (Lenes et al. 2001). An average of 71 % of the atmospheric dust loadings along the south Florida coast occur during June through August (Prospero 1999), the same period when 15 of the 16 of the most intense *K. brevis* blooms were initiated in the last 50 years (Walsh and Steidinger 2001). Although these patterns are intriguing, further research is needed to verify cause-and-effect linkages among atmospheric iron deposition and blooms of *Trichodesmium* and toxic dinoflagellates in the southeastern United States, and possibly in other tropical and subtropical coastal environments.
Much of the limited and inconclusive earlier evidence for trace metal effects in HABs has been previously reviewed (Boyer and Brand 1998). Because of limited space here, we refer the reader to that review for a discussion of that material.

References

Anderson MA, Morel FMM, Guillard RRL (1978) Growth limitation of a coastal diatom by low zinc ion activity. Nature 276:70–71

Badger MR, Price GD (1994) The role of carbonic anhydrase in photosynthesis. Annu Rev Plant Physiol and Plant Mol Biol 45:369–392

Barbeau K, Rue EL, Bruland KW, Butler A (2001) Photochemical cycling of iron in the surface ocean mediated by microbial iron(III)-binding ligands. Nature 413:409–413

Bates SS (1998) Ecophysiology and metabolism of ASP toxin production. In: Anderson DM, Cembella AD, Hallegraeff GM (eds) Physiological ecology of harmful algal blooms. NATO ASI Series 41. Springer, Berlin Heidelberg New York, pp 405–426

Bates SS, Garrison DL, Horner RA (1998) Bloom dynamics and physiology of domoic-acid-producing *Pseudo-nitzschia* species. In: Anderson DM, Cembella AD, Hallegraeff GM (eds) Physiological ecology of harmful algal blooms. NATO ASI Series 41. Springer, Berlin Heidelberg New York, pp 267–292

Berman-Frank I, Cullen JT, Shaked Y, Sherrell RM, Falkowski PG (2001) Iron availability, cellular iron quotas, and nitrogen fixation in *Trichodesmium*. Limnol Oceanogr 46:1249–1260

Berman-Frank I, Zohary T, Erez J, Dubinsky Z (1994) CO_2 availability, carbonic anhydrase, and the annual dinoflagellate bloom in Lake Kinneret. Limnol Oceanogr 39:1822–1834

Boyd PW et al (2000) A mesoscale phytoplankton bloom in the polar Southern Ocean stimulated by iron fertilization. Nature 407:695–702

Boyer GL, Brand LE (1998) Trace elements and harmful algal blooms. In: Andersen DM, Cembella AD, Hallegraeff GM (eds) Physiological ecology of harmful algal blooms. NATO ASI Series 41. Springer, Berlin Heidelberg New York, pp 489–508

Boyle E, Edmond JM, Sholkovitz ER (1977) The mechanism of iron removal in estuaries. Geochim Cosmochim Acta 41:1313–1324
Brand LE, Sunda WG, Guillard RRL (1983) Limitation of marine phytoplankton reproductive rates by zinc, manganese and iron. Limnol Oceanogr 28:1182–1198
Brand LE, Sunda WG, Guillard RRL (1986) Reduction of marine phytoplankton reproduction rates by copper and cadmium. J Exp Mar Biol Ecol 96:225–250
Bruland KW (1980) Oceanographic distributions of cadmium, zinc, nickel and copper in the North Pacific. Earth Planet Sci Lett 47:176–198
Bruland KW (1989) Complexation of zinc by natural organic ligands in the central North Pacific. Limnol Oceanogr 34:267–285
Bruland KW, Rue EL, Smith GJ (2001) Iron and macronutrients in California coastal upwelling regimes: implications for diatom blooms. Limnol Oceanogr 46:1661–1674
Coale KH (1991) Effects of iron, manganese, copper, and zinc enrichments on productivity and biomass in the subarctic Pacific. Limnol Oceanogr 36:1851–1864
Coale KH, Bruland KW (1990) Spatial and temporal variation of copper complexation in the North Pacific. Deep Sea Res 47:317–336
Coale KH et al (1996) A massive phytoplankton bloom induced by an ecosystem-scale iron fertilization experiment in the equatorial Pacific Ocean. Nature 383:495–501
Coale KH et al (2004) Southern Ocean iron enrichment experiment: carbon cycling in high- and low-Si waters. Science 304:408–414
Crawford DW et al (2003) Influence of zinc and iron enrichments on phytoplankton growth in the northeastern subarctic Pacific. Limnol Oceanogr 48:1583–1600
Duce RA, Tindale NW (1991) Atmospheric transport of iron and its deposition in the ocean. Limnol Oceanogr 36:1715–1726
Evans JC, Prepas EE (1997) Relative importance of iron and molybdenum in restricting phytoplankton growth in high phosphorus saline lakes. Limnol Oceanogr 42:461–472
Falkowski PG (1997) Evolution of the nitrogen cycle and its influence on the biological sequestration of CO_2 in the ocean. Nature 387:272–275
Frank VM, Bruland KW, Hutchins DA, Brzezinski MA (2003) Iron and zinc effects on silicic acid and nitrate uptake kinetics in three high nutrient, low chlorophyll regions. Mar Ecol Prog Ser 252:15–33
Granéli E, Haraldsson C (1993) Can increased leaching of trace metals from acidified areas influence phytoplankton growth in coastal waters? Ambio 22:308–311
Granéli E, Risinger L (1994) Effects of cobalt and vitamin B12 on the growth of *Chrysochromulina polylepis* (Prymnesiophyceae). Mar Ecol Prog Ser 1994:177–183
Harrison GI, Morel FMM (1983) Antagonism between cadmium and iron in the marine diatom *Thalassiosira weissflogii*. J Phycol 19:495–507
Hutchins DA, Bruland KW (1998) Iron-limited diatom growth and Si:N uptake ratios in a coastal upwelling regime. Nature 393:561–564
Hutchins DA, DiTullio GR, Zhang Y, Bruland KW (1998) An iron limitation mosaic in the California upwelling regime. Limnol Oceanogr 43:1037–1054
Hutchins DA et al (2002) Phytoplankton iron limitation in the Humboldt Current and Peru Upwelling. Limnol Oceanogr 47:997–1011
Johnson KS, Gordon RM, Coale KH (1997) What controls dissolved iron concentrations in the world ocean? Mar Chem 57:137–161
Kustka AB, Sañudo-Wilhelmy S, Carpenter EJ, Capone D, Burns J, Sunda WG (2003) Iron requirements for dinitrogen and ammonium supported growth in cultures of *Trichodesmium* (IMS 101): comparison with nitrogen fixation rates and iron:carbon ratios of field populations. Limnol Oceanogr 48:1869–1884
LaFontaine S, Quinn JM, Nakamoto SS, Page MD, Göhre V, Moseley JL, Kropat J, Merchant S (2002) Copper dependent iron assimilation pathway in the model photosynthetic eukaryote *Chlamydomonas reinhardtii*. Eukaryot Cell 1:736–757

Lane TW, Morel FMM (2000) Regulation of carbonic anhydrase expression by zinc, cobalt, and carbon dioxide in the marine diatom *Thalassiosira weissflogii*. Plant Physiol 123:345–352
Lenes JM et al (2001) Iron fertilization and the *Trichodesmium* response on the West Florida shelf. Limnol Oceanogr 46:1261–1277
Lui X, Millero FJ (2002) The solubility of iron in seawater. Mar Chem 77:43–54
Maldonado MT, Hughes MP, Rue EL, Wells ML (2002) The effect of Fe and Cu on growth and domoic acid production by *Pseudo-nitzschia multiseries* and *Pseudo-nitzschia australis*. Limnol Oceanogr 47:515–526
Maldonado MT, Price NM (1996) Influence of N substrate on Fe requirements of marine centric diatoms. Mar Ecol Prog Ser 141:161–172
Maldonado MT, Price NM (2001) Reduction and transport of organically bound iron by *Thalassiosira oceanica*. J Phycol 37:298–310
Martin JH, Gordon RM, Fitzwater S, Broenkow WW (1989) VERTEX: phytoplankton/iron studies in the Gulf of Alaska. Deep-Sea Res 36:649–680
Mills MM, Ridame C, Davey M, La Roche J, Geider RJ (2004) Iron and phosphorus co-limit nitrogen fixation in the eastern tropical North Atlantic. Nature 429:292–294
Moffett JW, Brand LE, Croot PL, Barbeau KA (1997) Cu speciation and cyanobacterial distribution in harbors subject to anthropogenic Cu inputs. Limnol Oceanogr 42:789–799
Moore JK, Doney SC, Glover DM, Fung IY (2002) Iron cycling and nutrient-limitation patterns in surface waters of the World Ocean. Deep-Sea Res (II Top Stud Oceanogr) 49:463–507
Morel FMM, Reinfelder JR, Roberts SB, Chamberlain CP, Lee JG, Yee D (1994) Zinc and carbon co-limitation of marine phytoplankton. Nature 369:740–742
Peers G, Price NM (2004) A role for manganese in superoxide dismutases and growth of iron-deficient diatoms. Limnol Oceanogr 49:1774–1783
Peers G, Quesnet S, Price NM (2005) Copper requirements for iron acquisition and growth of coastal and oceanic diatoms. Limnol Oceanogr 50:1149–1158
Price NM, Ahner BA, Morel FMM (1994) The equatorial Pacific Ocean: grazer-controlled phytoplankton population in an iron-limited ecosystem. Limnol Oceanogr 39:520–534
Price NM, Morel FMM (1991) Colimitation of phytoplankton growth by nickel and nitrogen. Limnol Oceanogr 36:1071–1077
Prospero JM (1999) Long-term measurement of the transport of African mineral dust to the southeastern United States: implications for regional air quality. J Geophys Res 104:15917–15927
Raven JA (1988) The iron and molybdenum use efficiencies of plant growth with different energy, carbon and nitrogen sources. New Phytol 109:279–287
Raven JA (1990) Predictions of Mn and Fe use efficiencies of plant growth with different energy, carbon and nitrogen sources. New Phytol. 109:279–287
Raven JA, Evans MCW, Korb RE (1999) The role of trace metals in photosynthetic electron transport in O_2-evolving organisms. Photosyn Res 60:111–149
Rue E, Bruland K (2001) Domoic acid binds iron and copper: a possible role for the toxin produced by the marine diatom *Pseudo-nitzschia*. Mar Chem 76:127–134
Rue EL, Bruland KW (1995) Complexation of iron(III) by natural organic ligands in the Central North Pacific as determined by a new competitive ligand equilibration/adsorptive cathodic stripping voltammetric method. Mar Chem 50:117–138
Rueter JG (1983) Theoretical iron limitation of microbial N_2 fixation in the oceans. EOS 63:445
Saito MA, Moffet JW, Chisholm SW, Waterbury JB (2002) Cobalt limitation and uptake in *Prochlorococcus*. Limnol Oceanogr 47:1629–1636

Scholin CA et al (2000) Mortality of sea lions along the central California coast linked to a toxic diatom bloom. Nature 403:80–84
Stoecker DK, Sunda WG, Davis LH (1986) Effect of copper and zinc on two planktonic ciliates. Mar Biol 92:21–29
Sunda WG (1988/89) Trace metal interactions with marine phytoplankton. Biol Oceanogr 6:411–442
Sunda WG (2000) Trace metal-phytoplankton interactions in aquatic systems. In: Lovely DR (ed) Environmental microbe-metal interactions. ASM Press, Washington, DC, pp 79–107
Sunda WG (2001) Bioavailability and bioaccumulation of iron in the sea. In: Turner DR, Hunter KA (eds) The biogeochemistry of iron in seawater. Wiley, New York, pp 41–84
Sunda WG, Guillard RLL (1976) The relationship between cupric ion activity and the toxicity of copper to phytoplankton. J Mar Res 34:511–529
Sunda WG, Hanson AK (1987) Measurement of free cupric ion concentration in seawater by a ligand competition technique involving copper sorption onto C_{18} SEP-PAK cartridges. Limnol Oceanogr 32:537–551
Sunda WG, Huntsman SA (1995a) Cobalt and zinc interreplacement in marine phytoplankton: biological and geochemical implications. Limnol Oceanogr 40:1404–1417
Sunda WG, Huntsman SA (1995b) Iron uptake and growth limitation in oceanic and coastal phytoplankton. Mar Chem 50:189–206
Sunda WG, Huntsman SA (1996) Antagonisms between cadmium and zinc toxicity and manganese limitation in a coastal diatom. Limnol Oceanogr 41:373–387
Sunda WG, Huntsman SA (1997) Interrelated influence of iron, light and cell size on marine phytoplankton growth. Nature 390:389–392
Sunda WG, Huntsman SA (1998) Interactive effects of external manganese, the toxic metals copper and zinc, and light in controlling cellular manganese and growth in a coastal diatom. Limnol Oceanogr 43:1467–1475
Sunda WG, Huntsman SA (2005) The effect of CO_2 supply and demand on zinc uptake and growth limitation in a coastal diatom. Limnol Oceanogr 50:1181–1192
Sunda WG, Tester PA, Huntsman SA (1987) Effects of cupric and zinc ion activities on the survival and reproduction of marine copepods. Mar Biol 94:203–210
Takeda S (1998) Influence of iron availability on nutrient consumption ratio of diatoms in oceanic waters. Nature 393:774–777
Thomas WH, Holm-Hansen O, Siebert DLR, Azam F, Hodson R, Takahashi M (1977) Effects of copper on phytoplankton standing crop and productivity: controlled ecosystem pollution experiment. Bull Mar Sci 27:34–43
Trainer VL, Hickey BM, Homer RA (2002) Biological and physical dynamics of domoic acid production off the Washington coast. Limnol Oceanogr 47:1438–1446
Tsuda A et al (2003) A mesoscale iron enrichment in the western subarctic Pacific induces a large centric diatom bloom. Science 300:958–961
Walsh JJ, Steidinger KA (2001) Saharan dust and Florida red tides: the cyanophyte connection. J Geophys Res 106:11597–11612
Wells ML, Trick CG, Cochlan WP, Hughes MP, Trainer VL (2005) Domoic acid: the synergy of iron, copper, and the toxicity of diatoms. Limnol Oceanogr 50:1908–1917
Wu J, Sunda W, Boyle E, Karl, DM (2000) Phosphate depletion in the western North Atlantic Ocean. Science 289:759–762
Wu J, Luther III GW (1995) Complexation of Fe(III) by natural organic ligands in the Northwest Atlantic Ocean by a competitive ligand equilibration method and a kinetic approach. Mar Chem 50:159–178
Wurtsbaugh WA, Horne AJ (1983) Iron in eutrophic Clear Lake, California: its importance for algal nitrogen fixation and growth. Can J Fish Aquat Sci 40:1419–1429

17 Molecular Physiology of Toxin Production and Growth Regulation in Harmful Algae

A. CEMBELLA and U. JOHN

17.1 Introduction

Molecular physiology is a hybrid discipline whereby cellular metabolism and regulatory and functional interactions of metabolites within organisms are explored via technological tools and concepts of molecular biology. From this perspective, molecular physiology, and the corresponding field of metabolomics derived from genomics, is in a relative state of infancy as applied to marine protists (or eukaryotic microalgae). Yet recent access to genomic databases for a variety of lower eukaryotes, and targeted genome studies involving cDNA libraries offer enormous potential to unlock mysteries relating algal genomes to cellular and ecological function.

Among the more than 100 protistan taxa considered to be "harmful" and/or that produce harmful algal blooms (HABs), the phytoflagellates and especially free-living dinoflagellates are most prominent. A bloom can result from rapid proliferation of cells, i.e., high intrinsic growth rate, or the maintenance of a high cell concentration via behavioral adaptations (swimming, sinking), successful resource competition, life-history transitions or allelochemical interactions. These processes are not mutually exclusive, but all have a cell regulatory dimension reflecting gene function. Toxic microalgae, particularly bloom-forming taxa, are fruitful subjects for molecular physiological investigations. For example, compared to many co-existing diatoms or other flagellates, large toxic dinoflagellates tend to be rather slow-growing (typically $\mu<0.5\ d^{-1}$). Therefore, to understand how such species are capable of bloom formation and in some instances to dominate the phytoplankton, it is necessary to obtain molecular physiological insights into regulation of both growth and toxin synthesis.

Marine protists produce a bewildering array of secondary metabolites, and many of these compounds possess high biological activity. Among these bioactives, the phycotoxins feature prominently, but it would be erroneous to

Ecological Studies, Vol. 189
Edna Granéli and Jefferson T. Turner (Eds.)
Ecology of Harmful Algae

conclude that toxicity necessarily reflects their ecological and evolutionary significance. Phycotoxins are most often synthesized by marine phytoflagellates (and rarely by pennate diatoms); more than 80 % of the eukaryotic algal taxa that produce defined phycotoxins belong to the dinoflagellates (Cembella 2003). These bioactive secondary metabolites may play diverse roles (currently unknown) in intracellular regulation of cell growth and metabolism as well as in extracellular regulation of population growth via allelochemical interactions. The molecular physiological approach facilitates the generation of testable hypotheses regarding ecological niche differentiation defined by intrinsic as opposed to extrinsic factors (e.g., nutrient acquisition, turbulence, etc.).

Certain cyanobacteria (colloquially "blue-green algae") share with the dinoflagellates the capability of synthesizing saxitoxin and other toxic tetrahydropurines. The cyanobacteria also produce hepatotoxic cyclic peptides (microcystins, nodularins) and the potent neurotoxic anatoxins, which are included as "phycotoxins," but these are dealt with here only by analogy. Molecular genetics of cyclic peptide production and regulation in cyanobacteria are well advanced (comprehensively reviewed by Börner and Dittmann 2005). For example, a biosynthetic operon (*myc* gene) for microcystins has been sequenced from a number of cyanobacterial genera. Homologies and differences in molecular organization for such toxin genes among the cyanobacteria have been interpreted as an indication of a common ancestor. Subject to many caveats, such advanced work on cyanobacteria provides guidance for corresponding work on toxin genetics in eukaryotic algae with larger and less tractable genomes.

Complex transcriptional changes and genetic regulation of growth and bioactive secondary metabolite production occur via biotic interactions, life history transitions, and the mitotic cycle. Such gene expression and the consequent ecological responses can be described and quantified using tools of modern molecular biology. Because we have limited knowledge about the molecular physiology and gene regulation of harmful algae, we have biased this review to consider only the few highlights of recent findings, supplemented with emerging methodologies that are currently being applied or will be used in the near future.

17.2 Phycotoxin Biosynthesis

From a molecular physiological and chemical ecological perspective, toxin classification based upon structural homology presumably reflects shared elements of biosynthetic pathways (Wright and Cembella 1998). Biosynthetic evidence from stable isotope–labeling studies indicates that most if not all polyether phycotoxins are produced via polyketide pathways (Shimizu 1996, cited in Cembella 1998), in which acetate units are added sequentially from acetyl-CoA within a pathway regulated by polyketide synthases (PKS). The classical approach is to provide an isotopic enrichment of low-molecular-weight putative precursors (e.g., ^{13}C-acetate, ^{15}N-glycine) and then to follow the incorporation pattern into the target phycotoxin by NMR spectroscopy. This approach has provided structural elucidation and plausible biosynthetic schemes for key phycotoxins, including saxitoxin and analogues from amino acid precursors in *Alexandrium tamarense* and the cyanobacterium *Aphanizomenon flos-aquae* (Shimizu 2000, cited in Cembella 2003), brevetoxins from *Karenia brevis* (Shimizu 1996, cited in Cembella 1998), sulphated dinophysistoxins from *Prorocentrum maculosum* (Macpherson et al. 2003), and domoic acid from *Pseudo-nitzschia pungens* (Douglas et al. 1992, cited in Bates 1998).

In polyketide synthesis of the sulphate esters DTX5a and DTX5b, the amino acid in the sulfated side-chain originates from glycine, whereas oxygen insertion occurs after polyketide formation (Macpherson et al. 2003). Biosynthetic studies on *Alexandrium ostenfeldii* supplemented with [1,2-$^{13}C_2$]-acetate and [1,2-$^{13}C_2$, ^{15}N]-glycine demonstrated by NMR spectroscopy that 13-desmethyl spirolide C is also polyketide-derived (Cembella et al. 2004). A biosynthetic scheme for pinnatoxins, presumed to be of dinoflagellate origin as are other spiroimine toxins (Wright and Cembella 1998), proposes derivation from a single polyketide chain and bears resemblance to the spirolide pattern.

Molecular ecological studies on ichthyotoxins and/or haemolysins produced by eukaryotic microalgae, particularly by prymnesiophytes and raphidophytes, but also including some dinoflagellates, are hampered by poorly described links between toxicity and chemical structures. For example, Yasumoto et al. (1990, cited in Eschbach 2005) attributed haemolysis and ichthyotoxicity of the prymnesiophyte *Chrysochromulina polylepis* to galactolipids, 1-acyl-3-digalacto-glycerol and polyunsaturated fatty acids (PUFAs). Later evidence indicting no differences in lipid and PUFA profiles between toxic versus non-toxic strains has challenged this assumption (John et al. 2002, cited in Cembella 2003). Where toxin biosynthetic pathways and structures are unknown, bioassays may be used to target genetics of toxicity expression through the cell cycle and under varying environmental conditions.

17.3 Growth and Regulation of Toxin Production

Cellular growth rate of marine microalgae is a function of extrinsic abiotic factors – salinity, temperature, turbulence, light, nutrient concentrations, ratios and supply rates, and accumulation of allelochemical and autoinhibitory metabolites. Growth rate is typically a hyperbolic function of the limiting resource, with the rate increasing with increasing available resource up to saturation. A diminution of growth rate accompanies an increase in resource limitation or stress. Exogenous environmental variables can be altered under controlled conditions to yield differential growth rates and hence insights into growth-rate dependency of gene expression. The maximum growth rate for each strain or species is also regulated by intrinsic factors, e.g., synthesis and activation of bioenergetic metabolites (ATP, NADPH), nucleic acids, and cell cycle regulators.

Many biologically active secondary metabolites, such as antibiotics produced by bacteria and fungi, are subject to induction effects, or genetic "up-regulation," when cells are exposed to unfavorable environmental conditions. These metabolites are often presumed to function in cell defense; thus it makes evolutionary sense and is metabolically efficient to reserve genetic expression for when they are critically required.

With respect to effects of abiotic factors, the relationship between growth and toxin production has been extensively reviewed for the dinoflagellates that produce saxitoxin and analogues (Cembella 1998), polyether toxins (Wright and Cembella 1998), and diatoms that synthesize domoic acid (Bates 1998). Perhaps paradoxically, there are few cases of stress-dependent triggers in toxigenic eukaryotic microalgae, even for dinoflagellate polyethers derived from polyketide pathways that are analogous if not homologous to those in bacteria and fungi.

The most notable exceptional example of apparent stress metabolism is the induction of domoic acid synthesis in *Pseudo-nitzschia* spp. under phosphate- and silicate-limitation (Bates 1998). The gradual diminution and eventual loss of toxin production in "older" mitotically reproducing strains without sexual recombination in this toxic diatom (reviewed in Pan et al. 1998; Bates 1998) is not usually apparent in toxigenic strains of dinoflagellates, which are nominally haploid in the vegetative stage. Sexual recombination is not essential for maintenance of toxicity in flagellates (at least over several years), although loss of toxicity perhaps through random mutation and clonal selection in culture may rarely occur.

It is not reasonable to conclude that nutrient limitation "induces" toxin production, based only upon evidence of high cell toxin quota in growth-limited nutrient–deficient cells. An increase in toxin cell quota may result merely from a decrease in the *rate* of cell division relative to the *rate* of toxin synthesis (Cembella 1998). This would account for high cell toxicity in *Chrysochromulina polylepis* (Edvardsen et al. 1996, cited in Eschbach 2005) in stationary growth phase under P-limited conditions. Under non-limiting nutrient conditions, for dinoflagellates (Cembella 1998) and other phytoflagellates, e.g., *C. polylepis* (Eschbach et al. 2005), toxin content is typically highest in mid-exponential phase, when the growth rate is maximal.

17.4 Toxin Production Through the Cell Cycle

The eukaryotic cell cycle, divided into M (mitosis), G1 (Gap 1), S (DNA synthesis), and G2 (Gap 2) phases, is universal for microalgae, even for the dinoflagellates (Taroncher-Oldenburg et al. 1997; Pan and Cembella 1998, both cited in Eschbach et al. 2005), which have a unique mitosis. Reduction in growth rate is correlated to changes in the duration of the cell cycle, and is attributable to expansion of a single cell-cycle phase (G1).

Circadian rhythms among the phototrophic microalgae, e.g., in cell division, nutrient assimilation, bioluminescence, toxin production, onset of sexual reproduction, and vertical migration, are linked to both nutrients and light regime. The cell-division cycles are therefore phased or synchronized to photocycles. Mitotic division in the dark is typical, but there are exceptions, e.g., shade-adapted dinoflagellate species such as *Prorocentrum lima* (Pan and Cembella 1998, cited in Eschbach et al. 2005) and *Amphidinium operculatum* (Leighfield and Van Dolah 1999) may divide in the light. In the latter species, it is the time from dark-to-light transition that dictates occurrence of M-phase, not whether it is in the dark or light period.

Induction and regulation of toxin production is best studied by analysis of synchronized cells because biosynthesis then occurs essentially at the same time-point. Entrainment of microalgal cells towards synchronous division is often accomplished by the block/release method via deprivation of light and/or regulation of nutrient concentration and supply rate (Taroncher-Oldenburg et al. 1997; Pan et al. 1999, both cited in Eschbach et al. 2005), or use of metabolic inhibitors (Van Dolah et al. 1998, cited in Eschbach et al. 2005). Cells arrested in G1 phase (then termed G0 phase), when released from the block, proceed through the cell cycle as a homogeneous population. Alternatively, cells can be synchronized by exposure to a tightly defined light/dark cycle (Eschbach et al. 2005). In synchronized eukaryotic microalgae, toxin production is generally discontinuous, restricted to a defined period of the cell cycle, and is light-dependent, although not always confined to the light phase (Taroncher-Oldenburg et al. 1997; Pan et al. 1999, both cited in Eschbach et al. 2005).

Discriminating between toxin expression events that are truly cell-cycle dependent (e.g., linked to G1) and those which are merely light-dependent (entrained by photoperiod) can be very challenging, particularly when complete synchrony in cell division is not maintained. In *Alexandrium fundyense*, saxitoxin analogues are produced only during a restricted period in G1, with slight time shifts in synthesis of different derivatives (Taroncher-Oldenburg et al. 1997, cited in Eschbach et al. 2005). In *Prorocentrum lima*, DTX4 production is initiated early in the light period in G1 whereas other derivatives are synthesized during S and G2 phases (Pan et al. 1999, cited in Eschbach et al. 2005). Production of spirolides in *A. ostenfeldii*, occurs primarily during dark in G2+M phases (John et al. 2001, cited in Cembella 2003). These differences suggest that not even polyketide toxins share a universal function in regulation and expression in the cell division cycle.

17.5 Molecular Approaches to Growth and Toxin Expression

Current data are insufficient to resolve the genetics of cell-cycle regulation in eukaryotic microalgae, but homologues of cell-cycle regulators have been found. Gene sequence data are available from the chlorophyte *Dunaliella tertiolecta* (Lin and Carpenter 1999) and cyclin B homologues have been discovered from genome projects on the apicomplexan *Cryptosporidium parvum* and the prasinophyte *Ostreococcus tauri*. Universal cell-cycle regulators, cyclin-dependent kinases (CDKs) and cyclin B-like proteins, have been reported from the dinoflagellate cell cycle (Van Dolah et al. 1998 cited in Eschbach et al. 2005, Barbier et al. 2003). A mitotic cyclin from the (often toxigenic) *Lingulodinium polyedrum* was isolated by functional complementation in yeast, then cloned and sequenced (Bertomeu and Morse 2004); expression levels through the cell cycle indicated a typical eukaryotic pattern.

Other limited genomic approaches to gene expression in toxigenic flagellates, e.g., expressed-sequence-tag (EST) analyses of *A. tamarense*, *A. ostenfeldii*, *Karenia brevis*, *P. lima*, and *Chrysochromulina polylepis* have not revealed homologues of CDKs or cyclins. However, 18 ESTs for *C. polylepis* (John et al. 2004a), 6 ESTs for *A. ostenfeldii* (Cembella et al. 2004), and 16 ESTs from *K. brevis* (Lidie et al. 2005) are attributable to orthologous genes for cell division and DNA replication.

The molecular expression of the proliferating-cell-nuclear-antigen (PCNA) gene has been described from several microalgal species (Lin and Corstjens 2002, and references therein). This auxiliary protein for DNA polymerase is essential for nuclear DNA synthesis and thus can serve as a cell-cycle marker for S-phase.

Molecular techniques for toxic algae allow for the detection of genes that are "turned on" during toxin production, and not merely on the detection of the toxins themselves (Plumley 1997). Modified versions of subtractive-hybridization libraries, DNA microarrays, EST libraries, and real-time polymerase chain reaction (QPCR) permit identification of differentially expressed genes in organisms grown under different conditions. These methods are currently being applied to toxic eukaryotic microalgae, for example, to investigate expression differences between toxic versus non-toxic strains (John et al. 2004a).

There are only a few defined examples of genetic regulation of toxin production in marine protists. Differential display (DD) of mRNA was used to identify differential gene expression of synchronized batch cultures of *Alexandrium fundyense* during toxin production at time-points in G1 phase versus time-points in G1, S and G2 + M phases where toxin production did not occur (Taroncher-Oldenburg and Anderson 2000). Three differentially expressed genes were identified – for S-adenosyl-homocysteine hydrolase, methionine aminopeptidase, and a histone-like protein – but none were clearly related to PSP toxin synthesis or regulation.

The limited genomic approach to gene expression, involving creation of a cDNA library for coding genes, followed by EST sequencing has proven to be extremely fruitful as a complement or alterative to whole genome sequencing. This is particularly true for dinoflagellates where the genome size can exceed 200 gbp; in contrast, the entire human genome is about 3.2 gbp! The genome size and technical constraints on sequencing dinoflagellates, perhaps the richest source of "toxin genes" among lower eukaryotes, restricts availability of whole genomic information.

Since many (perhaps all) polyether phycotoxins are derived via polyketide metabolism regulated by PKS, this enzyme complex merits particular attention in gene expression studies. Given the complex structures of the polyethers produced by eukaryotic microalgae (e.g., maitotoxin congeners originating from the dinoflagellate *Gambierdiscus toxicus* are >3,000 Da and contain 32 transfused ether rings), Type I modular PKS synthases are the most likely candidates for polyether synthesis (see models for structure and function of modular PKS in Börner and Dittmann 2005). In *Cryptosporidium parvum*, a close protistan relative of the dinoflagellates, a large modular PKS gene has been identified, with up to 40 kbp open reading frames (ORFs), and containing no introns in genomic DNA (Zhu et al. 2002). Putative PKS genes have been reported from the dinoflagellates *Gymnodinium catenatum*, *Amphidinium operculum*, *Prorocentrum lima*, *Karenia brevis* (Snyder et al. 2005, and references therein) and *Alexandrium ostenfeldii* (Cembella et al. 2004), and from the prymnesiophyte *C. polylepis* (John et al. 2004a). At least for *K. brevis*, the PKS genes have been confirmed to be of dinoflagellate as opposed to bacterial origin (Snyder et al. 2005). Nevertheless, no full-length PKS genes have been fully characterized and validated with published sequence data from toxic flagellates, and thus no definitive attribution can be made regarding their role in biosynthesis of polyketide toxins. We cannot rule out that an entirely new class of PKS may be involved in biosynthesis of the highly unusual polycyclic polyketides in dinoflagellates.

In *A. ostenfeldii*, a normalized cDNA library was used to generate ESTs (approximately 5,300), representing 2,400 unique sequences (updated from Cembella et al. 2004). Only 9% of the total sequences were homologues to known genes. Nevertheless, annotation revealed at least three putative PKS genes with sequences consistent with Type 1 modular PKS. Putative modular PKS genes also correlate in their expression profile with toxicity in synchronized *C. polylepis* (John et al. 2004a).

The low degree of sequence homology in *A. ostenfeldii* and perhaps other dinoflagellates presents a challenge for determining the expression and regulatory function of biosynthetic genes for dinoflagellate secondary metabolites. Developments in understanding the structure and function of operons for PKS and NRPS (non-ribosomal protein synthase) in toxic cyanobacteria (Börner and Dittmann 2005, and references therein) indicate that it may be difficult to identify "toxin genes" for polyketides in eukaryotic flagellates, even with high overall sequence homology.

Proteome analysis can elucidate which proteins are expressed in different growth and cell division phases under various growth conditions, during toxin synthesis and allelochemical interactions. Further information is provided on how these proteins and processes are post-translationally modified and how variation is expressed due to epigenetic phenomena.

Application of proteomics to harmful algae is thus far very limited and has been primarily targeted towards "molecular fingerprinting" of bloom populations. Recent characterization of microalgal populations, including *Prorocentrum* and other dinoflagellates (Chan et al. 2004) involved 2D-GE profiles for species identification and as potential protein biomarkers for bloom description and prediction. Existing protein databases have enough sequence data that a few distinctive proteins for HAB species can be identified from 2D-GE and/or peptide profiles, but the genomic databases necessary for protein profile screening of key toxic species are lacking.

Cryptic genetic diversity may be an important feature in the selection and viability of particular strains responsible for the formation of natural bloom populations. Molecular methods, such as rDNA sequencing (Scholin et al. 1994), amplified-fragment-length polymorphism (AFLP) (John et al. 2004b) and micro-satellite analysis (Evans et al. 2005), are being used for genotyping toxic microalgae, to differentiate genetically distinct sub-populations within species. Determination of genetic heterogeneity is an essential preliminary step in establishing the expected degree of variability in gene expression among populations and species.

In the *Alexandrium tamarense* species complex, the best-studied group of toxic dinoflagellates, the global expression of toxin phenotype is diverse, characterized by the production of more than two dozen saxitoxin analogues. The toxin profile of *Alexandrium* strains is relatively stable under favorable culture conditions but can differ markedly among strains. Unknown allelopathic compounds with biological activity against a wide range of planktonic organisms are also produced by members of the *A. tamarense* complex and by *A. ostenfeldii*, but these appear to be unrelated to the potent saxitoxin analogues or spirolides, respectively, produced by the dinoflagellates (Tillmann and John 2002, cited in Cembella 2003).

Toxin expression studies on *A. tamarense* populations from eastern Canada (Cembella and Destombe 1996, cited in Cembella 1998) and the Scottish east coast (Alpermann et al. 2004) indicated that multiple isolates (clones) from the same geographical area can be highly heterogeneous in their toxin profiles. Analysis of other phenotypic properties, including the expression of fatty acid profiles, has also revealed a high level of cryptic genetic diversity within populations. Studies of genetic structure based upon genetic markers such as AFLP analysis for the *Alexandrium tamarense* species complex (John et al. 2004b) and microsatellite DNA from *Pseudo-nitzschia pungens* from the North Sea (Evans et al. 2005) indicated high levels of genetic diversity, but low genetic and morphological differentiation. Preliminary genetic analysis of North Sea populations of *A. tamarense* using both microsatellites and AFLP as genetic markers, and toxin profiles as phenotypic markers, demonstrated similarly high genetic variation among isolates from discrete populations, but no correlation of genotypic markers with toxin expression patterns (Alpermann et al. 2004). A correlation between genetic markers and toxin expression or growth rate is not necessarily expected, but such studies indicate that dependence upon single or few clones in autecological investigations may lead to erroneous conclusions regarding regulation of growth and toxin biosynthesis.

17.6 Current and Future Perspectives

The way forward in molecular ecological studies on growth regulation and toxin production of harmful algae is to unite classical approaches to structural elucidation and metabolic pathway determination with genomic and proteomic technologies for gene expression. Analytical and physical-chemical methods such as tandem mass-spectrometry with liquid-chromatographic separation (LC-MS/MS) and NMR can resolve structures and provide basic biosynthetic pathways to phycotoxins, but provide little direct evidence on specific biosynthetic enzymes and regulatory functions. A combined approach linked to genomics and proteomics is required to fully describe toxin biosynthesis and regulation.

Many analytical technologies (e.g., LC-MS) have already been applied on board ship for field measurements and toxin profiling in near real-time in natural populations (Scholin et al. 2005). Furthermore, species-level taxonomic probes are available for field use even on deployable buoy systems in automated mode. As hypotheses or laboratory techniques are developed, these must be rapidly applied to field populations to evaluate the predictability, variability and role of environmental cues on phenotypic expression. Traditional autecological approaches to studies of gene expression, growth and toxin production have often relied on individual geographically "representative" clones of uncertain provenance that are maintained in long-term culture on artificially enriched seawater. As a first step, a large number of isolates from various geographical populations must be cultured under rigidly defined conditions and subjected to genotyping (DNA probing) and phenotyping (toxin characterization).

Total genome sequencing is desirable for gene-expression studies. The rapid advent of novel genomic technologies and associated databases for a host of prokaryotes and eukaryotes offers straightforward alternatives for sensitive and stringent gene screening. Whole-genome sequencing of a few eukaryotic microalgae (*Thalassiosira pseudonana*, *Phaeodactylum tricornutum*, *Ostreococcus tauri*) has been completed, is in progress (*Emiliana huxleyi*) or will soon be underway (e.g., toxigenic *Pseudo-nitzschia multiseries*). Nevertheless, no complete sequences are yet available for toxic eukaryotic algal species.

The DNA microarray approach is also a promising strategy to screen for genes that are induced or repressed as a result of intra- and inter-specific interactions and response to other environmental triggers in marine protists (Taroncher-Oldenburg et al. 2003; Lidie et al. 2005). DNA microarrays have been used to identify differentially expressed genes during toxin synthesis in the fungus *Aspergillus* (O'Brian et al. 2003). The basis for microarray development is availability of a substantial sequence data-base for a given species. From sequence data, gene-specific oligonucleotides can be designed or cDNA fragments can be generated using PCR. DNA microarrays loaded with many putative genes for cell cycle regulation and toxin synthesis can then screen for expression of toxin-specific genes in toxic versus non-toxic strains within a species or across a wide spectrum of taxa from field populations and laboratory cultures.

For full-length sequences of large gene candidates, such as modular PKS genes, as well as information on the 5' and 3' untranslated regions (UTRs), a cosmid library, or a bacterial artificial chromosome (BAC) library, must be constructed. Cosmids are useful for cloning large segments of foreign DNA (up to 50 kbp). For large genomes and for large target genes, BAC libraries of vectors containing inserts of high molecular-weight DNA from 75–150 kbp can be produced. Given the huge genome size of dinoflagellates, 7.5×10^6 clones may be required to achieve reasonable coverage of the genome with a mean insert size of ~150 kbp.

Identification of genes or primary gene products (enzymes) involved in cell growth and/or toxin synthesis is facilitated by comparison with strains of the same taxon, which lack the capability to express the particular trait. For some toxigenic species, e.g., *A. tamarense*, non-toxic analogues of the morphospecies are available. Mutations induced in a toxic strain by exposure to UV radiation or chemical mutagens also provide an alternative for comparison. Chemical mutagens can induce point mutations resulting in base mispairing and substitution or even deletions. Small deletions are desirable because they are more likely to abolish gene function but also to affect only a single gene, thus avoiding generation of complex phenotypes.

Proteomic approaches involving large-scale systematic analysis of proteins of harmful algae and their post-translational modifications can be accomplished by combining two-dimensional gel electrophoresis (2D-GE) and protein microanalysis with bioinformatics. Two-dimensional difference gel electrophoresis (2D-DIGE) allows co-separation of different CyDye DIGE fluor-labelled protein samples from toxic and non-toxic strains on the same gel. Protein spots of interest can be selected and excised from the gels for further characterization by matrix-assisted laser-desorption ionization time-of-flight mass spectrometry (MALDI-TOF MS) with post-source decay, or by Edman sequencing. After accurate determination of peptide masses, databases and EST libraries of investigated taxa must be searched for identification of original proteins.

Once cell-based molecular diagnostics (microarray analyses, Northern blots, QPCR) have been established, the next stage is to explore functional genetics of allelochemistry and toxin expression on target organisms, such as predators and putative competitors of toxic microalgae. The hypothesis that these bioactive substances may serve as quorum-sensing compounds should be evaluated with respect to the formation, maintenance and senescence of HABs of eukaryotes and cyanobacteria.

Cyanobacteria offer an advanced model for application of molecular techniques for toxin gene detection (cited in Börner and Dittmann 2005). Unfortunately, for eukaryotic microalgae, no definitive candidates for toxin genes are characterized, and transfection methods and expression systems are described for only a few non-toxic eukaryotes (such as *Ostreococcus* and *Phaeodactylum*). For cyanobacteria, a QPCR approach that is specific for the *myc* gene has been used to evaluate the relative contribution of toxic and non-toxic *Microcystis* in natural blooms, as well as to identify the microcystin producers among several potentially toxic species. Biosynthetic promoters for microcystins in cyanobacteria have also been fused to *lux* genes to evaluate environmental effects on toxin production. The cyanobacterial literature suggests that molecular tools should be taxon- rather than toxin-specific, but this model cannot be simply extrapolated to the eukaryotes. Genome organization in cyanobacteria and fungi is quite different from that of eukaryotic microalgae (especially dinoflagellates). Toxic cyclic peptides in cyanobacteria are not derived from pathways similar to those of the dinoflagellate polyketides. The vast diversity of eukaryotic toxins would defy diagnosis by a single genetic marker. Nevertheless, phylogenetic analysis suggests that focus on polyketide toxins is not unwarranted.

Further insight into the role of allelochemicals in the "watery arms race" sensu Smetacek is required to understand the capability of these species to form persistent, recurrent, dense, and on occasion nearly monospecific blooms. Knowledge of the cues and mechanisms that induce, regulate, and/or stimulate production and exudation of specific toxins, and yet-undefined extracellular lytic compounds, will significantly contribute to understanding the ecological significance of these allelochemicals in dynamics of key species and consequent effects on marine trophic webs.

References

Alpermann T, John U, Tillmann U, Evans K, Cembella A (2004) Genotypic and phenotypic variability in allelochemical potential within populations of *Alexandrium tamarense*. XI Inter Conf on Harmful Algal Blooms, Cape Town, South Africa (Abstract), p 57

Barbier M, Leighfield TA, Soyer-Gobillard M-O, Van Dolah FM (2003) Expression of a cyclin B homologue in the cell cycle of the dinoflagellate, *Karenia brevis*. J Eukaryot Microbiol 50:123–131

Bates SS (1998) Ecophysiology and metabolism of ASP toxin production. In: Anderson DM, Cembella AD, Hallegraeff GM (eds) Physiological ecology of harmful algal blooms. NATO ASI Series 41. Springer, Berlin Heidelberg New York, pp 405–426

Bertomeu T, Morse D (2004) Isolation of a dinoflagellate mitotic cyclin by functional complementation in yeast. Biochem Biophys Res Comm 323:1172–1183

Börner T, Dittmann E (2005) Molecular biology of cyanobacterial toxins. In: Huisman J, Matthijs HCP, Visser PM (eds) Harmful cyanobacteria. Springer, Berlin Heidelberg New York, pp 25–40

Cembella AD (1998) Ecophysiology and metabolism of paralytic shellfish toxins in marine microalgae. In: Anderson DM, Cembella AD, Hallegraeff GM (eds) Physiological ecology of harmful algal blooms. NATO ASI Series 41. Springer, Berlin Heidelberg New York, pp 381–403

Cembella AD (2003) Chemical ecology of eukaryotic microalgae in marine ecosystems. Phycologia 42:420–447

Cembella AD, John U, Singh R, Walter J, MacKinnon S (2004) Molecular genetic analysis of spirolide biosynthesis in the toxigenic dinoflagellate *Alexandrium ostenfeldii*. XI Inter Conf on Harmful Algal Blooms, Cape Town, South Africa (Abstract), p 82

Chan LL, Hodgkiss IJ, Lu S, Lo SC-L (2004) Use of two-dimensional gel electrophoresis proteome reference maps of dinoflagellates for species recognition of causative agents of harmful algal blooms. Proteomics 4:180–192

Eschbach E, John U, Reckermann M, Cembella AD, Edvardsen B, Medlin LK (2005). Cell cycle dependent expression of toxicity by the ichthyotoxic prymnesiophyte *Chrysochromulina polylepis*. Aquat Microb Ecol 39:85–95

Evans KM, Kühn SF, Hayes PK (2005) High levels of genetic diversity and low levels of genetic differentiation in North Sea *Pseudo-nitzschia pungens* (Bacillariophyceae) populations. J Phycol 41:506–514

John U, Groben R, Beszteri B, Medlin L (2004b) Utility of amplified fragment length polymorphisms (AFLP) to analyse genetic structures within the *Alexandrium tamarense* species complex. Protist 155:169–179

John U, Tillmann U, Cembella A, Medlin L (2004a) Expression of polyketide synthases (PKS) during induction of toxicity in the ichthyotoxic prymnesiophyte *Chrysochromulina polylepis*. XI Inter Conf on Harmful Algal Blooms, Cape Town, South Africa (Abstract), p 147

Leighfield TA, Van Dolah FM (1999) Cell cycle regulation in a dinoflagellate, *Amphidinium operculatum*: identification of the diel entraining cue and a possible role of cyclic AMP. J Exp Mar Biol Ecol 262:177–197

Lidie KB, Ryan JC, Barbier M, Van Dolah, FM (2005) Gene expression in Florida red tide dinoflagellate *Karenia brevis*: analysis of an expressed sequence tag library and development of DNA microarray. Mar Biotechnol Oct 2005, p 46

Lin S, Carpenter EJ (1999) A PSTTLRE-form of *cdc2*-like gene in the marine microalga *Dunaliella tertiolecta*. Gene 239:39–48

Lin S, Corstjens PLAM (2002) Molecular cloning and expression of the proliferating cell nuclear antigen from the coccolithophorid *Pleurochrysis carterae* (Haptophyceae). J Phycol 38:164–173

Macpherson GR, Burton IW, LeBlanc P, Walter JA, Wright JLC (2003) Studies of the biosynthesis of DTX-5a and DTX-5b by the dinoflagellate *Prorocentrum maculosum*: regiospecificity of the putative Baeyer-Villigerase and insertion of a single amino acid in a polyketide chain. J Org Chem 68:1659–1664

O'Brian GR, Fakhoury AM, Payne GA (2003) Identification of genes differentially expressed during aflatoxin biosynthesis in *Aspergillus flavus* and *Aspergillus parasiticus*. Fungal Genet Biol 39:118–127

Pan Y, Bates S, Cembella AD (1998) Environmental stress and domoic acid production by *Pseudo-nitzschia*: a physiological perspective. Nat Tox 6:127–135

Plumley FJ (1997) Marine algal toxins: biochemistry, genetics and molecular biology. Limnol Oceanogr 42:1252–1264

Scholin CA, Doucette GJ, Cembella AD (2006) Prospects for developing automated systems for in situ detection of harmful algae and their toxins. In: Babin M, Roesler C, Cullen J (eds) HABwatch: monographs on oceanographic methodology. UNESCO, Paris, (in press)

Scholin CA, Herzog M, Sogin M, Anderson DM (1994) Identification of group-and strain-specific genetic markers for globally distributed *Alexandrium* (Dinophyceae). 2. Sequence analysis of a fragment of the LSU rRNA gene. J Phycol 30:999–1011

Snyder VR, Guerrero MA, Sinigalliano CD, Winshell J, Perez R, Lopez JV, Rein KS (2005) Localization of polyketide synthase encoding genes to the toxic dinoflagellate *Karenia brevis*. Phytochemistry 66:1767–1780

Taroncher-Oldenburg G, Anderson DM (2000) Identification and characterization of three differentially expressed genes, encoding S-adenosylhomocysteine hydrolase, methionine aminopeptidase, and a histone-like protein, in the toxic dinoflagellate *Alexandrium fundyense*. Appl Environ Microbiol 66:2105–2112

Taroncher-Oldenburg G, Griner EM, Francis CA, Ward BB (2003) Oligonucleotide microarray for the study of functional gene diversity in the nitrogen cycle of the environment. Appl Environ Microbiol 69:1159–1171

Wright JLC, Cembella AD (1998) Ecophysiology and biosynthesis of polyether marine biotoxins. In: Anderson DM, Cembella AD, Hallegraeff GM (eds) Physiological ecology of harmful algal blooms. NATO ASI Series 41. Springer, Berlin Heidelberg New York, pp 427–451

Zhu G, LaGier MJ, Stejskal F, Millership JJ, Cai X, Keithly JS (2002) *Cryptosporidium parvum*: the first protist known to encode a putative polyketide synthase. Gene 298:79–89

18 Chemical and Physical Factors Influencing Toxin Content

E. GRANÉLI and K. FLYNN

18.1 Introduction

Many isolates of the same species display qualitatively similar toxin profiles supporting the view that toxin production is heritable, at least for PSP toxin producers (see Cembella and John, Chap. 17). It is not known, however, why some phytoplankton species produce toxins while most do not, or perhaps more perplexing, why some strains of the same species are toxic while others are not. One may expect that if it were advantageous, all strains of the same species would display similar levels of toxicity and they may also be expected to be toxic most of the time. It is even possible that at least some algae do not produce toxins themselves, but that toxin production is caused by intracellular bacteria (Kodama 1990, cited in Granéli et al. 1998, see also Chap. 19). Toxin production can also be stimulated by the presence of grazers (Pohnert et al. 2002).

Algal toxins are secondary metabolites that vary in structure, atomic composition, and functional activity. Thus, it is not surprising that factors stimulating toxin production in one algal species/group may have a different impact on another. This chapter considers the factors affecting the accumulation of toxins. In Chap. 17, Cembella and John consider this topic at the molecular and genetic level, and phase of cell cycle, while here we focus on interactions between the growth stage (lag, exponential and stationary phases) and physico-chemical environmental factors.

18.2 Growth Stage and Toxin Production

In general, changes in toxin content are associated with disturbed (unbalanced) physiology, with the up-shock or down-shock of cells exiting or

Ecological Studies, Vol. 189
Edna Granéli and Jefferson T. Turner (Eds.)
Ecology of Harmful Algae

entering stationary phases (Anderson 1994; Flynn et al. 1994, 1996; Johansson and Granéli 1999a, 1999b; Granéli et al. 1998; Edvardsen et al. 1990, cited in Granéli et al. 1998). Toxin production in normal experiments can be difficult to measure against a background of changing cell biomass, cell size and indeed toxin leakage or catabolism. Thus changes in toxin content per cell may simply reflect changes in cell size accompanying changes in nutrient status, temperature, or salinity. For this reason, here we have usually referred to changes in toxin content, rather than specifically to changes in production. Nonetheless, depending on the chemical composition of the toxin, different treatments may be expected to have different impacts. Thus N-rich PSP-toxins are synthesized during N-upshock and P-stress (which gives relatively high N-status) and not during N-downshock (Flynn et al. 1994; Granéli et al. 1998).

Table 18.1 illustrates the importance of up/down shock events in affecting toxicity. Thus, domoic acid (DA) has been shown to be produced primarily in stationary phase, with little or none in exponential phase, in several strains of *Pseudo-nitzschia multiseries* (Pan et al. 1996a, 1996 c, cited in Bates 1998). If toxin-containing cells from the late stationary phase are re-inoculated into fresh, nutrient-replete medium, cellular DA levels decrease as cells re-enter exponential phase (Douglas et al. 1993, cited in Bates 1998); the existing toxin produced during the stationary phase becomes more and more diluted by division of the exponentially growing cells. PSP content in the freshwater cyanobacterium *Anabaena circinalis* is highest in stationary phase (Negri et al. 1997) similar to the pattern of *P. multiseries* DA toxin production. The dinoflagellate *Prorocentrum minimum* is toxic only sporadically, but a higher toxicity always occurs when the cells are in the declining phase of the growth cycle (Wikfors 2005). Construction of models describing such activity helps us understand these conflicting and at times counter-intuitive processes (John and Flynn 2002; Flynn 2002a), but very few experiments yield suitable data sets for modeling.

18.3 Physical Factors Influencing Toxin Content

Low temperature decreased the growth rate of the PSP-producing dinoflagellates *Alexandrium catenella*, *A. cohorticula* and *Gymnodinium catenatum*, whereas it increased toxin content per cell (Ogata et al. 1989, cited in Granéli et al. 1998). In contrast, a low growth rate due to light inhibition did not cause an increase in toxin content, compared to cells growing at optimum illumination. Salinity had a greater effect on toxin production than did temperature, for the prymnesiophyte *Chrysochromulina polylepis* (Anderson 1994). The raphidophyte *Fibrocapsa japonica* strongly increased its hemolytic activity at low salinities (de Boer et al. 2004). Flynn et al. (1996) suggested that salinity may influence PSP toxin synthesis by affecting the distribution of metabolites toward osmotic regulation.

Toxin production in marine and freshwater cyanobacteria species has also been shown to be affected by temperature (e.g., Sivonen 1990; Lehtimäki et al. 1994), illumination (e.g., Sivonen 1990) and salinity (Lehtimäki et al. 1994). High pH can induce higher toxicity in *Chrysochromulina polylepis* and *Oscillatoria laetevirens* (Ray and Bagchi 2001; Schmidt and Hansen 2001). This has also been suggested to be the case for *Pseudo-nitzschia* species and *Nitzschia navis-varingica* (Lundholm et al. 2004).

A major problem with all these studies is that as cell size varies with physical environmental changes, toxin per cell will obviously vary to some extent. Many physical and biological factors also interact, promoting increases in toxicity for different phytoplankton species and toxin groups in different ways. pH will inevitably change during consumption of nutrients. Thus, a link between pH and toxin content may not be a cause, but an effect. Quite likely these interactions will also be synergistic. Flynn (2002a) simulated the synergistic impacts of nutrient stress, nutrient refeeding, and variable illumination encountered by vertically migrating *Alexandrium*; the combined impacts on toxin production exceeded those of single factors.

18.4 Inorganic Nutrients and Toxin Content

Changes in the relative and absolute discharges of phosphorus (P) and silicon (Si), and especially the increases of nitrogen (N) entering coastal waters, are not in balance with the demands from algae (Conley 1999). The extent to which changes in nutrient concentrations and ratios are linked to toxic/harmful phytoplankton blooms and their toxin production is poorly understood and highly controversial. These changed ratios affect not only algal physiology but also the structure of phytoplankton communities (Smayda 1997). Stress caused by rate limitation of these nutrients differs greatly. Si stress, which directly affects only diatoms and silicoflagellates, prevents cell division while not directly affecting carbon, nitrogen and phosphorus (CNP) physiology. N stress impacts primarily on somatic growth. P stress, however, has the potential for not only affecting growth but also for having numerous insidious effects through disturbing metabolic regulation; phosphorylation of intermediates is the most common form of biochemical regulation. That P stress should impact so frequently on the production of toxins is thus not surprising (Table 18.1). Neither is it surprising that these nutrient-stress should interact.

Phosphorus deficiency has been shown to increase PSP toxin levels 3- to 4-fold in *Alexandrium tamarense* in comparison to nutrient-sufficient or N-deficient cells (Boyer et al. 1987, cited in Cembella 1998). Also, the dinoflagellate *G. catenatum* has been shown to increase its intracellular toxin concentration under P deficiency (Boyer et al. 1987, cited in Cembella 1998; Flynn et al. 1996). Synthesis of N-containing toxins (such as DA and PSP) declines under N stress (Bates et al. 1991, cited in Bates 1998; Boyer et al. 1987, cited in Cembella 1998), and increases after reintroduction of N (Bates et al. 1991, cited in Bates 1998; Flynn et al. 1994). Toxin content in *A. excavatum* (Levasseur et al. 1995, cited in Granéli et al. 1998) and *P. multiseries* (Bates et al. 1993, cited in Bates 1998) has also been shown to increase when N is available in high concentration, but even more so if ammonium is used as a N source rather than nitrate (Levasseur et al. 1995, cited in Granéli et al. 1998).

Si-limited growth of *Pseudo-nitzschia* promotes synthesis of DA (a non-protein amino acid). For the pennate diatom *P. multiseries*, DA production is triggered by silicate (Si) and/or phosphate (P) limitation (Bates et al. 1991 and Pan et al. 1996a, 1996 c, cited in Bates 1998; Pan et al. 1998). By successively adding back these two limiting nutrients, Pan et al. 1996a, 1996 c, cited in Bates 1998) were able to gradually decrease DA production as stress decreased. The increase in the production of DA under such conditions is favored because, when primary metabolic activity is decreased, necessary precursors, as well as high energy compounds and several co-factors are made available (Pan et al. 1998). Toxic blooms of *P. multiseries* have been associated with nitrate pulses (Smith et al. 1990, cited in Granéli et al. 1998). Bates et al. (1993, cited in Bates 1998) found that this species required a high concentration of nitrate to produce DA. The toxin content of the cells increased from 0.2–5 pg DA $cell^{-1}$ when nitrate in the growth medium was increased from 55 to 880 μM and inorganic P was held constant at 33 μM. Whether the impact of N:P supply is due to different levels of N-P stress, or to the 'forced' entry of nitrate through a second nitrate transporter that operates only at high nitrate concentrations (see Flynn 2001) is not known.

In the cyanobacterium *Nodularia spumigena*, toxin production increases under P limitation. Stolte et al. (2002) grew *N. spumigena* under P limitation but with different sources of N (nitrate, ammonium and N_2-fixation); nitrate addition produced the highest amounts of biomass (chlorophyll-*a*) and nodularin. However, similar amounts of nodularin per amount of chlorophyll or carbon were produced during the stationary phase, independent of the addition of ammonium or nitrate. In contrast, the freshwater cyanobacterium *Oscillatoria agardhii* has been shown to increase toxin production proportionally to the increase of N in the medium (Sivonen 1990). Phosphorus concentrations of between 12.9 and 177 μM had no effect on toxin production, and concentrations below 3.3 μM could not support good toxin production (Sivonen 1990). In *Microcystis aeruginosa* removal of N from the medium correlated with the amount of microcystin synthesized (Orr and Jones 1998). Thus the same pattern of toxin production is seen in cyanobacteria as for the dinoflagellates mentioned previously; the cellular N status correlates with toxin content or production.

For the above groups of organisms under P limitation, the availability of nitrogenous compounds within the cell increases relative to cells grown under N limitation while cellular regulation controlling synthesis and growth slows or ceases. Accordingly, secondary metabolites accumulate and some of these are what we describe as toxins. Cells under different levels of N stress (deficiency) will have lower synthesis rates of secondary metabolites; strain and species differences in physiology will affect the balance between toxin synthesis and cell growth, which together affect the net rate of accumulation of toxin within the cell (Fig. 18.1). Because the availability of nutrients within the cell establishes the degree of nutrient stress, levels of illumination will affect C fixation and energy production, interacting with nutrient availability to set these levels of stress. The net rate of accumulation of toxins will also vary. For example, Flynn et al. (1994) found that neither short-term N nor P limitation promoted toxin increase in *A. minutum* cells whereas Guisande et al. (2002) showed an increase in toxin content when the same organism was grown under P limitation (see also Fig. 18.1). For *A. fundyense* N:P supply ratios did also affect toxin production (John and Flynn 2000) until nutrient exhaustion occurred.

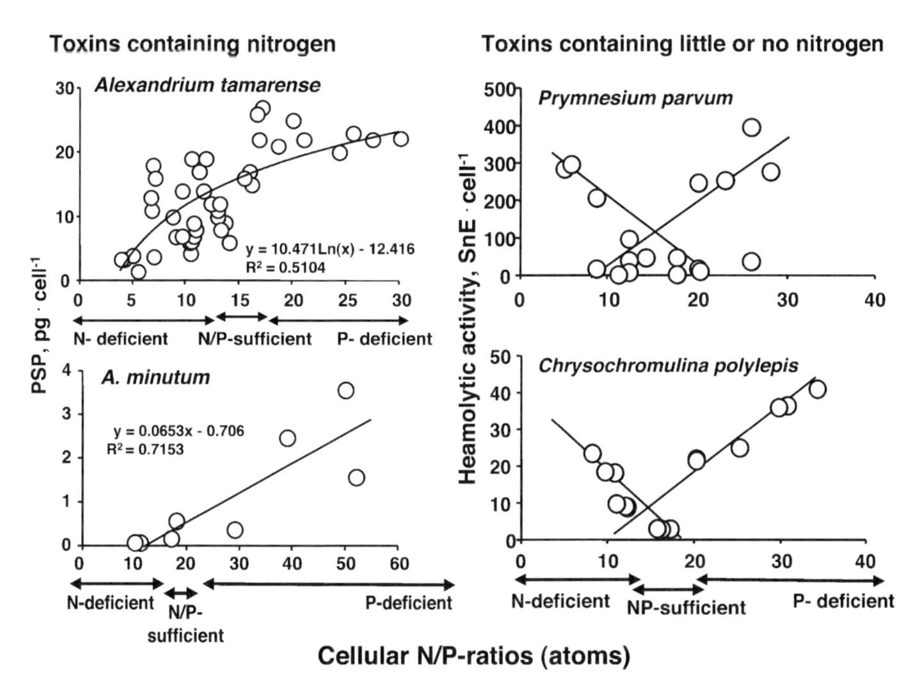

Fig. 18.1. Relation between intracellular nitrogen and phosphorus ratios and toxin concentrations for the toxin-producing dinoflagellates *Alexandrium tamarense* and *A. minutum* (data from the NUTOX project, see Acknowledgements) and flagellates *Prymnesium parvum* and *Chrysochromulina polylepis* grown under different N and P conditions (data from Johansson and Granéli 1999a, 1999b)

For diarrhetic shellfish poisoning (DSP)-producing *Prorocentrum lima*, both N and P deficiency have been found to promote high okadaic acid (OA) toxin accumulations (McLachlan et al. 1994 and Sohet et al. 1995, cited in Granéli et al. 1998). For *Dinophysis* species, Johansson et al. (1996) have shown that under N limitation, OA levels in *D. acuminata* cells increased (from 9 to 23 pg OA cell^{-1}), whereas in *D. acuta* there was an increase in OA only when the cells were grown in sufficient conditions with a N:P atomic ratio of 16:1. Under the same NP-sufficient conditions, *D. acuta* cells had high quantities of dinophysistoxin-1, whereas the DTX-1 content of *D. acuminata* was close to the analytical detection limit, independent of nutrient conditions.

For ichthyotoxic harmful algal species, it seems more the rule that limitation by either N or P increases toxin content (Fig. 18.1). The prymnesiophyte *Chrysochromulina polylepis* growing in a medium with a high N:P ratio increased its toxin content (Edvardsen et al. 1990, cited in Granéli et al. 1998; Johansson and Granéli 1999b). This may appear somewhat surprising considering that (so far) the only identified *C. polylepis* toxin is a N-poor compound; the implication is that disturbed physiology under P stress, rather than a surplus of N, is the trigger (noting that nitrate-grown cells are by definition more N stressed than ammonium-grown cells; Flynn 2001).

The prymnesiophyte *Prymnesium parvum* increased its toxin content under either N- or P-limited steady-state conditions, although toxin content was more accentuated under the latter (Fig. 18.1) (Johansson and Granéli 1999a). Similar results were also found by (Edvardsen et al. 1990, cited in Granéli et al. 1998) for *C. polylepis* growing in semi-continuous cultures; both P and N limitation increased intracellular toxin concentration in this flagellate. However, several other species and strains of *Chrysochromulina* sp. were tested and found not to contain high (if any) amounts of toxin under P-limiting conditions (Edvardsen 1993, cited in Granéli et al. 1998). *Chrysochromulina leadbeateri* cells were only slightly toxic under P limitation, whereas this species had been shown to be extremely toxic in the field, when it bloomed in northern Norway during May–June 1991, causing the death of tons of caged fish (Edvardsen 1993, cited in Granéli et al. 1998). Why this species did not produce high quantities of toxin in experiments might be explained by differences in the species strain and/or by differences in the nutrient concentration as well as its N:P (Edvardsen 1993, cited in Granéli et al. 1998). It should also be noted that a high N:P ratio at low ambient N concentrations will not cause the same level of P stress to a cell as will the same ratio at high N concentrations; in the former, dual (N and P) rate limitation will occur.

Thus, there is a group of phytoplankton species that can increase cellular toxin concentrations when grown under N- or P-deficiency. Both conditions will result in an imbalanced production of C metabolites. A dual limitation, however, will result in a rapid deregulation of C metabolism; toxin production under such conditions may be expected to be low. The low N and P threshold

Table 18.1. Toxic species and the factors affecting toxin production

Toxic species	Factors increasing cell toxin	Toxin[a]	References
Toxins containing nitrogen			
Alexandrium catenella *A. cohorticula*	Low temperature, low growth rate	PSP	Granéli et al. 1998
A. fundyense	N and P- limitation, temperature	PSP	Flynn 2002a; Flynn and Flynn 1995; John and Flynn 2000
A. minutum	P-limitation	PSP	Guisande et al. 2002
A. tamarense	P-limitation, N-excess, organic matter	PSP	Cembella 1998; Flynn et al. 1994; Flynn and Flynn 1995; John and Flynn 1999
Anabaena sp. *Anabaena flos-aquae*	High P, temperature 19–25 °C, moderate light	ANA, MIC, LPS	Chorus and Bartram 1999
Anabaena circinalis	Growth phase	PSP, LPS	Chorus and Bartram 1999; Negri et al. 1997
Aphanizomenon sp.	High P, temperature 20 °C, high light	ANA, PSP, CYL, LPS	Chorus and Bartram 1999
Aphanizomenon flos-aquae	High P	ANA, PSP, LPS	Chorus and Bartram 1999
Microcystis sp.	N excess, moderate to high light, 18–25 °C, high pH, high Fe, moderate pH	MIC, LPS	Chorus and Bartram 1999; Orr and Jones 1998
Nodularia sp.	P-limitation/high NO_3, growth phase, temperature 19–20 °C, high light, moderate salinity	Nodularin, LPS	Chorus and Bartram 1999; Granéli et al. 1998; Stolte et al. 2002
Oscillatoria agardhii	N excess + low light	MIC	Chorus and Bartram 1999
Oscillatoria sp.,	High N	ANA, LPS	Chorus and Bartram 1999; Ray et al. 2001
O. laetevirens	High ph	ANA, LPS	Schmidt and Hansen 2001

Table 18.1. (*Continued*)

Toxic species	Factors increasing cell toxin	Toxin[a]	References
Pseudo-nitzschia australis P. delicatissima, P. multi-iseries, P. pungens, P. seriata	P/Si limitation, N excess, growth phase, high light, high pH, high or low temperature, organic matter, Fe-limitation or Cu-stress	DA	Bates 1998; Lundholm et al. 2004; Maldonado et al. 2002
Toxins with little or no nitrogen			
Chrysochromulina leadbeateri, C. polylepis	P-limitation, salinity, high pH, N or P-limitation	ICHT	Anderson 1994; Granéli et al. 1998; Johansson and Granéli 1999b
Dinophysis sp., *D. acuminata, D. acuta*	N or P-limitation, N-limitation, N/P-sufficiency	OA, DTX, PTX	Granéli et al. 1998; Johansson et al. 1996
Fibrocapsa japonica	Low salinity	PUFA	de Boer et al. 2004
Gymnodinium catenatum	Low temperature, P-limitation	GYM	Cembella 1998; Flynn et al. 1996; Granéli et al. 1998
Heterosigma akashiwo	High temperature, Fe-limitation	ROS	Twiner and Trick 2000
Karenia brevis	Urea	Brevetoxin	Glibert and Terlizzi 1999
K. mikimotoi	N and P-limitation	ICHT	NUTOX project
Ostreopsis ovata	High temperature	Ostreocin	Granéli 2004
Prymnesium parvum P. patelliferum	N or P-limitation	Prymnesin	Johansson and Granéli 1999a; Legrand et al. 2001
Prorocentrum lima	N or P-limitation, organic matter	OA, DTX	Granéli et al. 1998
P. minimum	Growth phase	Unknown	Wikfors 2005

[a]*PSP* (saxitoxin (STX), neo-STX, gonyautoxin (GTX), C-toxins), *GYM* gymnodimine, *ANA* anatoxins, *MIC* microcystins, *LPS* lipopolysaccharide, *CYL* cylindrospermopsins, *ICHT* ichthyotoxins, *DA* domoic acid, *OA* okadaic acid, *DTX* Dinophysis toxin, *PTX* pectenotoxin, *PUFA* polyunsaturated fatty acids, *ROS* reactive oxygen species

concentrations that cause toxin levels to increase can only be achieved in waters where these nutrients are found in imbalanced conditions i.e., where N:P ratios are under or above 16.

Figure 18.1 and Table 18.1 summarize the evidence for the impact of environmental and nutrient availability on toxin production. It is important to appreciate that nutrient supply ratios are only part of the story. The impact of N:P ratios (for example) on algal growth and nutrient stress will depend greatly on the residual nutrient concentration in the seawater. Unfortunately, many studies of algal physiology have not considered the impact of such events, and models of algal physiology are usually poorly configured to reflect the interaction (Flynn 2001, 2002b). Further, while steady-state (discontinuous or chemostat) cultures may be favored for use by experimentalists, they do not reflect the dynamic situation in nature, which has the potential for significantly changing secondary- metabolite and toxin-production patterns (e.g., Flynn 2002a). The only nutrient levels and ratios that ultimately matter are internal, not external. Thus, interpretation of studies without cellular C:N:P(:Si) ratios plus concurrent C-specific growth determination and C-specific toxin content is difficult and potentially misleading.

18.5 Organic Matter and Toxin Content

Organic matter can affect the production of several toxins in different ways. Care must be taken in interpreting the ecological significance of some studies. While the uptake of organic nutrients at natural concentrations has been demonstrated (e.g., amino acids by *Alexandrium*; John and Flynn 1999), the impacts of substrates applied at concentrations orders-of-magnitude greater than those in nature, while perhaps demonstrating important biochemical pathways, may be of little or no ecological consequence.

The diatom *P. multiseries* has been shown to increase its DA content when grown with a mixture of gluconic acid/gluconolactone (produced by the bacterium *Alteromonas* sp., or added directly) (Osada and Stewart 1997, cited in Bates 1998). While *A. tamarense* decreases PSP toxin content under N limitation, it can increase it if yeast is added to the cultures (Ogata et al. 1996, cited in Cembella 1998); the implication is that this species can use organic nitrogen. *Gymnodinium breve* can increase brevetoxin content by up to six times when grown with urea at 0.5 to 1.0 mM nitrogen (cited in Glibert and Terlizzi 1999).

Very little is known about the connection between other forms of algal nutrition, e.g., ingestion of dissolved organic material (DOM) or particles (mixotrophy), and toxin production. Okadaic acid concentrations in *P. lima* cells were higher when organic P (glycerol) was supplied instead of inorganic P to the growth medium (Tomas and Baden 1993, cited in Granéli et al. 1998). However, Sohet et al. (1995, cited in Granéli et al. 1998) could not find a correlation between organic P or humic acid addition and the toxin content of *P. lima*. Phosphorus-limited *N. spumigena* cells growing in media containing additional humic substances produced half as much toxin as did P-limited cultures; the implication is that humics were a source of P, although this was not sufficient to decrease toxin content to the same low levels as found in the cultures growing under nutrient-balanced conditions (Granéli et al. 1998).

Okadaic acid content in *D. acuminata* correlates with the occurrence of inclusion bodies in cells under N-deficient conditions. It was not possible, however, to determine if the higher toxin levels found in the N-deficient treatment were associated directly with phagotrophy or with nutrient deficiency per se; nutrient deficiency may have triggered phagotrophy. For the haptophyte *Prymnesium parvum*, high toxin levels resulting from P deficiency decreased after the cells ingested bacteria; presumably the bacteria provided the missing P (Legrand et al. 2001). Thus, the hypothesis of toxin accumulation by active ingestion of these bacteria, in this case, can be ruled out.

18.6 Conclusions

There is a wide variation, not only between algal groups but also species and strains, with respect to their toxin-content response to different nutrient conditions. Phosphorus limitation often seems to enhance cellular toxin content. N limitation does so to a lesser extent, or may, in the cases of DA and PSP, do the opposite and lead to a fall in toxin content. In most instances, toxin content is relatively low under nutrient-balanced conditions (N:P ~16:1). Thus, it seems that physiological stress, rather than solely P or N limitation, can be associated with an increase in toxin content. However, because carbon fixation requires significant investment in terms of N- and P-containing compounds, even here, some interaction may be expected, depending on the timescale of stress induction.

A major problem in HAB research, in general, is the reliance on cell number as the basis for describing toxin content. Because cell size also varies with changes in environmental conditions such as temperature and light, it becomes less clear whether toxin production per se alters with these conditions. Biomass [CNP(Si)] determinations, indicating the internal nutrient status, are also necessary for an improved interpretation of the results. Further, in the absence of CNP(Si) data, most studies are also woefully inadequate for supporting modeling efforts. It is therefore not possible to unequivocally identify definitive common causal relationships between stressors (physical and/or nutrient) and toxin production.

Toxin accumulation occurs primarily under unbalanced-nutrient conditions, and when cells are not growing optimally. This suggests that production is only sometimes advantageous, even though production costs appear minor (John and Flynn 2002). If it were of universal value then one would expect toxin production to be common and to occur during normal exponential growth. This limited functionality, essentially being selection-neutral, may explain why toxin production ability varies so greatly among and within species and strains. However, it is quite likely that in both harmful algae and other phytoplankton, many other secondary metabolites are also produced that is at the least indigestible (if not unpalatable) to grazers thus affecting predator-prey interactions. Our understanding of what promotes toxin production in harmful algae has important ramifications for our general understanding of phytoplankton ecophysiology.

Acknowledgements. Edna Granéli thanks Christina Esplund-Lindquist for preparing the figures and tables and Roseni Carvalho for the reference list. Thanks to the Swedish Natural Research Council and the European Commission MAST-NUTOX (contract nr: MAS-3-CT97-0103) and FP 6, THRESHOLDS projects (contract nr: GOCE 003933) for financial support to EG. Kevin Flynn was supported by the Royal Society/Leverhulme Trust, and by the Natural Environment Research Council (UK).

References

Anderson DM (1994) Red tides. Sci Am 271:52–58

Bates SS (1998) Ecophysiology and metabolism of ASP toxin production. In: Anderson DM, Cembella AD, Hallegraeff GM (eds) Physiological ecology of harmful algal blooms. NATO ASI Series 41. Springer, Berlin Heidelberg New York, pp 405–426

Cembella AD (1998) Ecophysiology and metabolism of paralytic shellfish toxins in marine microalgae. In: Anderson DM, Cembella AD, Hallegraeff GM (eds) Physiological ecology of harmful algal blooms. NATO ASI Series 41. Springer, Berlin Heidelberg New York, pp 381–403

Chorus I, Bartram J (eds) (1999) Toxic cyanobacteria in water: a guide to their public health consequences, monitoring and management. © 1999, WHO http://www.who.int/docstore/water_sanitation_health/toxicyanobact/begin.htm#Contents

Conley DJ (1999) Biogeochemical nutrient cycles and nutrient management strategies. Hydrobiologia 410:87–96

de Boer MK, Tyl MR, Vrieling EG, van Rijssel M (2004) Effects of salinity and nutrient conditions on growth and haemolytic activity of *Fibrocapsa japonica* (Raphidophyceae). Aquat Microb Ecol 37:171–181

Flynn KJ (2001) A mechanistic model for describing dynamic multi-nutrient, light, temperature interactions in phytoplankton. J Plankton Res 23:977–997

Flynn KJ (2002a) Toxin production in migrating dinoflagellates: a modelling study of PSP producing *Alexandrium*. Harmful Algae 1:147–155

Flynn KJ (2002b) How critical is the critical N:P ratio? J Phycol 38:961–970

Flynn KJ, Flynn K (1995) Dinoflagellate physiology, nutrient stress and toxicity. In: Lassus P, Arzul G, Le Denn EE, Gentien P, Marcaillou C (eds) Harmful marine algal blooms. Lavoisier Intercept, Paris, pp 541–550

Flynn K, Franco JM, Fernández P, Reguera B, Zapata M, Wood G, Flynn KJ (1994) Changes in toxin content, biomass and pigments of the dinoflagellate *Alexandrium minutum* during nitrogen refeeding and growth into nitrogen and phosphorus stress. Mar Ecol Prog Ser 111:99–109

Flynn KJ, Flynn K, John EH, Reguera B, Reyero MI, Franco JM (1996) Changes in toxins, intracellular and dissolved free amino acids of the toxic dinoflagellate *Gymnodinium catenatum* in response to changes in inorganic nutrients and salinity. J Plankton Res 18:2093–2111

Glibert PM, Terlizzi DE (1999) Co-occurrence of elevated urea levels and dinoflagellate blooms in temperature estuarine aquaculture ponds. Appl Environ Microbiol 65:5594–5596

Granéli E (2004) Toxic algae – a global problem. HavsUtsikt 2:12–13 (in Swedish)

Granéli E, Johansson N, Panosso R (1998) Cellular toxin contents in relation to nutrient conditions for different groups of phycotoxins. In: Reguera B, Blanco J, Fernandez ML, Wyatt T (eds) Harmful algae. Xunta de Galicia and IOC-UNESCO, Grafisant, Santiago de Compostela, Spain, pp 321–324

Guisande C, Frangópulos M, Maneiro I, Vergara AR, Isabel Riveiro I (2002) Ecological advantages of toxin production by the dinoflagellate *Alexandrium minutum* under phosphorus limitation Mar Ecol Prog Ser 225:169–176

Johansson N, Granéli E (1999a) Influence of different nutrient conditions on cell density, chemical composition and toxicity of *Prymnesium parvum* (Haptophyta) in semi-continuous cultures. J Exp Mar Biol Ecol 239:243–258

Johansson N, Granéli E (1999b) Cell density, chemical composition and toxicity of *Chrysochromulina polylepis* (Haptophyta) in relation to different N:P supply ratios. Mar Biol 135:209–217

Johansson N, Granéli E, Yasumoto T, Carlsson P, Legrand C (1996) Toxin production by *Dinophysis acuminata* and *D. acuta* cells grown under nutrient sufficient and deficient conditions. In: Yasumoto T, Oshima Y, Fukuyo Y (eds) Harmful and toxic algal blooms. IOC-UNESCO, Paris, pp 277–280

John EH, Flynn KJ (1999) Amino acid uptake by the toxic dinoflagellate *Alexandrium fundyense*. Mar Biol 133:11–20

John EH, Flynn KJ (2000) Growth dynamics and toxicity of *Alexandrium fundyense* (Dinophyceae): the effect of changing N:P supply ratios on internal toxin and nutrient levels. Eur J Phycol 35:11–23

John EH, Flynn KJ (2002) Modelling changes in paralytic shellfish toxin content of dinoflagellates in response to nitrogen and phosphorus supply. Mar Ecol Prog Ser 225:147–160

Legrand C, Johansson N, Johnsen G, Borsheim KY, Granéli E (2001) Phagotrophy and toxicity variation in the mixotrophic *Prymnesium patelliferum* (Haptophyceae). Limnol Oceanogr 46:1208–1214

Lehtimäki J, Sivonen K, Luukkainen R, Niemela SI (1994) The effects of incubation-time, temperature, light, salinity, and phosphorus on growth and hepatotoxin production by *Nodularia* strains. Arch Hydrobiol 130:269–282

Lundholm N, Hansen PJ, Kotaki Y (2004) Effect of pH on growth and domoic acid production by potentially toxic diatoms of the genera *Pseudo-nitzschia* and *Nitzschia*. Mar Ecol Prog Ser 273:1–15

Maldonado MT, Hughes MP, Rue EL, Wells ML (2002) The effect of Fe and Cu on growth and domoic acid production by *Pseudo-nitzschia multiseries* and *Pseudo-nitzschia australis*. Limnol Oceanogr 47:515–526

McLachlan JI, Marr JC, Conolon-Kelly A, Adamson A (1994) Effects of nitrogen concentration and cold temperature on DSP-toxin concentration in the dinoflagellate *Prorocentrum lima* (Prorocentrales, Dinophyceae). Natural Toxins 2:263–270

Negri AP, Jones GJ, Blackburn SI, Oshima Y, Onodera H (1997) Effect of culture and bloom development and of sample storage on paralytic shellfish poisons in the cyanobacterium *Anabaena circinalis*. J Phycol 33:26–35

NUTOX project European Commission MAST-NUTOX (contract nr: MAS-3-CT97-0103)

Orr PT, Jones GJ (1998) Relationship between microcystin production and cell division rates in nitrogen-limited *Microcystis aeruginosa* cultures. Limnol Oceanogr 43:1604–1614

Pan Y, Bates SS, Cembella, AD (1998) Environmental stress and domoic acid production by *Pseudo-nitzschia*: a physiological perspective. Natural Toxins 6:127–135

Pohnert G, Lumineau O, Cueff A, Adolph S Codevant C, Lang M, Poulet S (2002) Are volatile unsaturated aldehydes from diatoms the main line of chemical defence against copepods? Mar Ecol Prog Ser 245:33–45

Ray S, Bagchi SN (2001) Nutrients and pH regulate algicide accumulation in cultures of the cyanobacterium *Oscillatoria laetevirens*. New Phytol 149:455–460

Schmidt LE, Hansen PJ (2001) Allelopathy in the prymnesiophyte *Chrysochromulina polylepis*: effect of cell concentration, growth phase and pH. Mar Ecol Prog Ser 216:67–81

Sivonen K (1990) Effects of light, temperature, nitrate, orthophosphate, and bacteria on growth of and hepatotoxin production by *Oscillatoria agardhii* strains. Appl Environ Microbiol 56:2658–2666

Smayda TJ (1997) Harmful algal blooms: their ecophysiology and general relevance to phytoplankton blooms in the sea. Limnol Oceanogr 42:1137–1153

Sohet K, Pereira A, Braeckman JC, Houvenaghel G (1995) Growth and toxicity of *Prorocentrum lima* (Ehrenberg) Dodge in different culture media: effects of humic acids and organic phosphorus. In: Lassus P, Arzul G, Le Denn EE, Gentien P, Marcaillou C (eds) Harmful marine algal blooms. Lavoisier Intercept Paris, pp 669–674

Stolte W, Karlsson C, Carlsson P, Granéli E (2002) Modeling the increase of nodularin content in Baltic Sea *Nodularia spumigena* during stationary phase in phosphorus-limited batch cultures. FEMS Microbiol Ecol 41:211–220

Twiner MJ, Trick CG (2000) Possible physiological mechanisms for production of hydrogen peroxide by the ichthyotoxic flagellate *Heterosigma akashiwo*. J Plankton Res 22:1961–1975

Wikfors GH (2005) A review and new analysis of trophic interactions between *Prorocentrum minimum* and clams, scallops, and oysters. Harmful Algae 4:585–592

19 Relationships Between Bacteria and Harmful Algae

M. Kodama, G.J. Doucette and D.H. Green

19.1 Introduction

The interactions of harmful algal (HA) species with their physico-chemical and biological environments ultimately determine their abundance and distribution. While an algal taxon's physiological tolerance limits (e.g., temperature, salinity, light) and intrinsic phenotypic traits (e.g., growth rate, nutrient uptake, vertical migration) largely define its ecological niche, relationships with the biological components of an ecosystem (e.g., grazers, microbes, pathogens, competing algal species) play a critical role in its ability to achieve concentrations that lead to the many negative impacts characterizing harmful species. A biological factor of potentially great significance in regulating the population and even toxin dynamics of HA that has received increasing yet comparatively little attention to date is their relationship with the ubiquitous and diverse bacterial community. These microbes, which exist as free-living forms as well as securely attached to algal cells, occasionally with a high degree of taxonomic specificity, have now been demonstrated to modulate (either positively or negatively) algal growth rates and transitions between life history stages, influence toxin production, and even induce the rapid lysis of algal cells.

Previous reviews of this topic (e.g., Doucette 1995; Doucette et al. 1998) described an exciting, emerging field in which molecular-based approaches were just beginning to supplement traditional bacteriological techniques, yielding new insights into the nature and ecological implications of bacterial-algal interactions. During the intervening time, continued application of molecular techniques, novel experimental methods, and targeted studies of natural bacterial communities have contributed valuable details of the relationships between bacteria and HA that will be the focus of this chapter.

Ecological Studies, Vol. 189
Edna Granéli and Jefferson T. Turner (Eds.)
Ecology of Harmful Algae

19.2 Diversity of Algal-Associated Bacteria

Through the last decade, molecular biological methods have increasingly been used to investigate algal-bacterial diversity due to their finer level of discrimination as compared with commonly used bacteriological identification protocols (e.g., biochemical tests). The culture-independent nature of most molecular techniques allows the importance of the numerous non-cultivable bacteria present in marine ecosystems to be evaluated. Moreover, the coupling of molecular-based detection strategies to automation represents a real potential for the development of in situ real-time monitoring as is starting to be realized for HA species (Babin et al. 2005). Nonetheless, the process of isolating and culturing bacteria continues to be a prerequisite for examining bacterial physiology, and although genomics and metagenomics will to a certain extent minimize the reliance on pure cultures, close partnering of genomics and pure culture studies will likely be the most valuable route to elucidating the functional significance of bacteria associated with harmful algae.

19.2.1 Bacteria Associated with Harmful Algal Species

There is now a growing list of studies of bacteria associated with HA taxa, especially within the dinoflagellate group. Most represent 'snapshots' of the bacteria present in the culture at a given point in time and the emerging picture suggests that these bacterial communities are more notable for their similarities than their differences (e.g., Fig. 2 of Hold et al. 2001). The bacteria associated with these dinoflagellates resemble an ordered and structured community rather than a random assemblage of species recruited from the marine bacterial metacommunity (Curtis and Sloan 2004). Figure 19.1 illustrates one such phylogenetic 'micro-cluster' of bacteria associated with a range of dinoflagellate taxa originating from distinct regions around the globe. One interpretation of this seemingly conserved community structure is that it may reflect a physiological requirement of the dinoflagellates (Alavi et al. 2001; Green et al. 2004). Nonetheless, most of these studies involve laboratory cultures that may not accurately reflect field populations and many of these micro-clusters are in common to both toxic and non-toxic algal species.

The bacterial group most often associated with dinoflagellates and diatoms, irrespective of their toxicity, is the alpha Proteobacteria, and most frequently, the *Roseobacter* clade and its relatives (e.g., Hold et al. 2001; Allgaier et al. 2003; Green et al. 2004). The *Roseobacter* clade alone can dominate the bacterial diversity of microalgal cultures (Alavi et al. 2001), but can also be an abundant feature of field blooms (Fandino et al. 2001, cited in

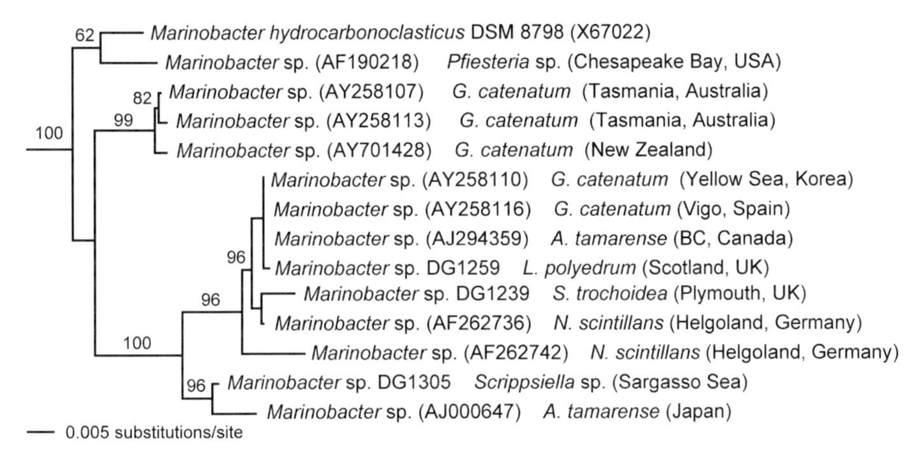

Fig 19.1. Specificity of association: the genus *Marinobacter*. Neighbor-joining tree (with bootstrap support for the branching order) depicting known dinoflagellate-associated *Marinobacter* spp. The dinoflagellate host and its isolation origin are noted in parentheses to the right of each bacterial strain

Long and Azam 2001). One suggested explanation of their predominant association with dinoflagellates is a growth promoting capability (Jasti et al. 2003).

Gamma Proteobacteria, beta Proteobacteria, and Cytophaga-Flavobacteria-Bacteroides (CFB) are the other bacterial groups reported to be associated with HA species. Within the gamma Proteobacteria, *Marinobacter* spp. (Fig. 19.1) and *Alteromonas* spp. appear to have an association with dinoflagellates (Hold et al. 2001; Ferrier et al. 2002). *Marinobacter* have no known function other than currently unpublished data (D. Green and C. Bolch) indicating that they stimulate dinoflagellate growth. Alteromonads have seemingly ambiguous roles, including stimulating dinoflagellate growth (Ferrier et al. 2002) as well as inhibiting *Alexandrium* cyst formation (Adachi et al. 2002). The presence of *Pseudoalteromonas* (gamma Proteobacteria) in algal cultures is more sporadic, which may reflect their propensity toward algicidal activity (Mayali and Azam 2004). Yet, *Pseudoalteromonas* spp. are frequently encountered in the field (Skerratt et al. 2002, cited in Mayali and Azam 2004), where they are a potentially significant factor in bloom decline. Representatives of the beta Proteobacteria are generally considered to be rare in marine systems, but were observed by Alverca et al. (2002) to dominate the intracellular bacterial flora of the dinoflagellate, *Gymnodinium instriatum*. Another significant bacterial phylum associated with many microalgal cultures and field populations is the CFB group. This group is associated with algal surfaces and in the free-living fraction of algal blooms (e.g., Rooney-Varga et al. 2005), where they are important in the degradation of high molecular weight dissolved organic matter (Kirchman 2002). More importantly, they are frequently associated with the exhibition of algicidal activity (e.g., Doucette et al. 1999), and like *Pseudoalteromonas*, may be important in bloom termination.

Apart from the ecological implications of bacteria-HAB associations, it is important to note that algal blooms can serve as reservoirs for human pathogens such as *Vibrio cholerae*. In this case, a viable, but unculturable form of *V. cholerae* associated with plankton can become infectious under conditions favoring bloom development. Moreover, this pathogen has been reported to survive for extended periods in the presence of cyanobacteria, diatoms, and dinoflagellates (Rose et al. 2001). Investigations of this relationship as it relates to the growth potential of *V. cholerae* are ongoing (e.g., Mouriño-Pérez et al. 2003).

19.2.2 Spatio-Temporal Relationships Between Bacteria and Algae

Aside from a need to identify bacteria and understand their physiological function, a pressing aim in algal-bacteria-HAB research is to understand the role bacteria play in field population dynamics – the 'how' and 'what' interactions are important in HAB ecology (Doucette et al. 1998). Are there 'key' or 'signature' bacteria that define the progress of a HAB event? Several recent studies have begun to provide insights into how bacterial communities can change during the growth of algal populations. Among the more interesting concepts to emerge from such work is that changes in attached (versus free-living) bacterial communities are linked closely with both successional changes in phytoplankton assemblages (Rooney-Varga et al. 2005) and with algal physiological status (Grossart et al. 2005). Moreover, members of the Bacteroidetes phylum tend to predominate the bacteria attached to algal cells, although the alpha Proteobacteria are also represented frequently in both attached and free-living communities over time. These findings strongly suggest that there are very specific, two-way interactions between phytoplankton and their attached microbial flora that can influence such important processes as bacterial-algal succession and the cycling of organic matter in the ocean. Notably, there are relatively few descriptions dealing specifically with the dynamics of bacterial communities associated with HABs (e.g., see Doucette et al. 1998; Töbe et al. 2001, cited in Green et al. 2004; Wichels et al. 2004). Preliminary data from a study of bacterial clone libraries associated with blooms of the dinoflagellate, *Karenia brevis*, on the west Florida shelf show a clear dominance of the alpha Proteobacteria and Bacteroidetes groups, as noted above for non-HA species (K. Jones, C. Mikulski, and G. Doucette unpubl. data). Clearly, additional efforts to describe both the spatio-temporal variability of bacterial communities associated with HABs and the functional attributes of dominant taxa are needed to better define how bacteria may influence algal bloom dynamics.

19.3 Bacterial Influences on Algal Growth, Metabolism, and Toxins

19.3.1 Bacterial Effects on Algal Growth

To what extent bacteria might be involved in regulating transitions between stages of HAB events (i.e., initiation, development, maintenance, decline) is an often neglected issue, despite multiple scenarios suggesting the potentially important influence of bacteria on algal growth. Note that the role of bacteria as prey for phagotrophic algal taxa is considered elsewhere (see Stoecker et al. Chap. 14). As discussed by Doucette et al. (1998), bacteria may contribute to the supply of trace elements such as iron, and recent work has further highlighted the potential significance of bacterially produced siderophores in defining iron chemistry in the coastal zone (Soria-Dengg et al. 2001, cited in Green et al. 2004). Interestingly, the unique chemistry of marine bacterial siderophores (Martinez et al. 2000, cited in Barbeau et al. 2001) can potentially increase the availability of Fe to phytoplankton (Barbeau et al. 2001). Whether bacteria play a fundamental role in mediating the biological availability of Fe and other essential trace elements to phytoplankton, remains poorly understood; however, this becomes an important question within the framework of increasing eutrophication and anthropogenic modification of the coastal zone (e.g., dumping of metal-rich sludge, increased run-off/erosion, etc.), which may lead to increased availability of trace elements, corresponding to an increased frequency of HABs.

Bacteria are generally thought to have an indirect role in phytoplankton growth through re-mineralization of excreted dissolved organic matter (Azam 1998). However, some evidence would suggest that bacteria have more specific and direct effects on algal growth, such as the supply of vitamins (Haines and Guillard 1974). More recently, specific bacterial taxa have been linked with growth stimulation of dinoflagellates such as *Gambierdiscus toxicus* (Sakami et al. 1999), *Pfiesteria* spp. (Alavi et al. 2001), and *A. fundyense* (Ferrier et al. 2002). It is interesting to speculate whether bacterial release of metabolites that may lead to enhanced algal growth is, in fact, a mutualistic strategy capable of maintaining the health of associated algal cells that in turn excrete organic matter required for bacterial growth or, more simply, one of altruism? Perhaps the most compelling evidence of a direct relationship between bacteria and algal growth comes from recent work demonstrating that the dinoflagellate *G. catenatum* has an obligatory requirement for bacteria following excystment from the sexual resting stage (D. Green and C. Bolch unpubl. data). Moreover, it was determined that individual bacterial strains could meet the dinoflagellate's bacterial requirement. Notably, *Marinobacter* was one of these 'growth-promoting' taxa, highlighting the remarkable phylogenetic relatedness observed amongst the dinoflagellate-associated *Mari-*

nobacters (see Fig. 19.1), and suggesting that other dinoflagellates may also have an obligatory growth requirement for certain bacteria. The mechanistic basis for this requirement remains uncertain, but examples of bacteria employing sophisticated secretion systems to establish mutualistic relationships with eukaryotes have been reported (e.g., Dale et al. 2002).

Initiation of sexuality in dinoflagellates has largely been thought of as a function of physical or chemical stresses, such as nitrogen or phosphorus deficiency commonly used to induce formation of sexual life history stages (e.g., Blackburn et al. 1989, cited in Green et al. 2004). An involvement of bacteria in dinoflagellate sexuality would not seem immediately obvious. Yet, Adachi et al. (1999) reported that a certain fraction of the naturally occurring bacteria within *Alexandrium* blooms promoted the formation of cysts. This group also observed the co-occurrence of bacteria in natural blooms that inhibited the formation of *Alexandrium* cysts (Adachi et al. 2002). It remains unclear if these bacteria were producing specific compounds that would promote or inhibit cyst formation or whether an indirect mechanism affecting the algal cells' physiology was involved.

19.3.2 The Role of Bacteria in Toxin Production

Historically, the bacteria associated with HA species have been studied through a causal link between algal toxin production and bacterial presence (e.g., Silva 1982, cited in Doucette 1995). This link was reinforced by findings that bacteria were capable of producing paralytic shellfish poisoning-like toxins (PST) (e.g., Kodama et al. 1988; Gallacher et al. 1997). Although a direct bacterial involvement in PST production and PSP events has been questioned (Sato and Shimizu 1998; Baker et al. 2003, cited in Martins et al. 2003; Martins et al. 2003; Green et al. 2004; Wichels et al. 2004), there is evidence that domoic acid (DA) production by *Pseudo-nitzschia* spp. may be stimulated indirectly by bacterial presence.

Bates et al. (1995) showed that non-axenic *P. multiseries* produced high DA levels at late-exponential phase coinciding with growth limiting conditions. The opposite trend in DA production was observed in axenic *P. multiseries* cultures; however, this pattern was reversed upon addition of bacteria isolated from the non-axenic cultures, providing compelling evidence that environmental bacteria can enhance DA production by *P. multiseries*. Recently, polymerase chain reaction (PCR) primers were used to amplify bacterial 16S rDNA associated with cells of axenic *P. multiseries* cultures. Among the PCR products obtained, several were of identical sequence to bacteria isolated from non-axenic *P. multiseries* (Kobayashi et al. 2003).

Work by Bates et al. (2004) has indicated that free-living bacteria associated with *P. multiseries* cultures are not capable of autonomous DA production, even in the presence of algal exudates, suggesting that any bacterial direct involvement in toxin production likely involves bacteria attached to the diatom cells. Kobayashi and co-workers (unpubl. data) have also examined this bacterial-algal relationship by measuring DA levels in axenic and non-axenic *P. multiseries* strains co-cultured with bacteria derived from the latter separated by cellophane membrane. Their results showed that while concentrations of dissolved DA were similar between cultures, maximum intracellular toxin levels were at least three-fold higher (1.1 pg cell^{-1}) in the non-axenic cultures, again indicating that direct contact with bacteria was essential for *P. multiseries* to produce large amounts of domoic acid. More recently, Kaczmarska et al. (2005) has proposed that a specific bacterial community composition or its density may be important to explaining the variable DA levels associated with some *P. multiseries* cultures. They hypothesized that some bacteria were antagonistic to the alga, which provoked an increase in DA production as a specific response to the bacteria's presence.

19.3.3 Bacterially-Mediated Release and Metabolism of Algal Toxins

In addition to their role as modulators of algal toxin production, as well as putatively autonomous sources of some toxins as discussed above, certain bacteria can mediate the release of phycotoxins into marine ecosystems through the lysis of algal cells. In the case of two such algicidal bacteria targeting the brevetoxin (PbTx)-producing dinoflagellate, *Karenia brevis*, disruption of the algal cell wall exposed previously intracellular toxin to the surrounding seawater (Roth 2005). By tracking the toxin present in several size fractions (>5.0, 0.22–5.0, <0.22 μm) following treatment of *K. brevis* cultures with these algicidal bacteria, a 50 ng mL^{-1} pulse of dissolved (<0.22 μm) brevetoxin was observed co-incident with the onset of cell lysis and a rapid decline in the >5.0-μm fraction, which initially accounted for 50 to 90% of the total toxin present (~125 ng PbTx-3 equiv mL^{-1}). The peak in dissolved toxin (initially ca. 60 % of total toxin) dissipated in about 5 days, likely due to natural and/or bacterial degradation. Nonetheless, it was clear that attack of *K. brevis* cells by algicidal bacteria resulted in a marked redistribution of brevetoxin to different size fractions and would be expected to alter the routes and efficacy of toxin trophic transfer.

It is well established that many bacteria are capable of metabolizing algal toxins to yield either known congeners or novel, chemically modified derivatives within a given toxin class. The sources of such bacterial isolates range from laboratory cultures of both toxic and non-toxic algae (Sakamoto et al. 2000; Smith et al. 2002) to those originating from natural collections of shellfish (Smith et al. 2001) or other invertebrates (Kotaki et al. 1985) known to accumulate phycotoxins. Interestingly, Stewart et al. (1998) reported that the capacity for domoic acid metabolism was higher in bacteria isolated from shellfish species that rapidly depurate toxins as compared to those known to retain toxins for extended periods, suggesting a linkage between such bacteria and toxin clearance rates in shellfish. In addition, the products of bacterial toxin metabolism may be either more or less potent than the starting compound(s) (e.g., PSP toxins; Kotaki 1989; Smith et al. 2001), with the former leading potentially to increased public health risks for commercially harvested seafood above that expected based solely on the algal toxin source (Bricelj and Shumway 1998).

In terms of actual metabolic transformations of PSP toxins, Sakamoto et al. (2000) reported evidence for the involvement of glutathione (GSH) in certain reactions, leading to production of stable toxin-GSH conjugates by covalent linkage of the GSH cystein moiety (involving the sulfur atom) at the toxin C-11 position (Sato et al. 2000). Such conjugates are formed between PSP toxins with 11-O-sulfate such as gonyautoxins (GTXs) and various thiol compounds, including those of biological origin. Notably, when these conjugates are treated with an excess of thiols, bound toxins are released without the 11-O-sulfate group, yielding more potent congeners such as saxitoxin. PSP toxins could bind with the cystein moiety of various proteins or amino acids and be co-metabolized in toxic organisms as well as in bacteria. The latter suggests a possible explanation for the difficulty in demonstrating autonomous bacterial production of these toxins and a potential justification for re-visiting this issue.

19.4 Potential Implications of Interactions Among Bacteria

The majority of this chapter has focused on characterizing relationships between bacteria and species of HA. However, interactions between members the microbial community (e.g., production of antibiotics; Holmström and Kjelleberg 1999) may play an important role in determining composition and thus functional attributes of the bacterial assemblage occurring in association with algal cells. This, in turn, can have a direct effect on the nature of algal-bacterial interactions. For example, Long and Azam (2001) reported that greater than 50% of 86 pelagic marine bacterial isolates tested exhibited antagonistic properties against other bacteria via the production of growth inhibiting compounds. Perhaps, not surprisingly, particle-associated isolates, which must compete for available surfaces to colonize, were far more likely than free-living forms to produce such substances and the gamma Proteobacteria contained the most representatives either producing inhibitory materials or resistant to their effects. These findings have important biogeochemical implications (e.g., carbon cycling), since changes in species composition resulting from antagonistic interactions between bacteria will affect both the quality and quantity of particulate organic material degraded by microbial communities.

Inhibition of a bacterial taxon by other bacteria may also indirectly modulate the population dynamics of HA species. Indeed, Mayali and Doucette

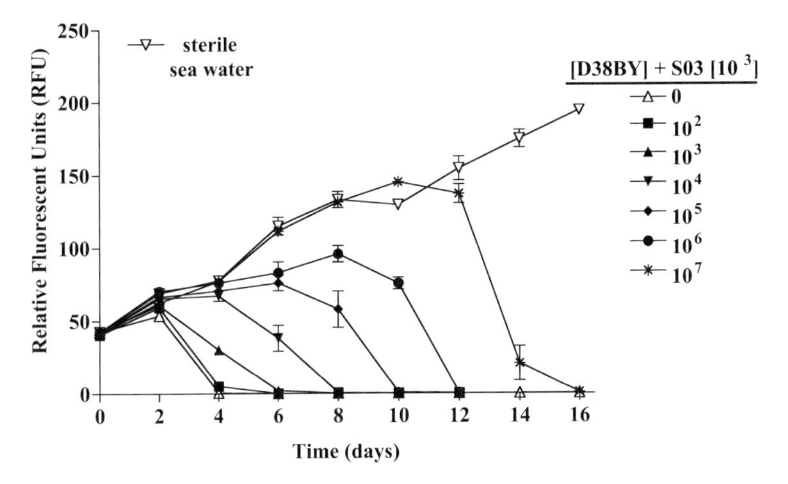

Fig 19.2. A 'protective' bacterium (strain D38BY, Flavobacteriaceae) was added at concentrations ranging from 0–10^7 cells mL^{-1} to bacteria-free *Karenia mikimotoi* co-cultures. D38BY inoculations followed immediately the addition of algicidal strain S03 at 10^3 cells mL^{-1} on Day 0. A positive control consisted of sterile seawater and culture growth and was monitored by in vivo fluorescence (after Roth 2005)

(2002) established that resistance of a *Karenia brevis* culture to attack by an algicidal bacterium was not an intrinsic property of the algal cells but instead was mediated by the associated bacterial community and that this resistance could be conferred to susceptible *K. brevis* cultures simply by switching ambient bacteria with a resistant culture. Recently, a representative of this "protective" bacterial community belonging to the Flavobacteriaceae was isolated and shown to inhibit the killing activity of this algicidal bacterium in a concentration-dependent manner (Fig. 19.2, Roth 2005), although the antagonistic mechanism involved remains to be determined. Nevertheless, given the potentially important role of algicidal bacteria in regulating algal population growth, it would appear that this algal-bacterial interaction may ultimately be regulated by antagonistic relationships between bacterial taxa. These findings, in turn, raise questions of whether algal species are able to selectively promote the growth of bacteria that may inhibit the killing activity of algicidal strains and if such ecological relationships are common in the marine environment.

19.5 Future Directions/Research Needs/Critical Questions

Discussion of algal-bacterial communication was outside the scope of the present summary. Nonetheless, it is evident from numerous examples throughout this chapter that bacterial-algal, cell-to-cell interactions are likely to be pivotal factors regulating HAB ecology, whether via cell signaling or micronutrient exchange (e.g., Miller et al. 2004). Parallel strategies employing model algal-bacterial communities in conjunction with classical physiological as well as gene expression (i.e., transcriptomics/proteomics/metabolomics) approaches are needed to reveal the micro-scale processes operating between cells that have meso- and meta-scale consequences (e.g., Moran et al. 2004). There does not appear to be any question that bacteria are involved in HAB ecology, yet the challenge we now face is to produce compelling, quantifiable evidence of their role in natural systems. This effort will require identification of specific mechanisms of action coupled with measurements of the chemicals being exchanged, regulated, or modified, in order to establish their significance throughout a bloom event.

Acknowledgements. The authors are grateful to the following colleagues for making available their unpublished data: C. Bolch, K. Jones, K. Kobayashi, C. Mikulski, P. Roth, M. Twiner. Partial support was provided by the ECOHAB Inter-Agency Funding Initiative and NOAA/NOS operational funds (GJD) and a New Zealand Foundation for Research, Science and Technology postdoctoral fellowship (DHG). This is contribution 197 of the U.S. ECOHAB Program, sponsored by NOAA, NSF, EPA, NASA, and ONR.

References

Adachi M, Kanno T, Matsubara T, Nishijima T, Itakura S, Yamaguchi M (1999) Promotion of cyst formation in the toxic dinoflagellate *Alexandrium* (Dinophyceae) by natural bacterial assemblages from Hiroshima Bay, Japan. Mar Ecol Prog Ser 191:175–185

Adachi M, Matsubara T, Okamoto R, Nishijima T, Itakura S, Yamaguchi M (2002) Inhibition of cyst formation in the toxic dinoflagellate *Alexandrium* (Dinophyceae) by bacteria from Hiroshima Bay, Japan. Aquat Microb Ecol 26:223–233

Alavi M, Miller T, Erlandson K, Schneider R, Belas R (2001) Bacterial community associated with *Pfiesteria*-like dinoflagellate cultures. Environ Microbiol 3:380–396

Allgaier M, Uphoff H, Felske A, Wagner-Dobler I (2003) Aerobic anoxygenic photosynthesis in *Roseobacter* clade bacteria from diverse marine habitats. Appl Environ Microbiol 69:5051–5059

Alverca E, Biegala IC, Kennaway GM, Lewis J, Franca S (2002) In situ identification and localization of bacteria associated with *Gyrodinium instriatum* (Gymnodiniales, Dinophyceae) by electron and confocal microscopy. Eur J Phycol 37:523–530

Azam F (1998) Microbial control of oceanic carbon flux: the plot thickens. Science 280:694–696

Babin M, Cullen JJ, Roesler CS, Donaghay PL, Doucette GJ, Kahru M, Lewis MR, Scholin CA, Sieracki ME, Sosik HM (2005) New approaches and technologies for observing harmful algal blooms. Oceanography 18:210–227

Barbeau K, Rue EL, Bruland KW, Butler A (2001) Photochemical cycling of iron in the surface ocean mediated by microbial iron (III)-binding ligands. Nature 413:409–413

Bates SS, Douglas DJ, Doucette GJ, Leger C (1995) Enhancement of domoic acid production by reintroducing bacteria to axenic cultures of the diatom *Pseudo-nitzschia multiseries*. Nat Toxins 3:428–435

Bates SS, Gaudet J, Kaczmarska I, Ehrman JM (2004) Interaction between bacteria and the domoic acid-producing diatom *Pseudo-nitzschia multiseries* (Hasle) Hasle; can bacteria produce domoic acid autonomously? Harmful Algae 3:11–20

Bricelj VM, Shumway SE (1998) Paralytic shellfish toxins in bivalve mollusks: occurrence, transfer kinetics, and biotransformation. Rev Fish Sci 6:315–383

Curtis TP, Sloan WT (2004) Prokaryotic diversity and its limits: microbial community structure in nature and implications for microbial ecology. Curr Opin Microbiol 7:221–226

Dale C, Plague GR, Wang B, Ochman H, Moran NA (2002) Type III secretion systems and the evolution of mutualistic endosymbiosis. PNAS 99:12397–12402

Doucette GJ (1995) Interactions between bacteria and harmful algae: a review. Natural Toxins 3:65–74

Doucette GJ, Kodama M, Franca S, Gallacher S (1998) Bacterial interactions with harmful algal bloom species: bloom ecology, toxigenesis, and cytology. In: Anderson DM, Cembella AD, Hallegraeff GM (eds) Physiological ecology of harmful algal blooms. NATO ASI Series 41. Springer, Berlin Heidelberg New York, pp 619–648

Doucette GJ, McGovern ER, Babinchak JA (1999) Algicidal bacteria active against *Gymnodinium breve* (Dinophyceae). I. Bacterial isolation and characterization of killing activity. J Phycol 35:1447–1454

Ferrier M, Martin JL, Rooney-Varga JN (2002) Stimulation of *Alexandrium fundyense* growth by bacterial assemblages from the Bay of Fundy. J Appl Microbiol 92:706–716

Gallacher S, Flynn KJ, Franco JM, Brueggemann EE, Hines HB (1997) Evidence for production of paralytic shellfish toxins by bacteria associated with *Alexandrium* spp. (Dinophyta) in culture. Appl Environ Microbiol 63:239–245

Green DH, Llewellyn LE, Negri AP, Blackburn SI, Bolch CJS (2004) Phylogenetic and functional diversity of the cultivable bacterial community associated with the paralytic shellfish poisoning dinoflagellate *Gymnodinium catenatum*. FEMS Microbiol Ecol 47:345–357

Grossart H-P, Levold F, Allgaier M, Simon M, Brinkhoff T (2005) Marine diatom species harbour distinct bacterial communities. Environ Microbiol 7:860–873

Haines KC, Guillard RRL (1974) Growth of vitamin B_{12}-requiring marine diatoms with vitamin B_{12}-producing marine bacteria. J Phycol 10:245–252

Hold GL, Smith EA, Rappe MS, Maas EW, Moore ERB, Stroempl C, Stephen JR, Prosser JI, Birkbeck TH, Gallacher S (2001) Characterisation of bacterial communities associated with toxic and non-toxic dinoflagellates: *Alexandrium* spp. and *Scrippsiella trochoidea*. FEMS Microbiol Ecol 37:161–173

Holmström C, Kjelleberg S (1999) Marine *Pseudoalteromonas* species are associated with higher organisms and produce biologically active extracellular agents. FEMS Microbiol Ecol 30:285–293

Jasti S, Sieracki ME, Poulton NJ, Giewat MW, Rooney-Varga JN (2003) Phylogenetic diversity of bacteria associated with *Alexandrium* spp. and other phytoplankton from the Gulf of Maine. Second Symposium on Harmful Marine Algae in the United States, Woods Hole, MA. In: Harmful Algae 3:216 (Abstract)

Kaczmarska I, Ehrman JM, Bates SS, Green DH, Leger C, Harris J (2005) Diversity and distribution of epibiotic bacteria on *Pseudo-nitzschia multiseries* (Bacillariophyceae) in culture, and comparison with those on diatoms in native seawater. Harmful Algae 4:725–741

Kirchman DL (2002) The ecology of *Cytophaga-Flavobacteria* in aquatic environments. FEMS Microbiol Ecol 39:91–100

Kobayashi K, Kobiyama A, Kotaki Y, Kodama M (2003) Possible occurrence of intracellular bacteria in *Pseudo-nitzschia multiseries*, a causative diatom of amnesic shellfish poisoning. Fish Sci 69:974–978

Kodama M, Ogata T, Sato S (1988) Bacterial production of saxitoxin. Agric Biol Chem 52:1075–1077

Kotaki Y (1989) Screening of bacteria which convert gonyautoxin 2,3 to saxitoxin. Nippon Suisan Gakkaishi 55:1239

Kotaki Y, Oshima Y, Yasumoto T (1985) Bacterial transformation of paralytic shellfish toxins in coral reef crabs and a marine snail. Bull Jpn Soc Sci Fish 51:1009–1013

Long RA, Azam F (2001) Microscale patchiness of bacterioplankton assemblage richness in seawater. Aquat Microb Ecol 26:103–113

Martins CA, Alvito P, Tavares MJ, Pereira P, Doucette GJ, Franca S (2003) Re-evaluation of paralytic shellfish toxin production by bacteria associated with dinoflagellates of the Portuguese coast. Appl Environ Microbiol 69:5693–5698

Mayali X, Azam F (2004) Algicidal bacteria in the sea and their impact on algal blooms. J Eukaryot Microbiol 51:139–144

Mayali X, Doucette GJ (2002) Microbial community interactions and population dynamics of an algicidal bacterium active against *Karenia brevis* (Dinophyceae). Harmful Algae 1:277–293

Miller TR, Hnilicka K, Dziedzic A, Desplats P, Belas R (2004) Chemotaxis of *Silicibacter* sp. strain TM1040 toward dinoflagellate products. Appl Environ Microbiol 70:4692–4701

Moran MA, Buchan A, González JM et al (2004) Genome sequence of *Silicibacter pomeroyi* reveals adaptations to the marine environment. Nature 432:910–913

Mouriño-Pérez RR, Worden AZ, Azam F (2003) Growth of *Vibrio cholerae* O1 in red tide waters off California. Appl Environ Microbiol 69:6923–6931

Rooney-Varga JN, Giewat MW, Savin MC, Sood S, Legresley M, Martin JL (2005) Links between phytoplankton and bacterial community dynamics in a coastal marine environment. Microb Ecol 49:163–175

Rose JB, Epstein PR, Lipp EK, Sherman BH, Bernard SM, Patz JA (2001) Climate variability and change in the United States: potential impacts on water- and foodborne diseases caused by microbiologic agents. Environ Health Perspect 109 2:211–221

Roth P (2005) The microbial community associated with the Florida red tide dinoflagellate *Karenia brevis*: algicidal and antagonistic interactions. College of Charleston, Charleston, SC. MSc Thesis, pp 161

Sakamoto S, Sato S, Ogata T, Kodama M (2000) Formation of intermediate conjugates in the reductive transformation of gonyautoxins to saxitoxins by thiol compounds. Fish Sci 66:136–141

Sakami T, Nakahara H, Chinain M, Ishida Y (1999) Effects of epiphytic bacteria on the growth of the toxic dinoflagellate *Gambierdiscus toxicus* (Dinophyceae). J Exp Mar Biol Ecol 233:231–246

Sato S, Shimizu Y (1998) Purification of a fluorescent product of the bacterium *Moraxella* sp.: a neosaxitoxin impostor. In: Reguera B, Blanco J, Fernandez ML, Wyatt T (eds) Harmful algae. Xunta de Galicia and IOC-UNESCO, Santiago de Compostela, Spain, pp 465–467

Sato S, Sakai R, Kodama M (2000) Identification of thioether intermediates in the reductive transformation of gonyautoxins into saxitoxins by thiols. Bioorg Med Chem 10:1787–1789

Smith EA, Grant F, Ferguson CMJ, Gallacher S (2001) Biotransformations of paralytic shellfish toxins by bacteria isolated from bivalve molluscs. Appl Environ Microbiol 67:2345–2353

Smith EA, Mackintosh FH, Grant F, Gallacher S (2002) Sodium channel blocking (SCB) activity and transformation of paralytic shellfish toxins (PST) by dinoflagellate-associated bacteria. Aquat Microbial Ecol 29:1–9

Stewart JE, Marks LJ, Gilgan MW, Pfeiffer E, Zwicker BM (1998) Microbial utilization of the neurotoxin domoic acid: blue mussels (*Mytilus edulis*) and soft shell clams (*Mya arenaria*) as sources of the microorganisms. Can J Microbiol 44:456–464

Wichels A, Hummert C, Elbrächter M, Luckas B, Schutt C, Gerdts G (2004) Bacterial diversity in toxic *Alexandrium tamarense* blooms off the Orkney Isles and the Firth of Forth. Helgoland Mar Res 58:93–103

Part D
Harmful Algae and the Food Web

20 Harmful Algae Interactions with Marine Planktonic Grazers

J.T. Turner

20.1 Introduction

Toxins from harmful algae (HA) can cause shellfish toxicity and become accumulated in and transported through benthic and pelagic marine food webs by trophic interactions. In pelagic food webs, this often causes harmful effects on upper-trophic-level consumers such as marine mammals, seabirds and humans (Turner and Tester 1997; Turner et al. 1998). Effects of HA toxins on their primary grazers, zooplankton and planktivorous fish are variable, in some cases causing deleterious effects on grazers, but in others having little or no obvious effects. Such variability likely relates to different concentrations of different toxins having different effects on different grazers. Just because some phytoplankters are toxic does not mean that toxins necessarily evolved to repel grazers. The numerous documented cases where phycotoxins have minimal effects on grazers suggest that toxicity may be coincidental, or that these chemicals may have evolved for other reasons, such as nitrogen storage, bioluminescence, chromosome structural organization, pheromones for induction of sexuality, or they may be vestigial remnants of archaic pathways for nucleic acid biosynthesis (Cembella 2003).

Previous reviews (Turner and Tester 1997; Turner et al. 1998) summarized information on interactions between toxic phytoplankton and their zooplanktonic grazers published prior to about the mid 1990s. Also, Turner et al. (2002) reviewed information on grazer interactions with *Phaeocystis*, published prior to 2002. Most previous information on HA grazer interactions dealt with copepods. There have been considerable advances within the last decade on grazer interactions with harmful algae, including not only copepods and other mesozooplankton (metazoans >200 μm in longest dimension), but also microzooplankton (mostly protists <200 μm in longest dimension). This recent information will be the focus of the present review, with citation of the aforementioned reviews for most studies published prior to

Ecological Studies, Vol. 189
Edna Granéli and Jefferson T. Turner (Eds.)
Ecology of Harmful Algae

1995. Due to space constraints, many papers that should be cited cannot be, so reference will be primarily to more recent papers that contain citations to others.

There has been a conceptual shift during the last decade in understanding of planktonic trophodynamics. It now appears that most predation mortality on marine phytoplankton is due to microzooplankton, rather than mesozooplankton grazing. Microzooplankton have been estimated to consume approximately 60–70 % of phytoplankton primary production (Calbet and Landry 2004), whereas mesozooplankton, particularly copepods may consume only 10–40 % (Calbet 2001). Thus, it has recently been suggested that phytoplankton blooms, including harmful ones, may be in response to "loopholes" in the normal microzooplankton grazing controls exerted on bloom development by the microbial loop (Irigoien et al. 2005). Microzooplankton protists that have been studied in terms of their predation on HA include heterotrophic dinoflagellates, tintinnids, and aloricate ciliates. Although some HA toxins appear to serve as grazer deterrents to some microzooplankton such as heterotrophic dinoflagellates and ciliates (Skovgaard and Hansen 2003), for other HA blooms, microzooplankton community grazing appears to be an important factor regulating bloom development. There have also been further studies of grazing on HA by mesozooplankton, including rotifers, copepods, euphausiids, and others. These studies, as well as those of zooplankton grazing impact on HA blooms, will be summarized below.

20.2 Planktonic Grazers

20.2.1 Heterotrophic Dinoflagellates and other Flagellates

Approximately half of the known species of dinoflagellates do not contain chlorophyll-*a* and appear to feed heterotrophically (Hansen and Calado 1999; Jacobson 1999; Jeong 1999). Several heterotrophic dinoflagellates have been shown to feed upon various species of harmful algae (Matsuyama et al. 1999; Matsuoka et al. 2000; Jeong et al. 2001, 2003; Olseng et al. 2002; Stoecker et al. 2002; Clough and Strom 2005, and references therein), and such predation may influence development or termination of HA blooms (Nakamura et al. 1996; Matsuyama et al. 1999).

Some autotrophic dinoflagellates are also mixotrophic (Stoecker 1999; Smalley et al. 2003), being capable of ingesting other dinoflagellates. Several heterotrophic dinoflagellates have been shown to ingest other demonstrably toxic dinoflagellates, but apparently only the studies of Jeong et al. (2001, 2003) show that heterotrophic dinoflagellates can retain dinoflagellate toxins and make them available for vectorial intoxication of their consumers. Jeong et al. (2001) found that copepods ingested fewer cells of a heterotrophic dinoflagellate that itself had been feeding on a toxic dinoflagellate, compared to diets of the same heterotrophic dinoflagellate that had been feeding on a non-toxic dinoflagellate. Jeong et al. (2003) found that toxicity declined in a heterotrophic dinoflagellate that had been initially fed another toxic dinoflagellate and then starved, showing that the heterotrophic dinoflagellate had retained toxins from its diet, but that it depurated toxins over time.

Toxic haptophycean flagellates of the genus *Prymnesium* (which are not dinoflagellates) use a combination of chemical warfare and predation to successfully compete against other phytoplankton. These algae produce chemicals with allelopathic effects which can kill other phytoplankton, which these flagellates then eat (Tillmann 2003).

20.2.2 Tintinnids and Aloricate Ciliates

Several species of tintinnids and aloricate ciliates (Maneiro et al. 2000; Jakobsen et al. 2001; Stoecker et al. 2002; Gransden and Lewitus 2003; Rosetta and McManus 2003; Kamiyama and Matsuyama 2005; Setälä et al. 2005, and references therein) are known to ingest toxic dinoflagellates or other harmful phytoplankton, and may be important in controlling their blooms (Montagnes and Lessard 1999). However, in other cases, toxic dinoflagellates or other harmful phytoplankton appear to have deleterious effects on ciliates such as changes in swimming behavior, reduced ingestion, inability to support growth or even causing ciliate mortality (Kamiyama and Arima 1997; Liu and Buskey 2000; Jakobsen et al. 2001; Granéli and Johansson 2003; Rosetta and McManus 2003; Clough and Strom 2005).

In what is apparently the first unambiguous indication of HA toxin accumulation in tintinnids, Maneiro et al. (2000) found that concentrations of okadaic acid in smaller zooplankton size fractions (100–200, 200–300 μm) showed good correlation with abundance of the tintinnid *Favella serrata*, indicating that the tintinnid had accumulated okadaic acid from feeding on blooms of the dinoflagellate *Dinophysis acuminata*. However, we suspect that levels of PSP toxins in the small zooplankton size fraction (64–100 μm) from Massachusetts Bay (Turner et al. 2000) and Casco Bay, Gulf of Maine (Doucette et al. 2005; Turner et al. 2005), during periods when *Alexandrium* was not abundant in that size fraction, but tintinnids were, may also reflect vectorial intoxication of tintinnids from consuming these dinoflagellates.

20.2.3 Rotifers

Rotifers are abundant components of the spring zooplankton in various temperate coastal waters. A combination of requirements for low salinity, high concentrations of phytoplankton food, and shallow water to facilitate the benthic resting egg portion of their life cycle limits marine rotifers to estuarine or coastal waters, but together with parthogenetic reproduction, allows explosive population growth, high biomass and high grazing impact on phytoplankton where such conditions are met (Park and Marshall 2000, and references therein).

Marine rotifers feed upon a variety of autotrophic and heterotrophic flagellate prey. The rotifer *Brachionus plicatilis* fed upon the toxic dinoflagellate *Pfiesteria piscicida* with no apparent adverse effects in terms of reduced fecundity or increased mortality, but the same rotifer species failed to sustain growth on a diet of the Texas brown tide alga (references in Turner and Tester 1997). Barreiro et al. (2005) found that ingestion of the toxic haptophyte *Prymnesium parvum* by *B. plicatilis* resulted in lower growth and higher mortality of the rotifer than on non-toxic prey. Wang et al. (2005) found that when *B. plicatilis* was exposed to ten different strains of *Alexandrium* spp., some were ingested with no adverse effects, whereas others were lethal.

There is apparently no information as to whether marine rotifers can concentrate and retain PSP toxins from toxic dinoflagellates, but results of Doucette et al. (2005) from the Gulf of Maine in which toxins were recorded for microplankton size fractions dominated by rotifers suggests that they do.

20.2.4 Copepods and other Mesozooplankton

Various species of copepods have been shown to feed upon a variety of species of toxic phytoplankton. In many recent studies, ingestion of toxic phytoplankton induced adverse effects on copepods, such as reduced ingestion, fecundity, egg hatching success or survival (Dutz 1998; Delgado and Alcaraz 1999; Koski et al. 1999; Teegarden 1999; Frangópulos et al. 2000; Maneiro et al. 2000; Jeong et al. 2001; Tang et al. 2001; Colin and Dam 2002; Guisande et al. 2002; Kozlowsky-Suzuki et al. 2003; da Costa et al. 2005; Dutz et al. 2005, and references therein). However, in other studies with different HA diets or copepod species, there were no apparent adverse effects associated with eating toxic algae (Lincoln et al. 2001; Teegarden et al. 2001; Colin and Dam 2002; Koski et al. 2002, 2005; Maneiro et al. 2002; Wexels Riser et al. 2003; da Costa et al. 2005; Kozlowsky-Suzuki et al. 2006; Turner and Borkman 2005).

Much of this variability likely relates to different species or concentrations of HA, with different toxin levels offered as food to various copepods. Generally, diets with more toxins cause greater effects. Also, as suggested by Turner and Tester (1989), there may be a biogeographic aspect to effects of HA toxins on copepods. Colin and Dam (2002) found that copepods from areas where they are routinely exposed to natural HA blooms were less affected by ingesting these toxic phytoplankters then copepods of the same species that were from areas where they were not naturally exposed to HA toxins.

20.3 HAB Toxin Accumulation in Zooplankton

Some copepods also accumulate phytoplankton toxins through ingestion. However, recent studies (Guisande et al. 2002; Lehtiniemi et al. 2002; Hamasaki et al. 2003; Teegarden et al. 2003) suggest that copepods may be inefficient in retaining ingested PSP toxins, and may either metabolize, and/or disperse considerable proportions of ingested PSP toxins into the environment where they are less likely to contribute to vectorial intoxication of higher trophic levels. Nonetheless, Teegarden et al. (2003) concluded that even if copepods do not retain much of the large amounts of toxins calculated to have been consumed, that body burdens of PSP toxins in copepods in the Gulf of Maine may still represent a threat of transfer of toxins to higher trophic levels.

Other mesozooplankters (>200 μm in length) also ingest HA and in some cases accumulate HA toxins. Included are euphausiids from California (Bargu et al. 2002) feeding on domoic-acid producing species of the diatom *Pseudo-nitzschia*, and euphausiids from the Antarctic and the Arctic feeding on colonial *Phaeocystis* (references in Turner et al. 2002). Planktonic crab larvae may exhibit adverse effects to exposure to toxic phytoplankton, including reduced respiration and locomotory anomalies and reduced ingestion (Perez and Sulkin 2005).

20.4 Selective Grazing and Feeding Deterrence by Harmful Algae

It is often assumed that HA toxins evolved to repel grazers, and that selective grazing against these species, or poisoning of grazers after ingestion of toxic algae contributes to HA bloom development. However, there is surprisingly little unambiguous evidence for this scenario, at least for the zooplankton, which are the primary grazers of HA microalgae.

Selective feeding implies disproportionate ingestion of a given food item. Thus, in order to properly address selective feeding, one must quantify the relative proportions of different food items in the ingested, as well as the relative proportions of the taxa comprising the natural food assemblage (see Turner and Borkman 2005).

Many studies addressing selection against HA have used laboratory diets of unialgal or mixed phytoplankton cultures at cell concentrations higher than those of most natural HA levels (papers published prior to approximately the mid-1990s reviewed by Turner et al. 1998). Many such studies revealed that some grazers appear to select against some toxic algae, but many others do not.

Laboratory studies within the last decade reveal similar selective feeding variability, for a variety of grazers, including protists (Granéli and Johansson 2003) and copepods (Teegarden 1999; Tang et al. 2001; Dutz et al. 2005). Increasingly, there have been studies of zooplankton grazing on HA taxa as well as other components of natural mixed plankton assemblages during natural blooms (Teegarden et al. 2001; Calbet et al. 2003; Caron et al. 2004; Campbell et al. 2005; Jansen et al. 2006; Koski et al. 2005; Kozlowsky-Suzuki et al. 2006; Turner and Borkman 2005). Despite some variability, an emerging pattern seems to be that HA species usually comprise a relatively small proportion of the total phytoplankton abundance in natural assemblages, even during HA blooms, and grazers unselectively ingest HA taxa, along with concurrent ingestion of other phytoplankton. This would dilute potential adverse effects of HA toxins on grazers, such as those seen in less diverse laboratory diets with artificially elevated concentrations of HA cells and toxins.

20.5 Impact of Zooplankton Grazing on Formation and Termination of HA Blooms

Turner et al. (1998) reviewed earlier studies of zooplankton community grazing impact on natural HA blooms. In some cases, grazing is incapable of preventing development of HA blooms, but may contribute to their termination (Nakmura et al. 1996; Buskey et al. 2001). Conversely, grazing may retard initial development a HA bloom, but once developed, the bloom becomes immune to grazing pressure. Recent studies indicate that copepod grazing impact on HA blooms is variable, but usually quite low (Campbell et al. 2005; Jansen et al. 2006; Kozlowsky-Suzuki et al. 2006; Turner and Borkman 2005).

Microzooplankton grazing impact on HA blooms appears to exceed that of mesozooplankton. Grazing by protists such as heterotrophic dinoflagellates and ciliates has been shown to cause considerable mortality (> 50 %) of even collapse of dinoflagellate blooms (Nakamura et al. 1996; Matsuyama et al. 1999; Kamiyama et al. 2001; Kamiyama and Matsuyama 2005). Jeong et al. (2005) estimated that the grazing impact of populations of the heterotrophic dinoflagellate *Stoeckeria algicida* on the harmful alga *Heterosigma akashiwo* could remove as much as 13 % of the prey population per minute. Caron et al. (2004) found that mortality rates of the brown tide alga *Aureococcus anophagefferens* due to microprotistan community grazing closely matched growth rates of this alga, during the later stages of blooms in Long Island, NY, embayments. In other brown tide blooms in New York (Gobler et al. 2004) and Texas estuaries (Buskey et al. 2001), it appeared that the microzooplankton communities could not effectively graze the brown-tide algae, allowing development of blooms. Johnson et al. (2003) estimated that potential microzooplankton grazing, primarily by heterotrophic dinoflagellates and ciliates, frequently exceeded growth rates of HA dinoflagellates during summer blooms in Chesapeake Bay.

In an elegant comparison of grazing impact by microzooplankton and mesozooplankton during a bloom of the toxic dinoflagellate *Alexandrium minutum* from the Mediterranean, Calbet et al. (2003) found that microzooplankton grazing impact on the bloom was substantial, but that of copepods was trivial. Grazing by the microzooplankton community (0.84 d^{-1}), and was equal to, or exceeded *A. minutum* growth rates (0.79 d^{-1}). However, the grazing impact of the community of copepods (*Acartia grani* and *Oithona davisae*) caused daily reductions of the *A. minutum* standing stock of only 0.007 and 0.003 %, respectively.

20.6 Conclusions

In the decade since the previous reviews of Turner and Tester (1997) and Turner et al. (1998), several important points have emerged from studies of interactions between pelagic grazers and HA species. These are summarized below.

Many planktonic grazers ingest toxic phytoplankton, often with no deleterious effects. When such adverse effects are noted, they are often at unrealistically high HA concentrations typical of laboratory cultures, rather than at the lower concentrations typically recorded for HA natural blooms. Thus, it is questionable whether HA toxins are effective grazer deterrents in nature. Even if HA toxins potentially discourage grazers from sustained ingestion of HA species, in natural mixed phytoplankton assemblages, and at the low concentrations of HA algae typically found in blooms, effects of HA toxins may be minimized or diluted to the irrelevant by concurrent ingestion of non-toxic co-occurring prey by non-selective omnivorous grazing.

The last decade has seen a growing realization that HA grazers include microzooplankton such as heterotrophic dinoflagellates and ciliates, in addition to traditionally viewed mesozooplankton grazers such as copepods, rotifers and euphausiids. It is now becoming apparent that the grazing impact of microzooplankton on HA blooms likely exceeds that of the mesozooplankton in many if not most cases (Calbet et al. 2003).

The zooplankton community grazing impact on HA blooms is highly variable. In some cases, grazing may retard initial development of HA blooms, whereas in other cases, grazing may contribute to termination of blooms. Breakdowns of community grazing at various trophic levels may also allow blooms such as the Texas brown tide to respond to anomalous hydrographic conditions, and once developed, to sustain themselves. There are also interactions between different components of grazer communities. An example would be when copepod predation on ciliates that eat single cells of *Phaeocys-*

tis might actually contribute to perpetuation of a *Phaeocystis* bloom (references in Turner et al. 2002). Variability in such interactions makes each HA bloom unique, and defies universal generalizations.

While microzooplankton grazing on HA blooms is now generally viewed as more important than mesozooplankton grazing, grazing by mesozooplankton (and phytoplanktivorous fish) is still viewed as the initial pathway through which HA toxins become vectored into pelagic food webs. Subsequent accumulation and trophic transferal can intoxicate higher-trophic-level consumers such as fish, sea birds, and marine mammals, often with catastrophic consequences (Scholin et al. 2000). There is also increasing evidence that HA toxins enter pelagic pathways through the microzooplankton, in addition to the mesozooplankton (Doucette et al. 2005).

Even if HA taxa are considerably grazed, the grazing impact might not be enough to prevent blooms. Since most HA blooms are coastal phenomena, they usually occur in well-mixed nutrient-replete, and shallow waters, where the euphotic zone may extend to near-bottom depths. Thus, even though such waters may be grazer-rich, they are optimal for phytoplankton growth, which may surpass the ability of the grazer community to control bloom development. Since many HA algae typically bloom in the temperate zone during spring, at the time when water temperatures are still cold, and seasonal zooplankton populations are just beginning to develop, it may be that low zooplankton community abundance, rather than toxin-associated grazer deterrence explains why community grazing cannot keep up with spring bloom development. Conversely, the causes of the decline of most HA blooms cannot usually separate effects of a robust grazer community from those of a senescent HA population in nutrient-depleted water. Further, intra-zooplankton predation patterns may actually perpetuate some HA blooms if microzooplankton are the major HA grazers, and their populations are reduced by mesozooplankton omnivory.

The importance of grazing on HA blooms becomes apparent by its absence. For whatever reasons, and despite a plethora of detailed complications, a HA bloom implies that HA phytoplankters have outgrown, or been physically accumulated beyond the ability of zooplankton grazers to regulate them. Thus, grazing on HA blooms is an important aspect of the ecology of HA blooms.

References

Bargu S, Powell CL, Coale SL, Busman M, Doucette GJ, Silver MW (2002) Krill: a potential vector for domoic acid in marine food webs. Mar Ecol Prog Ser 237:209–216

Barreiro A, Guisande C, Maneiro I, Lien TP, Legrand C, Tamminen T, Lehtinen S, Uronen P, Granéli E (2005) Relative importance of the different negative effects of the toxic haptophyte *Prymnesium parvum* on *Rhodomonas salina* and *Brachionus plicatilis*. Aquat Microb Ecol 38:259–267

Buskey EJ, Liu H, Collumb C, Bersano JGF (2001) The decline and recovery of a persistent Texas brown tide algal bloom in the Laguna Madre (Texas USA). Estuaries 24:337–346

Calbet A (2001) Mesozooplankton grazing effect on primary production: a global comparative analysis in marine ecosystems. Limnol Oceanogr 46:1824–1830

Calbet A, Landry M (2004) Phytoplankton growth, microzooplankton grazing, and carbon cycling in marine ecosystems. Limnol Oceanogr 49:51–57

Calbet A, Vaqué D, Felipe J, Vila M, Sala MM, Alcaraz M, Estrada M (2003) Relative grazing impact of microzooplankton and mesozooplankton on a bloom of the toxic dinoflagellate *Alexandrium minutum*. Mar Ecol Prog Ser 259:303–309

Campbell RG, Teegarden GJ, Cembella AD, Durbin EG (2005) Zooplankton grazing impacts on *Alexandrium* spp. in the near-shore environment of the Gulf of Maine. Deep-Sea Res II 52:2817–2833

Caron DA, Gobler CJ, Lonsdale DJ, Cerrato RM, Schaffner RA, Rose JM, Buck NJ, Taylor G, Boissonneault KR, Mehran R (2004) Microbial herbivory on the brown tide alga, *Aureococcus anophagefferens*: results from natural ecosystems, mesocosms and laboratory experiments. Harmful Algae 3:439–457

Cembella AD (2003) Chemical ecology of eukaryotic microalgae in marine ecosystems. Phycologia 42:420–447

Clough J, Strom S (2005) Effects of *Heterosigma akashiwo* (Raphidophyceae) on protist grazers: laboratory experiments with ciliates and heterotrophic dinoflagellates. Aquat Microb Ecol 39:121–134

Colin SP, Dam H (2002) Latitudinal differentiation in the effects of the toxic dinoflagellate *Alexandrium* spp. on the feeding and reproduction of populations of the copepod *Acartia hudsonica*. Harmful Algae 1:113–125

da Costa RM, Franco J, Cacho E, Fernández F (2005) Toxin content and toxic effects of the dinoflagellate *Gyrodinium corsicum* (Paulmier) on the ingestion and survival rates of the copepods *Acartia grani* and *Euterpina acutifrons*. J exp Mar Biol Ecol 322:177–183

Delgado M, Alcaraz M (1999) Interactions between red tide microalgae and herbivorous zooplankton: the noxious effects of *Gyrodinium corsicum* (Dinophyceae) on *Acartia grani* (Copepoda: Calanoida). J Plankton Res 21:2361–2371

Doucette GJ, Turner JT, Powell CL, Keafer BA, Anderson DM (2005) Trophic accumulation of PSP toxins in zooplankton during *Alexandrium fundyense* blooms in Casco Bay, Gulf of Maine, April–June, 1998. I. Toxin levels in *A. fundyense* and zooplankton size fractions. Deep-Sea Res II 52:2764–2783

Dutz J (1998) Repression of fecundity in the neritic copepod *Acartia clausi* exposed to the toxic dinoflagellate *Alexandrium lusitanicum*: relationship between feeding and egg production. Mar Ecol Prog Ser 175:97–107

Dutz J, Klein Breteler WCM, Kramer G (2005) Inhibition of copepod feeding by exudates and transparent exopolymer particles (TEP) derived from a *Phaeocystis globosa* dominated phytoplankton community. Harmful Algae 4:929–940

Frangópulos M, Guisande C, Maneiro I, Riveiro I, Franco J (2000) Short-term and long-term effects of the toxic dinoflagellate *Alexandrium minutum* on the copepod *Acartia clausi*. Mar Ecol Prog Ser 203:161–169

Gobler CJ, Deonarine S, Leigh-Bell J, Gastrich MD, Anderson OR, Wilhelm SW (2004) Ecology of phytoplankton communities dominated by *Aureococcus anophagefferens*: the role of viruses, nutrients, and microzooplankton grazing. Harmful Algae 3:471–483

Granéli E, Johansson N (2003) Effects of the toxic haptophyte *Prymnesium parvum* on the survival and feeding of a ciliate: the influence of different nutrient conditions. Mar Ecol Prog Ser 254:49–56

Gransden SG, Lewitus AJ (2003) Grazing of two euplotid ciliates on the heterotrophic dinoflagellates *Pfiesteria piscicida* and *Cryptoperidiniopsis* sp. Aquat Microb Ecol 33:303–308

Guisande C, Frangópulos M, Carotenuto Y, Maneiro I, Riveiro I, Vergara AR (2002) Fate of paralytic shellfish poisoning toxins ingested by the copepod *Acartia clausi*. Mar Ecol Prog Ser 240:105–115

Hamasaki K, Takahashi T, Uye S-I (2003) Accumulation of paralytic shellfish poisoning toxins in planktonic copepods during a bloom of the toxic dinoflagellate *Alexandrium tamarense* in Hiroshima Bay, western Japan. Mar Biol 143:981–988

Hansen PJ, Calado AJ (1999) Phagotrophic mechanisms and prey selection in free-living dinoflagellates. J Eukaryot Microbiol 46:382–389

Irigoien X, Flynn KJ, Harris RP (2005) Phytoplankton blooms: a 'loophole' in microzooplankton grazing impact? J Plankton Res 27:313–321

Jacobson DM (1999) A brief history of dinoflagellate feeding research. J Eukaryot Microbiol 46:376–381

Jakobsen HH, Hyatt C, Buskey EJ (2001) Growth and grazing on the 'Texas brown tide' alga *Aureoumbra lagunensis* by the tintinnid *Amphorides quadrilineata*. Aquat Microb Ecol 23:245–252

Jansen S, Wexels Riser C, Wassmann P, Bathmann U (2006) Copepod feeding behaviour and egg production during a dinoflagellate bloom in the North Sea. Harmful Algae 5:102–112

Jeong HJ (1999) The ecological roles of heterotrophic dinoflagellates in marine planktonic community. J Eukaryot Microbiol 46:390–396

Jeong HJ, Kang H, Shim JH, Park JK, Kim JS, Song JY, Choi H-J (2001) Interactions among the toxic dinoflagellate *Amphidinium carterae*, the heterotrophic dinoflagellate *Oxyrrhis marina*, and the calanoid copepods *Acartia* spp. Mar Ecol Prog Ser 218:77–86

Jeong HJ, Park KH, Kim JS, Kang H, Kim CH, Choi H-J, Kim YS, Park JY, Park MG (2003) Reduction in the toxicity of the dinoflagellate *Gymnodinium catenatum* when fed on by the heterotrophic dinoflagellate *Polykrikos kofoidii*. Aquat Microb Ecol 31:307–312

Jeong HJ, Kim JS, Kim JH, Kim ST, Seong KA, Kim TH, Song JY, Kim SK (2005) Feeding and grazing impact of the newly described heterotrophic dinoflagellate *Stoeckeria algicida* on the harmful alga *Heterosigma akashiwo*. Mar Ecol Prog Ser 295:69–78

Johnson MD, Rome M, Stoecker DK (2003) Microzooplankton grazing on *Prorocentrum minimum* and *Karlodinium micrum* in Chesapeake Bay. Limnol Oceanogr 48:238–248

Kamiyama T, Arima S (1997) Lethal effect of the dinoflagellate *Heterocapsa circularisquama* upon the tintinnid ciliate *Favella taraikaensis*. Mar Ecol Prog Ser 160:27–33

Kamiyama T, Matsuyama Y (2005) Temporal changes in the ciliate assemblage and consecutive estimates of their grazing effect during the course of a *Heterocapsa circularisquama* bloom. J Plankton Res 27:303–311

Kamiyama T, Takayama H, Nishii Y, Uchida T (2001) Grazing impact of the field ciliate assemblage on a bloom of the toxic dinoflagellate *Heterocapsa circularisquama*. Plankton Biol Ecol 48:10–18

Koski M, Rosenberg M, Viitasalo M, Tanskanen S, Sjölund U (1999) Is *Prymnesium patelliferum* toxic for copepods? -Grazing, egg production, and egestion of the calanoid copepod *Eurytemora affinis* in mixtures of "good" and "bad" food. ICES J Mar Sci 56:131–139

Koski M, Schmidt K, Engström-Öst J, Viitasalo M, Jónasdóttir SH, Repka S, Sivonen K (2002) Calanoid copepods feed and produce eggs in the presence of toxic cyanobacteria *Nodularia spumigena*. Limnol Oceanogr 47:878–885

Koski M, Dutz J, Klein Breteler WCM (2005) Selective grazing of *Temora longicornis* in different stages of a *Phaeocystis globosa* bloom – a mesocosm study. Harmful Algae 4:915–927

Kozlowsky-Suzuki B, Karjalainen M, Lehtiniemi M, Engström-Öst J, Koski M, Carlsson P (2003) Feeding, reproduction and toxin accumulation by the copepods *Acartia bifilosa* and *Eurytemora affinis* in the presence of the toxic cyanobacterium *Nodularia spumigena*. Mar Ecol Prog Ser 249:237–249

Kozlowsky-Suzuki B, Carlsson P, Rühl A, Granéli E (2006) Food selectivity and grazing impact on toxic *Dinophysis* spp. by copepods feeding on natural plankton assemblages. Harmful Algae 5:57–68

Lehtiniemi M, Engström-Öst J, Karjalainen M, Kozlowsky-Suzuki B, Viitasalo M (2002) Fate of cyanobacterial toxins in the pelagic food web: transfer to copepods or to faecal pellets? Mar Ecol Prog Ser 241:13–21

Lincoln JA, Turner JT, Bates SS, Léger C, Gauthier DA (2001) Feeding, egg production, and egg hatching success of the copepods *Acartia tonsa* and *Temora longicornis* on diets of the toxic diatom *Pseudo-nitzschia multiseries* and the non-toxic diatom *Pseudo-nitzschia pungens*. Hydrobiologia 453/454:107–120

Liu H, Buskey EJ (2000) The extracellular polymeric substance (EPS) layer surrounding *Aureoumbra lagunensis* cells reduces grazing by protozoa. Limnol Oceanogr 45:1187–1191

Maneiro I, Frangópulos M, Guisande C, Fernández M, Reguera B, Riveiro I (2000) Zooplankton as a potential vector of diarrhetic shellfish poisoning toxins through the food web. Mar Ecol Prog Ser 201:155–163

Maneiro I, Guisande C, Frangópulos M, Riveiro I (2002) Importance of copepod faecal pellets to the fate of the DSP toxins produced by *Dinophysis* spp. Harmful Algae 1:333–341

Matsuoka K, Cho H-J, Jacobson DM (2000) Observations of the feeding behavior and growth rates of the heterotrophic dinoflagellate *Polykrikos kofoidii* (Polykrikaceae, Dinophyceae). Phycologia 39:82–86

Matsuyama Y, Miyamoto M, Kotani Y (1999) Grazing impacts of the heterotrophic dinoflagellate *Polykrikos kofoidii* on a bloom of *Gymnodinium catenatum*. Aquat Microb Ecol 17:91–98

Montagnes DJS, Lessard EJ (1999) Population dynamics of the marine planktonic ciliate *Strombidinopsis multiauris*: its potential to control phytoplankton blooms. Aquat Microb Ecol 20:167–181

Nakamura Y, Suzuki S, Hiromi J (1996) Development and collapse of a *Gymnodinium mikimotoi* red tide in the Seto Inland Sea. Aquat Microb Ecol 10:131–137

Olseng CD, Naustvoll L-J, Paasche E (2002) Grazing by the heterotrophic dinoflagellate *Protoperidinium steinii* on a *Ceratium* bloom. Mar Ecol Prog Ser 225:161–167

Park GS, Marshall HG (2000) The trophic contributions of rotifers in tidal freshwater and estuarine habitats. Est Coast Shelf Sci 51:729–742

Perez MF, Sulkin SD (2005) Palatability of autotrophic dinoflagellates to newly hatched larval crabs. Mar Biol 146:771–780

Rosetta CH, McManus GB (2003) Feeding by ciliates on two harmful algal bloom species, *Prymnesium parvum* and *Prorocentrum minimum*. Harmful Algae 2:109–126

Scholin CA et al (2000) Mortality of sea lions along the central California coast linked to a toxic diatom bloom. Nature 403:80–84

Setälä O, Autio R, Kuosa H (2005) Predator-prey interactions between a planktonic ciliate *Strombidium* sp. (Ciliophora, Oligotrichida) and the dinoflagellate *Pfiesteria piscicida* (Dinamoebiales, Pyrrophyta). Harmful Algae 4:235–247

Skovgaard A, Hansen PJ (2003) Food uptake in the harmful alga *Prymnesium parvum* mediated by excreted toxins. Limnol Oceanogr 48:1161–1166

Smalley GW, Coats DW, Stoecker DK (2003) Feeding in the mixotrophic dinoflagellate *Ceratium furca* is influenced by intracellular nutrient concentrations. Mar Ecol Prog Ser 262:137–151

Stoecker DK (1999) Mixotrophy among dinoflagellates. J Eukaryot Microbiol 46:397–401

Stoecker DK, Parrow MW, Burkholder JM, Glasgow HB Jr (2002) Grazing by microzooplankton on *Pfiesteria piscicida* cultures with different histories of toxicity. Aquat Microb Ecol 28:79–85

Tang KW, Jakobsen HH, Visser AW (2001) *Phaeocystis globosa* (Prymnesiophyceae) and the planktonic food web: feeding, growth, and trophic interactions among grazers. Limnol Oceanogr 46:1860–1870

Teegarden GJ (1999) Copepod grazing selection and particle discrimination on the basis of PSP toxin content. Mar Ecol Prog Ser 181:163–176

Teegarden GJ, Campbell RG, Durbin EG (2001) Zooplankton feeding behavior and particle selection in natural plankton assemblages containing toxic *Alexandrium* spp. Mar Ecol Prog Ser 218:213–226

Teegarden GJ, Cembella AD, Capuano CL, Barron SH, Durbin EG (2003) Phycotoxin accumulation in zooplankton feeding on *Alexandrium fundyense* – vector or sink? J Plankton Res 25:429–443

Tillmann U (2003) Kill and eat your predator: a winning strategy of the planktonic flagellate *Prymnesium parvum*. Aquat Microb Ecol 32:73–84

Turner JT, Borkman DG (2005) Impact of zooplankton grazing on *Alexandrium* blooms in the offshore Gulf of Maine. Deep-Sea Res II 52:2801–2816

Turner JT, Tester PA (1989) Zooplankton feeding ecology: copepod grazing during an expatriate red tide. In: Cosper EM, Bricelj VM, Carpenter EJ (eds) Novel phytoplankton blooms: causes and impacts of recurrent brown tides and other unusual blooms. Springer, Berlin Heidelberg New York, pp 359–374

Turner JT, Tester PA (1997) Toxic marine phytoplankton, zooplankton grazers and pelagic food webs. Limnol Oceanogr 42:1203–1214

Turner JT, Tester PA, Hansen PJ (1998) Interactions between toxic marine phytoplankton and metazoan and protozoan grazers. In: Anderson DM, Cembella AD, Hallegraeff GM (eds) Physiological ecology of harmful algal blooms. NATO ASI Series 41. Springer, Berlin Heidelberg New York, pp 453–474

Turner JT, Doucette GJ, Powell CL, Kulis DM, Keafer BA, Anderson DM (2000) Accumulation of red tide toxins in larger size fractions of zooplankton assemblages from Massachusetts Bay, USA. Mar Ecol Prog Ser 203:95–107

Turner JT, Ianora A, Esposito F, Carotenuto Y, Miralto A (2002) Zooplankton feeding ecology: does a diet of *Phaeocystis* support good copepod grazing, survival, egg production and egg hatching success? J Plankton Res 24:1185–1195

Turner JT, Doucette GJ, Keafer BA, Anderson DM (2005) ECOHAB-Gulf of Maine. Trophic accumulation of PSP toxins in zooplankton during *Alexandrium fundyense* blooms in Casco Bay, Gulf of Maine, April–June, 1998. II. Zooplankton abundance and size-fractionated community composition. Deep-Sea Res II 52:2784–2800

Wang L, Yan T, Yu R, Zhou M (2005) Experimental study on the impact of dinoflagellate *Alexandrium* species on populations of the rotifer *Brachionus plicatilis*. Harmful Algae 4:371–382

Wexels Riser C, Jansen S, Bathmann U, Wassmann P (2003) Grazing of *Calanus helgolandicus* on *Dinophysis norvegica* during bloom conditions in the North Sea: evidence from investigations of faecal pellets. Mar Ecol Prog Ser 256:301–304

21 Pathogens of Harmful Microalgae

P.S. Salomon and I. Imai

21.1 Introduction

Pathogens are any organisms that cause disease to other living organisms. Parasitism is an interspecific interaction where one species (the parasite) spends the whole or part of its life on or inside cells and tissues of another living organism (the host), from where it derives most of its food. Parasites that cause disease to their hosts are, by definition, pathogens. Although infection of metazoans by other metazoans and protists are the more frequently studied, there are interactions where both host and parasite are single-celled organisms. Here we describe such interactions involving microalgae as hosts. The aim of this chapter is to review the current status of research on pathogens of harmful microalgae and present future perspectives within the field. Pathogens with the ability to impair and kill microalgae include viruses, bacteria, fungi and a number of protists (see reviews by Elbrächter and Schnepf 1998; Brussaard 2004; Park et al. 2004; Mayali and Azam 2004; Ibelings et al. 2004). Valuable information exists from non-harmful microalgal hosts, and these studies will be referred to throughout the text. Nevertheless, emphasis is given to cases where hosts are recognizable harmful microalgae.

21.2 Viruses

Viruses and virus-like particles (VLPs) have been found in more than 50 species of eukaryotic microalgae, and several of them have been isolated in laboratory cultures (Brussaard 2004; Nagasaki et al. 2005). These viruses are diverse both in size and genome type, and some of them infect harmful algal bloom (HAB)-causing species (Table 21.1).

Studies of viral impact on harmful microalgal blooms in coastal waters include host species such as *Heterosigma akashiwo*, *Heterocapsa circular-*

Ecological Studies, Vol. 189
Edna Granéli and Jefferson T. Turner (Eds.)
Ecology of Harmful Algae

Table 21.1. Viruses infectious to marine eukaryotic microalgae (adapted from Brussaard 2004; Nagasaki et al. 2005)

Virus	Host microalga	Virus size (nm)	Virus genome
BtV	*Aureococcus anophagefferens* (Ochrophyta)*	140	dsDNA
CbV	*Chrysochromulina brevifilum* (Haptophyta)	145–170	dsDNA
CeV	*Chrysochromulina ericina* (Haptophyta)	160	dsDNA, 510 kbp
CsNIV	*Chaetoceros salsugineum* (Ochrophyta)	38	(ss+ds)DNA, 6.0 kb
EhV	*Emiliania huxleyi* (Haptophyta)*	170–200	dsDNA, 410–415 kbp
HaNIV	*Heterosigma akashiwo* (Ochrophyta)*	30	No report
HaV	*Heterosigma akashiwo* (Ochrophyta)*	202	dsDNA, 294 kbp
HaRNAV	*Heterosigma akashiwo* (Ochrophyta)*	25	ssRNA, 9.1 kb
HcRNAV	*Heterocapsa circularisquama* (Dinophyta)*	30	ssRNA, 4.4 kb
HcV	*Heterocapsa circularisquama* (Dinophyta)*	197	dsDNA, 356 kb
MpRNA	*Micromonas pussila* (Chlorophyta)	50–60	dsRNA, 24.6 kb
MpV	*Micromonas pussila* (Chlorophyta)	115	dsDNA, 200 kb
PoV	*Pyramimonas orientalis* (Chlorophyta)	180–220	dsDNA, 560 kb
PpV	*Phaeocystis pouchetii* (Haptophyta)*	130–160	dsDNA
PgV	*Phaeocystis globosa* (Haptophyta)*	No report	dsDNA

* Bloom-forming microalgae

isquama, *Phaeocystis pouchetii*, *Emiliania huxleyi*, *Aureococcus anophagefferens* and *Micromonas pusilla* (Nagasaki et al. 1994, 2004; Jacobsen et al. 1996; Gastrich et al. 1998; Zingone et al. 1999, cited in Brussaard 2004 and Nagasaki et al. 2005). Viral infections of *H. akashiwo* and *H. circularisquama* have received special attention due to their negative impacts on aquaculture and aquatic environments.

VLP-containing cells were found at the very final stage of *Heterosigma akashiwo* blooms (Nagasaki et al. 1994). Algicidal agents were also detected in the size fraction smaller than 0.2 μm in the same area and at the same stage of *H. akashiwo* bloom (Imai et al. 1998). These facts imply that virus-induced mortality is an important factor in termination of red tides. Moreover, a sudden collapse of a *H. akashiwo* bloom coinciding with an increase in HaV (*H. akashiwo* virus) was observed in a coastal sea (Tarutani et al. 2000). However, the fate of HaV's produced during the bloom and the origins of viruses that initiate infections are unclear.

The dynamics of the bivalve-killing dinoflagellate *Heterocapsa circularisquama* and its virus (HcRNAV) were studied in Ago Bay, Japan (Nagasaki et al. 2004). The abundance of infectious HcRNAV was high from the bloom peak and throughout the post-bloom period, falling to undetectable levels few weeks later. The proportion of VLP-harboring *H. circularisquama* cells reached 88 %, indicating the significance of viral mortality for termination of the red tides. An increase in viral abundance in sediments was noticed during the bloom period, where sediments stayed infective for at least three months, highlighting the importance of sediments as a reservoir for viruses.

21.2.1 Host Specificity

Viruses infecting microalgae are usually host specific. Tomaru et al. (2004a) investigated the virus sensitivity of *Heterosigma akashiwo* and the host specificity of HaV by using 90 *H. akashiwo* clones and 65 virus clones isolated from the same bloom in the Seto Inland Sea. These authors demonstrated the coexistence of different types of HaV during the bloom, and concluded that viral infections affected both biomass and clonal composition of *H. akashiwo* populations. In the case of *Heterocapsa circularisquama* and its virus HcRNAV, Tomaru et al. (2004b) showed that infection could be strain-specific rather than species-specific. Consequently, a single virus clone cannot exterminate a specific algal species that is composed of different ecotypes. Such strain-specificity can be a drawback for the use of viruses to exterminate harmful algal blooms.

21.3 Algicidal Bacteria

During the last two decades, many algicidal bacteria have been identified and isolated from marine coastal areas and have received considerable attention as terminators of HABs (Doucette et al. 1998; Mayali and Azam 2004). Analyses of small subunit ribosomal DNA has shown that the most common algicidal bacteria are gram-negatives that belong to the gamma-Proteobacteria line (mainly the genera *Alteromonas* and *Pseudoalteromonas*) or the phylum Bacteroides (mainly the genera *Cytophaga* and *Saprospira*). New members of the alpha-Proteobacteria line with algicidal effects have been isolated from a seaweed bed in the Seto Inland Sea, Japan (Imai et al. 2006). Gram-positive algicidal bacteria belonging to the phylum Firmicutes have been reported from Australian waters (Skerratt et al. 2002).

21.3.1 Modes of Algicidal Activity and Specificity

Gliding bacteria such as the genera *Cytophaga* and *Saprospira* are usually direct-attack types, requiring physical contact with the hosts to elicit the algicidal effect, whereas members of the gamma-Proteobacteria and Firmicutes release algicidal compounds into the water (Imai et al. 1993, 1995; Skerratt et al. 2002). Direct-attack type bacteria tend to display a wide range of target microalgae. An interesting role of dissolved organic matter was reported in *Alteromonas* E401, which kills the harmful dinoflagellate *Karenia mikimotoi* (Yoshinaga et al. 1995). A high molecular weight (>10 kD), heat-labile compound showing algicidal activity was produced by the bacterium in response to excreted organic matter (EOM) from *K. mikimotoi*. The algicidal activity was restricted to *K. mikimotoi* and *Gymnodinium catenatum*, with no effects on other dinoflagellates, raphidophytes or diatoms. If this is the case, species-specific algicidal activity against blooming algal species might be induced by the EOM from each microalgal species.

Ectoenzymes, particularly ectoproteases, are suspected to be the algicidal compounds released by certain bacteria. Mitsutani et al. (2001) found that cell extracts from a stationary culture of *Pseudoalteromonas* A25 showed both algicidal and high protease activities.

Gram-negative algicidal bacteria appear to use the AI-2 mechanism (quorum sensing) at mid- to late-stage of log-growth phase (Skerratt et al. 2002). The AI-2 mechanism, rather than acetylated homoserinelactones (AHL), is believed to be involved in inter-species bacteria communication. There is a metabolic benefit for these mechanisms to be activated individually or as bacteria requires, rather than simultaneously and continuously.

The swarming capacity of certain bacteria toward target algal cells is thought to be an advantageous strategy for algicidal strains. Imai et al. (1995) observed the swarming of the bacterium *Alteromonas* S to the valve face of the diatom *Ditylum brightwellii* in early stage of algicidal attack. Skerratt et al. (2002) also reported the swarming of algicidal bacteria to the target microalgae. Swarming might be the result of a signal that targets the cells to the prey. Whether algicidal activity is related to swarming and quorum sensing remains an interesting subject.

21.3.2 Ecology of Algicidal Bacteria and Harmful Microalgae

In northern Hiroshima Bay, the Seto Inland Sea, the dynamics of *Heterosigma akashiwo*-killer bacteria was closely related with that of *H. akashiwo* populations (Imai et al. 1998; Yoshinaga et al. 1998). Gamma-Proteobacteria represented 70–80 % of algicidal bacteria isolated during the termination of red tides (Yoshinaga et al. 1998).

At a station in Harima-Nada, also in the Seto Inland Sea area, the cell density of algicidal *Cytophaga* sp. J18/M01 (originally isolated from the same station in Harima-Nada) increased just after the peak of a small bloom of *Chattonella* spp. (Imai et al. 2001). This bacterium has a wide range of targets, which is reflected in a close relationship between its abundance and the change in total microalgal biomass.

Bacteria heavily colonize macroaggregates, being important for biochemical processes as “hot spots” (Simon et al. 2002). Notably, *Cytophaga* and gamma-Proteobacteria are among the dominant bacteria attached to macro-aggregates or marine snow (Doucette et al. 1998). Swarming was observed in the algicidal process by *Alteromonas* sp. and *Cytophaga* sp. (Imai et al. 1995; Skerratt et al. 2002). When a small number of algicidal bacteria cells aggregate around a single microalgal cell or on a macroaggregate, such microscale patchiness can create algicidal hot spots in the sea, and may play an important role in the dynamics of algicidal bacteria and microalgae (Doucette et al. 1998; Imai et al. 1998; Mayali and Azam 2004).

21.3.3 Seaweed Beds as Prevention of HABs

A hitherto-unexplored aspect on the ecology of algicidal bacteria is the huge number of such microorganisms attached onto the surface of seaweeds such as *Ulva* sp. (Chlorophyta) and *Gelidium* sp. (Rhodophyta) (Imai et al. 2002). Maximum numbers of about 10^5~10^6 bacteria g seaweed wet weight^{-1} were detected for bacteria targeting *Karenia mikimotoi, Fibrocapsa japonica* (Ochrophyta), and *Heterosigma akashiwo*. Algicidal bacteria belonged to alpha- and gamma-Proteobacteria and the phylum Bacteroides and were also abundant in the water around seaweed beds.

Based on these studies, a new prevention strategy for red tides in aquaculture area is proposed. Co-culturing of *Gelidium* sp. or *Ulva* sp. and finfish is proposed to be effective in cage cultures (Imai et al. 2002). Many algicidal bacteria will be continuously released from the surface of seaweeds to the surrounding water, contributing to prevent HABs. This strategy may be effective in enclosed and small-scale inlets. Artificial restoration of seaweed beds over large coastal areas should work in a similar way to prevent HAB events.

21.4 Parasitic Fungi

Fungi commonly involved in parasitic associations with microalgae are uniflagellated Chytridiomycetes, usually called chytrids (Ibelings et al. 2004). To a lesser extent, biflagellated forms belonging to the Oomycetes and other types of fungi have also been observed (Mountfort et al. 1996; Elbrächter and Schnepf 1998). Parasite dispersal is via free-swimming, flagellate zoospores. Chytrid zoospores penetrate their hosts' cells using their flagella, and then form an intracellular rhizoid through which the zoospore outside the host is nourished. The zoospore becomes a sporangium that matures and produces new zoospores (Van Donk 1989).

Freshwater diatoms and their chytrid parasites are probably the best-studied planktonic host-parasite systems (Ibelings et al. 2004). Fungal infection causes high mortality of freshwater diatom populations, influencing succession of phytoplankton assemblages (Van Donk 1989). Fungal pathogens of marine microalgae include uni- and biflagellated forms infecting diatoms (Wetsteyn and Peperzak 1991; Elbrächter and Schnepf 1998) and dinoflagellates (Mountfort et al. 1996). Whether the dynamics of marine microalgal assemblages are influenced by pathogenic fungi to the same extent as freshwater assemblages is not yet known.

There is a substantial lack of information about the incidence of parasitic fungi on harmful microalgal species, most likely reflecting low interest by the scientific community rather than the absence of such a phenomenon in nature. Canter (1972) noted that a few freshwater cyanobacteria, including the toxin-producing genera *Microcystis* and *Anabaena*, were susceptible to fungal parasites, but apparently no further studies were done. In the marine environment, diatoms of the genus *Coscinodiscus*, which occasionally form deleterious blooms, are hosts for *Lagenisma coscinodisci* (Schnepf et al. 1978). *Verticillium lecanii*, a fungus isolated from marine waters releases algicidal substances that kill dinoflagellates (Montfort et al. 1996).

21.4.1 Host Specificity

Fungal parasites usually show a narrow host range. The chytrid *Zygorhizidium planktonicum* was reported to infect only two diatom genera, *Asterionella* and *Synedra* (Canter et al. 1992). Further studies demonstrated the co-existence of species-specific variants of *Z. planktonicum* (Doggett and Porter 1995). *Verticillium lecanii* kills the red tide-forming dinoflagellates *Karenia mikimotoi* (formerly *Gymnodinium mikimotoi*) and *Akashiwo sanguinea* (formerly *G. sanguineum*), but does not harm *Heterocapsa triquetra* and three species of *Alexandrium* (Montfort et al. 1996).

21.5 Parasitic Protists

Protists have long been described as parasites of planktonic microalgae (see review by Elbrächter and Schnepf 1998). Included are various zooflagellates, amoebae, and euglenas that infect mostly diatoms (Schweikert and Schnepf 1997; Tillmann et al. 1999; Elbrächter and Schnepf 1998). Dinoflagellates are also known to parasitize planktonic microalgae in marine waters (Park et al. 2004). For example, *Dubosquella melo* infects the red tide-forming dinoflagellate *Noctiluca scintillans* (Cachon 1964), and species of *Paulsenella* have been observed infecting diatoms (Drebes and Schnepf 1988). However, the genus *Amoebophrya* is noticeably the most widespread parasitic dinoflagellate of marine microalgae, infecting several free-living dinoflagellates (Park et al. 2004). In addition, two newly described perkinsozoan flagellates, *Parvilucifera infectans* and *Cryptophagus subtilis*, have been reported to infect marine dinoflagellates and cryptophytes, respectively (Norén et al. 1999; Brugerolle 2002). Together, *Amoebophrya* spp. and *P. infectans* infect over 50 marine dinoflagellates, among them several harmful species (Park et al. 2004). These two parasites have received considerable attention due to their worldwide occurrence and, at times, high virulence against harmful dinoflagellates.

Members of the genus *Amoebophrya* infecting several dinoflagellates were initially described as one species, *Amoebophrya ceratii*, but are now regarded as a cluster of closely related strains or species (Coats et al. 1996; Gunderson et al. 2002; Salomon et al. 2003; Coats and Park 2002). Dispersal is by small (8–10 µm long), free-swimming, biflagellate zoospores (called dinospores) (Cachon and Cachon 1987), which penetrate the host becoming a trophont (Fig. 21.1). The parasite leaves the host as a vermiform stage that differentiates into new dinospores. Prevalence of up to 80 % was reported during epidemic outbreaks of *Amoebophrya* sp. infecting *Akashiwo sanguinea* (Coats et al. 1996). In such cases, *Amoebophrya*-induced mortality might offset in situ growth rates of host dinoflagellates, being relevant to bloom termination.

Infection by *Parvilucifera infectans* is mediated by free-living, 2–4 µm long, biflagellate zoospores that enter the host and develop into thick-walled, round bodies called sporangia (Delgado 1999; Norén et al. 1999) (Fig. 21.2). When mature, zoospores inside the sporangium become active and escape to the water, completing the cycle. A single sporangium can harbor as many as 200 zoospores (Erard-Le Denn et al. 2000). *P. infectans* can cause high mortality when inoculated in cultures of compatible dinoflagellate hosts.

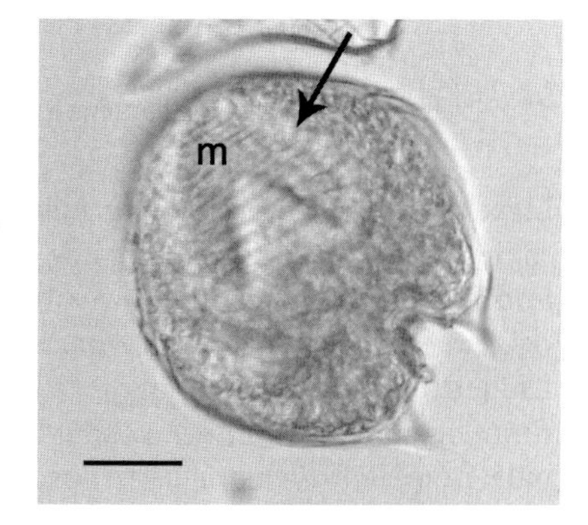

Fig. 21.1. *Protoperidinium* sp. from the Southern Atlantic (Brazilian coast of Arraial do Cabo, Rio de Janeiro state) infected with a trophont of the parasitic dinoflagellate *Amoebophrya* sp. Note the "beehive" aspect of the trophont (*arrow*) and the mastigocoel (*m*), typical of this kind of parasite. Specimen fixed with paraformaldehyde and ethanol. Bright field, 40×. *Scale* 20 µm

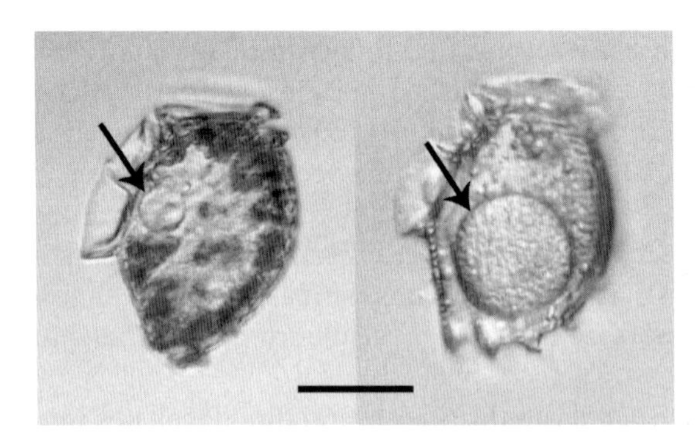

Fig. 21.2. Two cells of *Dinophysis norvegica* from the Baltic Sea infected with *Parvilucifera*. Sporangia (*arrows*) at early (*left cell*) and advanced (*right cell*) developmental stages are shown. Live specimens. Bright field with DIC, 40X. *Scale* 20 μm

21.5.1 Host Specificity

Host specificity for parasitic protists of microalgae is variable. Cross-infection experiments with six species of *Pirsonia* and more than 20 species of diatoms revealed parasites with both narrow and wide host ranges (Kühn et al. 1996). Various strains of *Amoebophrya* have been maintained in culture with their original hosts for successive generations, but attempts to artificially establish uninterrupted infection on dinoflagellate hosts other than the ones from which they were isolated have been unsuccessful, indicating high host specificity for the group (Coats et al. 1996; Park et al. 2004). Notable exceptions are two *Amoebophrya* strains recently isolated from *Alexandrium* spp., which are infectious to several other thecate dinoflagellates (Sengco et al. 2003). *Parvilucifera infectans* appears to be one of the less selective parasitic protists infecting microalgae. Although infection seems to be restricted to dinoflagellates, mostly thecate forms, zoospores originating from a given host are normally infectious to a large array of other dinoflagellates in culture (Delgado 1999; Norén et al. 1999; Erard-Le Denn 2000).

21.5.2 Host Avoidance of Parasitic Infection

Little is known about defense mechanisms elicited by microalgae to escape their parasites. Few insights come from studies of dinoflagellate hosts. Coats and Park (2002) observed that a fraction of cells in cultures of *Karlodinium micrum* inoculated with *Amoebophrya* sp. were resistant to infection. Inducible defense against parasite infection was reported in the PSP-causative dinoflagellate *Alexandrium ostenfeldii*, which responded to water-borne signals from *Parvilucifera infectans* by shifting from motile to resting stages (Toth et al. 2004). Moreover, behavioral features also seem to protect microalgal populations from heavy parasitism. *Amoebophrya* infection in *Akashiwo sanguinea* populations leads to a physical separation between diel-vertical-migrating, non-infected cells and metabolically impaired, infected cells. This in turn helps to prevent parasite epidemics in healthy populations (Coats and Bockstahler 1994).

21.6 Conclusions and Future Perspectives

The influence of pathogens as controlling mechanism for microalgal populations is still a relatively neglected subject (if compared to nutrient limitation and grazing). Infections by freshwater fungi and marine protists are the most conspicuous exceptions to this pattern. This generalized lack of studies might be, at least in part, due to methodological difficulties to recognize the parasites on and inside microalgal cells.

High host specificity and virulence of certain pathogens highlights their potential in controlling host populations, and lends credibility to previous suggestions of their use as mitigation to harmful blooms (Taylor 1968). Uncertainties about host specificity, pathogen stability, and fate within the aquatic food web, and possible negative impacts to other members of the aquatic biota are issues that still must be examined. The use of pathogens as mitigation for harmful algal blooms is thoroughly discussed in Chap. 25.

Bloom termination by viruses and algicidal bacteria releases huge amounts of organic matter that must enter the aquatic food web mainly via the microbial food web component (Kamiyama et al. 2000). If not recycled, this organic matter will most likely contribute to the deterioration of aquatic environments through anoxia. On the other hand, attack of harmful microalgae by parasites like chytrids and protists reorganizes much of the algal biomass into smaller packages – in the form of parasite zoospores – readily available to grazers (Johansson and Coats 2002). Blooming microalgae released from grazing due to toxin production or size refuge weakens matter and energy fluxes to higher trophic levels. Parasitism might re-establish these fluxes via alternative, temporary routes mediated by zoospores (Park et al. 2004).

Finally, it is widely accepted that release from natural enemies favors the success of plants and animals introduced into new territories (Torchin et al. 2003). Whether or not a release from pathogens influences the success of harmful microalgae introduced in coastal areas (e.g., via cargo-vessel ballast water) is still an open question.

References

Brugerolle G (2002) *Cryptophagus subtilis*: a new parasite of cryptophytes affiliated with the Perkinsozoa lineage. Eur J Protistol 37:379–390

Brussaard CPD (2004) Viral control of phytoplankton populations – a review. J Eukaryot Microbiol 51:125–138

Cachon J (1964) Contribuition à l'étude des Péridiniens parasites. Cytologie, cycles évolutifs. Ann Sci Nat Zool 6 (12e Sèrie):1–158

Cachon J, Cachon M (1987) Parasitic dinoflagellates. In: Taylor FJR (ed) The biology of dinoflagellates. Botanical Monographs, vol 21. Blackwell Sci Publ London, pp 571–610

Canter HM (1972) A guide to the fungi occurring on planktonic blue-green algae. In: Desikachary TV (ed) Taxonomy and biology of blue-green algae. Univ of Madras, India, pp 145–148

Canter HM, Jaworski GHM, Beakes GW (1992) Formae speciales differentiation in the chytrid *Zygorhizidium planktonicum* Canter, a parasite of the diatoms *Asterionella* and *Synedra.* Nova Hedwigia 55:437–455

Coats DW, Adam EJ, Gallegos CL, Hedrick S (1996) Parasitism of photosynthetic dinoflagellates in a shallow subestuary of Chesapeake Bay, USA. Aquat Microb Ecol 11:1–9

Coats DW, Bockstahler KR (1994) Occurrence of the parasitic dinoflagellate *Amoebophrya ceratii* in Chesapeake Bay populations of *Gymnodinium sanguineum.* J Eukaryot Microbiol 41:586–593

Coats DW, Park MG (2002) Parasitism of photosynthetic dinoflagellates by three strains of *Amoebophrya* (Dinophyta): parasite survival, infectivity, generation time, and host specificity. J Phycol 38:520–528

Delgado M (1999) A new "diablillo parasite" in the toxic dinoflagellate *Alexandrium catenella* as a possibility to control harmful alga blooms. Harmful Algal News 19:1–3

Doggett MS, Porter D (1995) Further evidence for host-specific variants in *Zygorhizidium planktonicum.* Mycologia 87:161–171

Doucette GJ, Kodama M, Franca S, Gallacher S (1998) Bacterial interactions with harmful algal bloom species: bloom ecology, toxigenesis, and cytology. In: Anderson DM, Cembella AD, Hallegraeff GM (eds) Physiological ecology of harmful algal blooms. NATO ASI Series 41. Springer, Berlin Heidelberg New York, pp 619–647

Drebes G, Schnepf E (1988). *Pausenella* Chatton (Dinophyta), ectoparasites of marine diatoms: development and taxonomy. Helgol Meersunters 42:563–581

Elbrächter M, Schnepf E (1998) Parasites of harmful algae. In: Anderson DM, Cembella AD, Hallegraeff GM (eds) Physiological ecology of harmful algal blooms. NATO ASI Series 41. Springer, Berlin Heidelberg New York, pp 351–369

Erard-Le Denn E, Chrétiennot-Dinet M-J, Probert I (2000) First report of parasitism on the toxic dinoflagellate *Alexandrium minutum* Halim. Est Coast Shelf Sci 50:109–113

Gunderson JH, John SA, Boman II C, Coats DW (2002) Multiple strains of the parasitic dinoflagellate *Amoebophrya* exist in Chesapeake Bay. J Eukaryot Microbiol 49:469–474

Ibelings BW, De Bruin A, Kagami M, Rijkeboer M, Brehm M, van Donk E (2004) Host parasite interactions between freshwater phytoplankton and chytrid fungi (Chytridiomycota). J Phycol 40:437–453

Imai I, Ishida Y, Hata Y (1993) Killing of marine phytoplankton by a gliding bacterium *Cytophaga* sp., isolated from the coastal sea of Japan. Mar Biol 116:527–532

Imai I, Ishida Y, Sakaguchi K, Hata Y (1995) Algicidal marine bacteria isolated from northern Hiroshima Bay, Japan. Fisheries Sci 61:624–663

Imai I, Kim MC, Nagasaki K, Itakura S, Ishida Y (1998) Relationships between dynamics of red tide-causing raphidophycean flagellates and algicidal micro-organisms in the coastal sea of Japan. Phycol Res 46:139–146

Imai I, Sunahara T, Nishikawa T, Hori Y, Kondo R, Hiroishi S (2001) Fluctuations of the red tide flagellates *Chattonella* spp. (Raphidophyceae) and the algicidal bacterium *Cytophaga* sp. in the Seto Inland Sea, Japan. Mar Biol 138:1043–1049

Imai I, Fujimaru D, Nishigaki T (2002) Co-culture of fish with macroalgae and associated bacteria: a possible mitigation strategy for noxious red tides in enclosed coastal sea. Fisheries Sci 68 (Suppl):493–496

Imai I, Fujimaru D, Nishigaki T, Kurokawa M, Sugita H (2006) Algicidal bacteria against harmful microalgae isolated from the surface of seaweeds in the coast of Osaka Bay, Seto Inland Sea, Japan. Afr J Mar Sci (in press)

Johansson M, Coats DW (2002) Ciliate grazing on the parasite *Amoebophrya* sp. decreases infection of the red-tide dinoflagellate *Akashiwo sanguinea*. Aquat Microb Ecol 28:69–78

Kamiyama T, Itakura S, Nagasaki K (2000) Changes in microbial loop components: effects of a harmful algal bloom formation and its decay. Aquat Microb Ecol 21:21–30

Kühn S, Drebes G, Schnepf E (1996) Five new species of the nanoflagellate *Pirsonia* in the German Bight, North Sea, feeding on planktic diatoms. Helgol Meeresunters 50:205–222

Mayali X, Azam F (2004) Algicidal bacteria in the sea and their impact on algal blooms. J Eukaryot Microbiol 51:139–144

Mitsutani A, Yamasaki I, Kitaguchi H, Kato J, Ueno S, Ishida Y (2001) Analysis of algicidal proteins of a diatom-lytic marine bacterium *Pseudoalteromonas* sp. strain A25 by two-dimensional electrophoresis. Phycologia 40:286–291

Mountfort DO, Atkinson M, Ponikla K, Burke B, Todd K (1996) Lysis of *Gymnodinium* species by the fungus *Verticillium lecanii*. Botanica Marina 39:159–165

Nagasaki K, Ando M, Itakura S, Imai I, Ishida Y (1994) Viral mortality in the final stage of *Heterosigma akashiwo* (Raphidophyceae) red tide. J Plankton Res 16:1595–1599

Nagasaki K, Tomaru Y, Nakanishi K, Hata N, Katanozaka N, Yamaguchi M (2004) Dynamics of *Heterocapsa circularisquama* (Dinophyceae) and its viruses in Ago bay, Japan. Aquat Microb Ecol 34:219–226

Nagasaki K, Tomaru Y, Takao Y, Nishida K, Shirai Y, Suzuki H, Nagumo T (2005) Previously unknown virus infects marine diatom. Appl Environ Microbiol 71:3528–3535

Norén F, Moestrup Ø, Rehnstam-Holm A-S (1999) *Parvilucifera infectans* Norén et Moestrup gen. et. sp . nov. (Perkinsozoa phylum nov.): a parasitic flagellate capable of killing toxic microalgae. Eur J Protistol 35:233–254

Park MG, Yih W, Coats DW (2004) Parasites and phytoplankton, with special emphasis on dinoflagellate infections. J Eukaryot Microbiol 51:145–155

Salomon PS, Janson S, Granéli E (2003) Multiple species of the dinophagous dinoflagellate genus *Amoebophrya* infect the same host species. Environ Microbiol 5:1046–1052

Schnepf E, Deichgräber G, Drebes G (1978) Development and ultrastructure of the marine, parasitic Oomycete, *Lagenisma coscinodisci* Drebes (Lageniales). The infection. Arch Microbiol 116:133–139

Schweikert M, Schnepf E (1997) Light and electron microscopical observations on *Pirsonia punctigerae* spec nov., a nanoflagellate feeding on the marine centric diatom *Thalassiosira punctigera*. Eur J Protistol 33:168–177

Sengco MR, Coats DW, Popendorf KJ, Erdner DR, Gribble KE, Anderson DM (2003) Biological and phylogenetic characterization of *Amoebophrya* sp. ex *Alexandrium tamarense*. Abstract Second Symposium on Harmful Marine Algae in the US, Woods Hole, MA, p 57

Simon M, Grossart HP, Schweitzer B, Ploug H (2002) Microbial ecology of organic aggregates in aquatic ecosystems. Aquat Microb Ecol 28:175–211

Skerratt JH, Bowman JP, Hallegraeff G, James S, Nichols PD (2002) Algicidal bacteria associated with blooms of a toxic dinoflagellate in a temperate Australian estuary. Mar Ecol Prog Ser 244:1–15

Tarutani K, Nagasaki K, Yamaguchi M (2000) Viral impacts on total abundance and clonal composition of the harmful bloom-forming phytoplankton *Heterosigma akashiwo*. Appl Environ Microbiol 66:4916–4920

Taylor FJR (1968) Parasitism of the toxin-producing dinoflagellate *Gonyaulax catenella* by the endoparasitic dinoflagellate *Amoebophrya ceratii*. J Fish Res Bd Can 25:2241–2245

Tillmann U, Hesse K, Tillmann A (1999) Large-scale parasitic infection of diatoms in the North Frisian Wadden Sea. J Sea Res 42:255–261

Tomaru Y, Tarutani K, Yamaguchi M, Nagasaki K (2004a) Quantitative and qualitative impacts of viral infection on a *Heterosigma akashiwo* (Raphidophyceae) bloom in Hiroshima Bay, Japan. Aquat Microb Ecol 34:227–238

Tomaru Y, Katanozaka N, Nishida K, Shirai Y, Tarutani K, Yamaguchi M, Nagasaki K (2004b) Isolation and characterization of two distinct types of HcRNAV, a single-stranded RNA virus infecting the bivalve-killing microalga *Heterocapsa circularisquama*. Aquat Microb Ecol 34:207–218

Torchin ME, Lafferty KD, Dobson AP, McKenzie VJ, Kuris AM (2003) Introduced species and their missing parasites. Nature 421:628–630

Toth GB, Norén F, Selander E, Pavia H (2004) Marine dinoflagellates show induced life-history shifts to escape parasite infection in response to water-borne signals. Proc R Soc Lond B 271:733–738

Van Donk E (1989) The role of fungal parasites in phytoplankton succession. In: Sommer U (ed) Plankton ecology: succession in plankton communities. Springer, Berlin Heidelberg New York, pp 171–194

Wetsteyn LP, Peperzak L (1991) Field observations in the Oosterchelde (The Netherlands) on *Coscinodiscus concinnus* and *Coscinodiscus granii* (Bacillariophyceae) infected by the marine fungus *Lagenisma coscinodiscus* (Oomycetes). Hydrobiol Bull 25:15–21

Yoshinaga I, Kawai T, Ishida Y (1995) Lysis of *Gymnodinium nagasakiense* by marine bacteria. In: Lassus P, Arzul G, Le Denn EE, Gentien P, Marcaillou C (eds) Harmful marine algal blooms. Lavoisier Intercept, Paris, pp 687–692

Yoshinaga I, Kim MC, Katanozaka N, Imai I, Uchida A, Ishida Y (1998) Population structure of algicidal marine bacteria targeting the red tide forming alga *Heterosigma akashiwo* (Raphidophyceae), determined by restriction fragmental length polymorphism analysis of the bacterial 16S ribosomal RNA genes. Mar Ecol Prog Ser 170:33–44

22 Phycotoxin Pathways in Aquatic Food Webs: Transfer, Accumulation and Degradation

G. DOUCETTE, I. MANEIRO, I. RIVEIRO and C. SVENSEN

22.1 Introduction

Transfer of phycotoxins through aquatic food webs is an important aspect of harmful algal bloom (HAB) dynamics affecting multiple trophic levels. Contamination can be caused by toxins residing inside particles (phytoplankton cells, faeces, or tissues of non-toxin producers that have ingested toxin or toxic species) and dissolved toxins produced by either active excretion, passive leakage from senescent cells, or cell lysis via attack by pathogens or natural cell death (see Landsberg 2002). Flocculation and sedimentation allows phycotoxins to reach benthic organisms, which can also be exposed through ingestion of toxic organisms or detritus. Moreover, dissolution and/or degradation of toxins are mediated through the action of bacteria on dead organisms or excreted material. A diagram indicating the possible pathways of toxin trophic transfer is provided in Fig. 22.1 and the fate of algal toxins in each major trophic group is described below.

22.2 Bacteria

Contributions of bacteria to the pathways of phycotoxin transfer include the enhancement or inhibition of toxin production, the lysis of toxic algal cells, and the transformation or degradation of these compounds. Bacteria can affect the production of certain toxins by microalgae. This interaction can be positive or negative and has been observed recently for several toxin classes such as domoic acid (DA) and saxitoxin (STX) and its analogues (e.g., Su et al. 2003).

Bacteria can cause the disruption of toxic cyanobacteria (Nakamura et al. 2003) as well as the breakdown of cyanotoxins released from these cells (Saito

Ecological Studies, Vol. 189
Edna Granéli and Jefferson T. Turner (Eds.)
Ecology of Harmful Algae

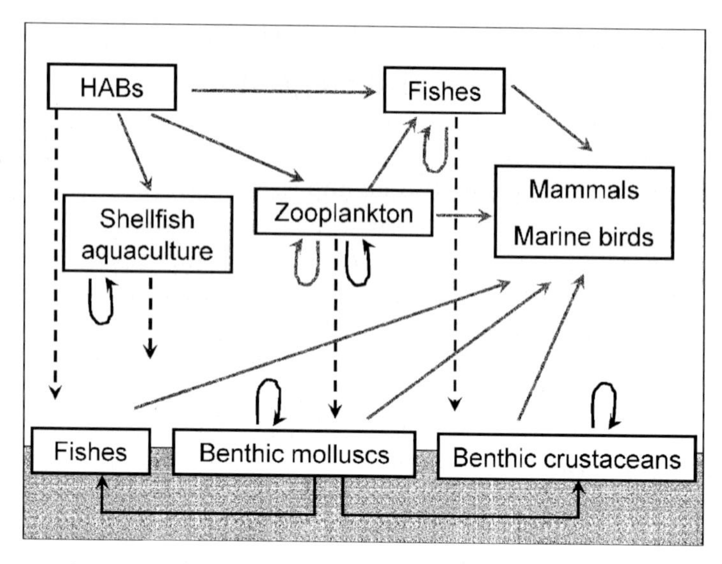

Fig. 22.1. Predominant pathways of phycotoxin trophic transfer, including organisms ingesting toxins or toxic algae (*grey lines*), ingestion of detritus (*black lines*), and sinking of particulate matter to the benthos (*dashed lines*)

et al. 2003). Bacterially mediated lysis has also been observed for a number of other toxic HAB species (see Salomon and Imai, Chap. 21; Kodama et al. Chap. 19). In many cases, such algicidal activity results in the release of dissolved toxins that can lead to the contamination of other aquatic organisms, as will be described below.

Some bacteria are capable of transforming PSP (paralytic shellfish poisoning) toxins into related compounds that may be more-or-less potent than the original toxins. Microbial taxa contained within a variety of shellfish, including *Pseudomonas* and *Vibrio* species, have the enzymatic capacity to chemically modify or degrade PSP toxins (Smith et al. 2001). Other bacterial communities present in crabs or snails can also transform PSP toxins. Interestingly, the rate of transformation by bacteria originating from animal tissues appears to be higher than that of bacteria isolated from the ambient marine environment (Kotaki 1989).

Stewart et al. (1998, cited in Smith et al. 2001) reported the isolation of bacteria that exhibited growth on and biodegradation of domoic acid, from bivalves that eliminate DA readily (e.g., blue mussel, *Mytilus edulis*; soft-shell clam, *Mya arenaria*). These findings suggest that autochthonous bacteria may be significant factors in the elimination of DA from these two shellfish species. However, in more recent studies, other bacteria were unable to degrade either dissolved DA or PSP toxins, while cyanobacterial toxins (e.g., nodularin) were biodegraded quickly when occurring in dissolved form (Granéli et al. 2003).

22.3 Zooplankton

Zooplankton can be contaminated by consumption of phycotoxins associated with particulate matter, including toxic algal cells. The resulting toxicity has been detected in many zooplankton groups, including tintinnids, copepods, cladocera, and larvae of crustaceans or krill (Turner and Tester 1997; Maneiro et al. 2000). Although dissolved forms of several algal toxins adversely affect zooplankton (Shaw et al. 1997, cited in Maneiro et al. 2006), there is as yet no evidence of toxin accumulation via this route of exposure.

Phycotoxins can be accumulated by organisms and/or released as particulate material and/or dissolved compounds. The assimilation of toxins by zooplankton is both toxin- and species-specific. For both saxitoxin and domoic acid, less than 10 % of the toxins acquired during feeding activity are retained by these grazers (Guisande et al. 2002; Teegarden et al. 2003; Maneiro et al. 2006). Despite the low efficiency of toxin retention, the accumulation of PSP toxins makes zooplankton an important source of toxicity to higher trophic levels (e.g., Doucette et al. 2006b). Turner et al. (2000, cited in Doucette et al. 2006b) observed that these toxins can be accumulated disproportionately by zooplankton in larger size fractions (e.g., *Calanus finmarchicus*) that are the preferred prey of certain whales. In fact, natural zooplankton populations can contain levels of PSP toxins (ca. 90 µg STX eq $(100\ g)^{-1}$) higher than the legal limit for human consumption (Durbin et al. 2002, cited in Doucette et al. 2006a). It has now been established through analysis of faecal material that concentrations as high as 1.0–0.5 µg STX eq g^{-1} can occur in the endangered North Atlantic right whale through toxin transfer by such zooplankton vectors (Doucette et al. 2006a).

In the case of PSP toxins, there are differences in toxin profiles (i.e., relative toxin composition) between contaminated biological tissues (e.g., shellfish) and the toxic algal source (Guisande et al. 2002). Differences in toxin profiles can be due to different assimilation efficiencies of toxins, transformations, and/or dissimilar toxin detoxification rates exhibited by exposed organisms.

It seems that while assimilation efficiencies vary among toxin classes (Maneiro et al. 2006), consistent trends generally occur among toxins of the same group (Teegarden et al. 2003). A similar relationship has been observed in the detoxification process, which seems to be a non-selective process among toxins within the same class (Guisande et al. 2002; Teegarden et al. 2003).

Evidence for metabolic transformation of PSP toxins within zooplankton was reported by several investigators and later confirmed by Teegarden and Cembella (1996, cited in Guisande et al. 2002), who observed epimerization of C2 toxin to C1 and desulfation of C1 toxin to the more toxic GTX2. These data suggest that PSP toxin transformations can enhance the toxicity of zooplankton vectors in marine food webs.

There is little information about the pathway of phycotoxin release in zooplankton, but regurgitation and "sloppy feeding" seem to be the most important mechanisms for the release of ingested PSP toxins (Teegarden et al. 2003), which would then be made available largely in dissolved form for potential uptake and/or degradation. These toxins can also be discharged into the surrounding water inside eggs (Frangópulos et al. 2000, cited in Maneiro et al. 2006) or within faecal pellets (Guisande et al. 2002). The quantity of PSP toxins relocated through any of these processes is reported to be less than that ingested (Guisande et al. 2002).

Contaminated faecal pellets can, nonetheless, be a potentially important vector for the delivery of algal cells and their toxins into benthic food webs, depending on a variety of factors (e.g., food concentration and composition, ingestion rate, etc.) that influence pellet size and sinking rate, which can range from <10 to 100s of meters per day (Turner 2002). However, faecal material that sinks at a slow rate is generally consumed and digested by pelagic organisms or fragmented/dissolved (chemically or biologically), causing its contents to leak into the water column before reaching the sediments. Toxins associated with faecal pellets may thus be cycled within the pelagic zone, exposing organisms for at least brief periods that extend past the termination of a toxic bloom.

22.4 Bivalves

In recent decades, human deaths and illness caused by consumption of bivalves contaminated with phycotoxins have been recorded, and the fate of toxins in these organisms has thus been well documented.

There are three possible pathways of contamination for filter-feeding bivalves: ingestion of toxic algal cells, ingestion of toxic faeces, and uptake of dissolved toxins. It has been established that mussels, clams, oysters, scallops, and cockles can feed on toxic microalgae (FAO 2004). Ingestion of toxic faeces by mussels was observed for cyanobacterial toxins (Svensen et al. 2005), and shellfish toxicity due to dissolved DA was demonstrated by Novaczek et al. (1991) for *M. edulis.*

After their incorporation into a biological matrix, phycotoxins can be accumulated, transformed, and/or eliminated. Due to the accumulation of toxins, bivalve molluscs act as a source of toxicity to other trophic levels. Particulate algal toxins are accumulated mainly in the viscera (e.g., PSP toxins, Bricelj and Shumway 1998; DSP (diarrhetic shellfish poisoning) toxins, Bauder et al. 2001; DA, see Landsberg 2002) or in the gills in the case of dissolved DA (Novaczek et al. 1991), becoming a serious risk to mammalian consumers, including humans and wildlife.

PSP toxin profiles are frequently different among toxic phytoplankton and bivalve grazers (see Bricelj and Shumway 1998). It has been suggested that selective retention of individual toxin congeners may explain these differences (Anderson et al. 1989, cited in Bricelj and Shumway 1998). However, recent studies indicate that the retention of these toxins is non-selective (Suzuki et al. 2003 and references therein). Changes in PSP toxin profiles within shellfish tissues may therefore arise from epimerisation, various other transformation processes, or possibly selective elimination of individual toxins.

Toxin biotransformation in bivalves has been observed for DSP (Moroño et al. 2003) and PSP toxins (Bricelj and Shumway 1998) and, as noted above, may be influenced by bacterial activity. Although the exact nature of biotransformation varies greatly among shellfish species and toxin class, this process generally leads to changes in net toxicity. For the PSP toxins, increased toxicity often reflects an enhancement in the proportion of the more potent carbamate derivates such as saxitoxin and decarbamoylsaxitoxin relative to the less toxic N-sulfocarbamate forms.

Following the termination of a HAB event, there is not a clear pattern in the detoxification of accumulated toxins. Several factors, including different detoxification rates among toxins (FAO 2004), bivalve species (Bricelj and Shumway 1998), and tissues (Bauder et al. 2001) contribute to this variability. The depuration of toxins from bivalves occurs predominantly through their release either in faecal material or in dissolved form. In the case of the mussel, *Mytilus galloprovincialis*, most of the ingested PSP toxins appear to be excreted in dissolved form (Suzuki et al. 2003).

Still, bivalve faeces can play an important role in the fate of algal toxins. Bivalves can produce toxic faeces (Amorim and Vasconcelos 1999; Svensen et al. 2005) that act as a source of toxins to other organisms in the pelagic and benthic food webs. In fact, the same bivalve can be contaminated by its own toxic faeces as has been observed for microcystins in *M. galloprovincialis* (Amorim and Vasconcelos 1999). However, for *M. edulis* grazing on toxic *Nodularia spumigena*, it was demonstrated that when this bivalve was fed faeces with a high nodularin content, the toxin concentration in the resulting faeces was reduced by 99 % and nodularin did not accumulate after ingestion of the toxic faeces, but was thought to be degraded either by metabolic or microbial activity, or actively excreted with the urine (Svensen et al. 2005).

22.5 Benthic Invertebrates (Non-Bivalves)

The initial transfer of algal toxins into aquatic food webs generally occurs in the water column (e.g., zooplankton, planktivorous fish) or via bivalve species in the benthos. However, there are other benthic organisms (e.g., deposit feeders and non-bivalve filter feeders) that are also exposed to these toxins as the algae flocculate and settle to the bottom over the course of a bloom (Trainer et al. 1998). Other processes such as the sinking of toxic faecal material (Guisande et al. 2002) and the ingestion of previously contaminated benthic animals by various scavengers may also introduce phycotoxins into benthic food webs. Contamination of non-bivalve benthic organisms has been reported for ASP (amnesic shellfish poisoning) toxins (Ferdin et al. 2002), DSP toxins (Castberg et al. 2004), PSP toxins (see Shumway 1995; Pitcher et al. 2001), and NSP (neurotoxic shellfish poisoning) toxins (see Landsberg 2002). In the case of some commercially important species (e.g., lobster, crab), toxicity values higher than the legal limit for human consumption have been reported (Wekell et al. 1994, cited in Landsberg 2002; Desbiens and Cembella 1995). Moreover, the abundance and distribution of a wide range of benthic fauna (e.g., brachiopods, echinoderms, gastropods, polychaetes) can be adversely affected by certain algal blooms (e.g., *Karenia* spp.; Wear and Gardner 2001; Landsberg 2002).

22.6 Fishes

The presence of phycotoxins has been observed frequently in many fish species. Toxicity can result from direct ingestion of toxic algal cells (Robineau et al. 1991; Lefebvre et al. 2002; Jos et al. 2005) or consumption of contaminated prey, mainly zooplankton and herbivorous fish (especially in the case of ciguatoxins; Lewis and Holmes 1993), that have fed previously on toxic algae (Gosselin et al. 1989; Tester et al. 2000; Lefebvre et al. 2002).

Fish may accumulate algal toxins in their tissues, which allows them to serve as vectors for toxin transfer to other organisms, including humans. For example, the intoxication syndrome known as ciguatera fish poisoning is caused by eating carnivorous tropical fish (e.g., barracuda, grouper, jack) that have accumulated high levels of ciguatoxins in their muscle tissue via biomagnification. The toxins are obtained by consuming many herbivorous prey species that have ingested toxic algal cells (mainly *Gambierdiscus toxicus*) growing on surfaces associated primarily with reef environments (Lewis and Holmes 1993; Landsberg 2002). Goldberg et al. (submitted) have demonstrated that a diverse array of benthic taxa can accumulate elevated concentrations of DA in the vicinity of *Pseudo-nitzschia australis* blooms, in some cases far exceeding levels generally reported for many filter-feeding bivalves (with maximum values in the predatory Pacific sanddab *Citharichthys sordidus*, of 514 μg DA g^{-1}). By comparison, the highest DA concentration measured in the viscera of the northern anchovy *Engraulis mordax*, associated with a mass mortality of California sea lions in 1998, was 220 μg DA g^{-1} (Lefebvre et al. 1999, cited in Lefebvre et al. 2002; Scholin et al. 2000).

The distribution of algal toxins accumulated in fish organs and tissues does not appear to be uniform. Many toxins, including saxitoxin, domoic acid, and nodularin, are localized more frequently in the liver and other viscera (Haya et al. 1990, cited in Castonguay et al. 1997; Sipiä et al. 2001; Costa and Garrido 2004) as compared to muscle and blood, thereby reducing the health risk to both the fish and human consumers. Nonetheless, toxin groups such as ciguatoxins (Lewis and Holmes 1993) and microcystins (Magalhães et al. 2001) are commonly detected in the muscle tissue of certain fish, which has important and adverse health implications for human consumption. Interestingly, certain toxins (e.g., PSP toxins) can remain sequestered and even accumulate over the life of the fish in organs such as the liver (Castonguay et al. 1997), frequently without harmful effects on the fish.

Finally, an intriguing interaction observed between fish and some toxic cyanobacterial species is an increase in microcystin production when exposed to certain fish. It appears that *Microcystis aeruginosa* can respond to a chemical signal related to feeding, even when the fish are not vigorously consuming the algae (Jang et al. 2004).

22.7 Seabirds and Marine Mammals

Information on the occurrence and impacts of phycotoxins in seabirds and marine mammals has been reviewed recently (Shumway et al. 2003; Van Dolah et al. 2003), but is somewhat limited, since data arise mainly from mortality events. Also, it is difficult to establish a clear connection between HABs and such mortalities due to a lack of knowledge on the sensitivity of these animals to algal toxins. However, Scholin et al. (2000) reported detection of domoic acid in dead sea lions, their planktivorous anchovy prey, and the toxin producing *Pseudo-nitzschia australis*, strongly implicating this toxin as the cause of the sea lion mortalities. Moreover, Lefebvre et al. (1999, cited in Lefebvre et al. 2002) detected frustules of the toxic genus *P. australis* in the faeces of these sea lions. Similar cases have been reported for seabirds (Shumway et al. 2003).

Among the species experiencing mortalities associated with toxic HAB events and/or showing evidence of phycotoxin exposure are whales (Lefebvre et al. 2002; Doucette et al. 2006a), manatees (Bossart et al. 1998, cited in Landsberg 2002), dolphins (Flewelling et al. 2005), and several seabirds including cormorants, penguins, pelicans, and ducks (Shumway et al. 2003). In addition, changes in feeding behaviour have been noted in the presence of toxic food, such as clear preference for non-toxic prey items by sea otters (Kvitek and Bretz 2004) and various shorebirds, including oystercatchers and Eider ducks (Kvitek and Bretz 2005; Shumway et al. 2003).

22.8 Summary and Conclusions

Table 22.1 summarises some of the values of assimilation, detoxification, and accumulation for algal toxins obtained from the literature. Phycotoxin assimilation by zooplankton is usually lower than for bivalves. From the information available, it is not possible to establish differences in assimilation among different toxin classes. Detoxification rates are generally higher in zooplankton than in bivalves and benthic invertebrates, and by toxin class, are higher for PSP toxins than for ASP or DSP toxins. The toxin levels accumulated by bivalves and benthic invertebrates are higher than in zooplankton, probably as a result of the higher assimilation and lower detoxification rates compared with zooplankton, as noted above.

Clearly, there are many pathways for particulate and dissolved phycotoxin trophic transfer and numerous processes regulating the direction and efficiency of toxin movement. Predictive models of toxin transfer incorporating such information are now needed to enable forecasting of the nature and intensity of adverse HAB impacts at levels up to and including the ecosystem. Moreover, these models must also account for variations in algal cell toxicity that, in addition to cell concentrations, can markedly influence the initial incorporation of toxins into aquatic food webs.

Table 22.1. Values of toxin assimilation, detoxification, and accumulation observed in different trophic levels

Species	Assimilation (% of toxin ingested)	Detoxification rate	Toxin accumulation	Toxic species	Reference
Zooplankton					
Acartia clausi	11.6	~0.12 (d^{-1})	19 µg STXeq 100 g^{-1} wet wt	*Alexandrium fundyense*	White (1981)
Barnacle nauplii	32	~0.25 (d^{-1})	54 µg STXeq 100 g^{-1} wet wt		
Eurytemora herdmani	1.25–14.61		0.22–3.14 ng STXeq ind^{-1}	*Alexandrium fundyense*	Teegarden and Cembella (1996, cited in Landsberg 2002)
Acartia tonsa	0.20–30.91		0.02–1.02 ng STXeq ind^{-1}		
Euphausia pacifica			0.1–44 µg DAeq g^{-1} wet wt	*Pseudo-nitzschia* spp.	Bargu et al. (2002)
Acartia clausi	3.8	0.58 (d^{-1})	2,398 µg STXeq 100 g^{-1} dry wt	*Alexandrium minutum*	Guisande et al. (2002)
Acartia clausi	4.8	1.01 (d^{-1})	40.84–97.47 µgDA g^{-1} dry wt	*Pseudo-nitzschia multiseries*	Maneiro et al. (2006)
Bivalves					
Mytilus edulis	78		5,000 µg STXeq 100 g^{-1}	*Alexandrium fundyense*	Bricelj et al. (1990)
Mytilus galloprovincialis		~80% in 2 weeks	10.7 mg microcystin g^{-1} dry wt	*Microcystis aeruginosa*	Amorim and Vasconcelos (1999)
Argopecten iridians	<1–10	0.088–1.137 (d^{-1})	1.8 µg DSP toxins g^{-1} wet wt	*Dinophysis* spp.	Bauder et al. (2001)
Mytilus galloprovincialis		0.58 (d^{-1})	153 µg DA g^{-1} wet wt	*Pseudo-nitzschia australis*	Blanco et al. (2002)
Mytilus edulis			244.9 µg g^{-1} dry wt	*Nodularia spumigena*	Svensen et al. (2005)

Table 22.1. Values of toxin assimilation, detoxification, and accumulation observed in different trophic levels

Species	Assimilation (% of toxin ingested)	Detoxification rate	Toxin accumulation	Toxic species	Reference
Benthic invertebrates					
Homarus americanus		several months of retention	42–1,512 µg STXeq 100 g^{-1} wet wt	*Alexandrium excavatum*	Desbiens and Cembella (1995)
Gasterosteus aculeatus			35–170 ng g–1 dry wt	*Nodularia spumigena*	Kankaanpää et al. (2002)
Cancer pagurus	3–30	14–18 d half-life	20–1,976 µg OA kg^{-1}	*Dinophysis* spp.	Castberg et al. (2004)
Telmessus acutidens	34.7	5 d half-life	12.8 MU g^{-1}	PSP producer	Oikawa et al. (2005)
Fishes					
Gadus morhua *Platichthys flesus*			25–400 ng g^{-1} dry wt	*Nodularia spumigena*	Sipiä et al. (2001)
Sardina pilchardus			128.5 µg DA g^{-1} wet wt	*Pseudo-nitzschia* spp.	Costa and Garrido (2004)
Scomber scombrus			112.4 µg STX 100 g^{-1} wet wt	PSP producer	Castonguay et al. (1997)
Birds and marine mammals					
Pelecanus occidentalis			37.2 µg DA g^{-1}	*Pseudo-nitzschia* sp.	Sierra-Beltrán et al. (1997, cited in Landsberg 2002)
Zalophus californianus			223 µg DA g^{-1} wet wt 136.5 µg DA g^{-1} faeces	*Pseudo-nitzschia australis*	Scholin et al. (2000)
Megaptera novaeangliae *Balaenoptera musculus*			10 µg DA g^{-1} faeces 25 and 207 µg DA g^{-1} faeces	*Pseudo-nitzschia australis*	Lefebvre et al. (2002)
Phoeniconaias minor			0.196 µg g^{-1} fresh weight	Several cyanobacteria	Krienitz et al. (2003)
Eubalaena glacialis			0.5–1.0 µg STX eq g^{-1} faeces	*Alexandrium fundyense*	Doucette et al. (2006a)

Acknowledgements. Partial support was provided by the ECOHAB Inter-Agency Funding Initiative and NOAA/NOS operational funds (GJD). This is contribution 195 of the U.S. ECOHAB Program, sponsored by NOAA, NSF, EPA, NASA, and ONR.

References

Amorim Á, Vasconcelos V (1999) Dynamics of microcystins in the mussel *Mytilus galloprovincialis*. Toxicon 37:1041–1052

Bargu S, Powell CL, Coale SL, Busman M, Doucette GJ, Silver MW (2002) Krill: a potential vector for domoic acid in marine food webs. Mar Ecol Prog Ser 237:209–216

Bauder AG, Cembella AD, Bricelj VM, Quilliam MA (2001) Uptake and fate of diarrhetic shellfish poisoning toxins from the dinoflagellate *Prorocentrum lima* in the bay scallop *Argopecten irradians*. Mar Ecol Prog Ser 213:39–52

Blanco J, Bermúdez de la Puente M, Arévalo F, Salgado C, Moroño A (2002) Depuration of mussels (*Mytilus galloprovincialis*) contaminated with domoic acid. Aquat Living Resour 15:53–60

Bricelj VM, Lee JH, Cembella AD, Anderson DM (1990) Uptake kinetics of paralytic shellfish toxins from the dinoflagellate *Alexandrium fundyense* in the mussel *Mytilus edulis*. Mar Ecol Prog Ser 63:181–188

Bricelj VM, Shumway SE (1998) Paralytic shellfish toxins in bivalve molluscs: occurrence, transfer kinetics, and biotransformation. Rev Fish Sci 6:315–383

Castberg T, Torgersen T, Aasen J, Aune T, Naustvoll LJ (2004) Diarrhoetic shellfish poisoning toxins in *Cancer pagurus* Linnaeus, 1758 (Brachyura, Cancridae) in Norwegian waters. Sarsia 89:311–317

Castonguay M, Levasseur M, Beaulieu JL, Gregoire F, Michaud S, Bonneau E, Bates SS (1997) Accumulation of PSP toxins in Atlantic mackerel: seasonal and ontogenetic variations. J Fish Biol 50:1203–1213

Costa PR, Garrido S (2004) Domoic acid accumulation in the sardine *Sardina pilchardus* and its relationship to *Pseudo-nitzschia* diatom ingestion. Mar Ecol Prog Ser 284:261–268

Desbiens M, Cembella AD (1995) Occurrence and elimination kinetics of PSP toxins in the American lobster (*Homarus americanus*) In: Lassus P, Arzul G, Erard-Le Denn E, Gentien P, Marcaillou C (eds) Harmful marine algal blooms. Lavoisier Intercept, Paris, pp 433–438

Doucette GJ, Cembella AD, Martin JL, Michaud J, Cole TVN, Rolland RM (2006a) PSP toxins in North Atlantic right whales (*Eubalaena glacialis*) and their zooplankton prey in the Bay of Fundy, Canada. Mar Ecol Prog Ser 306:303–313

Doucette GJ, Turner JT, Powell CL, Keafer BA, Anderson DM (2006b) ECOHAB-Gulf of Maine. Trophic accumulation of PSP toxins in zooplankton during *Alexandrium fundyense* blooms in Casco Bay, Gulf of Maine, April–June, 1998. I. Toxin levels in *A. fundyense* and zooplankton size fractions. Deep-Sea Res II 52:2764–2783

FAO (2004) Marine biotoxins. FAO Food and Nutrition Papers, Rome, 8 pp

Ferdin ME, Kvitek RG, Bretz CK, Powell CL, Doucette GJ, Lefebvre KA Coale S, Silver MW (2002) *Emerita analoga* (Stimpson) – possible new indicator species for the phycotoxin domoic acid in California coastal waters. Toxicon 40:1259–1265

Flewelling LJ, Naar JP, Abbott JP et al (2005) Brevetoxicosis: red tides and marine mammal mortalities. Nature 435:755–756

Magalhães VF de, Soares RM, Azevedo SMFO (2001) Microcystin contamination in fish from the Jacarepaguá Lagoon (Rio de Janeiro, Brazil): ecological implication and human health risk. Toxicon 39:1077–1085

Goldberg JD, Kvitek RG, Smith GJ, Doucette GJ, Silver MW (submitted) Domoic acid contamination within eight representative species from the benthic food web of Monterey Bay, California, USA. Mar Ecol Prog Ser

Gosselin S, Fortier L, Gagné JA (1989) Vulnerability of marine fish larvae to the toxic dinoflagellate *Protogonyaulax tamarensis*. Mar Ecol Prog Ser 57:1–10

Granéli E, Hagström JA, Sengco MR, Anderson DM, Tamminen T, Trinh PL (2003) Mitigation of *Prymnesium parvum* blooms by clay flocculation and bacterial degradation of dissolved and particulate algal toxins. EUROHAB Cluster Workshop, Amsterdam (NL), 17/18 March 2003

Guisande C, Frangópulos M, Carotenuto Y, Maneiro I, Riveiro I, Vergara AR (2002) Fate of paralytic shellfish poisoning toxins ingested by the copepod *Acartia clausi*. Mar Ecol Prog Ser 240:105–115

Jang MH, Ha K, Lucas MC, Joo GJ, Takamura N (2004) Changes in microcystin production by *Microcystis aeruginosa* exposed to phytoplanktivorous and omnivorous fish. Aquat Toxicol 68:51–59

Jos A, Pichardo S, Prieto AI, Repetto G, Vázquez CM, Moreno I, Cameán AM (2005) Toxic cyanobacterial cells containing microcystins induce oxidative stress in exposed tilapia fish (*Oreochromis* sp.) under laboratory conditions. Aquat Toxicol 72:261–271

Kankaanpää H, Vuorinen PJ, Sipiä V, Keinänen M (2002) Acute effects and bioaccumulation of nodularin in sea trout (*Salmo trutta* m. *trutta* L.) exposed orally to *Nodularia spumigena* under laboratory conditions. Aquat Toxicol 61:155–168

Kotaki Y (1989) Screening of bacteria which convert gonyautoxin 2,3 to saxitoxin. Bull Jap Soc Sci Fish 55:1293

Krienitz L, Ballot A, Kotut K, Wiegand C, Puetz S, Metcalf JS, Codd GA, Pflugmacher S (2003) Contribution of hot spring cyanobacteria to the mysterious deaths of Lesser Flamingos at Lake Bogoria, Kenya. FEMS Microbiol Ecol 43:141–148

Kvitek R, Bretz C (2004) Harmful algal bloom toxins protect bivalve populations from sea otter predation. Mar Ecol Prog Ser 271:233–243

Kvitek R, Bretz C (2005) Shorebird foraging behavior, diet, and abundance vary with harmful algal bloom toxin concentrations in invertebrate prey. Mar Ecol Prog Ser 293:303–309

Landsberg JH (2002) The effects of harmful algal blooms on aquatic organisms. Rev Fish Sci 10:113–390

Lefebvre KA, Bargu S, Kieckhefer T, Silver MW (2002) From sanddabs to blue whales: the pervasiveness of domoic acid. Toxicon 40:971–977

Lewis RJ, Holmes MJ (1993) Origin and transfer of toxins involved in ciguatera. Comp Biochem Physiol 106C:615–628

Maneiro I, Frangopulos M, Guisande C, Fernandez M, Reguera B, Riveiro I (2000) Zooplankton as a potential vector of diarrhetic shellfish poisoning toxins through the food web. Mar Ecol Prog Ser 201:155–163

Maneiro I, Iglesias P, Guisande C, Riveiro I, Barreiro A, Zervoudaki S, Granéli E (2005) Fate of domoic acid ingested by the copepod *Acartia clausi*. Mar Biol 148:123–130

Moroño A, Arévalo F, Fernández ML, Maneiro J, Pazos Y, Salgado C, Blanco J (2003) Accumulation and transformation of DSP toxins in mussels *Mytilus galloprovincialis* during a toxic episode caused by *Dinophysis acuminata*. Aquat Toxicol 62:269–280

Nakamura N, Nakano K, Sugiura N, Matsumura M (2003) A novel cyanobacteriolytic bacterium, *Bacillus cereus*, isolated from a eutrophic Lake. J Biosci Bioeng 95:179–184

Novaczek I, Madhyastha MS, Ablett RF, Johnson G, Nijjar MS, Sims DE (1991) Uptake, disposition and depuration of domoic acid by blue mussels (*Mytilus edulis*). Aquat Toxicol 21:103–118

Oikawa H, Satomi M, Watabe S, Yano Y (2005) Accumulation and depuration rates of paralytic shellfish poisoning toxins in the shore crab *Telmessus acutidens* by feeding toxic mussels under laboratory-controlled conditions. Toxicon 45:163-169

Pitcher GC, Franco JM, Doucette GJ, Powell CL Mouton A (2001) Paralytic shellfish poisoning in the abalone *Haliotis midae* on the west coast of South Africa. J Shellfish Res 20:895–904

Robineau B, Gagné JA, Fortier L, Cembella AD (1991) Potential impact of a toxic dinoflagellate (*Alexandrium excavatum*) bloom on survival of fish and crustacean larvae. Mar Biol 108:293–301

Saito T, Okano K, Park H, Itayama T, Inamori Y, Neilan BA, Burns BP, Sugiura N (2003) Detection and sequencing of the microcystin LR-degrading gene, mlrA, from new bacteria isolated from Japanese lakes. FEMS Microbiol Lett 229:271–276

Scholin CA, Gulland F, Doucette GJ et al (2000) Mortality of sea lions along the central California coast linked to a toxic diatom bloom. Nature 403:80–84

Shumway SE (1995) Phycotoxin-related shellfish poisoning: bivalve molluscs are not the only vectors. Rev Fish Sci 3:1–31

Shumway SE, Allen SM, Boersma PD (2003) Marine birds and harmful algal blooms: sporadic victims or under-reported events? Harmful Algae 2:1–17

Sipiä V, Kankaanpää H, Lahti K, Carmichael WW, Meriluoto J (2001) Detection of nodularin in flounders and cod from the Baltic Sea. Environ Toxicol 16:121–126

Smith EA, Grant F, Ferguson CMJ, Gallacher S (2001) Biotransformations of paralytic shellfish toxins by bacteria isolated from bivalve molluscs. Environ Microbiol 67:2345–2353

Su J, Zheng T, Yu Z, Song X (2003) Effect of marine bacteria on tile growth and PSP production of red-tide algae. Oceanol Limnol Sin/Haiyang Yu Huzhao 34:44–49

Suzuki T, Ichimi K, Oshima Y, Kamiyama T (2003) Paralytic shellfish poisoning (PSP) toxin profiles and short-term detoxification kinetics in mussels *Mytilus galloprovincialis* fed with the toxic dinoflagellate *Alexandrium tamarense*. Harmful Algae 2:201–206

Svensen C, Strogyloudi E, Riser CW, Dahlmann J, Legrand C, Wassmann P, Granéli E, Pagou K (2005) Reduction of cyanobacterial toxins through coprophagy in *Mytilus edulis*. Harmful Algae 4:329–336

Teegarden GJ, Cembella AD, Capuano CL, Barron SH, Durbin EG (2003) Phycotoxin accumulation in zooplankton feeding on *Alexandrium fundyense*: vector or sink? J Plankton Res 25:429–443

Tester PA, Turner JT, Shea D (2000) Vectorial transport of toxins from the dinoflagellate *Gymnodinium breve* through copepods to fish. J Plankton Res 22:47–61

Trainer VL, Adams NG, Bill BD, Anulación BF, Wekell JC (1998) Concentration and dispersal of a *Pseudo-nitzschia* bloom in Penn Cove, Washington, USA. Nat Toxins 6:113–126

Turner JT (2002) Zooplankton fecal pellets, marine snow and sinking phytoplankton blooms. Aquat Microb Ecol 27:57–102

Turner JT, Tester PA (1997) Toxic marine phytoplankton, zooplankton grazers, and pelagic food webs. Limnol Oceanogr 42:1203–1214

Van Dolah FM, Doucette GJ, Gulland FMD, Rowles TL, Bossart GD (2003) Impacts of algal toxins on marine mammals. In: Vos JG, Bossart GD, Fournier M, O'Shea TJ (eds) Toxicology of marine mammals. Taylor & Francis, London, pp 247–269

Wear RG, Gardner JPA (2001) Biological effects of the toxic algal bloom of February and March 1998 on the benthos of Wellington Harbour, New Zealand. Mar Ecol Prog Ser 218:63–76

White AW (1981) Marine zooplankton can accumulate and retain dinoflagellate toxins and cause fish kills. Limnol Oceanogr 26:103–109

Part E
Studying and Mitigating Harmful Algae: New Approaches

23 Molecular Approaches to the Study of Phytoplankton Life Cycles: Implications for Harmful Algal Bloom Ecology

R. W. Litaker and P. A. Tester

23.1 Introduction

Knowledge of life cycles is crucial for understanding the ecology of harmful algal bloom (HAB) species. The purpose of this chapter is to provide a brief overview of molecular and imaging techniques that can be used to unambiguously identify the life cycle stages of HAB species in cultured and field materials. This overview, in conjunction with key references, is designed to serve as a general guide for those wishing to incorporate these techniques into their research. For specific details of harmful algal bloom life cycles, see Chap. 4.

23.2 Identifying Life Cycle Stages Using Fluorescence In Situ Hybridization (FISH)

The asexual and sexual stages of many HAB species are often highly variable and can be difficult to distinguish using light microscopy (Cerino et al. 2005). These varying life-cycle stages can play distinctly different roles in the life history of phytoplankton, including the initiation and development of blooms (Anderson et al. 1983; Garcés et al. 2004). Consequently, accurate identification of all life cycle stages is important for understanding and modeling HAB bloom dynamics. In situ hybridization provides a molecular means for unambiguously identifying a species and its morphologically disparate life-cycle stages (Miller and Scholin 2000). This method employs fluorescently labeled nucleic acid probes that target unique species-specific sequences contained in genomic DNA or expressed RNAs (Litaker and Tester 2002). In a standard fluorescent in situ hybridization (FISH) assay, preserved cells are mixed with an oligonucleotide probe, which is complementary to the target sequence, and

Ecological Studies, Vol. 189
Edna Granéli and Jefferson T. Turner (Eds.)
Ecology of Harmful Algae

which has been labeled at the 5' or 3' end with a fluorochrome. After the probe has had time to penetrate the cell and bind, any remaining unbound probe is removed by washing, and the different life-cycle stages are identified using epifluorescence microscopy or flow cytometry (Scholin et al. 1996; Groben et al. 2004). The unique sequences most often targeted for in situ hybridization assays are located in ribosomal RNA (rRNA) genes. These genes make attractive targets because there are several hundred or more loci per cell and their transcripts constitute a major portion of total cellular RNA. This high number of genomic and RNA target sites provides enhanced assay sensitivity.

For unculturable species, the single-cell polymerase chain reaction (PCR) assay may be used to amplify sections of ribosomal genes, which can then be cloned and sequenced to identify species-specific target sites (Yamaguchi and Horiguchi 2005). Once unique target sites are identified, probes can be synthesized and tested for specificity. Those probes that do not cross-react can be used to identify the various morphologically distinct forms within the life cycles of the non-culturable species.

Important considerations when designing and using FISH probes include: (1) labeling the probe with a fluorochrome that fluoresces at wavelengths that do not overlap with the natural fluorescence ("autofluorescence") of preserved cells in the sample; (2) being certain that the probe is specific and does not cross-react with other algal species; (3) verifying that the preservative used to fix the cells does not significantly decrease hybridization efficiency; and (4) being aware that cells embedded in detrital material will not hybridize properly due to diffusion limitations (Miller and Scholin 2000; Litaker et al. 2002a). The extent to which this latter effect causes an underestimate of the life-cycle stages present in a sample can be obtained by extracting alga DNA from an aliquot of the sample, and independently determining cell density using quantitative PCR (Galluzzi et al. 2004) or sandwich hybridization assays (Scholin et al. 1996).

Another problem encountered with in situ hybridization is the failure of probes to penetrate thick-walled resting cysts crucial to bloom initiation in some HAB species (Anderson et al. 1983; Garcés et al. 1998). This problem can be overcome using peptide nucleic acid probes (PNAs), synthetic DNA analogues that have a peptide backbone rather than a phosphate backbone. PNAs are neutrally charged and penetrate cells more effectively than nucleic acid probes. In combination with other treatments to weaken the cell wall, these probes, have proved effective in identifying cysts stages (Litaker et al. 2002a).

A logical question that arises when one first undertakes these assays is how much starting material is required? As is the case for the other assays described in this chapter, the ideal cell densities must be determined empirically. Generalized starting concentrations often can be obtained from the cited papers. By bracketing those suggested concentrations in preliminary experiments it is possible to establish optimal cell densities. For example, we typically start with 10–50 ml at ~10^6 cell ml^{-1} from a bloom population or laboratory culture when doing an in situ assay. The initial result determines whether a small or larger volume of cells should be processed for the samples to be efficiently assayed. When developing new assays from an existing protocol, it is particularly advantageous to obtain the same organism used in the original study to serve as a positive control. This allows negative results to be properly interpreted and can greatly accelerate assay development.

In summary, despite some limitations, FISH can allow rapid identification of life cycle stages and can provide data relevant to evaluating the role of cysts in the development and maintenance of HAB species.

23.3 Nuclear Staining to Determine Ploidy and Growth Rates

Because mitotic divisions produce inherently faster population growth than meiotically dividing cells, quantifying the abundance of asexual and sexual stages has significant implications for understanding HAB population dynamics (Figueroa 2005). Given that ploidy levels can vary with life-cycle stage, a potential way of differentiating mitotic from meiotic cells is by measuring DNA content per cell. DNA content is typically determined using nuclear stains such as CyberGreen or propidium iodide. These dyes change confirmation upon binding double-stranded DNA, and fluorescence in proportion to the amount of DNA present when exposed to UV light. In stained cells, sufficiently cleared of natural pigments, the ploidy can be quantified using an epifluorescence microscope and image-analysis software or flow cytometer (Garcés et al. 1998). If part of the population is undergoing synchronized mitosis or meiosis, the distribution of fluorescence intensities per cell will show peaks that correspond with haploid, diploid and tetraploid cells. In laboratory cultures, the ploidy of particular morphological forms can be readily determined and a hypothetical life cycle constructed. In field samples, it may be necessary to simultaneously label the cells with both a species-specific FISH probe and a nuclear stain to determine the relationship between ploidy and the morphology of specific life-cycle stages. Occasionally, information about mitotic or meiotic stages can also be obtained by observing the structure of the nucleus itself (Figueroa and Bravo 2005). Unfortunately, for some species, specific mitotic or meiotic life-cycle stages have equivalent morphologies and DNA contents, thereby complicating the use of ploidy alone in constructing life cycles, or in estimating the proportion of mitotic or meiotic stages in field samples (Litaker et al. 2002a). FISH assays similarly fail to differentiate these asexual (mitotic) or sexual (meiotic) life stages. When this occurs, careful observations of life-cycle transformations or other techniques (see 23.4) must be used to distinguish mitotic from meiotic stages.

For synchronous, mitotically dividing cell populations, two-fold changes in ploidy can be used to estimate specific growth rates (Chang and Carpenter 1994; Garcés et al. 1998). However, this approach does not work well in populations undergoing asynchronous division (Gisselson et al. 1999) or where high levels of sexual reproduction are occurring (Reguera et al. 2003). When appropriate conditions prevail, however, growth rates can be estimated and used to model HAB bloom dynamics. These data can also be used to determine the extent to which changes in cell abundance occur from in situ growth or hydrodynamic concentration, such as in frontal zones (Litaker et al. 2002b).

23.4 Genomic Approaches to Identifying Mitotic and Meiotic Life Cycle Stages

As mentioned in section 23.3, the proportion of mitotic versus meiotic cells in field populations has important implications for modeling the population dynamics of HABs (Peperzak et al. 2005). Sexual (meiotic) reproduction is energetically expensive and contributes substantially less to population growth than asexual (mitotic) cell division (Williams 1975). Recent evidence suggests that sexual reproduction commonly occurs more frequently during blooms than previously suspected (Figueroa 2005). An ability to quantify the number of cells undergoing mitosis and meiosis would improve the accuracy of population models, and increase our understanding of how life-cycle dynamics enable specific HAB species to exploit temporally variable environments and bloom. Effective means of identifying these mitotic and meiotic life-cycle stages using oligonucleotide or antibody probes that respectively bind mRNAs or proteins expressed only during specific stages of mitosis or meiosis are currently being developed. Evolutionary studies have shown that many of these proteins are conserved among species and can be identified using a variety of techniques (Schwarzacher 2003). The most common of these techniques is to screen a complementary DNA (cDNA) library for genes homologous to those which code for proteins involved in stage-specific cell cycle regulation in other species. This approach is feasible because conserved genes involved in different stages of mitosis and meiosis are already known for many species (Schwarzacher 2003; Oliva et al. 2005).

A cDNA library is created by isolating expressed RNAs from a species at different life-cycle stages, converting those RNAs to cDNAs, and then cloning and sequencing the cDNAs. cDNA libraries have been constructed for several HAB species (Hackett et al. 2005; Yoon et al. 2005), as well as for closely related species that are likely to contain homologous genes (Mann et al. 1999; Armbrust et al. 2004; Tanikawa et al. 2004; Maheswari et al. 2005; Montsant et al. 2005). Once the expressed cDNA sequences are available, bioinformatics software programs can translate them into putative protein sequences that can be compared to known stage-specific proteins from other species. Once matching stage-specific proteins have been identified, their expression pattern during different phases of mitosis or meiosis can be verified in laboratory experiments. If the expression patterns prove diagnostic of a specific mitotic or meiotic life-cycle phase, then a FISH probe can be synthesized and used to identify that life-cycle stage in field samples.

Other means of identifying stage-specific proteins include functional complementation assays. In essence, these assays are designed to identify interactions where one protein binds to and alters the function of another protein. Many of the key enzymes involved in mitosis and meiosis are regulated in this manner. Specifically, those enzymes contain domain(s) that bind regulatory proteins that are only expressed during specific life-cycle stages (Corellou et al. 2005). The binding of a regulatory protein to the regulatory domain causes a conformational change in the enzyme, which alters its activity. It is possible to exploit this binding arrangement to discover novel mitosis and meiosis stage-specific regulatory proteins. This is most often accomplished using some variation of the yeast two-hybrid system.

The two-hybrid system is based on several engineered plasmids that are autonomous stands of DNA capable of independently replicating in yeast. In the case of mitosis or meiosis, stage-specific proteins, one of the plasmids is engineered to contain (1) a regulatory binding domain from an enzyme in yeast known to play a key role in mitosis or meiosis, and (2) part of a transcription factor that controls the production of a specific yeast protein capable of conferring a selectable phenotype. Fortunately, a number of the binding domains for key enzymes in yeast that are controlled by mitosis or meiosis stage-specific proteins have been characterized and can be used for construct development. Once the construct is complete, it is transformed into a haploid yeast cell where it replicates and expresses the hybrid protein containing both the yeast binding domain and partial transcription factor. This is often referred to as the "bait" protein.

The corresponding construct in the yeast two-hybrid is an engineered "prey" protein. In this case, the self-replicating plasmid is engineered to contain the remainder of the transcription factor not contained in the "bait" construct, and a cloning site designed to receive cDNAs from the library. The HAB cDNAs are then cloned into the "prey" construct and transformed into haploid yeast cells. The result is thousands of yeast cells, each containing a different cloned cDNA construct. The yeast cells containing the "bait" are then mated with the cells expressing the cDNA or "prey" proteins from the HAB library. These matings form diploid cells, which allow the proteins expressed by both constructs to come in contact. If the "bait" protein containing the known binding domain from yeast binds an expressed HAB protein, it will bring both portions of the transcription factor together in such a way that the transcription factor becomes capable of activating a specific target gene. Typically, the protein produced by the newly activated gene allows the diploid yeast cells containing the binding constructs to grow on selective media which would otherwise be lethal to diploid cells lacking the binding partners necessary to produce a functional transcription factor. This allows rapid elimination of cells expressing cDNA proteins incapable of specifically binding the bait protein. Approximately 50 % of all interacting proteins have been discovered in this manner. The genes producing the functional assay can then easily be isolated from the surviving yeast cells and sequenced to identify the putative mitosis or meiosis regulatory genes (Bertomeu and Morse 2004). Ito (2001) presents additional details concerning the yeast two-hybrid and other related systems.

Another related approach to finding conserved stage-specific proteins is through the use of degenerate nucleic-acid primers that correspond to conserved regions of homologous regulatory proteins from other species. Once synthesized, these primers can be used to amplify genomic DNA from the species of interest. The resulting amplicons can be cloned, sequenced and translated for comparison with other known mitosis or meiosis stage-specific proteins. If a potential match is found, further characterization can be undertaken.

Proteomics methods involving 2-D gel electrophoresis may also be employed to identify differentially expressed proteins involved in cell-cycle regulation (Chan et al. 2004). Once these proteins have been identified, they can be isolated and sequenced to determine if they are homologous to other proteins known to regulate mitosis or meiosis. Knowledge of the protein sequence facilitates design of degenerate PCR primers, which can be used to identify the gene that codes for that protein. Once the gene is known, its stage-specific expression can be verified in culture, and probes developed for use in physiological or environmental studies.

Antibody detection of stage-specific proteins in HAB species can also be used to identify specific mitotic or meiotic stages. Production of these antibodies, however, first requires that sufficient amounts of stage-specific protein be produced to immunize mice or rabbits. This protein can be obtained by isolating and purifying the protein directly from culture material or by cloning the stage-specific gene into a protein-expression vector system. Once produced, the antibodies can be labeled and used to bind the stage-specific proteins in preserved, permeablized cells in a manner analogous to that employed in the FISH assays (Barbier et al. 2003). If the antibody binds specifically, it can be used to determine the location of the expressed protein during various life-cycle stages and to unambiguously establish the life-cycle stages in cultured and field samples.

The stage-specific proteins that have been studied most extensively with FISH and antibody probes are the cyclins, proteins involved in the regulation of the cell cycle. There are numerous cyclins with concentrations that rise and fall during different phases of mitosis or meiosis, while the level of others remains constant. Cyclins have been identified for *Karenia brevis* (Barbier et al. 2003) and *Aureococcus anophagefferens* (Lin et al. 2000). Other, non-cylcin stage-specific proteins have been identified as well (Chan et al. 2002, 2004).

23.5 Measuring Genetic Recombination During Sexual Reproduction

Genetic recombination helps organisms adapt to changing environments. However, little is known about the extent of recombination that occurs during sexual reproduction in HAB species. Recombination can be assessed in the laboratory by establishing single cell isolates ("clones"), performing pair-wise matings, and assessing genetic variation at the individual level in the parental clones and their offspring. Genetic variation at the individual level is measurable using either RAPD (random amplification of polymorphic DNA) or AFLP (amplified fragment length polymorphisms) assays (Figueroa 2005). A detailed description of these techniques can be found in Litaker and Tester (2002).

23.6 Future Application of Reverse Transcriptase Assays and DNA Microarrays in Life Cycle Studies

As more genes specific for mitosis and meiosis are discovered, it will be possible to employ two other molecular biological tools in HAB life cycle research: quantitative reverse-transcriptase PCR (qRT-PCR) and DNA microarrays. RT-PCR involves the quantitative isolation of total RNA from a culture or field sample, and the conversion of the RNA to DNA by means of a reverse transcriptase (RT). Quantitative PCR is then used to measure the amount of DNA produced by the RT reaction, which is proportional to the amount of expressed RNA in the sample. The assay can be normalized on a per-cell basis by estimating HAB cell density in a replicate sample using a FISH, quantitative PCR or sandwich hybridization assay. This approach will allow changes in cell stage-specific gene expression to be tracked temporally or spatially to determine how environmental factors influence life cycle transitions within a given population (Armbrust and Galindo 2001; Chepurnov et al. 2004). Once the genes that regulate the switch from mitosis to meiosis have been more fully characterized, it will be possible to monitor activation of these genes in the field.

The viability of cysts, which allow many HAB species to escape periods of unfavorable growth conditions, can also be potentially monitored using RT-PCR. This approach is important because current microscopic counts and certain molecular methods cannot distinguish between viable and non-viable.

The reason is that cyst cell walls and DNA are relatively stable and can persist for a significant period of time after the cell dies. Direct microscopic counts or DNA-based PCR assays would identify these cells as viable. This would result in an overestimate of viable cyst densities. mRNA transcripts, however, are much less stable than DNA, and degrade rapidly, making RT-PCR measurement of specific RNA transcripts a potentially good proxy for cyst viability (Coyne and Cary 2005).

DNA microarrays have great promise for life cycle studies as well. They are based on the principle of affixing cDNAs, or more commonly, short, gene-specific oligonucleotides derived from cDNAs, onto very precise locations within a 2x2 matrix on the surface of a microchip. Depending on the method of construction, 10,000 or more cDNAs or oligonucleotide sequences can be placed on a single chip. Once the chip has been constructed, mRNA is isolated from the sample of interest and converted to single-stranded DNA in a reaction mixture containing reverse transcriptase, nucleotides and a specific or random oligonucleotide primer. Commonly, one of the four nucleic acids used to synthesize the single-stranded DNA is fluorescently labeled. The RNA in the resulting mRNA/DNA hybrid is then hydrolyzed using NaOH. Other schemes for labeling single-strand DNA have been developed as well. In each case, the resulting single-stranded DNA is allowed to bind the cDNA or gene-specific oligonucleotide sequences on the chip. After incubation, and wash steps to remove the unbound labeled DNA, the chip is scanned using a laser beam. The fluorescence intensity at each spot on the microarray is recorded. Fluorescence intensity from each location on the chip is proportional to the amount of labeled DNA that was bound to the target DNA. Because the imaging system is highly accurate, the fluorescence intensities in any location can be precisely matched with the specific DNA or oligonucleotide deposited in that location.

Microarrays allow examination of expression by multiple genes simultaneously. By extracting mRNA from different growth and life-cycle stages, changes in expression of the putative cell-cycle gene(s) previously identified from a cDNA library can be followed. Changes in the expression levels of other uncharacterized cDNAs, which correlate with life-cycle stages observed in culture, can also be used to identify potential meiosis-and mitosis-specific genes (Lidie et al. 2006). Genes exhibiting stage-specific expression can then be tested using qRT-PCR assays of samples from laboratory cultures exhibiting different life-cycle stages, or by using in situ hybridization assays to determine at what stage of the life cycle the gene(s) are expressed. This approach should also provide valuable information on the basic genetic pathways that regulate meiosis and mitosis as well as circadian rhythms (Okamoto and Hastings 2003).

HAB microarray construction is currently limited due to the small number of cDNA libraries that have been fully sequenced. The number of characterized libraries, however, is expected to increase over the next 5 years. Other concerns include the high cost of manufacturing microarrays and controversies over the most appropriate internal controls and how best to normalize the fluorescence data. Because this technique is being used extensively in medical research to follow gene expression in cancer and other diseases, considerable efforts are being made to reduce synthesis costs and to deal with internal control and data normalization issues. It is anticipated that these problems will be resolved by the time the next generation of HAB cDNA libraries become available for incorporation into the chip format.

23.7 Conclusions

The molecular methods described in this chapter offer an opportunity to identify specific life stages and to study the mitotic and meiotic cell cycles of harmful algal bloom species with unprecedented accuracy. Given that different life-cycle stages often interact in complex ways to influence the success of algal species, these techniques will provide critical insights into the factors that regulate HAB population dynamics.

Acknowledgements. Ed Noga and Bill Sunda reviewed this manuscript and provided constructive criticism.

References

Anderson DM, Chisholm SW, Watras CJ (1983) Importance of life cycle events in the population dynamics of *Gonyaulax tamarensis*. Mar Biol 76:179–190
Armbrust EV et al (2004) The genome of the diatom *Thalassiosira pseudonana*: ecology, evolution, and metabolism. Science 306:79–86
Armbrust EV, Galindo HM (2001) Rapid evolution of a sexual reproduction gene in centric diatoms of the genus *Thalassiosira*. Appl Environ Microbiol 67:3501–3513
Barbier M, Leighfield TA, Soyer-Gobillard MO, Van Dolah FM (2003) Permanent expression of a cyclin B homologue in the cell cycle of the dinoflagellate *Karenia brevis*. J Eukaryot Microbiol 50:123–131
Bertomeu T, Morse D (2004) Isolation of a dinoflagellate mitotic cyclin by functional complementation in yeast. Biochem Biophys Res Comm 323:1172–1183
Cerino F, Orsini L, Sarno D, Dell'Aversano C, Tartaglione L, Zingone A (2005) The alternation of different morphotypes in the seasonal cycle of the toxic diatom *Pseudo-nitzschia galaxiae*. Harmful Algae 4:33–48
Chan KL, New D, Ghandhi S, Wong F, Lam CMC, Wong JTY (2002) Transcript levels of the eukaryotic translation initiation factor 5A gene peak at early G_1 phase of the cell cycle in the dinoflagellate *Crypthecodinium cohnii*. Appl Environ Microbiol 68:2278–2284
Chan LL, Hodgkiss IJ, Wan JMF, Lum JHK, Mak ASC, Sit WH, Lo SCL (2004) Proteomic study of a model causative agent of harmful algal blooms, *Prorocentrum triestinum*

II: the use of differentially expressed protein profiles under different growth phases and growth conditions for bloom prediction. Proteomics 4:3214–3226
Chang J, Carpenter EJ (1994) Active growth of the oceanic dinoflagellate *Ceratium teres* in the Caribbean and Sargasso Seas estimated by cell-cycle analysis. J Phycol 30:375–381
Chepurnov VA, Mann DG, Sabbe K, Vyverman W (2004) Experimental studies on sexual reproduction in diatoms. Int Rev Cytol 237:91–15
Corellou F, Camasses A, Ligat L, Peaucellier G, Bouget F-Y (2005) Atypical regulation of a green lineage-specific B-type cyclin-dependent kinase. Plant Physiol 138:1627–1636
Coyne KJ, Cary SC (2005) Molecular approaches to the investigation of viable dinoflagellate cysts in natural sediments from estuarine environments. J Eukaryot Microbiol 52:90–94
Figueroa RI (2005) The significance of sexuality and cyst formation in the life-cycles of four marine dinoflagellate species. PhD Thesis, Dept Limnology, Lund Univ, Sweden, 153 pp
Figueroa RI, Bravo I (2005) Sexual reproduction and two different encystment strategies of *Lingulodinium polyedrum* (Dinophyceae) in culture. J Phycol 41:370–379
Galluzzi L, Penna A, Bertozzini E, Vila M, Garcés E, Magnani M (2004) Development of a real-time PCR assay for rapid detection and quantification of *Alexandrium minutum* (a dinoflagellate). Appl Environ Microbiol 70:1199–1206
Garcés E, Bravo I, Vila M, Figueroa RI, Masó M, Sampedro N (2004) Relationship between vegetative cells and cyst production during *Alexandrium minutum* bloom in Arenys de Mar harbour (NW Mediterranean). J Plankton Res 26:637–645
Garcés E, Delgado M, Masó M, Camp J (1998) Life history and in situ growth rates of *Alexandrium taylori* (Dinophyceae, Pyrrophyta). J Phycol 34:880–887
Gisselson LA, Granéli E, Carlsson P (1999) Using cell cycle analysis to estimate in situ growth rate of the dinoflagellate *Dinophysis acuminata*: drawbacks of the DNA quantification method. Mar Ecol Prog Ser 184:55–62
Groben R, John U, Eller G, Lange M, Medlin LK (2004) Using fluorescently labeled rRNA probes for hierarchical estimation of phytoplankton diversity – a mini-review. Nova Hedwigia 79:313–320
Hackett JD, Scheetz TE, Yoon HS, Soares MB, Bonaldo MF, Casavant TL, Bhattacharya D (2005) Insights into a dinoflagellate genome through expressed sequence tag analysis. BMC Genomics 6:80
Ito T, Chiba T, Yoshida M (2001) Exploring the protein interactome using comprehensive two-hybrid projects. Trends Biotechnol 19:S23–S27
Lidie KL, Ryan JC, Barbier M, Van Dolah FM (2006) Gene expression in the Florida red tide dinoflagellate *Karenia brevis*: analysis of an expressed sequence tag (EST) library and development of a DNA microarray. Mar Biotechnol (in press)
Lin S, Magaletti E, Carpenter EJ (2000) Molecular cloning and antiserum development of cyclin box in the brown tide alga *Aureococcus anophagefferens*. Mar Biotechnol 2:577–586
Litaker RW, Tester PA (2002) Molecular methods for detecting and characterizing harmful phytoplankton. In: Hurst CJ, Crawford RL, Knudsen GR, McInerney MJ, Stetzenbach LD (eds) Manual of environmental microbiology. ASM Press, Washington, DC, pp 342–353
Litaker RW, Vandersea MW, Kibler SR, Madden VJ, Noga EJ, Tester PA (2002a) Life cycle of the heterotrophic dinoflagellate *Pfiesteria piscicida* (Dinophyceae). J Phycol 38: 442–463
Litaker RW, Warner VE, Rhyne C, Duke CS, Kenney BE, Ramus J, Tester PA (2002b) Effect of diel and interday variations in light on the cell division pattern and in situ growth rates of the bloom-forming dinoflagellate *Heterocapsa triquetra*. Mar Ecol Prog Ser 232:63–74

Maheswari U, Montsant A, Goll J, Krishnasamy S, Rajyashri KR, Patell VM, Bowler C (2005) The diatom EST database. Nucleic Acids Res 33:D344–D347

Mann DG, Chepurnov VA, Droop SJM (1999) Sexuality, incompatibility, size variation, and preferential polyandry in natural populations and clones of *Sellaphora pupula* (Bacillariophyceae). J Phycol 35:152–170

Miller PE, Scholin CA (2000) On detection of *Pseudo-nitzschia* (Bacillariophyceae) species using whole cell hybridization: sample fixation and stability. J Phycol 36:238–250

Montsant A, Jabbari K, Maheswari U, Bowler C (2005) Comparative genomics of the pennate diatom *Phaeodactylum tricornutum*. Plant Physiol 137:500–513

Okamoto OK, Hastings JW (2003) Novel dinoflagellate clock-related genes identified through microarray analysis. J Phycol 39:519–526

Oliva A, Rosebrock A, Ferrezuelo F, Pyne S, Chen H, Skiena S, Futcher B, Leatherwood J (2005) The cell cycle-regulated genes of *Schizosaccharomyces pombe*. PLoS Biol 3:1239–1260

Peperzak L, Garcés E, Kremp A, Montresor M (2005) Model-aided quantification of dinoflagellate sexual cycles reveals unprecedented species-specific life cycle variability. ASLO Summer Meeting, Santiago de Compostela, Spain, (abstract), p 118

Reguera B, Garcés E, Pazos Y, Bravo I, Ramilo I, González-Gil S (2003) Cell cycle patterns and estimates of in situ division rates of dinoflagellates of the genus *Dinophysis* by a postmitotic index. Aquat Microb Ecol 24:117–131

Scholin C, Buck K, Britschgi T, Cangelosi G, Chavez F (1996) Identification of *Pseudo-nitzschia australis* (Bacillariophyceae) using rRNA-targeted probes in whole cell and sandwich hybridization formats. Phycologia 35:190–197

Schwarzacher T (2003) Meiosis, recombination and chromosomes: a review of gene isolation and fluorescent in situ hybridization data in plants. J Exp Bot 54:11–23

Tanikawa N, Akimoto H, Ogoh K, Chun W, Ohmiya Y (2004) Expressed sequence tag analysis of the dinoflagellate *Lingulodinium polyedrum* during dark phase. Photochem Photobiol 80:31–35

Williams GC (1975) Sex and evolution. Princeton Univ Press, Princeton, 210 pp

Yamaguchi A, Horiguchi T (2005) Molecular phylogenetic study of the heterotrophic dinoflagellate genus *Protoperidinium* (Dinophyceae) inferred from small subunit rRNA gene sequences. Phycol Res 53:30–42

Yoon HS, Hackett JD, Van Dolah FM, Nosenko T, Lidie KL, Bhattacharya D (2005) Tertiary endosymbiosis driven genome evolution in dinoflagellate algae. Mol Biol Evol 22: 1299–1308

24 Laboratory and Field Applications of Ribosomal RNA Probes to Aid the Detection and Monitoring of Harmful Algae

K. Metfies, K. Töbe, C. Scholin, and L.K. Medlin

24.1 Introduction

As discussed elsewhere, harmful algal blooms are steadily increasing worldwide. Given the numerous threats posed by them, early warning of harmful blooms and rapid detection of the species that cause them is highly desirable. Many countries have launched monitoring programs to serve as warning systems. Samples are typically transported to a centralized laboratory and examined for harmful or toxic species using traditional methods based on microscopy (e.g. Todd 2003). This approach has proven very successful, but becomes more difficult to manage as the number of samples and frequency of their collection increases, reaching a point where analysis of a sample can take days. This time lag can increase further when more elaborate methods are required to identify positively species that cannot be resolved using light microscopy alone. Such time lags impair our ability to provide an early warning of bloom events. For this reason, on-site, near real-time water analysis is desirable. This requirement can be met in part by equipping trained observers (paid staff or volunteers) with field microscopes to speed up assessments concerning the relative abundance and distribution of numerous species (Conrad et al. 2003), and that information can be used in an adaptive fashion to "flag" specific samples for expert examination. Whereas this can help alleviate some time lag associated with identifying problem species, there still remains a need to quantify them and accommodate the need for detailed analyses when required. Towards that end, we consider here application of ribosomal RNA (rRNA) targeted probes in fluorescent in situ hybridization (FISH) and DNA probe array formats. Reviews of other techniques that also have proven useful for identifying harmful algae in cultured and natural samples as well as the prospects of using those methods aboard in-water autonomous sensor systems are found elsewhere (Scholin et al. 2003).

Ecological Studies, Vol. 189
Edna Granéli and Jefferson T. Turner (Eds.)
Ecology of Harmful Algae

24.2 Ribosomal RNA Sequences as Markers for Phylogenetic Studies and Species Identification

Molecular biological techniques have greatly enhanced our ability to understand phylogenetic relationships among organisms and to develop means to detect specific species, genes and gene products. Although a number of genetic markers are used for this purpose (Paul et al. 1999), ribosomal RNA (rRNA) genes have historically figured most prominently in studies of harmful algae. Ribosomal RNA molecules have a number of attributes that make them excellent molecular markers (Woese 1987). The conserved and variable regions of the molecule can be used to develop oligonucleotide probes of varying specificity, making it possible to identify phytoplankton at various taxonomic levels from classes down to species or strains using whole-cell and cell-free formats (Scholin et al. 2003; John et al. 2003, 2005; Metfies and Medlin 2004).

24.3 Fluorescent In Situ Hybridization (FISH) for Identifying Intact Cells

Fluorescence hybridization (FISH) allows the rapid detection of algal groups and even the separation of closely related, morphologically similar species or strains (Lim et al. 1993; Miller and Scholin 1996, 2000; Simon et al. 1997, 2000; Peperzak et al. 2001; Rhodes et al. 2001; John et al. 2003, 2005; Groben et al. 2004; Groben and Medlin 2005; Anderson et al. 2005). FISH techniques rest on hybridization of fluorescently labelled oligonucleotide probes to rRNA within intact cells, thus "labelling" target species when appropriate reaction conditions are met.

The FISH technique begins with a fixed sample to preserve overall cell morphology, reduce autofluorescence, and permeabilise the cell wall to exchange probes and hybridization solutions. Several FISH protocols are in use, but the diversity of organisms targeted makes it difficult to find a method that works well for all or at least most species. Many different fixation and sample processing protocols have emerged in recent years. For example, Anderson et al. (2005) have used a two-step fixation where cells are initially treated with formaldehyde, and within 36 h resuspended in cold methanol and then stored refrigerated. For *Alexandrium*, preserved cells were stored for at least 1 year without signal loss. Others have explored alternative means of preserving samples for FISH. Medlin and co-workers compared different fixation protocols and found that saline ethanol treatment (Scholin et al. 1996) gave good results with most species tested, but it must be made fresh and is stable only for a few hours. "Modified saline ethanol fixative" has a reduced ethanol concentration and is stable for several months at room temperature, Cells preserved using this solution can be stored for at least 1 month at room temperature without signal loss (Miller and Scholin 1998, 2000). Its ease of preparation, low toxicity, and stability make it an attractive choice when working outside of a laboratory, but samples must be processed within several weeks after collection. Some species will autofluoresce even with prolonged exposure to it. Here 50 % dimethylformamide (DMF) treatment can help (Groben and Medlin 2005).

FITC (fluorescein-5- isothiocyanate) is often used to label probes in FISH experiments. With an excitation maximum of 494 nm and an emission maximum of 517 nm labelled algal cells are coloured light green and are normally easily distinguished from non-labelled and autofluorescent cells appearing orange to red in colour depending on the filter set (Scholin et al. 2003). Negative controls are always recommended in FISH experiments, e.g. "no probe" (hybridization buffer only) and/or application of a labelled probe that does not react with the target species (NON-EUK 1209R; Amann et al. 1990; Scholin et al. 2003). Another common FISH fluorochrome is Cy5 (Cyanin5), a derivative from indodicarbocyanine (Shapiro 2003), which has an excitation maximum at 649 nm and emission maximum at 670 nm in the far red/close infrared region. Visualizing Cy5 requires use of infrared sensitive cameras connected to the microscope.

Hybridization reactions for phytoplankton are often performed on polycarbonate filter membranes with one or two fluorescently labelled probes. In addition to flow-through methods where a sample may be captured in a filtration apparatus with an entire hybridization process following (e.g. Miller and Scholin 1998, 2000; Anderson et al. 2005), filters can also be cut into pieces and treated individually for the detection of more algal species (e.g. Miller and Scholin 1998; John et al. 2003). Where target and non-target sequences are very similar, stringent hybridization conditions must be used. Groben and Medlin (2005) found that addition of formamide in the hybridization buffer (up to 20 %) and salt reduction in washing steps permitted discrimination of single-base mismatches. Sodium dodecylsulfate (SDS), the detergent commonly used in FISH hybridization buffers (Amann 1995) can destroy fragile cells like unarmoured dinoflagellates. In contrast, IGEPAL-CA630 (or the chemically identical NONIDET-P40) maintains cell integrity while giving a good probe penetration into the cell. Fading of the fluorescence signal is reduced by applying an antifade mounting solution to the filter before sealing it with a coverslip on a slide (e.g. SlowFade Lite™, Molecular Probes, Inc.). The sealed filter can be examined microscopically directly or stored at −20 °C for several days without a loss of the fluorescence signal. Finally, DAPI (4′, 6-diamidino-2-phenylindole) counterstains cells' DNA (Shapiro 2003). DAPI is excited with UV light at 365 nm and stains DNA bright blue.

24.3.1 TSA-FISH for Flow Cytometry

Enumerating labelled phytoplankton cells manually using epifluorescence microscopy has proven a viable means of conducting a variety of research and monitoring programs (e.g. Rhodes et al. 2004; Scholin et al. 2003; Anderson et al. 2005), but the approach is time consuming and demanding when dealing with large numbers of samples. For this reason a number of workers have explored the use of flow cytometry. Flow cytometry (FCM) detects microalgal cells in liquid suspension based on their optical characteristics. Phytoplankton can be rapidly counted and sized with FCM by analyzing cells autofluorescence (Veldhuis and Kraay 2000) but limited discrimination of taxonomic groups and species is possible (e.g. Jonker et al. 2000; Marie et al. 2005). Application of the FISH technique enhances resolution afforded by FCM. Most flow cytometers equipped with a single light source use an argon laser emitting blue-green (excitation 488 nm), and dual-laser instruments possess a red (excitation 633–640 nm) helium-neon or diode laser as well. Larger benchtop and sorting cytometers often contain a UV laser (excitation 325–365 nm). Fluorescein and Cy5 have been used in FCM protocols (Shapiro 2003) and dual labelling of phytoplankton in one sample is possible. FISH for flow cytometry is typically performed in suspension. The cells are processed in a tube with reagent exchange generally achieved by centrifugation. Careful and precise removal of the supernatant is required to minimize the cell loss. Additionally, the adhesion of cells to the tube surface can lead to cell loss. Treating the tubes with surfactants, adding surfactants to the cell suspension and sonification can remedy this problem (Biegala et al. 2003). High autofluorescence, low rRNA content and poor accessibility of probe target sites can result in weakly labelled cells using a traditional FISH technique (Fuchs et al. 2001). The tyramide signal amplification (TSA), or synonymously called catalyzed reporter deposition (CARD), method can overcome this problem and can be used with FCM (Biegala et al. 2003). The enzyme horseradish peroxidase (HRP) is linked to the 5'- end of the oligonucleotide probe, and in the presence of small amounts of hydrogen peroxide it converts its labelled substrate, tyramide, into short-lived, extremely reactive intermediates that covalently link to electron rich regions of adjacent proteins, such as tyrosine residues. This reaction only occurs adjacent to the probe target sites (Schönhuber et al. 1997, 1999; Pernthaler et al. 2002). The fluorochrome that is bound to the tyramide can be FITC, Cy5 or Alexa fluor conjugates (Shapiro 2003). This method greatly enhances signal intensity relative to a fluorescent label directly attached to a probe. However, naturally occurring peroxidases can lead to non-specific binding of the tyramide and therefore must be quenched (Pougnard et al. 2002). Addition of dextran sulphate to the fluorescence tyramide substrate solution improves localization of the fluorescently labelled tyramide and prevents its free diffusion before it is immobilized (Schönhuber et al. 1999). Hybridization reactions with TSA must be performed between 35 and 37 °C to prevent HRP denaturation, so higher formamide concentrations in the hybridization buffer are required to ensure probe specify. Negative control reactions also include a "no probe" treatment with the addition of the labelled tyramide. When proper care is taken, the TSA method greatly improves the signal-to-noise ratio, particularly for bacteria (Schönhuber et al. 1997), cyanobacteria (Schönhuber et al. 1999; West et al., 2001), picoplankton (Not et al. 2002, 2004; Biegala et al. 2003) and bacteria associated with microalgae (Biegala et al. 2002; Alverca et al. 2002).

24.3.2 TSA-FISH for Solid-Phase Cytometry

Solid-phase cytometry (SPC) combines the advantages of FCM with image analysis (Kamentsky 2001; Lemarchand et al. 2001). In SPC, a laser is moved over cells immobilised on a solid support (Vives-Rego et al. 2000). SPC allows the rapid enumeration of several thousand cells with similar accuracy to FCM and is reported useful for the detection of rare events as compared to epifluorescence microscopy and flow cytometry (Lemarchand et al. 2001). The Chem*Scan*™ (Chemunex, France) is a SPC (Mignon-Godefroy 1997; Reynolds and Fricker 1999), initially developed for industrial and environmental microbiology (Vives-Rego et al. 2000). Recently, it was adapted for the detection of toxic microalgae using antibodies (West et al. 2006) and oligonucleotides probes (Töbe et al. 2006). The Chem*Scan* ™ with a 488-nm argon-ion laser is suited for FITC labelled probes. Samples are collected by filtration onto membranes, treated with the probe, and subsequently scanned. The optical system records emissions at three different wavelengths: green (500–530 nm), green-yellow (540–570 nm) and yellow-amber (570–585 nm) (Bauer et al. 1996; Roubin et al. 2002). Fluorescent particles are detected and the Chem*Scan* computer program applies discrimination criteria to discriminate between "true" and "false" events (Roubin et al. 2002), while calculating the signal ratios of the positive signals detected at the three wave-length intervals. The oligonucleotide probe-labelled cells ("true" signals) can automatically be distinguished from particles or auto-fluorescent cells ("false signals"). Expected cell size and shape can be defined for more discrimination. Positively identified cells are coloured spots on a scan map, a display of the membrane. The cells are visualized by transferring the membrane holder to an epifluorescence microscope equipped with a computer controlled motorized stage connected to the Chem*Scan*. Each positive point is validated by the user as desired (Reynolds and Fricker 1999; Roubin et al. 2002). TSA enhancement is strongly recommended for reliable detection of target cells with the Chem*Scan* (Fig. 24.1), because FITC-labelled cells give insufficient intensity for sufficient discrimination between labelled and non-labelled cells. Although promising, this method is only adequate for counting spherical microalgae, not long filamentous cells. However, improved software should eliminate this problem. Thus, validation of the positive counted cells, at least for a subset of the filter, is recommended.

24.4 Detecting Many Species Simultaneously Using DNA Probe Arrays

FISH and TSA-FISH only allow the identification of one or a few organisms at a time (Metfies and Medlin 2004; Groben et al. 2004). Cells must remain intact throughout those procedures, so careful sample handling and efforts to minimize cell loss are needed. Thus, many researchers changed to cell-free methods, such as those that rely on inferring presence of organisms based on detecting sequences of nucleic acid in solution. DNA probe arrays offer more possibilities to identify numerous "signature sequences" simultaneously in a single sample (DeRisi et al. 1997; Lockhart et al. 1996; Brown and Botstein 1999). According to the "Taxonomic Reference List of Toxic Plankton Algae" from 2002 (IOC 2002), there are approximately 85 algal species that can form blooms and produce potent toxins. Many are cosmopolitan. Many areas are threatened by multiple toxic algal species. Thus, DNA probe arrays could be developed to monitor harmful species, as well as many other organisms of interest. In this section we present several methods based on this concept wherein probes are attached to a solid support and target sequences are detected using fluorescent, electrochemical and chemiluminescent reporting.

24.4.1 Microarrays on Glass Slides and Fluorescence Detection

DNA microarray technology is based on an ordered array of probes attached to a solid support (Lockhart and Winzeler 2000; Shena 2000; Rampal 2001). Deposition of probes onto "DNA-microchips" is achieved in two ways: direct synthesis on the chip-surface (Singh-Gasson et al. 1999) or deposition on the surface with a high precision robot. Glass slides are the most commonly used solid support for DNA-microchips because they have low auto fluorescence.

Traditional microarray experiments employ a step to label fluorescently the target DNA/RNA prior to hybridization with the chip. The labelling step can either happen directly by an incorporation of a fluorescent dye into the target or indirectly by the incorporation of some other moiety (e.g. biotin) that is detected with another fluorescent label (Southern et al. 1999; Cheung et al. 1999). Once the hybridizations are completed, the chip is scanned with a device containing a laser or a polychromatic light source, and the fluorescence pattern recorded (DeRisi et al. 1997).

Probe development for arrays is based on known sequences. Although the number of ribosomal RNA sequences is continually growing (Maidak et al.

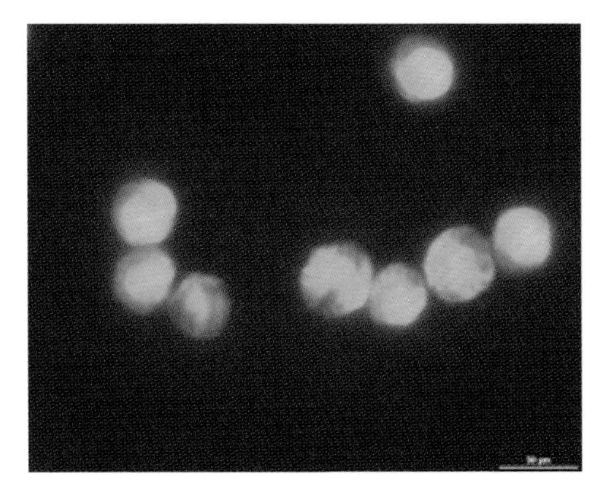

24.1. Whole-cell hybridization coupled with tyramide signal amplification and *Alexandrium fundyense* (CA 28) TSA-FISH with HRP-labelled probe NA 1 (Miller and Scholin 1998). Bar = 30 :m

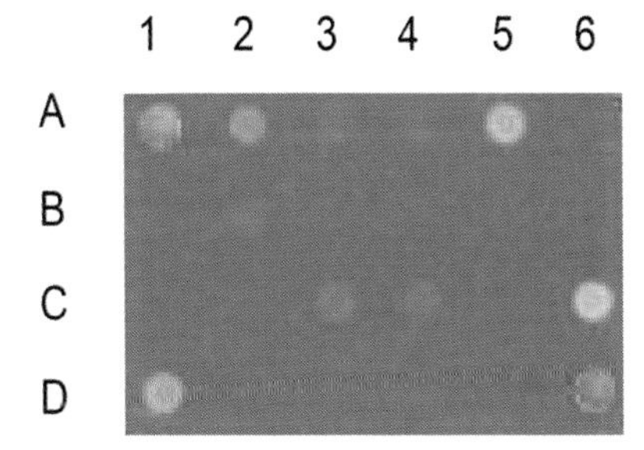

A1: Positive Control
A2: Euk 328 (all Eukaryotes)
A5: Chlo 02 (Class - Chlorophyta)
C6: Pras 04 (Family - Mamelliaceae)
D1: Bathy 01 (Genus - Bathycoccus)
D6: Positive Control

24.2. DNA-microarray that contains a hierarchical set of probes for identification of the genus *Bathycoccus*. The microarray was hybridized to a PCR-product that was amplified from a clone assigned to the genus *Bathycoccus*. The *dots* result from a hybridization of the immobilized probe with the target DNA

2001), it is widely accepted that the majority of microbes are unknown at the rRNA level. Consequently, the probe specificity has to be updated continually with respect to new sequences and results of empirical tests. Cross-reactions are always possible, particularly when dealing with environmental samples. The application of hierarchical probes, ones that detect that target species at different taxonomic levels, is possible given the large number of probes that can be spotted on a single chip. This approach makes identification of harmful species more accurate because a "positive" detection would depend on multiple probes all reacting with each sample. Hierarchical probes were implemented on a preliminary DNA-microarray dedicated to the assessment of phytoplankton composition (Fig. 24.2). That array contained probes limited to higher taxonomic levels of algae. Environmental samples taken in the North Sea at Helgoland were analysed using that chip. The data obtained were consistent with clone library results from the same samples (Medlin et al. 2006).

Despite the obvious potential of DNA-microarrays to facilitate monitoring and identification of microbes and harmful algae (Guschin et al. 1997; Loy et al. 2002; Call et al. 2003; Metfies and Medlin 2004), some limitations and pitfalls should be noted. Developing and evaluating DNA-microarrays is time-consuming and costly. Cross-reactions are always possible. Second, current hybridization reactions require all probes to have the same optimal hybridization temperature. This is not a trivial task when dealing with many probes, and intensive optimisation experiments are required to find a comprehensive set of probes that work well under the same conditions (Feriotto et al. 2002; Boireau et al. 2005). Third, whereas it is possible to quantify the amount of target bound to probes on the DNA-chip, it remains to be determined whether those signals can be correlated reliably to cell numbers in natural samples (Anderson et al. 2005). Commercial DNA microarray processors and scanners are currently designed for use in the laboratory.

24.4.2 Handheld Array Device That Uses Electro-Chemical Detection

Electrochemical detection of DNA probe/target hybrids offers an alternative to fluorescence-based systems (Azek et al. 2000; Litaker et al. 2001; Baeumner et al. 2003; Metfies et al. 2005). Such applications have been explored at the AWI (Bremerhaven, Germany) for the detection of *Alexandrium ostenfeldii* (Metfies et al. 2005). The sensor consists of two major parts: a disposable sensor chip and a portable handheld device in which the chip is inserted for the measurement of the electrochemical signals. Detection of target sequences utilizes a sandwich-hybridization method that takes place on a carbon electrode of the chip (Zammatteo et al. 1995; Rautio et al. 2003, Fig. 24.3). Sand-

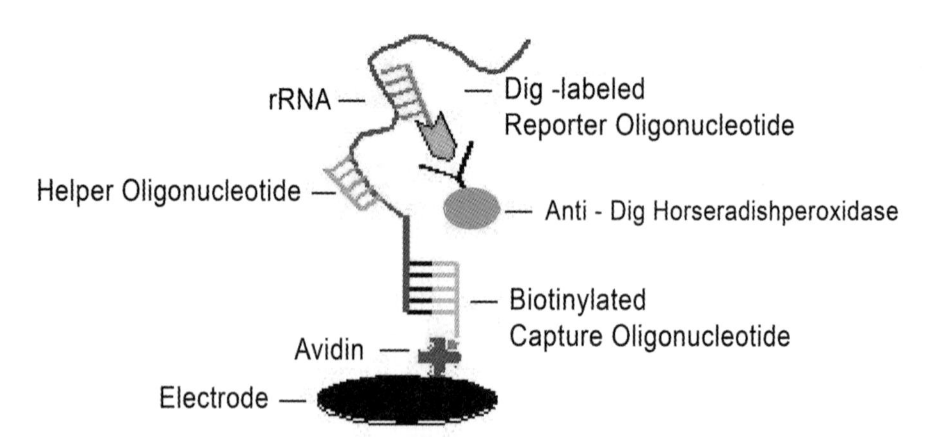

24.3. Principle of electrochemical detection of nucleic acids using sandwich hybridization

wich-hybridization reactions use a set of two probes that bind in close proximity to the target nucleic acid. In the current assay one probe is immobilised via biotin on a carbon electrode coated with avidin. The second probe signals the captured molecules via an antibody-HRP reaction that, in turn, catalyses the reduction of H_2O_2 to water. The resulting redox electron-transfer is measured as current and is proportional to the amount of target applied to the sensor. The device is currently limited to the detection of *A. ostenfeldii* and *A. tamarense* but is being expanded in the EU ALGADEC project to regional chips for 14 species. Manual isolation of the RNA is done prior to analysis; fully automated sample processing is planned.

24.4.3 DNA Probe Arrays for Autonomous Detection of Species Using the Environmental Sample Processor (ESP)

The ESP is an electromechanical/fluidic instrument system designed to collect discrete water samples from the ocean subsurface, concentrate cells (par-

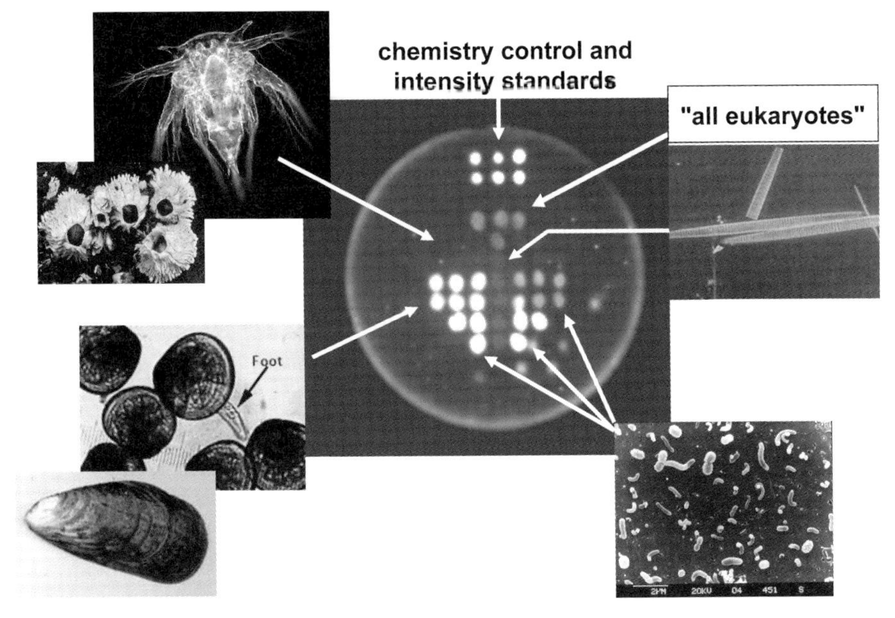

24.4. Example of a custom 25-mm DNA probe array developed automatically in the ESP using a natural seawater sample. Printed for demonstration purposes are (*top center*) probes for chemistry controls and array intensity standards, and (*clockwise, right to left*) small-subunit rRNA-targeted, probes for "universal eukaryote" "pennate diatoms," specific groups of marine bacteria, *Roseobacter* and *Cytophaga* , mussel larvae and barnacle larvae . This array illustrates the simultaneous detection of mussel larvae, marine bacteria including *Roseobacter* (weak) and *Cytophaga* (strong), and pennate diatoms (weak); barnacle larvae were not present in this sample. ©2005 MBARI

ticulates), and automate application of rRNA-targeted probes. In addition, the ESP archives discrete samples for a variety of nucleic acid analyses, microscopy and other types of analytical procedures after the instrument is recovered. The ESP has been applied to detect a broad range of marine planktonic organisms (Fig. 24.4). "First generation" ESP prototypes were deployed in Monterey Bay, CA, and Gulf of Maine, ME, USA (Goffredi et al. 2005; Scholin et al. 2005).

To develop a probe array, the ESP first collects a sample and removes seawater, then homogenizes material retained using a chaotrop and heat. A crude homogenate is applied to the array, followed by a sequence of reagents that reveal target molecules retained at specific locations on the array grid using sandwich hybridization and chemiluminescence (Scholin et al. 2005). An array image is captured by a CCD camera and transmitted to a remote location for interpretation (Fig. 24.4). The entire process, from collection of a live sample to broadcast of the imaged array takes about 2 h and occurs sub-surface. The reagents employed in the ESP assays are stable for extended periods (none require refrigeration), and the chemical reactions are amenable to microfluidic scaling. Different arrays can be tailored to specific groups of organisms. The ESP can support detection of many different rRNA target sequences using a common methodology, suite of reagents and core sample processing instrumentation.

24.5 Conclusions

Molecular techniques can be used to unequivocally and rapidly identify, and in some cases quantify, particular species and strains of harmful algae. Integrated instrument systems that are designed to automate sample preparation and distribution, as well as to employ molecular probe technologies for detecting a variety of cell markers and processing raw data to speed up and aid interpretation of assay results, are becoming increasingly common, although limited to use in a research laboratory. Much of that technological revolution is driven by the biomedical research and diagnostics industries, but parallel applications in environmental science have also firmly taken root. For example, "portable" sample preparation and analysis systems designed for use outside of a laboratory are gaining attention, but at this time are not routine in environmental research and monitoring practices. Prototypes of autonomous, in-water sensors that utilize molecular probe technology are also emerging (Scholin et al. 2005), but like the portable devices are still in the experimental stage and their use largely restricted to a handful of researchers studying a restricted set of target organisms. Nonetheless, the prospects of developing both the field portable and in-water systems to the point where they are robust and available commercially are bright, and it is well within the foreseeable future for such tools to be part of an "early warning system" in some areas and for certain species as an aid to invoke mitigation strategies to minimize the effects of harmful blooms. Numerous issues remain as to how such techniques and instruments will be tested, calibrated, validated, made available commercially, and ultimately used routinely in concert with the present-day, accepted methods for identifying and quantifying harmful species and their toxins (Scholin et al. 2005). In the meantime, those involved with harmful algae research and monitoring can expect a future that includes vigorous debate and intense innovation on both the analytical and instrumentation development fronts. The remaining issue is having these rapid techniques validated and accepted by the monitoring agencies worldwide to replace the more traditional methods.

References

Alverca E, Biegala IC, Kennaway GM, Lewis J, Franca S (2002) In situ identification and localization of bacteria associated with *Gyrodinium instriatum* (Gymnodiniales, Dinophyceae) by electron and confocal microscopy. Eur J Phycol 37:523–530

Amann RI (1995) In situ identification of micro-organisms by whole cell hybridization with rRNA-targeted nucleic acid probes. In: Akkermans ADL, van Elsas JD, de Bruijn FJ (eds) Molecular microbial ecology manual 3.3.6. Kluwer Academic Publishers, Dordrecht, pp 1–15

Amann RI, Binder BJ, Olson RJ, Chisholm SW, Devereux R, Satahl DA (1990) Combination of 16S rRNA-targeted oligonucleotide probes with flow cytometry for analysing mixed microbial populations. Appl Environ Microbiol 56:1919–1925

Anderson DM, Kulis DM, Keafer BA, Gribble KE, Marin R, Scholin CA (2005) Identification and enumeration of *Alexandrium* sp. from the Gulf of Maine using molecular probes. Deep-Sea Res II (in press)

Azek F, Grossiord C, Joannes M, Limoges B, Brossier P (2000) Hybridization assay at a disposable electrochemical biosensor for the attomole detection of amplified human cytomegalovirus DNA. Anal Biochem 284:107–113

Baeumner A J , Cohen RN, Miksic V, Min J (2003) RNA biosensor for the rapid detection of viable *Escherichia coli* in drinking water. Biosens Bioelectron 18:405–413

Bauer C, Butin M, Decrulle M, Foulon V, Gaubert S, Guyomard S, Lejuez V, Mandray C, Philippe L, Sofia T (1996) Alternative methods of microbiological control. Review of the major rapid techniques. Report of a SFSTP commission. S.T.P. Pharma Practiques 6:449–464

Biegala IC, Kennaway G, Alverca E, Lennon JF, Vaulot D, Simon N (2002) Identification of bacteria associated with dinoflagellates (Dinophyceae) *Alexandrium* spp. using tyramide signal amplification-fluorescent in situ hybridization and confocal microscopy. J Phycol 38:404–411

Biegala IC, Not F, Vaulot D, Simon N (2003) Quantitative assessment of picoeukaryotes in the natural environment by using taxon-specific oligonucleotide probes in association with tyramide signal amplification-fluorescence in situ hybridization and flow cytometry. Appl Environ Microbiol 69:5519–5529

Boireau W, Zeeh JC, Puig PE, Pompon D (2005) Unique supramolecular assembly of a redox protein with nucleic acids onto hybrid bilayer: towards a dynamic DNA chip. Biosens Bioelectron 20:1631–1637

Brown PO, Botstein D (1999) Exploring the new world of the genome with DNA microarrays. Nature Genet 21:33–37

Call DR, Borucki MK, Loge FJ (2003) Detection of bacterial pathogens in environmental samples using DNA microarrays. J Microbiol Meth 53:235–243

Cheung VG , Morley M, Aguilar F, Massimi A, Kucherlapati R, Childs G (1999) Making and reading microarrays. Nat Genet 21(1 Suppl):15–19

Conrad SM, Ethridge S, Hall S (2003) SEAPORT: improving biotoxin management through citizen involvement. Second Symposium on Harmful Marine Algae in the U.S. Abstracts, p 88

DeRisi JL, Iyer VR, Brown PO (1997) Exploring the metabolic and genetic control of gene expression on a genomic scale. Science 278:680–686

Feriotto G, Borgatti M, Mischiati C, Bianchi N, Gambari R (2002) Biosensor technology and surface plasmon resonance for real-time detection of genetically modified Roundup Ready soybean gene sequences. J Agric Food Chem 50:955–962

Fuchs BM, Syutsubo K, Ludwig W, Amann R (2001) In situ accessibility of *Escherichia coli* 23S rRNA to fluorescently labelled oligonucleotide probes. Appl Environ Microbiol 67:961–968

Goffredi SK, Jones W, Scholin C, Marin R, Hallam S, Vrijenhoek RC (2005) Molecular detection of marine larvae. Mar Biotechnol (in press)

Groben R, John U, Eller G, Lange M, Medlin LK (2004) Using fluorescently labelled rRNA probes for hierarchical estimation of phytoplankton diversity – a mini review. Nova Hedwigia 79:313–320

Groben R, Medlin LK (2005) In situ hybridization of phytoplankton using fluorescently labelled rRNA Probes. In: Zimmer EA, Roalson E (eds) Methods in enzymology. Elsevier, San Diego 395:299–310

Guschin DY, Mobarry BK, Proudnikov D, Stahl DA, Rittmann BE, Mirzabekov AD (1997) Oligonucleotide microchips as genosensors for determinative and environmental studies in microbiology. Appl Environ Microbiol 63:2397–2402 IOC 2002

John U, Cembella A, Hummert C, Ellbrächter M, Groben R, Medlin LK (2003) Discrimination of the toxigenic dinoflagellates *Alexandrium tamarense* and *A. ostenfeldii* in co-occurring natural populations from Scottish coastal waters. Eur J Phycol 38:25–40

John U, Medlin LK, Groben R (2005) Development of specific rRNA probes to distinguish between geographic clades of the *Alexandrium tamarense* species complex. J. Plankton Res 27:199–204

Jonker R, Groben R, Tarran G, Medlin L, Wilkins M, Garcia L, Zabala L, Boddy L (2000). Automated identification and characterization of microbial populations using flow cytometry: the AIMs project. In: Reckermann M, Colijn F (eds) Aquatic flow cytometry: achievements and prospects. Sci Mar (64)2:225–234

Kamentsky LA (2001) Laser scanning cytometry. Meth Cell Biol 63:51–87

Knauber DC, Berry ES, Fawley MW (1996) Ribosomal RNA-based oligonucleotide probes to identify marine green ultraphytoplankton. J Euk Microbiol 43:89–94

Lemarchand K, Parthuisot N, Catala P, Lebaron P (2001) Comparative assessment of epifluorescence microscopy, flow cytometry and solid-phase cytometry used in the enumeration of specific bacteria in water. Aquatic Microbial Ecol 25:301–309

Lim EL, Amaral LA, Caron DA, DeLong EF (1993) Application of rRNA-based probes for observing marine nanoplanktonic protists. Appl Environ Microbiol 59:1647–1655

Litaker W, Sundseth R, Wojciechowski M, Bonaventura C, Henkens R, Tester P (2001) Electrochemical detection of DNA or RNA from harmful algal bloom species. In: Hallegraeff GM, Blackburn SI, Bolch CJ, Lewis RJ (eds) Harmful algal blooms. Intergovernmental Oceanographic Commission of UNESCO, pp 242–249

Lockhart DJ, Dong H, Byrne MC, Follettie MT, Gallo MV, Chee MS, Mittmann M, Wang C, Kobayashi M, Horton H, Brown EL (1996) Expression monitoring by hybridization to high-density oligonucleotide arrays. Nat Biotechnol 14:1675–1680

Lockhart DJ, Winzeler EA (2000) Genomics, gene expression and DNA arrays. Nature 405:827–836
Loy A , Lehner A, Lee N, Adamczyk J, Meier H, Ernst J, Schleifer KH, Wagner M (2002) Oligonucleotide microarray for 16S rRNA gene-based detection of all recognized lineages of sulfate-reducing prokaryotes in the environment. Appl Environ Microbiol 68:5064–5081
Maidak BL, Cole JR, Lilburn TG, Parker CTJ, Saxman PR, Farris RJ, Garrity GM, Olson GJ, Schmidt TM, Tiedje JM (2001) The RDP-II (ribosomal Database Project). Nucleic Acids Res 29:173–174
Marie D, Simon N, Vaulot D (2005) Phytoplankton cell counting by flow cytometry. In: Andersen RA (ed) Algal culturing techniques. Academic Press, Elsevier (in press)
Medlin LK, Lange M, Noethig EV (2000) Genetic diversity of marine phytoplankton: a review and a look to Antarctic phytoplankton. Antarctic Sci 12:325–331
Medlin LK, Metfies K, Mehl H, Valentin K (2005) Picoplankton diversity at the Helgoland time series site as assessed by three molecular methods, Microb Ecol (in press)
Metfies K, Huljic S, Lange M, Medlin LK (2005) Electrochemical detection of the toxic dinoflagellate *Alexandrium ostenfeldii* with a DNA-biosensor. Biosens Bioelectron 20:1349–1357
Metfies K, Medlin L (2004) DNA microchips for phytoplankton: the fluorescent wave of the future. Nova Hedwigia 79:321–327
Metfies K, Medlin LK (2005) DNA ribosomal RNA probes and microarrays: their potential use in assessing microbial biodiversity. In: Zimmer EA, Roalson E (eds) Methods in enzymology. Elsevier, San Diego, 395:258–278
Mignon-Godefroy K, Guillet JC, Butor C (1997) Laser scanning cytometry for the detection of rare events. Cytometry 27:336–344
Miller PE, Scholin CA (1996) Identification of cultured *Pseudo-nitzschia* (Bacillariophyceae) using species specific LSU rRNA-targeted fluorescent probes. J Phycol 32:646–655
Miller PE, Scholin CA (1998) Identification and enumeration of cultured and wild *Pseudo-nitzschia* (Bacillariophyceae) using species specific LSU rRNA-targeted fluorescent probes and filter-based whole cell hybridization. J Phycol 34:371–382
Miller PE, Scholin CA (2000) On detection of *Pseudo-nitzschia* (Bacillariophyceae) species using whole cell hybridization: sample fixation and stability. J Phycol 36:238–250
Not F, Simon N, Biegala IC, Vaulot D (2002) Application of fluorescent in situ hybridization coupled with tyramide signal amplification (FISH-TSA) to assess eukaryotic picoplankton composition. Aquatic Microbial Ecol 28:157–166
Not F, Latasa M, Marie D, Cariou T, Vaulot D, Simon N (2004) A single species, *Micromonas pusilla* (Prasinophyceae), dominates the eukaryotic picoplankton in the western English channel. Appl Environ Microbiol 70:4063–4072
Paul JH, Pichard SL, Kang JB, Watson GMF, Tabita FR (1999) Evidence for a clade-specific temporal and spatial separation in ribulose bisphosphate carboxylase gene expression in phytoplankton populations off Cape Hatteras and Bermuda. Limnol Oceanogr 44:12–23
Peperzak LB, Sandee B, Scholin C, Miller P, Van Nieuwerburgh L(2001) Application and flow cytometric detection of antibody and rRNA probes to *Gymnodinium mikimotoi* (Dinophyceae) and *Pseudo-nitzschia multiseries* (Bacillariophyceae). In: Hallegraeff GM, Blackburn SI, Bolch CJ, Lewis J (eds) Harmful algal blooms 2000. IOC, UNESCO, pp 206–209
Pernthaler A, Pernthaler J, Amann R (2002) Fluorescence in situ hybridization hybridisation and catalyzed reporter deposition for the identification of marine bacteria. Appl Environ Microbiol 68:3094–3101

Pougnard C, Catala P, Drocourt JL, Legastelois S, Pernin P, Pringuez E, Lebaron P (2002) Rapid detection and enumeration of *Naegleria fowleri* in surface waters by solid-phase cytometry. Appl Environ Microbiol 68:3102–3107

Rampal JB (2001) DNA arrays, methods and protocols. Humana Press, Inc., Totowa, New Jersey, 264 pp

Rautio J, Barken KB, Lahdenera J, Breitenstein A, Molin S, Neubauer P (2003) Sandwich hybridization assay for quantitative detection of yeast RNAs in crude cell lysates. Microb Cell Fact 2:4

Reynolds DT, Fricker CR (1999) Application of laser scanning for the rapid and automated detection of bacteria in water samples. J Appl Microbiol 86:785–795

Rhodes L, Scholin CA, Tyrell JV, Adamson J, Todd K (2001) The integration of DNA probes into New Zealand's routine phytoplankton monitoring programs. In: Hallegraeff GM, Blackburn SI, Bolch CJ Lewis RJ (eds). Harmful Algal Blooms 2000. Intergovernmental Oceanographic Commission of UNESCO, pp 429–432

Rhodes L, Haywood A, Adamson J, Ponikla K, Scholin C (2004) DNA probes for the rapid detection of *Karenia* species in New Zealand's coastal waters. In: Steidinger KA, Landsberg JH, Tomas CR, Vargo GA (eds) Harmful algae 2002. Florida Fish and Wildlife Conservation Commission, Florida Institute of Oceanography, and Intergovernmental Oceanographic Commission of UNESCO. St. Petersburg, Florida, USA, pp 273–275

Roubin MR, Pharamond JS, Zanelli F, Poty F, Houdart S, Laurent F, Drocourt JL, Van Poucke S (2002) Application of laser scanning cytometry followed by epifluorescent and differential interference contrast microscopy for the detection and enumeration of Cryptosporidium and Giardia in raw and potable waters. J Appl Microbiol 93: 599–607

Scholin CA, Buck KR, Britschig T, Cangelosi G, Chavez FP (1996) Identification of *Pseudo-nitzschia australis* (Bacillariophyceae) using rRNA-targeted probes in whole cell and sandwich hybridization formats. Phycologia 35:190–197

Scholin CA, Doucette GJ, Cembella AD (2005) Prospects for developing automated systems for in situ detection of harmful algae and their toxins. In: Babin M, Roesler C, Cullen J (eds) Monographs on oceanographic methodology UNESCO (in press)

Scholin CA, Vrieling E, Peperzak L, Rhodes L, Rublee P (2003) Detection and quantification of HAB species using lectin, antibody and DNA probes: progress to date and future research objectives. In: Hallegraeff GM, Anderson DM, Cembella AD (eds) Manual on harmful marine microalgae, pp 131–164. Algae. Proc 8th Int Conf Harmful Algae, Paris, Intergovernmental Oceanographic Commission of UNESCO (vol 11, 2nd edn), pp 253–257

Schönhuber W, Fuchs B, Juretschko S, Amann R (1997) Improved sensitivity of whole-cell hybridization by the combination of horseradish peroxidase-labeled oligonucleotides and tyramide signal amplification. Appl Environ Microbiol 63:3268–3273

Schönhuber W, Zarda B, Eix S, Rippka R, Herdmann M, Ludwig W, Amann R (1999) In situ identification of cyanobacteria with horseradish peroxidase-labeled, rRNA-targeted oligonucleotide probes. Appl Environ Microbiol 65:1259–1267

Shapiro HM (2003) Excitation and emission spectra of common dyes. In: Robinson JP, Darzynkiewicz Z, Hyun W, Orfao A, Rabinovitch PS (eds) Current protocols in flow cytometry. Wiley, New York, Unit 1.19, Suppl. 26

Shena M (ed) (2000) Microarray biochip technology. Eaton Publishing, MA, USA, 297 pp and Addendum (32pp).

Simon N, Brenner J, Edvardsen B, Medlin LK (1997) The identification of *Chrysochromulina* and *Prymnesium* species (Haptophyta, Prymnesiophyceae) using fluorescent or chemiluminescent oligonucleotide probes: a means for improving studies on toxic algae. Eur J Phycol 32:393–401

Simon N, Campbell L, Ornolfsdottir E, Groben R, Guillou L, Lange M, Medlin LK (2000) Oligonucleotide probes for the identification of three algal groups by dot blot and fluorescent whole-cell hybridization. J Euk Microbiol 47:76–84

Singh-Gasson S, Green RD, Yue Y, Nelson C, Blattner F, Sussman MR, Cerrina F (1999) Maskless fabrication of light-directed oligonucleotide microarrays using a digital micromirror array. Nat Biotechnol 17:974–978

Southern E, Mir K, Shchepinov M (1999) Molecular interactions on microarrays. Nat Genet 21: (1 Suppl.) pp 5–9

Töbe K, Eller G, Medlin LK (2006) Automated detection and enumeration for toxic algae by solid-phase cytometry and the introduction of a new probe for *Prymnesium parvum* (Haptophyta: *Prymnesiophyceae*) J Plankton Res (in press)

Todd K (2003) Role of phytoplankton monitoring in marine biotoxins programmes. In: Hallegraeff GM, Anderson DM, Cembella AD (eds) Manual on harmful marine microalgae. Intergovernmental Oceanographic Commission, UNESCO. (vol 11, 2nd edn), pp 131–164

Veldhuis MJW, Kraay G (2000) Application of flow cytometry in marine phytoplankton research: current applications and future perspectives. In: Reckermann M, Colijn F (eds) Aquatic flow cytometry: achievements and prospects. Sci Mar (64)2:121–134

Vives-Rego J, Lebaron P, Nebe von C (2000) Current and future applications of flow cytometry in aquatic microbiology. FEMS Microbiol Revs 24:429–444

West NJ, Schönhuber WA, Fuller N, Amann RI, Rippka R, Post A, Scanalan D (2001) Closely related *Plochlorococcus* genotypes show remarkably different depth distributions in two oceanic regions as revealed by in situ hybridization using 16SrRNA-targeted oligonucleotides. Microbiology 147:1731–1744

West NJ, Bacchieri R, Hansen G, Tomas C, Lebaron P, Moreau H (2006) Rapid quantification of the toxic alga *Prymnesium parvum* in natural samples by use of a specific monoclonal antibody and solid-phase cytometry. Appl Environ Microbiol 72:860–868

Woese CR (1987) Bacterial evolution. Microbiol Rev 51:221–271

Zammatteo N, Moris P, Alexandre I, Vaira D, Piette J, Remacle J (1995) DNA probe hybridisation in microwells using a new bioluminescent system for the detection of PCR-amplified HIV-1 proviral DNA. J Virol Meth 55:185–197

25 Mitigation and Controls of HABs

H.G. Kim

25.1 Introduction

The impacts of harmful algal blooms (HABs) on public health and fisheries economics can be severe and are increasing along many coasts of the world. Other ecological impacts include the alteration of marine habitats and devastation of the coastal amenities. Commercial oysters and mussels along the Atlantic and Pacific coasts are exposed to paralytic, diarrhetic and amnesic shellfish poisoning (Oshima 1982; Yasumoto 1982; Sundström et al. 1990; Anderson et al. 2000; Landsberg 2002). Fish-killing HABs have become major threats to aquaculture industries in Asian countries (Chen 1993; Okaichi 1997; Dickman 2001; Kim 2005). Approximately 150 species of aquatic microalgae are harmful and toxic (Landsberg 2002). Such microalgae as *Chattonella antiqua* (Okaichi 1997), *Cochlodinium polykrikoides* (Kim 1998), *Heterosigma akashiwo* (Honjo 1992), *Karenia mikimotoi* (Iizuka and Irie 1966), *Heterocapsa circularisquama* (Matsuyama et al. 1995), and *Karenia brevis* (Ingersoll 1882) cause mass mortalities of wild and cultured fish almost every year.

In the USA during the 1987–1992 period, the total cost of HABs averaged $49 million per year (Anderson et al. 2000). In Korea, from 1995–2004, the estimated fisheries damage approximated 1.31 million US$ per year (Kim 2005). The average economic loss in Japan is over a billion yen per year (Imai 2005). Hong Kong's worst red tides, in 1998, caused fish kills valued at 10.3 million US$ (Dickman 2001). Nonetheless, predictive models and mitigation techniques are not developed enough to secure the safety of seafood and to minimize the impacts of HABs on commercial fisheries. The goal of mitigation and control of HABs is to protect public health, fisheries resources, and marine ecosystems. Present approaches focus on mitigation to prevent the impacts of HABs on fisheries resources and the aquaculture industry rather than public health.

Ecological Studies, Vol. 189
Edna Granéli and Jefferson T. Turner (Eds.)
Ecology of Harmful Algae

25.2 Mitigation Strategies and Control of HABs

A variety of mitigation strategies can be employed directly or indirectly to affect the size of a HAB population or its impacts. These can be classified into two categories: precautionary impact preventions, and bloom controls (Table 25.1). Bloom control can be categorized as either direct or indirect, depending upon whether the effort targets an existing bloom or strives to reduce future blooms, such as through alteration of pollution inputs and consumption of dissolved nutrient by beneficial diatom blooms (SCOR/GEOHAB 1998; Kim 2005).

25.2.1 Precautionary Impact Preventions

Precautionary impact prevention includes HAB monitoring, prediction, and emergent actions. The role of monitoring is to detect HABs and their associated toxins in algae or fish and shellfish. Prediction involves more scientific approaches based on the oceanography and ecology. Precautionary emergent actions are described in contingency plans for fish culture later in this chapter.

Monitoring for HABs requires identification of target species, determination of toxins, understanding oceanographic properties underlying population dynamics, and analysis of environmental and meteorological changes to build integrated prediction models. Sometimes, international and regional links are recommended to compare regional biodiversity and biogeography of harmful species. The Intergovernmental Oceanographic Commission

Table 25.1. Current mitigation and control strategies for HABs

Category	Before HAB	After HAB
Precautionary impact preventions	Monitoring Prediction Precautionary actions - Early harvesting - Provide less feed to fish - Prepare mitigation facilities - Clay, pump/aspirator	Emergent actions - Move pens to refuge site - Enclosure of fish cages - Water circulation - Oxygenation - Aeration - Ozonization
Bloom controls	Indirect controls - Reduce nutrient inputs - Modification of water circulation - Bio-remediation	Direct controls - Physical control - Chemical control - Biological control

(IOC) recommends the use of the Global Ocean Observing System (GOOS), whose broad portfolio of products and services allow the design of good cooperative monitoring and prediction systems. To make timely predictions, better monitoring tools must be developed.

Accurate forecasting of the timing and transport pathway of HABs can help fish farmers and other affected parties to take emergency actions. Some predictive models such as numerical models (Eilertsen and Wyatt 1998) and multiparameter ecosystem models (Allen 2004) have already been developed for some dinoflagellates blooms.

Thus far, however, truly practical predictive models of population and transport dynamics of important species have not yet been developed. Prediction models should be established on the basis of detailed understanding of the factors controlling the oceanographic and bloom dynamics (Allen 2004).

To maximize timeliness essential to protect the safety of cultured animals, all HAB information has to be distributed on a real-time basis. For example, the timing and transport pathways of HABs must be distributed immediately to all fish farmers through public communication networks such as television broadcasting, the Internet, facsimile and newsletter, automated telephone response systems (ARS), and SMS.

Audio-visual image transport systems are sometimes very efficient for exchanging oral and visual information. This provides the exchange of microscopic images between local laboratories and a central institute. This will permit overcoming the shortage of identification experts and budgetary restrictions, because this framework connects all the laboratories and terminals within one network system.

25.2.2 Direct and Indirect Bloom Controls

Since the first use of copper sulphate (Rounsefell and Evans 1958), a variety of chemicals, flocculants, and biological and physical techniques have been used in attempts to directly control HABs (Table 25.2).

Biological Control. Biological techniques apply top-down grazing and bottom-up bacterial decomposition. Some copepods and ciliates can graze dinoflagellates such as *C. polykrikoides* and *K. mikimotoi* (Jeong et al. 1999; Kim et al. 1999), but in reality, it is difficult to use them as HAB control agents due to a lack of facilities for mass culture and logistical problems. Viruses, parasites, and bacteria are also promising bottom-up control agents, as they can be abundant in marine systems, replicate rapidly, and sometimes are host-specific. Flagellates such as *Amoebophrya ceratii* and *Parvilucifera infectans* are well-known intracellular parasites of free-living dinoflagellates, including the genus *Alexandrium* (Taylor 1968). A possible use of a virus to

Table 25.2. Harmful algal bloom controls and available agents

Controls	Mechanisms	Available agent
Biological	Grazing (top-down)	Copepods, ciliates, bivalves
	Algicidal agents	Bacteria, viruses
	Parasites	*Amoebophrya, Parvilucifera*
	Enzymes	Mannosidase
Physical	Destruction	Ultrasound
	Electrolysis	NaOCl
	Removal	Skimmer, screen filter
	Isolation	Shield curtain, perimeter skirt
Chemical	Flocculants	Clays and long-chain polymers
	Surfactants	Sophorolipid, aponin
	Mucolytic coagulants	Cysteine compounds
	Metals and liquids	Copper, $Mg(OH)_2$, H_2O_2

control *H. akashiwo* and *H. circularisquama* has been reported (Nagasaki et al. 2005). Investigations of algicidal bacteria targeting a dinoflagellate *K. brevis* were conducted in the Gulf of Mexico (Doucette 2002). Algicidal microorganisms affecting red tides caused by *C. antiqua* and *H. akashiwo* were studied in northern Hiroshima Bay in the Seto Inland Sea (Imai et al. 1998).

Physical Control. Physical control involves removing the harmful algae cells using physical treatments such as skimming, isolation, ultrasonic destruction, and electrolyzation of seawater. Centrifugal separation equipment was used to remove *C. polykrikoides* cells in a land-based tank (Chang, pers. comm.), but the treatment of a number of different harmful algae was a difficult problem. Ultrasound was only effective over 50 cm in depth and was not useful at low cell concentrations (Shirota 1989).

Chemical Control. Copper sulphate was applied to control a red tide in Florida (Rounsefell and Evans 1958). However, it was found that dusting the sea with copper sulfate was too expensive as a control method for red tides, and copper was lethal to sensitive marine organisms. A chemical called "aponin", a sterol surfactant produced by the blue-green alga *Gomphosphaeria aponina*, was once suggested as a control agent for toxic *K. brevis* blooms (McCoy and Martin 1977). Microbial surfactant sophorolipids have also been used to control *C. polykrikoides* blooms (Choi et al. 2002). Other chemicals such as cysteine compounds (Jenkinson and Arzul 2001), octadecatetraenoic acid extracted from brown algae (Kakizawa et al. 1988), sodium hypochlorite produced by electrolysis of natural sea water (Kim et al. 2000), ozone (Rosenthal 1981), hydrogen peroxide as an extermination agent against cysts (Ichikawa et al. 1993), and magnesium hydroxide (Maeda et al. 2005) are in trial tests for the possibility of urgent suppression of red tides. Another study screened 4,300 chemicals against Florida's red tide algae but did not find one that was sufficiently potent and did not adversely affect other organisms (Marvin and Proctor 1964). Thereafter, chemical control options have received little attention, because most of the chemicals are likely to be non-specific and thus will kill co-occurring algae and other organisms indiscriminately. Each candidate chemical will require extensive testing for lethality, specificity, and general safety, and each must surmounts significant regulatory hurdles (Anderson et al. 2001).

Flocculant Clays. One promising non-chemical strategy involves the treatment of blooms with flocculant clays, which scavenge particles (including algal cells) from seawater and carry them to bottom sediments. The properties of cation exchange and plasticity are the most important characteristics for their use as flocculants for harmful microalgae. Clay minerals include kaolinite, illite, and montmorillonite. The first clay experiments were done in Japan to control *Olisthodiscus* sp. (Shirota and Adachi 1976), and then in Korea to control *Prorocentrum triestinum* (Kim 1987). Recently, clay has been widely applied to control HABs in Japan (Shirota and Adachi 1976), Korea (Kim 1987), China (Yu et al. 1994), and the USA (Sengco et al. 2000). As for the specific mechanisms, Shirota (1989) suggested that the clay not only flocculates and removes cells by sedimentation, but that some cells are killed by aluminum eluted from the clay by ion exchange. It was found that the concentrations of iron, aluminum and TiO_2 in clay were proportional to the removal rate of *C. polykrikoides*, such that the higher the concentration of iron and aluminum, the higher the removal rate (Kim et al. 2001).

Cochlodinium polykrikoides Blooms and Clay Control in Korea. In Korea, yellow clay was first applied to control a dinoflagellate *P. triestinum* bloom in 1986 (Kim 1987). Numerous of clays have been subsequently applied in fish farms to control blooms of the fish-killing dinoflagellate *C. polykrikoides*. When montmorillonite clay was sprinkled on a concentration of 1,000 cells/ml of *C. polykrikoides*, 74 % of the living cells were removed at a clay concentration of 2 g/ml, and 98 % at 4 g/ml, within 30 min (Kim 1998). In a field survey of *C. polykrikoides* blooms, the removal rates ranged from 74 to 85 % in 30 min after dispersion (Table 25.3), and the more the iron and aluminum, the higher the removal rate (NFRDI 1999). The removal rate of *C. polykrikoides* according to the density and elapsed time is summarized in the following table (NFRDI 1999).

In terms of toxicity of yellow clay to fish and shellfish, including abalone and flatfish, there were no significant impacts at a clay concentration of 20 g/l within 24 h (NFRDI 1999). A 5-year survey of benthic fauna at the clay dispersal site near Tongyong city, Korea, where clay has been distributed

Table 25.3. Removal rates (%) of *C. polykrikoides* according to the bloom density and elapsed time after clay dispersions

Bloom density (cells/ml)	Removal rate (%) of clay dispersion (10g/l) Just after	After 30 min	After 60 min
100–500	40–50	74–76	80–83
500–1,000	50–55	76–78	85–88
1,000–3,000	55–65	74–85	84–92

every year since 1996, showed no changes in the species composition, diversity and abundance of benthos (NFRDI 1999). Thus, clay has been widely applied by local governments for the control of fish-killing *C. polykrikoides* blooms.

A guidebook, "HABs and Clay Dispersion" has been published and distributed to all stakeholders for adequate clay dispersion in Korea and Japan (NFRDI 1997; JFA 1982). This guidebook recommends that clay minerals be powdered to a particle size of less than 50 μm, and dispersed at concentrations of 100–400 g/m^2 by mixing with seawater at mid-day (because the *C. polykrikoides* cells migrate to subsurface layers in mid-day). Taking into account the diffusion and sinking rate of clay minerals, the surface area of clay dispersion at fish cages would be about three times that of the area of the cages in order to protect fish staying at the bottom of the cage in mid-day. The interval for dispersion time is 30–40 min, taking into account the sinking rate of clay and the 10-m depth of the fish pens. The clay is dispersed in the tidal currents so that it drifts in the direction of the fish farm. If HABs are already inside of the fish cages, clay suspensions are dispersed in a "merry-go-round" fashion (Fig. 25.1). Acknowledging that the higher the density of *C. polykrikoides*, the better the removal efficiency of the clay, the local government recommends dispersing the clay when the density exceeds 1,000 cells /ml, the level of a "Red Tide Alert", taking into account the expenses and manpower for clay dispersion.

As mentioned above, Japan, Korea, China, and the USA use dispersed clays and/or have performed pilot studies for controlling dinoflagellate blooms. Clay mitigation seems to be promising, but environmental risk assessments should be carefully implemented to clarify the chronic impacts of clays on the marine ecosystem.

Eutrophication is a major cause of the initiation of HABs. Thus, enclosed or semi-enclosed bays are highly susceptible to harmful and toxic blooms due to their relatively shallow depths, restricted circulation and input of anthropogenic nutrients from land runoff. Therefore, reduction of terrestrial nutrients is the best way to reduce the outbreaks of HABs in such locations (SCOR/GEOHAB 1998).

Fig. 25.1. Photos showing clay dispersion **a** using an application ship equipped with a seawater electrolization system and **b** an aerial view of clay dispersion around the fish cages

Nutrients/Eutrophication. A case has been made that increases in nutrients are linked to increases in the frequency and abundance of HABs (Granéli et al. 1999). It follows that a reduction in pollution would lead to a decrease in bloom frequency. There are two classic examples; one is that the number of red tides began to decrease by reduction of industrial and domestic effluents in the Seto Inland Sea of Japan (Okaichi 1997), and the other is the diminishing of picoplanktonic green tides by opening of a channel to the ocean and by the gradual demise of the duck-farming business in Great South Bay and Moriches Bay on Long Island, New York, USA (SCOR/GEOHAB 1998).

In some cases, changes in N:P:Si ratios influence species composition (Smayda and Borkman 2005) and beneficial species might be encouraged; therefore, nutrient ratios and their effects on the growth of harmful algae and species composition should be considered as important indirect controls.

Bio-Manipulation. Human modification of physical structure of the ecosystem may be considered to conserve, establish or re-establish a biological community that may prevent HABs (SCOR/GEOHAB 1998). This is termed „bio-remediation", and is applicable as one component of integrated coastal zone management. An example might be the establishment of populations of benthic filter feeders to control populations of HABs. Another might be artificial aeration to mix the water column, favoring species that thrive in well-mixed waters over those requiring stratification. Another possibility is to stimulate blooms of diatoms instead of harmful dinoflagellates in summer by inducing germination of diatoms (Itakura 2002).

The design and evaluation of bio-manipulation strategies requires a fundamental understanding of associated processes, such as grazing losses, or the influence of water-column mixing on species succession.

Modification of Water Circulation. In some semi-enclosed areas, HABs linked to either local eutrophication or restricted circulation can be minimized by changing the circulation of water masses to optimize flushing of nutrient-rich water and HAB species (SCOR/GEOHAB 1998). Tidal dams that are designed to open the upper gate to allow entry of flood tides into a bay and open the lower gate to flush bottom eutrophicated waters during the ebb tides are examples of how tidal circulation may be enhanced. Again, this requires understanding of linkages between hydrography, nutrient loading, and bloom dynamics, of which little is known for most HAB species. Aeration may be useful in mitigating such small-scale blooms in fish farms by breaking down stratification.

25.2.3 Contingency Plans for Fish Culture

The goal of precautionary actions is to minimize the impacts on commercial fish and shellfish by taking mitigation actions prior to the development of high-density HABs. In Korea and Japan, early harvesting, reducing densities of cultured fish, quantities of fish food and metabolic substances, and disease diagnosis are highly recommended for marine fish culture. For land-based fish culture, compressed oxygen and pumps, filtering apparatus, and HAB warning machines are essential facilities (Anderson et al. 2001; Kim 1999). Clay stock and clay application ships near the areas of predicted HABs are important standby measures of local governments.

Mechanisms of mass mortalities of fish include ichthyotoxins produced by harmful algae, suffocation by gill-clogging and lack of dissolved oxygen especially at night, and reactive oxygen substances (ROS) on the gill membranes (Kim et al. 1999). Therefore, emergent measures are important to alleviate the direct cause of mass mortalities of cultured fish.

One of these emergent measures in Korea is to disperse yellow clay on fish farms, and another is to protect cultured animals either by transportation of fish cages into a zone free of HABs, or by enclosure of fish cages to prevent exposure of fish to HABs. Transportation of fish cages away from an area affected by fish-killing blooms to a known offshore refuge site has been widely practiced by mariculturists in Japan and Korea. The primary equipment for the transportation is a towing tugboat and a refuge container to accommodate transferred fish. Wave breakers can serve as refuge containers. Enclosure of fish cages has been done by shield curtains or perimeter skirts whose shape and deployment resembles purse seines. Enclosed fish can survive for 1–2 weeks and sometimes up to a month if no food is provided. Aeration is needed to reduce fish mortality by increasing dissolved oxygen concentrations in water exposed to fish gills. Atmospheric air can be introduced into the fish cages via aspiration with air-blowers or compressors. Pumping bottom waters and dispersing them on the surface of fish cages helps supply oxygen and dilutes high surface densities of harmful algae.

A final choice of releasing culture fish into the wild sea just before the invasion of HABs can be considered, due to the likelihood of fish survival and adaptability in the natural coastal environment. It is very difficult to determine the timing of the release and appropriate compensation.

25.3 Conclusions

The goals of management and mitigation of HABs is to secure public health and to protect aquaculture producers against economic losses. These goals can be accomplished by direct control of HABs and reducing terrestrial pollutants in order to reduce eutrophication that leads to frequent HABs.

Real-time monitoring and prediction is the first precautionary action to be implemented to minimize damage caused by HABs. In periods of high bloom risk, aircraft visual surveys, satellite remote sensing, and vessel and shore-based observations are proving to be effective tools for early warning and bloom tracking. Acknowledging that international cooperation enhances the mitigation efficiency, regional and international cooperative monitoring systems and mitigation strategies should be established for widespread HABs in the global ocean. The SEAWATCH™ system is a good example for those who want a complete operational monitoring and information system. According to the scale of the impacts and the increasing trends in incidence, HAB mitigation and control should be a prime research topic. Many chemicals and flocculants have been used as HAB mitigation agents. Clay is one of the promising agents for HAB mitigation and control, if its environmental effects are minimized.

Ballast water has long been recognized as a major vector for the introduction of non-indigenous and harmful organisms (Rosenthal 1981). Invasions of exotic HAB species are causing significant ecological and economic damage in various parts of the world. At present, restrictions on ballast water discharges should also be considered, as well as the manner in which live fish and shellfish are transported and dispersed (Hallegraeff et al. 1990).

References

Allen JC (2004) Multiparameter ecosystem models as tools for process modeling and prediction of HABs in fjords and coastal embayments. In: GEOHAB Open Sci Meeting. HABs in fjords and coastal embayments. Valparaiso, Chile. Abstracts, pp 20–21

Anderson DM, Andersen P, Bricelj VM, Cullen JJ, Rensel JE (2001) Monitoring and management strategies for harmful algal blooms in coastal waters. APEC #201-MR-01.1, Asia Pacific Economic Program, Singapore, and IOC-UNESCO, Tech Series 59, Paris, pp 174–264

Anderson DM, Kaoru Y, White AM (2000) Estimated annual economic impacts from harmful algal blooms (HABs) in the United States. Tech Rep, Woods Hole Oceanogr Inst, WHOI-200-11, Woods Hole, MA, USA, pp 3–8

Chen YQ, Gu XG (1993) An ecological study of red tides in the East China Sea. In: Smayda TJ, Shimizu Y (eds) Toxic phytoplankton blooms in the sea. Elsevier, Amsterdam, pp 217–221

Choi JK, Sun X, Lee YJ, Kim EK (2002) Synergistic effects of sophorolipid and loess combination in harmful algal blooms mitigation. In: Proc 3rd Int Symp Harmful Algal Blooms and Control, Pusan, Korea, pp 85–90

Dickman MD (2001) Hong Kong's worst red tide induced fish kill (March–April, 1998). In: Hallegraeff GM, Blackburn SI, Bolch CJ, Lewis RJ (eds) Harmful algal blooms 2000. IOC-UNESCO, Paris, pp 58–61

Doucette G (2002) Investigations of algicidal bacteria targeting the red tide dinoflagellate *Karenia brevis* in the Gulf of Mexico, USA. In: Proc 3rd Int Sym Harmful Algal Blooms and Control, Pusan, Korea, pp 57–64

Eilertsen HC, Wyatt T (1998) A model of *Alexandrium* population dynamics. In: Reguera B, Blanco J, Fernandez ML, Wyatt T (eds) Harmful algae. Xunta de Galicia and IOC-UNESCO, Grafisant, Santiago de Compostela, Spain, pp 196–199

Granéli E, Carlsson P, Legrand C (1999) The role of C, N and P in dissolved and particulate organic matter as a nutrient source for phytoplankton growth, including toxic species. Aquat Ecol 33:37–54

Hallegraeff GM, Bolch CJ, Bryan J, Koerbin B (1990) Microalgal spores in ship's ballast water: a danger to aquaculture. In: Granéli E, Sundström B, Edler L, Anderson DM (eds) Toxic marine phytoplankton. Elsevier, New York, pp 475–480

Honjo T (1992) Harmful red tides of *Heterosigma akashiwo*. NOAA Tech Rep NMFS, 111:27–32

Ichikawa S, Wakao y, Fukuyo Y (1993) Hydrogen peroxide an extermination agent against cysts of red tide and toxic dinoflagellates. In: Smayda TJ, Shimizu Y (eds) Toxic phytoplankton blooms in the sea. Elsevier, Amsterdam, pp 133–138

Iizuka S, Irie H (1966) The hydrographic conditions and the fisheries damage by the red tide that occurred in Omura Bay in summer 1965. II. Bull Fac Fish Nagasaki Univ 21:67–101

Imai I (2005) Mitigation strategies against harmful red tides by use of algicidal bacteria: application of seedweed beds as huge sources of algicidal bacteria. Proc 1st Int Work HAB Northwest Pacific region, Toyama, Japan, pp 99–105

Imai I, Kim MC, Nagasaki K, Itakura S, Ishida Y (1998) Relationship between dynamics of red tide-causing raphidophycean flagellates and algicidal microorganisms in the coastal sea of Japan. Phycol Res 46:139–146

Ingersoll E (1882) On the fish mortality in the Gulf of Mexico. Proc US Nat Mus 4:74–80

Itakura S (2002) Prevention of harmful flagellate blooms by diatoms. In: Hiroshi S, Imai I, Ishimaru T (eds) Prevention and extermination strategies of harmful algal blooms. Nippon Suisan Gakkai. Fisheries Series 134:9–18

Jenkinson IR, Arzul G (2001) Mitigation by cysteine compounds of rheotoxicity, cytotoxicity and fish mortality caused by the dinoflagellates, *Gymnodinium mikimotoi* and *G.* cf. *maguelonnense*. In: Hallegraeff GM, Blackburn SI, Bolch CJ, Lewis RJ (eds) Harmful algal blooms 2000. IOC-UNESCO, Paris, pp 461–464

Jeong HJ, Shim JH, Lee CW, Kim JS, Koh SM (1999) Growth and grazing rates of the marine planktonic ciliate *Strombidinopsis* sp. on red tide and toxic dinoflagellates. J Eukaryot Microbiol 46:69–76

JFA (1982) Clay dispersion. A manual for the mitigation of HABs impacts. Japan Fisheries Agency, 31 pp

Kakizawa H, Asari F, Kusumi T, Toma T, Sakurai T, Oohusa T, Hara Y, Chihara M (1988) An allelopathic fatty acid from the brown alga *Chadosiphon okamuransus*. Phytochemistry 27:731–735

Kim CS, Lee SG, Kim HG, Jung J (1999) Reactive oxygen species as causative agents in the ichthyotoxicity of the red tide dinoflagellate, *Cochlodinium polykrikoides*. J Plankton Res 21:2105–2115

Kim CS, Bae HM, Cho YC (2001) Control of harmful algal blooms by clay via photochemical reactions. J Korea Soc Phycol, Algae 16:67–73

Kim HG (1987) Ecological study of dinoflagellates responsible for red tide. 1. The population growth and control of *Prorocentrum triestinum* Schiller. Bulletin NFRDI 39:1–6

Kim HG (1998) *Cochlodinium polykirkoides* blooms in Korean waters and their mitigation. In: Reguera B, Blanco J, Fernandez ML, Wyatt T (eds) Harmful algae. Xunta de Galicia and IOC-UNESCO, Grafisant, Santiago de Compostela, Spain, pp 227–228

Kim HG (2005) Harmful algal blooms in the sea. Dasom Publ Co. 467 pp (in Korean)

Kim HG, Jeong HJ, Hahn SD et al (1999) Management and mitigation techniques to minimize the impacts of harmful algal blooms. Korean Ministry of Maritime Affairs and Fisheries, special rep 527 pp

Kim HR, Kim KI, Kim DS, Park KS, Hong TH, Jeong HJ (2000) Developing a method of controlling the outbreak and maintenance of red tides using NaOCl produced by electrolysis of natural seawater. In: 9th Int Conf Harmful Algal Blooms, Hobart, Australia, 2000. (Abstract), p 28

Landsberg JH (2002) The effects of harmful algal blooms on aquatic organisms. Reviews Fish Sci 10:113–390

Maeda H, Sasaki H, Nishino H, Taoka Y, Dung N, Hidaka M (2005) Suppression of red tide by magnesium hydroxide. In: Proc 1st Int Workshop on HAB Northwest Pacific region, Toyama, Japan, pp 95–98

Marvin KT, Proctor RR Jr (1964) Preliminary results of the systematic screening of 4,306 compounds as „red-tide“ toxicants. USFWS Bureau of Commercial Fisheries report, Galveston, Texas. Contribution no 184, 85 pp

Matsuyama Y, Nagai K, Mizuguchi T, Fujiwara M, Ishimula M, Yamaguchi M, Uchida T, Honjo T (1995) Ecological features and mass mortality of pearl oysters during the red tide of *Heterocapsa* sp in Ago Bay in 1992. Nippon Suisan Gakkaishi 61:35–41 (In Japanese)

McCoy JLF, Martin DF (1977) The influence of *Gomphosphaeria aponina* on the growth of *Gymnodinium breve* and the effects of aponin on the ichthyotoxicity of *Gymnodinium breve*. Chem Biol Interactions 17:17–24

Nagasaki T, Tomaru Y, Shirai Y, Mizumoto H, Nishida K, Takao Y (2005) Possible use of viruses as a microbiological agent to control HAB. Proc 1st Int Workshop on HAB Northwest Pacific region, Toyama, Japan, pp 106–108

NFRDI (1999) The clay and HABs. Korean Nat Fish Res Dev Inst Tech Rep 19 pp

Okaichi T (1997) Red tides in the Seto Inland Sea. In: Okaichi T, Yanagi Y (eds) Sustainable development in the Seto Inland Sea – from the viewpoint of fisheries. Tera Sci Publ Co, Tokyo, Japan, pp 251–304

Oshima Y (1982) Occurrence of paralytic shellfish poison in Japan and tropical areas. In: Toxic phytoplankton-occurrence, mode of action and toxins. Nippon Suisan Gakkai, Fish Series 42:73–87

Rosenthal H (1981) Ozonation and sterilization. In: World symposium on new developments in the utilization of heated effluents and recirculation systems. Stavanger, Norway, 511 pp

Rounsefell GA, Evans JE (1958) Large-scale experimental test of copper sulfate as a control for the Florida red tide. US Fish Wildlife Serv Spec Sci Rep, 270 pp

SCOR/GEOHAB (1998) The global ecology and oceanography of harmful algal blooms. A plan for co-ordinated scientific research and co-operation to develop international capabilities for assessment, prediction and mitigation, Joint SCOR-IOC Workshop Rep. Havreholm, Denmark 13–17 Oct 1998, 22 pp

Sengco MR, Li A, Henry MS, Petersen B, Anderson DM, Pierce R (2000) Preliminary mesocosm studies of clay dispersal in Sarasota Bay, Florida (USA). In: Proc 1st Inter-state Symp: The evaluation on the mitigation capability of clay and yellow loess at the HABs and their impacts on marine ecosystem, Pusan, Korea, pp 53–59

Shirota A (1989) Red tide problem and countermeasures. Int J Aquat Fish Technol 1:195–293

Shirota A, Adachi M (1976) Abst Annual Meeting of Japan Oceanogr Soc, p 15

Smayda T, Borkman D (2005) Multidecadal changes in the diatom: flagellate ratio and Si:N and Si:P ratios in Narragansett Bay, and influence of Si:N supply ratios on diatom species competition. In: GEOHAB Open Sci Meeting HABs and Eutrophication, Baltimore, Maryland USA, (abstract), p 56

Sundström B, Edler L, Granéli E (1990) The global distribution of harmful effects of phytoplankton. In: Granéli E, Sundström B, Edler L, Anderson DM (eds) Toxic marine phytoplankton. Elsevier, New York, pp 537–541

Taylor FJR (1968) Parasitism of the toxin-producing dinoflagellate *Gonyaulax catenella* by the endoparasitic dinoflagellate *Amoebophrya ceratii.* J Fish Res Board Can 25:2241–2245

Yasumoto T (1982) Occurrence and chemistry of diarrhetic shellfish poison. In: Toxic phytoplankton-occurrence, mode of action and toxins. Nippon Suisan Gakkai, Fisheries Series 42:102–111

Yu ZM, Zou JZ, Ma X (1994) Application of clays to removal of red tide organisms 1. Coagulation of red tide organisms with clays. Chin J Oceanol Limnol 12:193–200

Part F
Human Impact on Harmful Algae and Harmful Algae Impact on Human Activity

26 The Complex Relationships Between Increases in Fertilization of the Earth, Coastal Eutrophication and Proliferation of Harmful Algal Blooms

P.M. GLIBERT and J.M. BURKHOLDER

26.1 Introduction

The past five decades have witnessed a dramatic increase in the availability of nutrients on land, in the atmosphere, and in the oceans. This change has occurred largely due to the development of industrial fertilizers, changing practices in the raising of animals for consumption on land and in the sea, and increased consumption of fossil fuels (Vitousek 1997; Smil 2001). Over-enrichment of coastal waters by nutrients is considered a major pollution problem worldwide (Vitousek et al. 1997; Howarth et al. 2002) and one of the important factors contributing to global habitat change, including the geographic and temporal expansion of some harmful algal bloom (HAB) species (Smayda 1990; Anderson et al. 2002; Glibert et al. 2005a, 2005b). In this chapter, major trends impacting global nutrients are first reviewed, followed by a review of the concept of eutrophication and a synopsis of some of the evidence that HABs are increasing in frequency or extent in parallel with these trends. Lastly, these patterns are placed in context with other factors that must also be considered when understanding the relationship between eutrophication and HABs.

26.2 Global Trends in Population, Agricultural Fertilizer Usage and Implications for Export to Coastal Waters

The global population has expanded nearly five-fold in past half-century (Fig. 26.1A). It has been suggested that the "expansion from 1.6 billion people in 1900 to today's 6 billion...would not have been possible without the synthe-

Ecological Studies, Vol. 189
Edna Granéli and Jefferson T. Turner (Eds.)
Ecology of Harmful Algae

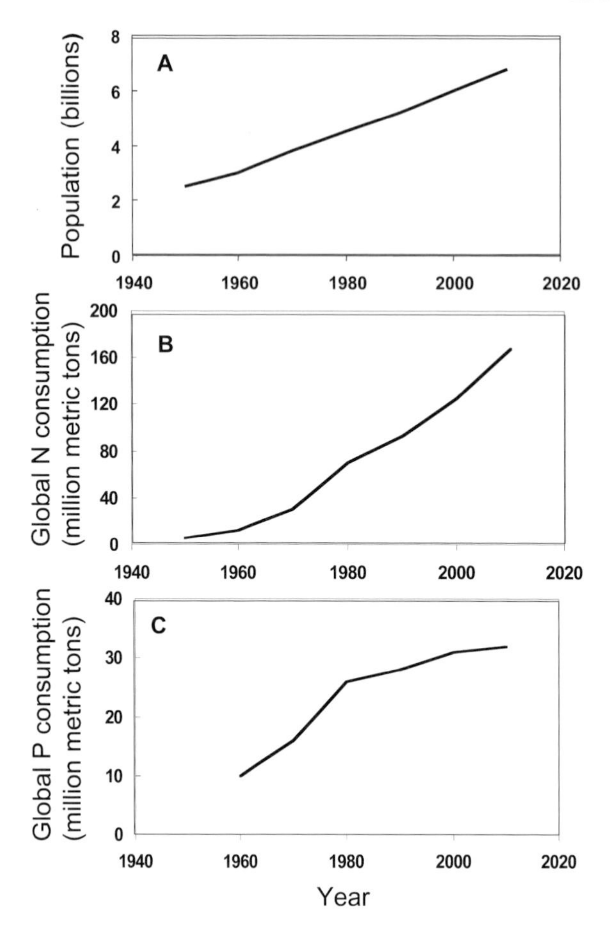

Fig. 26.1. A The change in world human population (in billions) from 1950 through 2000, projected to the year 2010. Data and projection are from www.census.gov/ipc/www/img/ worldpop.gif. **B** The change in global consumption of total synthetic nitrogen fertilizers (in million metric tons of N) for the same years as in panel A. **C** The change in global consumption of phosphate (in million metric tons of P) for the same years as in panel A. Data for panels B and C are from the Global Fertilizer Industry (International Fertilizer Industry 2005) and the projections through 2010 are based on an increase of 3 %

sis of ammonia" (Smil 2001). Industrial synthesis of ammonia permitted enormous expansion in agriculture, from improved yields per crop to improved protein diets of the global population (Smil 2001; Dalgaard et al. 2003). This process has also transformed the global nitrogen cycle and budget, however, with impacts ranging from atmospheric emissions to coastal pollution. Global use of nitrogen fertilizers has increased roughly 20-fold over the past five decades, while global use of phosphorus has increased only about 4-fold over the same period (Fig. 26.1B, C; International Fertilizer Industry 2005). In addition, the dominant form of synthetic nitrogen used worldwide has changed from inorganic nitrogen to organic nitrogen, primarily urea (Glibert et al. 2006). Urea is projected to represent nearly 70 % of global nitrogen fertilizer within the next decade (Glibert et al. 2006). Surface waters across the Earth are also sustaining impacts from anthropogenic nutrient enrichment from many other sources. Sewage (still untreated or poorly treated in many countries) has increased from the growing population. The development of concentrated animal feed operations (CAFOs), as well as large-scale aquaculture operations, have led to large amounts of organic feed and waste concentrated in small areas of the land, with high concentrations of nutrients and other pollutants added to adjacent rivers and estuaries (e.g., Mallin 2000). Atmospheric deposition of nitrogen has also grown in significance, from local emissions near large-animal operations (U.S. Environmental Protection Agency 1998) and from the production and consumption of fossil fuels (e.g., Driscoll et al. 2003).

Estimating nutrient export to the coastal zone thus remains a challenge because nutrient discharge is highly variable both globally and locally in time and space; nutrient forms and ratios vary with land use; and the composition of nutrient discharges is rapidly changing, reflecting changes in land use patterns and agricultural practices. Despite these challenges, models of both inorganic and organic nutrient export are now available (Seitzinger and Kroeze 1998; Seitzinger et al. 2002; Harrison et al. 2005a, 2005b). Such models show that nitrogen and phosphorus exports are greatest from European and Asian lands, followed by the United States. As explained below, not only the total load but also the nutrient form affects its fate in global nutrient transport and cycling, and its likelihood to stimulate HABs.

26.3 Nutrient Limitation versus Eutrophication: Basic Conceptual Framework

A long-standing paradigm in aquatic science is that primary production in freshwater systems typically is limited by phosphorus, whereas the primary limiting nutrient in marine systems is nitrogen (Hecky and Kilham 1988). Although many have questioned this paradigm as too simplistic, the fundamental principle remains: the nutrient in least availability relative to the needs of the organisms, that is, the nutrient that algae deplete first, will limit total production (e.g., Leibig's Law of the Minimum, reviewed in Wetzel 2001). Total algal production cannot increase infinitely, however; a maximum is attained as other factors begin to play a more important controlling role (Cloern 2001).

The concept of limiting nutrients in a eutrophic system actually is very complex. Ecosystem response to nutrient enrichment, or eutrophication, is a continual process rather than a static condition or a trophic state (Cloern 2001; Smayda 2005). There are several published formal definitions of eutrophication, all of which consider the changes accompanying nutrient enrichment as a continuum or a process. Nixon (1995), for example, defined eutrophication as an increase in the rate of organic carbon production in an ecosystem. Historically, the term was mostly applied to the natural aging of lakes, from deeper waters to a marsh (Wetzel 2001); more recently, the terms "accelerated" or "cultural" eutrophication have been used in recognition of major human influences. As relationships are explored between eutrophication and HAB species, it is important to recognize that different systems occur at different points along the eutrophication continuum.

26.4 Nutrient Loading, Nutrient Composition, and HABs

Nutrient enrichment of the globe has impacted the proliferation of harmful algal species in several major ways. Increases in total nutrient load can support higher HAB biomass, and alterations in nutrient form can lead to a nutrient regime favoring HAB growth relative to other algal species. Numerous direct relationships between nutrient load and HAB abundance have been established and recently reviewed (e.g., Cloern 2001; Anderson et al. 2002; Sellner et al. 2003; Glibert et al. 2005b); here, only a few examples can be highlighted. In Northern European waters, blooms of the mucus-forming HAB species *Phaeocystis globosa* have been shown to be directly related to the excess nitrate content of riverine and coastal waters, that is, the nitrate remaining after other species of algae deplete silicate (Fig. 26.2A; Lancelot 1995). In the USA, a strong positive relationship has been documented between increased nitrate loading from the Mississippi River to the Louisiana shelf and increased abundance of the toxigenic diatom *Pseudo-nitzschia pseudodelicatissima,* based on the geological record of the siliceous cell walls of

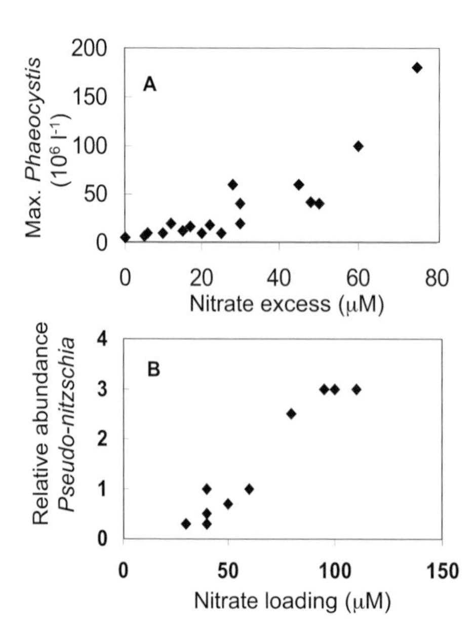

Fig. 26.2. Two examples of the relationship between nitrate and the occurrences of specific HAB outbreaks. **A** Maximum observed cells of the colony-forming prymnesiophyte, *Phaeocystis,* as a function of the excess nitrate (that remaining upon silicate depletion) in the Eastern Channel and Southern Bight of the North Sea for 1980–1990 (redrawn from Lancelot 1995). **B** Average abundance (%) of the diatom, *Pseudo-nitzschia,* in the sedimentary record as a function of the nitrate loading in the northern Gulf of Mexico (redrawn from Turner and Rabalais 1991; Parsons et al. 2002)

this species found in sediment cores (Fig. 26.2B; Parsons et al. 2002). In Puget Sound, Washington, USA, a striking correlation has been found between the growth in documented cases of paralytic shellfish toxins over four decades and the growth in the human population, based on USA census statistics, strongly indicative of nutrient loading and eutrophication as the causative agent of change (Trainer et al. 2003). Most notable are the blooms off the Chinese coast that have expanded in recent years in geographic extent (from km^2 to tens of km^2), in duration (days to months), species and harmful impacts (Fig. 26.3). These blooms parallel the increase in fertilizer use that has occurred during the past two decades (Fig. 26.3). The Baltic, Aegean, northern Adriatic, and the Black Sea have all experienced increases in HABs coincident with increases in nutrient loading (e.g., Larsson and Elmgren 1985; Bodenau 1993; Moncheva et al. 2001). On a shorter time scale, Beman et al. (2005) recently found a strong positive relationship between nitrogen-rich agricultural runoff to the Gulf of California and the development, 3–5 days later, of massive (~50–580 km^2) phytoplankton blooms. The broad relationship between nitrogen loading and HAB proliferation is also evident from the comparison of the global distribution of nitrogen export (based on the models of Seitzinger and Kroeze 1998) and the documented occurrences of several HABs (Fig. 26.4). Globally, harmful algal species that are responsible for paralytic shellfish poisoning (PSP), and the high biomass, toxigenic species *Proro-*

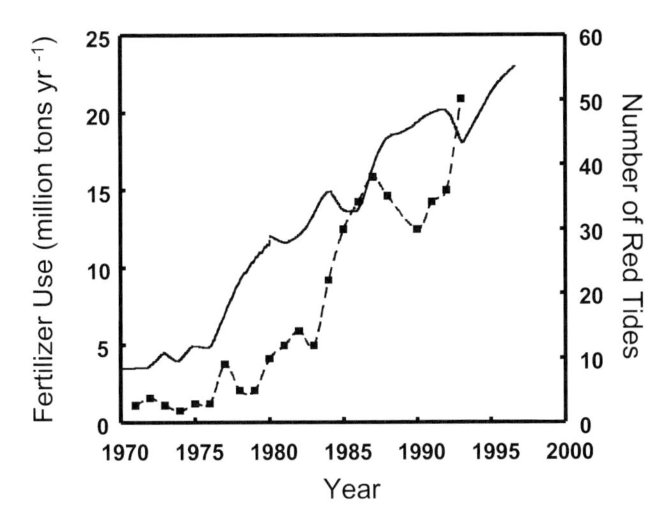

Fig. 26.3. Trends in nitrogen fertilizer use (*solid line*) and the number of red tides (*dashed line*) reported for Chinese coastal waters through the mid-1990s. Data were derived from Smil (2001) for fertilizer use and Zhang (1994) for red tide abundance (reproduced from Anderson et al. 2002 with permission of the Estuarine Research Federation). A review of species, the extent of the blooms and their impacts can be found at www.pices.int/publications/ scientific_reports/Report 23/HAB_china.pdf

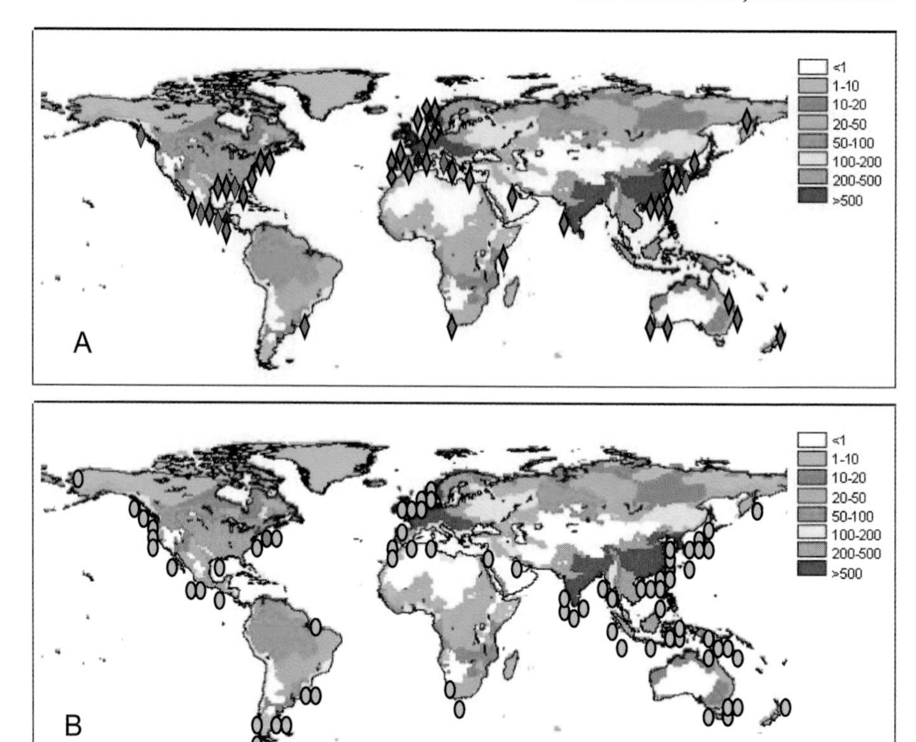

Fig. 26.4. Global distribution of recorded incidences of two major HABs types superimposed on a global map of modeled nitrogen export (base map from Seitzinger and Kroeze 1998; reproduced with permission of American Geophysical Union). Nitrogen export is calculated as kg N km^{-2} watershed year^{-1}. **A** Documented occurrences of *Prorocentrum minimum*, based on the review by Heil et al. (2005). **B** Documented occurrences of those HABs that produce paralytic shellfish poisoning (PSP, e.g., the dinoflagellates *Alexandrium tamarense, A. minutum, Gymnodinium catenatum* and *Pyrodinium bahamense* var. *compressum*), modified from the synthesis of GEOHAB (2001)

centrum minimum, are most common along the coasts of Asia, Europe, and North America where fertilizer use is high and consequently nitrogen export is also high (Heil et al. 2005).

Whereas total nitrogen loads are associated with some blooms, and total nitrogen loading exceeds phosphorus loading on a global basis, increased occurrences of other HABs have been associated with alterations in nutrient composition, especially decreases in the N:P ratio of the nutrient pool that reflect regional, disproportionate phosphorus loading relative to nitrogen. For example, in Tolo Harbor, Hong Kong, where phosphorus loading increased in parallel with increasing human population growth in the late 1980s, a distinct shift from diatoms to dinoflagellates was observed, coincident with a decrease in the ambient N:P ratio from roughly 20:1 to <10:1

(Hodgkiss and Ho 1997; Hodgkiss 2001). Along the eastern seaboard of the USA, outbreaks of the toxigenic dinoflagellate *Pfiesteria* spp. have been associated with low N:P ratios due to high phosphorus loading from CAFOs (Mallin 2000). Blooms of *Karenia brevis* on the western Florida shelf are found in waters with low N:P ratios due to phosphorus loading from local phosphate mining (Vargo et al. 2006).

The importance of organic nutrients in the nutrition of many HABs is also increasingly recognized, and some HABs have been related to organic nutrient load (see Glibert and Legrand, Chap. 13). For example, *Aureococcus anophagefferens* preferentially uses organic nitrogen over nitrate, and *Prorocentrum minimum*, *Lingulodinium polyedrum*, *Alexandrium catenella* and various other species are stimulated by organic nutrients (Glibert and Legrand, Chap. 13). Many harmful algal species prefer some forms of organic nutrients over inorganic forms, may have multiple acquisition mechanisms to obtain them, or may exhibit faster rates of growth on these nutrients (e.g., Berg et al. 1997; Berman and Chava 1999; Mulholland et al. 1999). Moreover, some species are obligate mixotrophs or heterotrophs, requiring organic forms of nutrients (Jones 1994).

Relationships between eutrophication and HABs extend beyond correlations between total nutrient load and changes in nutrient composition, however, as nutrients additionally can stimulate or enhance the impact of toxic or harmful species in more subtle ways. Nutrient availability or composition may also alter the toxin content of individual species without altering their total abundance, or may impact other members of the food web, such as bacteria and viruses, which in turn directly or indirectly impact the abundance or toxicity of harmful algal species (e.g., Carlsson et al. 1998; Anderson et al. 2002).

26.5 Factors Complicating the Relationship Between Eutrophication and HABs

Both eutrophication and the frequency and duration of many HABs have increased, but finding direct links between eutrophication and HABs has often been difficult because not all eutrophic waters support HABs, and not all HABs occur in waters rich in nutrients. One of the difficulties in linking nutrient loading to HABs is the multiplicity of factors contributing to HAB species responses to nutrient loading. The inability to universally apply a single criterion, such as total N concentration or N:P (or other nutrient) ratio, or organic:inorganic nutrient ratio, to determine whether eutrophication is stimulating HABs does not negate the utility of this approach. Rather, it underscores the interdependence of conducive environmental factors, physiological factors, and trophic interactions in the outcome of any species succession, as well as the importance of appropriately scaled temporal and spatial data (Glibert et al. 2005b). Thus, a given suite of nutrients may be insufficient to yield an outbreak of a harmful algal species. Nutrient availability must be matched with the preferences of the cells and their physiological condition, and with the physical and trophic structure of the water column at the time of nutrient delivery.

A classic example of differences in absolute nutrient requirements by specific species groups is that of diatoms, which, unlike most HAB species, require silica to construct their cell walls. If nutrient loading causes a proportional enrichment of nitrogen or phosphorus relative to hydrated silica, then a proportional shift away from a diatom-dominated community might be expected. Nitrogen-fixing cyanobacteria represent another example. Their ability to fix gaseous nitrogen can allow them to dominate under conditions in which dissolved inorganic or organic nitrogen is depleted but other nutrients, such as phosphorus and iron, are sufficiently available. Thus, high N:Si ratios (due to hydrated Si depletion) may favor flagellate or dinoflagellate abundance, while low N:P ratios may favor noxious cyanobacteria (Tilman 1977; Smayda 1990).

While some species have an absolute requirement for a particular nutrient, other species seem to have lost the ability to acquire specific nutrients. For example, the brown tide species, *Aureoumbra lagunensis*, apparently is incapable of assimilating nitrate (DeYoe and Suttle 1994). In such a case, nutrient ratios based solely on inorganic nutrients may not yield insights into the regulation of this species by nutrients.

A recent conceptual framework of the nutritional responses by HAB species included nine categories of dinoflagellates, each having distinctive morphological and habitat preferences, ranging from invasive species that dominate in habitats with enriched nutrient loading, to those that thrive in more oligotrophic, stratified systems (Smayda and Reynolds 2001). Large blooms of *P. minimum*, found in many regions affected by anthropogenic nutrient inputs (Heil et al. 2005), are an example of the former. In contrast, *Karenia brevis* and *K. mikimotoi* bloom in open coastal waters, aggregate in fronts, and are transported by coastal currents (Dahl and Tangen 1993; Walsh et al. 2001). As noted by Smayda (2005, p. 96), "It may be that as eutrophication progresses through its various stages, changes in life-form conditions occur which determine which life-form type of phytoplankter will predominate."

As a further, more fundamental complication, different strains within the same species often respond to nutrients differently. It is commonly assumed that the characteristics of one strain, maintained for years under highly artificial laboratory conditions, are representative of all strains of that species in the natural environment. This assumption overlooks the fact that for nearly all algal species studied, different strains within the same species commonly have shown marked differences in fundamental traits ranging from growth characteristics, toxicity, and bloom-forming behavior to responses to nutrients and other environmental conditions (Wood and Leatham 1992; Burkholder et al. 2001, 2005; Burkholder and Glibert 2006).

Nutritional history also affects the affinity of an algal cell for a particular form of nutrient, the immediate fate of that nutrient once taken up and, in the case of some harmful algal species, the degree of toxicity (e.g., Johansson and Granéli 1999; Skovgaard et al. 2003; Leong et al. 2004). For a nutrient-replete cell, the rate of incorporation of newly acquired nutrient may be slower than the rate of incorporation by a nutrient-deficient cell. Several decades of research on short-term physiological responses by various phytoplankton species have demonstrated that nitrogen-limited cells enhance nutrient uptake capabilities by orders of magnitude, relative to their rates of nutrient uptake when nitrogen-sufficient (e.g., McCarthy and Goldman 1979; Goldman and Glibert 1982). Thus, a nutrient pulse will be assimilated by species at different rates depending on whether they are nutrient-limited or nutrient-sufficient. Moreover, the same strain, as well as a group of strains with the same species, can show a different response to the same nutrient pulse at different phases of growth.

Physiological rate processes such as uptake or growth also vary in response to other environmental factors, such as temperature or salinity. Most physiological rates tend to increase with increasing temperature. This generalization appears to hold for the uptake of ammonium (e.g., Paasche and Kristiansen 1982), but not for uptake of all forms of nitrogen (Fan et al. 2003). During estuarine spring diatom blooms, for example, it has been shown that nitrate uptake is inversely related to temperature: uptake rates decrease with increasing temperature (Lomas and Glibert 1999). Differences in the temperature response of the enzyme responsible for nitrate uptake, nitrate reductase, in some diatoms compared to flagellates may help to explain such observations (Lomas and Glibert 2000). Nutrient pulses delivered at different times of year potentially can stimulate different species groups that are each adapted to specific temperature regimes. Meteorological conditions are also important, as the impact of nutrient delivery depends on seasonality and the extent to which nutrients may be delivered in a pulsed fashion, such as following rainfall, or in more continuous fashion, such as may occur via groundwater flow.

The relationship between nutrient availability/composition and HAB species is also constrained by the extent to which grazers assimilate select fractions of the plankton community (e.g., Neuer and Cowles 1995; Polis and Winemiller 1996). Thus, the nutrient pool may select for growth of a particular group of species, but if the rate of grazing exceeds the rate of growth of those species, such a relationship will not be observed. Some bloom species have predation avoidance mechanisms, such as spines or toxic compounds, that allow then to escape predation (Irigoien et al. 2005), which, in combination with nutrient availability, will aid bloom formation. Selective feeding by microzooplankton on certain species can alter the structure of a community that developed under a specific nutrient regime. Micro-predators can include some heterotrophic dinoflagellates that have been shown to graze [other] harmful algal species (e.g., Nakamura et al. 1995; Jeong et al. 2001). The close coupling of production and consumption by the microbial community, and the blurred distinction of autotrophic and heterotrophic nutrition of many flagellates, can make it difficult to differentiate the role of dissolved nutrients in bloom development from other mechanisms that structure populations.

26.6 Conclusions

For many HAB species, the effects of nutrient enrichment are complex. Nevertheless, fundamental physiological relationships have permitted some generalizations to be drawn about the role of nutrient quality and harmful algal species composition. Organic as well as inorganic nutrient forms are important in controlling HAB species responses to nutrient enrichment. Mixotrophy, via direct uptake of dissolved organic nutrients, cell surface oxidation, or phagotrophy, is common among flagellates, especially dinoflagellates (Glibert and Legrand, Chap. 13). Determination of the overall role of nutrient quantity and quality in affecting HAB species composition requires a fundamental understanding of physiological differences within and between species groups, their nutritional history, and intraspecific differences in response. While necessary, such information still can be insufficient to predict species outcomes in natural systems because physiological capabilities are also modulated by environmental conditions such as temperature, and by the community composition at the time of nutrient delivery. Thus, the same nutrient load – in quantity and quality – can have different impacts in different sites or at different times because of the ambient environmental conditions at the time of delivery. The ultimate success of a given species, and its response to nutrient enrichment, will depend on its ability to exploit both quantity and quality of available nutrients, the timing and intensity of the nutrient supply, and the interaction of other environmental factors and competitor or consumer species.

Acknowledgements. The authors thank E. Granéli for the invitation to write this chapter. PMG was funded by the NOAA MERHAB Program, and JMB was funded by the U.S. EPA. This is contribution number 3896 from the University of Maryland Center for Environmental Science.

References

Anderson DA, Glibert PM, Burkholder JM (2002) Harmful algal blooms and eutrophication: nutrient sources, composition and consequences. Estuaries 25:562–584

Beman JM, Arrigo KR, Matson PA (2005) Agricultural runoff fuels large phytoplankton blooms in vulnerable areas of the ocean. Nature 434:211–214

Berg GM, Glibert PM, Lomas MW, Burford M (1997) Organic nitrogen uptake and growth by the chrysophyte *Aureococcus anophagefferens* during a brown tide event. Mar Biol 129:377–387

Berman T, Chava S (1999) Algal growth on organic compounds as nitrogen sources. J Plankton Res 21:1423–1437

Bodenau N (1993) Microbial blooms in the Romanian area of the Black Sea and contemporary eutrophication conditions. In: Smayda TJ, Shimizu Y (eds) Toxic phytoplankton blooms in the sea. Elsevier, Amsterdam, pp 203–209

Burkholder JM, Glasgow HB, Deamer-Melia NJ, Springer J, Parrow MW, Zheng C, Cancellieri P (2001) Species of the toxic *Pfiesteria* complex, and the importance of functional type in data interpretations. Environ Health Perspect 109:667–679

Burkholder JM, Glibert PM (2006) Intraspecific variability: An important consideration in forming generalizations about toxigenic algal spcies. S Africa J Mar Sci 28 (in press)

Burkholder JM, Gordon AS, Moeller PD, Law JM, Coyne KJ, Lewitus AJ, Ramsdell JS, Marshall HG, Deamer NJ, Cary SC, Kempton JW, Morton SL, Rublee PA (2005) Demonstration of toxicity to fish and mammalian cells by *Pfiesteria* species: comparison of assay methods and multiple strains. Proc Nat Acad Sci USA 102:3471–3476

Carlsson P, Edling H, Béchamin C (1998) Interactions between a marine dinoflagellate (*Alexandrium catenella*) and a bacterial community utilizing riverine humic substances. Aquat Microb Ecol 16:65–80

Cloern JE (2001) Our evolving conceptual model of the coastal eutrophication problem. Mar Ecol Prog Ser 210:223–253

Dahl E, Tangen K (1993) 25 years experience with *Gyrodinium aureolum* in Norwegian waters. In: Smayda TJ, Shimizu Y (eds) Toxic phytoplankton blooms in the sea. Elsevier, Amsterdam, pp 15–22

Dalgaard T, Hutchings NJ, Porter JR (2003) Agroecology, scaling and interdisciplinarity. Agric Ecosys Environ 100:39–51

DeYoe HR, Suttle CA (1994) The inability of the Texas "brown tide" alga to use nitrate and the role of nitrogen in the initiation of a persistent bloom of this organism. J Phycol 30:800–806

Driscoll CT, Whitall D, Aber JD, Boyer E, Castro M, Cronan C, Goodale C, Groffman P, Hopkinson C, Lambert K, Lawrence G, Ollinger SV (2003) Nitrogen pollution in the Northeastern United States: sources, effects and management options. BioScience 53:357–374

Fan C, Glibert PM, Burkholder JM (2003) Characterization of the nitrogen uptake kinetics of *Prorocentrum minimum* in natural blooms and laboratory cultures. Harmful Algae 2:283–299

GEOHAB (2001) Global ecology and oceanography of harmful algal blooms. Glibert P, Gentien G (eds) SCOR and IOC, Baltimore and Paris, 86 pp

Glibert PM, Anderson DA, Gentien P, Granéli E, Sellner KG (2005a) The global, complex phenomena of harmful algal blooms. Oceanography 18:136–147

Glibert PM, Seitzinger S, Heil CA, Burkholder JM, Parrow MW, Codispoti LA, Kelly V (2005b) The role of eutrophication in the global proliferation of harmful algal blooms. Oceanography 18:198–209

Glibert PM, Harrison J, Heil C, Seitzinger S (2006) Escalating worldwide use of urea – a global change contributing to coastal eutrophication. Biogeochemistry 77:441–463

Goldman JC, Glibert PM (1982) Comparative rapid ammonium uptake by four species of marine phytoplankton. Limnol Oceanogr 27:814–827

Harrison JA, Caraco NF, Seitzinger SP (2005a) Global distribution and sources of dissolved organic matter export by rivers: results from a spatially explicit, global model (NEWS-DOM). Global Biogeochem Cycles 19:GB4S04, DOI:10.1029/2005GB002480

Harrison JA, Seitzinger SP, Bouwman AF, Caraco NF, Beusen AHW, Vorosmarty C (2005b) Dissolved inorganic phosphorus export to the coastal zone: results from a spatially explicit, global model (NEWS-DIP). Global Biogeochem Cycles 19:GB4S03, DOI:10.1029/2004GB002357

Hecky RE, Kilham P (1988) Nutrient limitation of phytoplankton in freshwater and marine environments: a review of recent evidence on the effects of enrichment. Limnol Oceanogr 33:796–822

Heil CA, Glibert PM, Fan C (2005) *Prorocentrum minimum* (Pavillard) Schiller – a review of a harmful algal bloom species of growing worldwide importance. Harmful Algae 4:449–470

Hodgkiss IJ (2001) The N:P ratio revisited. In: Ho KC, Wang ZD (eds) Prevention and management of harmful algal blooms in the South China Sea. School Sci Techn, the Open Univ Hong Kong

Hodgkiss IJ, Ho KC (1997) Are changes in N:P ratios in coastal waters the key to increased red tide blooms? Hydrobiologia 852:141–147

Howarth RW, Sharpley A, Walker D (2002) Sources of nutrient pollution to coastal waters in the United States: implications for achieving coastal water quality goals. Estuaries 25:656–676

International Fertilizer Industry (2005) www.fertilizer.org/ifa/statistics

Irigoien X, Flynn KJ, Harris RP (2005) Phytoplankton blooms: a 'loophole' in microzooplankton grazing impact? J Plankton Res 27:313–321

Jeong HJ, Kim SK, Kim JS, Kim ST, Yoo YD, Yoon JY (2001) Growth and grazing rates of the heterotrophic dinoflagellate *Polykrikos kofoidii* on red-tide and toxic dinoflagellates. J Eukaryot Microbiol 48:298–308

Johansson N, Granéli E (1999) Influence of different nutrient conditions on cell density, chemical composition and toxicity of *Prymnesium parvum* (Haptophyta) in semicontinuous cultures. J Exp Mar Biol Ecol 239:243–258

Jones RI (1994) Mixotrophy in planktonic protists as a spectrum of nutritional strategies. Mar Microb Food Webs 8:87–96

Lancelot C (1995) The mucilage phenomenon in the continental coastal waters of the North Sea. Sci Total Env 165:83–102

Larsson U, Elmgren R Wulff E (1985) Eutrophication and the Baltic Sea. Ambio 14:9–14

Leong SCY, Murata A, Nagashima Y, Taguchi S (2004) Variability in toxicity of the dinoflagellate *Alexandrium tamarense* in response to different nitrogen sources and concentration. Toxicon 43:407–415

Lomas MW, Glibert PM (1999) Temperature regulation of NO_3^- uptake: a novel hypothesis about NO_3^- uptake and cool-water diatoms. Limnol Oceanogr 44:556–572

Lomas MW, Glibert PM (2000) Comparisons of nitrate uptake, storage, and reduction in marine diatoms and flagellates. J Phycol 36:903–913

Mallin MA (2000) Impacts of industrial animal production on rivers and estuaries Am Sci 88:2–13

McCarthy JJ, Goldman JC (1979) Nitrogenous nutrition of marine phytoplankton in nutrient depleted waters. Science 203:670–672

Moncheva S, Gotsis-Skretas O, Pagou K, Krastev A (2001) Phytoplankton blooms in Black Sea and Mediterranean coastal ecosystems subjected to anthropogenic eutrophication: similarities and differences. Estuar Coast Shelf Sci 53:281–295

Mulholland MR, Ohki K, Capone DG (1999) Nitrogen utilization and metabolism relative to patterns of N_2 fixation in cultures of *Trichodesmium* NIBB1967. J Phycol 35:977–988

Nakamura Y, Suzuki S, Hiromi J (1995) Population dynamics of heterotrophic dinoflagellates during a *Gymnodinium mikimotoi* red tide in the Seto Inland Sea. Mar Ecol Prog Ser 125:269–277

Neuer S, Cowles TJ (1995) Comparative size-specific grazing rates in field populations of ciliates and dinoflagellates. Mar Ecol Prog Ser 125:259–267

Nixon SW (1995) Coastal marine eutrophication: a definition, social causes, and future concerns. Ophelia 41:199–219

Paasche E, Kristiansen S (1982) Nitrogen nutrition of the phytoplankton in the Oslofjord. Est Coastal Shelf Sci 14:237–249

Parsons ML, Dortch Q, Turner RE (2002) Sedimentological evidence of an increase in *Pseudo-nitzschia* (Bacillariophyceae) abundance in response to coastal eutrophication. Limnol Oceanogr 47:551–558

Polis GA, Winemiller KO (eds) (1996) Food webs: integration of patterns and dynamics. Chapman and Hall, New York, 475 pp

Seitzinger SP, Kroeze C (1998) Global distribution of nitrous oxide production and N inputs in freshwater and coastal marine ecosystems. Global Biogeochem Cycles 12:93–113

Seitzinger SP, Kroeze C, Bouwman AF, Caraco N, Dentener F, Styles RV (2002) Global patterns of dissolved inorganic and particulate nitrogen inputs to coastal systems: recent conditions and future projections. Estuaries 25:640–655

Sellner KG, Doucette GJ, Kirkpatrick GJ (2003) Harmful algal blooms: causes, impacts and detection. J Ind Microbiol Biotechnol 30:383–406

Skovgaard A, Legrand C, Hansen PJ, Granéli E (2003) Effects of nutrient limitation on food uptake in the toxic haptophyte *Prymnesium parvum*. Aquat Microb Ecol 31:259–265

Smayda TJ (1990) Novel and nuisance phytoplankton blooms in the sea: evidence for a global epidemic. In: Granéli E, Sundstrom B, Edler L, Anderson DM (eds) Toxic marine phytoplankton. Elsevier, New York, pp 29–40

Smayda TJ (2005) Eutrophication and phytoplankton. In: Wassmann P, Olli K (eds) Drainage basin nutrient inputs and eutrophication: an integrated approach. Univ Tromsoe, Norway, www.ut.ee/~olli/eutr/

Smayda TJ, Reynolds CS (2001) Community assembly in marine phytoplankton: application of recent models to harmful dinoflagellate blooms. J Plankton Res 23:447–461

Smil V (2001) Enriching the Earth: Fritz Haber, Carl Bosch, and the transformation of world food. MIT Press, Cambridge, 338 pp

Tilman D (1977) Source competition between plankton algae: an experimental and theoretical approach. Ecology 58:338–348

Trainer VL, Le Eberhart B-T, Wekell JC, Adams NG, Hanson L, Cox F, Dowell J Dowell (2003) Paralytic shellfish toxins in Puget Sound, Washington. J Shellfish Res 22:213–223

Turner RE, Rabalais NN (1991) Changes in Mississippi River water quality this century. BioScience 41:140–147

United States Environmental Protection Agency (1998) Environmental impacts of animal feeding operations. Office of Water, US EPA, Office of Water, Standards and Applied Sciences Division, 81 pp

Vargo GA, Heil CA, Fanning KA, Dixon LK, Neely MB, Lester K, Ault D, Musasko S, Havens J, Walsh J, Bell S (2006) Nutrient availability in support of *Karenia brevis* on the central West Florida Shelf: what keeps *Karenia* blooming? Cont Shelf Res (in press)

Vitousek PM, Aber J, Howarth RW, Likens GE, Matson PA, Schindler DW, Schlesinger WH, Tilman GD (1997) Human alteration of the global nitrogen cycle: causes and consequences. Ecol Appl 7:737–750
Walsh JJ, Penta B, Dieterle DA, Bissett WP (2001) Predictive ecological modeling of harmful algal blooms. Human Ecol Risk Assessment 7:1369–1383
Wetzel RG (2001) Limnology: lake and river ecosystems, 3rd edn. Academic Press, San Diego, 1006 pp
Wood AM, Leatham T (1992) The species concept in phytoplankton ecology. J Phycol 28:723–729

27 "Top-Down" Predation Control on Marine Harmful Algae

J.T. Turner and E. Granéli

27.1 Introduction

Removal of upper-trophic-level predators by industrialized fishing appears capable of disrupting and restructuring some marine ecosystems through a series of trophic cascades (Frank et al. 2005). There is the potential that such trophic cascades might favor harmful algal bloom (HAB) species of phytoplankton.

There is extensive literature on „top-down" control of phytoplankton by predation-induced trophic cascades, whereby predation on herbivorous zooplankton leads to increases in phytoplankton and cyanobacteria. Such studies come primarily from freshwater ecosystems (reviewed by Carpenter et al. 1985; Carpenter 1988), and are rare in marine ecosystems. However, there are suggestions that predation on marine grazers such as on copepods by gelatinous predators such as scyphomedusae and ctenophores, or by fish may decrease zooplankton grazing pressure, and increase the biomass or alter species composition of phytoplankton, including HAB species.

Inverse field abundances of copepods and medusae (Möller 1980; Greve and Reiners 1988; Greve 1994; Schneider and Behrends 1998), copepods and ctenophores of the genus *Mnemiopsis* (Reeve and Walter 1978; Kremer 1979; Deason and Smayda 1982; Turner et al. 1983; French and Smayda 1995), or studies from mesocosms using ctenophores of the genus *Pleurobrachia* (Olsson et al. 1992; Granéli et al. 1993a), suggest that predation impact of gelatinous carnivores on copepods can be high. In contrast, there are indications of a relatively small predatory impact of the ctenophore *Pleurobrachia pileus* on copepods in European coastal waters (Miller and Daan 1989; Kuipers et al. 1990; Båmstedt 1998). However, such conclusions are usually equivocal, based upon correlations of field abundances, which might be due to advection or other factors, rather than predation effects.

Selective grazing by mesozooplankton may change phytoplankton species composition. Selective grazing on diatoms by copepods (Turner and Tester

Ecological Studies, Vol. 189
Edna Granéli and Jefferson T. Turner (Eds.)
Ecology of Harmful Algae

1989) or euphausiids (Granéli et al. 1993b) could reduce grazing pressure on phytoflagellates and other non-diatom phytoplankters (Granéli et al. 1989). This could possibly contribute to events such as the ichthyotoxic Scandinavian bloom of *Chrysochromulina polylepis,* which began in the Skagerrak in late-spring of 1988 (Dahl et al. 1989; Maestrini and Granéli 1991). Alternatively, if ctenophore predation removes copepods, this could contribute to blooms of phytoplankters normally grazed by copepods.

We will review evidence for "top-down" influences on harmful algal blooms in the sea. These will include possible effects of gelatinous predators such as ctenophores and medusae, as well as fish. One hypothesized scenario is that predation by ctenophores, medusae and/or fish on mesozooplankton grazers such as copepods would reduce mesozooplankton grazing pressure on phytoplankton, possibly contributing to differential growth of certain phytoplankton taxa, some of which may include harmful algae. Also, global depletion of large piscivorous fishes by industrialized fishing (Myers and Worm 2003) may promote increases in smaller zooplanktivorous fish, increasing predation pressure on zooplankton, and diminishing zooplankton grazing on phytoplankton, including HAB species. Similar cascades might result from human "fishing down" marine food webs (Pauly et al. 2000), whereby industrialized fishing sequentially exploits organisms at lower trophic levels (such as small planktivorous fish), after severe depletion of larger apex predators

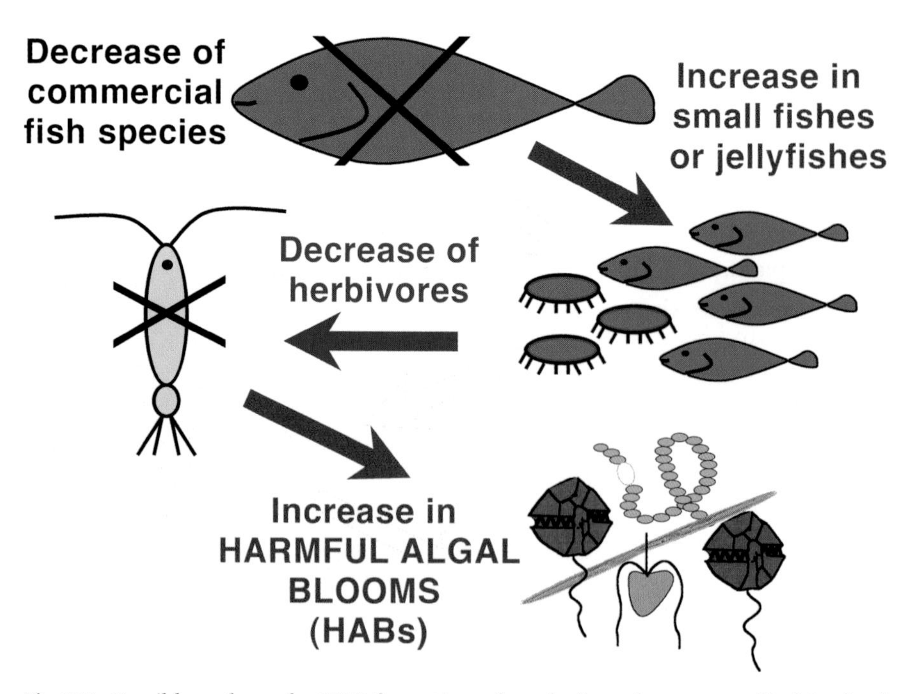

Fig 27.1. Possible pathway for HAB formation when the "top-down control" of the food chain is disrupted, as e.g., by overfishing. (Redrawn from Granéli 2004)

(Fig. 27.1). We will examine evidence for top-down predation-induced cascades in the Black Sea, and in mesocosm experiments designed to elucidate such effects. We will conclude that much more must be learned in order to clarify the extent of "top-down predation" effects on HABs, by either gelatinous predators such as medusae or ctenophores, or by fish.

27.2 "Top-down" Predators

27.2.1 Medusae

Greve (1994) reported that in July, 1989 in the German Bight of the North Sea, an increased population of the siphonophore *Muggiaea atlantica* was associated with lower than long-term mean levels of copepods, and higher than long-term mean levels of phytoplankton biomass. Greve concluded that siphonophore predation on copepods reduced grazing pressure on phytoplankton.

Schneider and Behrends (1998) found that abundant summer appearances of the scyphomedusa *Aurelia aurita* in Kiel Bight (western Baltic Sea) were associated with declines in mesozooplankton. This was particularly true for filter-feeding copepods such as *Pseudocalanus* spp., *Paracalanus* spp. and *Oithona similis*, compared to the copepods *Centropages hamatus* and *Acartia* spp. The latter two copepods reproduce throughout the summer in these waters, whereas *Pseudocalanus* reproduces only in the spring. The *Aurelia* increases were associated with increases in ultraplankton (<15 µm), which comprised 85 % of the summer chlorophyll. There was no change in the abundance of larger phytoplankton such as diatoms and dinoflagellates. Schneider and Behrends concluded that medusae exerted "top-down" predation on mesozooplankton copepods, causing reduced copepod grazing pressure on the most abundant phytoplankton.

Lindahl and Hernroth (1983) concluded that in Gullmar Fjord (west coast of Sweden) that spring and summer diatom blooms were related to low zooplankton grazing pressure because of predation on copepods and marine cladocerans by *Aurelia aurita*. When *Aurelia* declined in the fall, copepods increased in abundance and there were fall blooms of dinoflagellates such as *Gyrodinium aureolum* (=*Karenia mikimotoi*) and *Ceratium* spp. which may be toxic or difficult for zooplankton to graze. However, the hypothesis that these blooms were caused by top-down effects by jellyfish was later rejected by Lindahl (1987) who concluded that water exchange processes were more probable causes of the blooms (Granéli et al. 1993a).

Huntley and Hobson (1978) found that in Saanich Inlet, a fjord on Vancouver Island in British Columbia, Canada, that there was an initial spring diatom bloom (*Thalassiosira nordenskioldii*) in April–May, followed by a second spring diatom bloom (*Skeletonema costatum* and *Chaetoceros* spp.) in June. The second bloom may have been caused by predation of the medusa *Phialidium gregarium* on mesozooplankton, thereby reducing zooplankton grazing pressure on the diatoms.

Other studies suggest a lesser role for medusae predation impact on other zooplankton, and by inference, phytoplankton. Barz and Hirche (2005) found that in the Bornholm Basin of the central Baltic Sea, predation by *Aurelia aurita* did not regulate abundances of copepods and marine cladocerans, consuming averages of only 0.1 % of the copepods and 0.5 % of the cladocerans per day. Maximal impact was only 7.9 % of the cladocerans.

27.2.2 Ctenophores

Various field sampling programs have revealed precipitous declines in abundance or biomass of mesozooplankton such as copepods coincident with increases in ctenophores (reviewed by Reeve and Walter 1978; Kremer 1979; Deason and Smayda 1982; Turner et al. 1983; Frank 1986). In some of these cases (Deason and Smayda 1982; Turner et al. 1983), these patterns were coincident with increases in total phytoplankton abundance or biomass. Deason and Smayda (1982) found that for 4 of the 6 years studied in Narragansett Bay, Rhode Island, summer/fall pulses of the ctenophore *Mnemiopsis leidyi* were accompanied by rapid declines in zooplankton abundance and phytoplankton blooms. Yearly variations in the summer abundance of the diatom *Skeletonema costatum* were positively related to ctenophore abundance. Deason and Smayda (1982) concluded that *M. leidyi* regulated summer dynamics of phytoplankton and zooplankton in Narragansett Bay, and that such influence on phytoplankton dynamics through predation by a carnivore two trophic levels above the phytoplankton was analogous to other predation-induced trophic cascades reported for benthic marine communities.

In an extension of these studies in Narragansett Bay, French and Smayda (1995) proposed that the toxic flagellate *Heterosigma akashiwo* (formerly *Olisthodiscus luteus*, and *Heterosigma carterae*) can bloom in late spring even though the diatom *S. costatum* can outcompete it in growth rate. This can occur if copepod grazing on *H. akashiwo* is impaired by toxicity (Tomas and Deason 1981), but absence of ctenophore predation allows copepods to eat substantial amounts of *S. costatum*.

27.2.3 Fishes

There is little convincing evidence for putative top-down effects on marine phytoplankton communities through fish predation on zooplankton grazers.

As with most of the papers on "top-down" predation by gelatinous carnivores such as ctenophores and medusae, such effects due to fish remain more speculated than demonstrated. The primary reason for this is that there is very little information on the impact of predation by planktivorous fish on zooplankton communities.

Several attempts to estimate the importance of ichthyoplankton predation on zooplankton concur that such predation has little impact on the zooplankton. Dagg and Govoni (1996) estimated that at three localities in the northern Gulf of Mexico, predation by larval fish induced daily mortality of <1 % of the abundance of copepod eggs, nauplii or copepodites within the size range ingested by the fish larvae. Maximum potential predation by unusually high numbers of fish larvae concentrated in the plume of the Mississippi River was only 2–18 % of the copepodites per day, but most values were <2.5 % per day. Munk and Nielsen (1994) estimated that ichthyoplankton in frontal zones in the North Sea ingested only 2.3–3.5 % of copepod biomass per day. Pepin and Penney (2000) calculated that daily predation by a coastal Newfoundland larval fish community on the zooplankton community was <0.1 % of zooplankton abundance.

In estuarine enclosure studies, the predation effects of fish on zooplankton communities can range from weak (Granéli et al. 1993a) to substantial (Horsted et al. 1988; Riemann et al. 1988). Granéli et al. (1993a) found in mesocosm experiments, that the introduction of zooplanktivorous fish decreased the copepod biomass by half in enclosures where copepods were at natural abundance levels, as well as those where copepods were added at 10x their natural abundance. The resulting effect was that *Gyrodinium aureolum* (=*Karenia mikimotoi*) increased in abundance in these cylinders where grazing on dinoflagellates by copepods was released by the predation on the copepods by fish. However, there were no such effects apparent for another large dinoflagellate, *Ceratium furca*. Thus, different phytoplankton species were affected differently by fish predation on zooplankton.

27.3 Case Studies

27.3.1 Black Sea

The Black Sea is thought to be a major site of intense top-down predation effects on marine food webs. There has been an overall deterioration in the pelagic ecosystems of the Black Sea since introduction of the North American ctenophore *Mnemiopsis leidyi*, first detected in 1982 (Zaitsev 1992). These changes include increases in harmful algal blooms (Kideys 1994, 2002; Moncheva et al. 2001a, 2001b), suggesting the possibility of links to top-down predation by ctenophores on mesozooplankton, reducing grazing pressure on phytoplankton. However, it is difficult to separate effects of ctenophore predation from those of concurrent eutrophication due to increases in nutrient loading and changes in nutrient ratios due to damming of the Danube (Humborg et al. 1997; Lancelot et al. 2002), overfishing (Daskalov 2002; Gucu 2002) and climate change (Daskalov 2003).

Several simulations using physical-biological models have investigated changes in the pelagic ecosystems of the Black Sea in relation to increased predation by the ctenophore *Mnemiopsis leidyi* and the medusa *Aurelia aurita* during the last two decades (Oguz et al. 2001; Daskalov 2002, 2003, and references therein). The model of Oguz et al. (2001) reproduced reasonably well the observed planktonic community and food web structures in a typical eutrophic coastal location in the late 1970s and early 1980s prior to introduction of ctenophores. Simulations for the ctenophore-dominated ecosystems of the late 1980s indicated that outbreaks of either medusae or ctenophores reduced mesozooplankton grazing and led to increased phytoplankton blooms, as actually observed throughout the late 1980s and early 1990s. However, except for the omnivorous non-toxic dinoflagellate *Noctiluca scintilans*, simulated linkages between ctenophore or medusae predation and harmful algal blooms in the Black Sea remain inconclusive. Also, simulated trophic "cascades" caused by overfishing are complicated by other factors such as eutrophication and climate change (Daskalov 2002, 2003), and eutrophication-related changes in phytoplankton are complicated by climate-related changes in hydrography (Yunev et al. 2005). Effects of the invasion of the Caspian Sea by *Mnemiopsis leidyi* in 1995, probably from the Black Sea, are similarly complicated, and blaming such effects on a ctenophore-induced trophic cascade is uncertain (Bilio and Niermann 2004).

27.3.2 Mesocosm Studies

The Role of Ciliates

Planktonic ciliates are both grazers of phytoplankton, including HAB species, as well as prey for larger omnivorous zooplankters such as copepods (Turner and Roff 1993). Thus, predation on ciliates by copepods may actually remove the major grazers of HAB phytoplankton, enhancing their blooms.

Granéli and Turner (2002) performed mesocosm experiments (west coast of Sweden) to investigate whether ctenophore predation caused a cascade of "top-down" effects on phytoplankton abundance and species composition. A major result, which emerged from these experiments, was the unanticipated importance of ciliates in such interactions. Although copepod biomass levels in treatments with and without added *Pleurobrachia pileus* were not significantly different except for the last day of the experiment, it appeared that ctenophore predation diminished copepod abundance or slowed copepod growth enough to reduce copepod predation pressure on ciliates, which were the major grazers of smaller phytoflagellates. In mesocosms with natural copepod abundances, ctenophore predation contributed to increased ciliate biomass by removal of copepods. Resultant ciliate increases led to reductions in small phytoflagellates due to ciliate grazing. These effects were particularly magnified in mesocosms with added zooplankton (primarily copepods) at 10x natural abundance. There, copepod predation decimated ciliate biomass, both in the presence and absence of ctenophores. Differences in ciliate biomass and the presence versus absence of ctenophores in these enclosures were minor compared to differences in ciliate biomass in mesocosms with and without added zooplankton. This was because copepods were dramatically effective in removing ciliates. Consequently, small phytoflagellates and monads actually increased in mesocosms with added zooplankton (mainly copepods) at 10x natural concentrations, either with or without ctenophores. The toxic dinoflagellate *Karenia mikimotoi* was somewhat reduced (presumably by grazing) in the containers with 10x natural copepod abundance, but in treatments with natural abundances of copepods and ctenophores, *K. mikimotoi* increased in abundance, as ctenophores reduced the grazing pressure on *K. mikimotoi* by preying on copepods, promoting growth of this dinoflagellate.

The implication that copepods in the mesocosms were substantial predators on ciliates and grazers on the diatom *Skeletonema costatum* was supported by independent grazing experiments (Turner and Granéli 1992), which were performed concurrently with the Granéli and Turner (2002) mesocosm studies. In the grazing experiments, the copepod *Acartia clausi* grazed *S. costatum* and was predatory on ciliates, whereas *Centropages hamatus* only fed upon ciliates. These experimental results matched the mesocosm observations that there were practically no *S. costatum* ever, and no ciliates present after 3 days in the mesocosms with added zooplankton. Similarly, Olsson et al. (1992) and Nejstgaard et al. (1997) found that copepods preferred eating ciliates to most phytoplankters in mesocosm experiments. Other studies on the importance of ciliates as a food source for copepods have been reviewed by Turner and Roff (1993), Levinsen et al. (2000), Calbet and Saiz (2005) and references therein.

The high copepod predation on ciliates in the mesocosms of Granéli and Turner (2002) allowed total phytoplankton and small phytoflagellates to increase. Ciliates may even compete with copepods for consumption of prey larger than 4 μm (Rassoulzadegan et al. 1988). While the ciliates appear to have been extensively consumed by copepods, it appears that small phytoflagellates were not. This was confirmed by independent experiments of Turner and Granéli (1992). The high copepod grazing on diatoms and predation on ciliates gave the small phytoflagellates release from copepod grazing pressure.

The increase in ungrazed small phytoflagellates in the presence of abundant copepods was similar to mesocosm experiments from the Adriatic (Turner et al. 1999), where small phytoflagellate growth proceeded relentlessly until terminated by nutrient depletion (Granéli et al. 1999). This suggests that trophic transfer from phytoplankton communities dominated by flagellates through copepods to higher consumers may be inefficient, ultimately leading to a limited production of fish (Legendre 1990).

The original hypothesis of Granéli and Turner (2002) that ctenophore predation might contribute to blooms of all non-diatoms (i.e., all sizes of dinoflagellates and phytoflagellates) by removing copepods that are selective feeders on diatoms was not upheld. There was a trophic cascade effect of ctenophore predation that was transmitted down the food chain to the level of the phytoplankton, but it was not the effect that was hypothesized. Although copepods quickly grazed down *Skeletonema costatum* in enclosures and feeding experiments, this was a sideshow. The primary copepod trophic interaction was predation on ciliates, and ciliate grazing on small phytoflagellates was a major determinant of phytoplankton abundance and composition. Copepod grazing was ineffective in controlling small phytoflagellate abundance. Most larger dinoflagellates (except for *K. mikimotoi*) and small phytoflagellates (which include most nuisance algal bloom taxa in the Skagerrak) were unaffected by copepod grazing and ctenophore predation. Thus, either direct or indirect top-down influences on harmful algal blooms appear to be complicated.

Since *Chrysochromulina polylepis*, the flagellate which caused the 1988 toxic bloom in the Skagerrak, is only 10 μm in size, results of Granéli and Turner (2002) suggest that it would likely be grazed primarily by ciliates rather than copepods. Thus, if ctenophore predation on copepods was sufficient to reduce copepod predation pressure on ciliates, this would increase ciliate grazing pressure on *Chrysochromulina*.

27.4 Conclusions

The notion that top-down predation cascade effects on marine phytoplankton result in harmful algal blooms is theoretically attractive, and has often been suggested in the literature. However, this notion has prompted surprisingly little experimental investigation resulting in data from marine ecosystems. To date, only a few mesocosm studies demonstrate that trophic cascade effects of predation by zooplanktivorous fishes or gelatinous predators can promote increases in some HAB species. In one mesocosm study (Graneli et al. 1993a), the introduction of zooplanktivorous fish appeared to decrease copepod biomass and increase that of the dinoflagellate *Karenia mikimotoi* (=*Gyrodinium aureolum*). This was presumably because predation on the copepods by fish reduced grazing on dinoflagellates by copepods. However, the large dinoflagellate *Ceratium furca* appeared unaffected even when copepods were present in high abundance. In another mesocosm study (Granéli and Turner 2002), there were surprises by the unanticipated importance of trophic interactions involving ciliates. Ctenophore predation on copepods appeared to promote increases in the dinoflagellate *K. mikimotoi,* concurrent with ciliate increases and decreases in small phytoflagellates such as *Chrysochromulina polylepis.* Thus, different trophic cascades appear to work differently for different HAB phytoplankton species. Other complications with marine trophic cascades are that poorly understood effects of water advection on plankton in open marine systems confuse inverse field correlations between different trophic levels, unlike the elegant trophic cascades that have been demonstrated in some closed lake ecosystems. Further, there is a growing realization that heterotrophic protists, rather than metazoans such as copepods, are the primary grazers of phytoplankton, including HAB species, in most marine systems (Calbet and Landry 2004; Landry and Calbet 2004). Thus, scenarios based on top-down predation, which assume that metazoans such as copepods are the primary grazers of phytoplankton are oversimplified, and may actually work in the reverse way as hypothesized. Demonstration of top-down predation effects on marine plankton by fish that extend down the food chain to the phytoplankton lack clarification because, with a few exceptions, the effects of fish predation on zooplankton populations thus far remain poorly quantified. In view of recent indications that fishing is substantially altering the structure of marine food webs (Frank et al. 2005), it seems that connections between human industrialized fishing, top-down predation and trophic cascades, and harmful algal blooms should be more intensely investigated.

References

Båmstedt U (1998) Trophodynamics of *Pleurobrachia pileus* (Ctenophora, Cydippida) and ctenophore summer occurrence off the Norwegian north-west coast. Sarsia 83:169–18

Barz K, Hirche H-J (2005) Seasonal development of scyphozoan medusae and the predatory impact of *Aurelia aurita* on the zooplankton community in the Bornholm Basin (central Baltic Sea). Mar Biol 147:465–476

Bilio M, Niermann U (2004) Is the comb jelly really to blame for it all? *Mnemiopsis leidyi* and the ecological concerns about the Caspian Sea. Mar Ecol Prog Ser 269:173–183

Calbet A, Landry MR (2004) Phytoplankton growth, microzooplankton grazing and carbon cycling in marine systems. Limnol Oceanogr 49:51–57

Calbet A, Saiz E (2005) The ciliate-copepod link in marine ecosystems. Aquat Microb Ecol 38:157–167

Carpenter SR (ed) (1988) Complex interactions in lake communities. Springer, Berlin Heidelberg New York, 283 pp

Carpenter SR, Kitchell JF, Hodgson JR (1985) Cascading trophic interactions and lake productivity. BioScience 35:634–639

Dagg MJ, Govoni JJ (1996) Is ichthyoplankton predation an important source of copepod mortality in subtropical coastal waters? Mar Freshw Res 47:137–144

Dahl E, Lindahl O, Paasche E, Throndsen J (1989) The *Chrysochromulina polylepis* bloom in Scandinavian waters during spring 1988. In: Cosper EM, Bricelj VM, Carpenter EJ (eds) Novel phytoplankton blooms: causes and impacts of recurrent brown tides and other unusual blooms. Springer, Berlin Heidelberg New York, pp 383–405

Daskalov GM (2002) Overfishing drives a trophic cascade in the Black Sea. Mar Ecol Prog Ser 225:53–63

Daskalov GM (2003) Long-term changes in fish abundance and environmental indices in the Black Sea. Mar Ecol Prog Ser 255:259–270

Deason EE, Smayda TJ (1982) Ctenophore-zooplankton-phytoplankton interactions in Narragansett Bay, Rhode Island, USA, during 1972–1977. J Plankton Res 4:203–217

Frank KT (1986) Ecological significance of the ctenophore *Pleurobrachia pileus* off southwestern Nova Scotia. Can J Fish Aquat Sci 43:211–222

Frank KT, Petrie B, Choi JS, Leggett WC (2005) Trophic cascades in a formerly cod-dominated ecosystem. Science 308:1621–162

French DP, Smayda TJ (1995) Temperature regulated responses of nitrogen limited *Heterosigma akashiwo*, with relevance to its blooms In: Lassus P, Arzul G, Le Denn EE, Gentien P, Marcaillou C (eds) Harmful marine algal blooms. Lavoisier Intercept, Paris, pp 585–590

Granéli E (2004) Eutrophication and harmful algal blooms. In: Wassmann P, Olli K (eds) Drainage basin nutrient inputs and eutrophication: an integrated approach. ISBN 82-91086-36-2, pp 99–112 (Also: http://www.ut.ee/~olli/eutr/)

Granéli E, Turner JT (2002) Top-down regulation in ctenophore-copepod-ciliate-diatom-phytoflagellate communities in coastal waters: a mesocosm study. Mar Ecol Prog Ser 239:57–68

Granéli E, Carlsson P, Olsson P, Sundström B, Granéli W, Lindahl O (1989) From anoxia to fish poisoning: the last ten years of phytoplankton blooms in Swedish marine waters. In: Cosper EM, Bricelj VM, Carpenter EJ (eds) Novel phytoplankton blooms: causes and impacts of recurrent brown tides and other unusual blooms. Springer, Berlin Heidelberg New York, pp 407–427

Granéli E, Olsson P, Carlsson P, Granéli W, Nylander C (1993a) Weak „top-down" control of dinoflagellate growth in the coastal Skagerrak. J Plankton Res 15:213–237

Granéli E, Granéli W, Rabbani MM, Daugbjerg N, Fransz G, Cuzin-Roudy J, Alder VA (1993b) The influence of copepod and krill grazing on the species composition of phytoplankton communities from the Scotia-Weddell Sea. Polar Biol 13:201–213

Granéli E, Carlsson P, Turner JT, Tester PA, Béchemin C, Dawson, R, Funari E (1999) Effects of N:P:Si ratios and zooplankton grazing on phytoplankton communities in the northern Adriatic Sea. I. Nutrients, phytoplankton biomass, and polysaccharide production. Aquat Microb Ecol 18:37–54

Greve W (1994) The 1989 German Bight invasion of *Muggiaea atlantica.* ICES J Mar Sci 51:355–358

Greve W, Reiners F (1988) Plankton time – space dynamics in German Bight – a system approach. Oecologia 77:487–496

Gucu AC (2002) Can overfishing be responsible for the successful establishment of *Mnemiopsis leidyi* in the Black Sea? Est Coast Shelf Sci 54:439–451

Horsted SJ, Nielsen TG, Riemann B, Pock-Steen J, Bjørnsen PK (1988) Regulation of zooplankton by suspension-feeding bivalves and fish in estuarine enclosures. Mar Ecol Prog Ser 48:217–224

Humborg C, Ittekkot V, Coclasu A, Von Bodungen B (1997) Effect of Danube River dam on Black Sea biogeochemistry and ecosystem structure. Nature 386:385–388

Huntley ME, Hobson LA (1978) Medusa predation and plankton dynamics in a temperate fjord, British Columbia. J Fish Res Bd Can 35:257–261

Kideys AE (1994) Recent dramatic changes in the Black Sea ecosystem: the reason for the sharp decline in Turkish anchovy fisheries. J Mar Syst 5:171–181
Kideys AE (2002) Fall and rise of the Black Sea ecosystem. Science 297:1482–1484
Kremer P (1979) Predation by the ctenophore *Mnemiopsis leidyi* in Narragansett Bay, Rhode Island. Estuaries 2:97–105
Kuipers BR, Gaedke U, Enserink L, Witte H (1990) Effect of ctenophore predation on mesozooplankton during an outburst of *Pleurobrachia pileus*. Neth J Sea Res 26:111–124
Lancelot C, Staneva J, Van Eeckhout D, Beckers J-M, Stanev E (2002) Modelling the Danube-influenced north-western continental shelf of the Black Sea. II: ecosystem response to changes in nutrient delivery by the Danube River after its damming in 1972. Est Coast Shelf Sci 54:473–499
Landry MR, Calbet A (2004) Microzooplankton production in the oceans. ICES J Mar Sci 61:501–507
Legendre L (1990) The significance of microalgal blooms for fisheries and for the export of particulate organic carbon in the oceans. J Plankton Res 12:681–699
Levinsen H, Turner JT, Nielsen TG, Hansen BW (2000) On the trophic coupling between protists and copepods in Arctic marine ecosystems. Mar Ecol Prog Ser 204:65–77
Lindahl O (1987) Plankton community dynamics in relation to water exchange in the Gullmar Fjord, Sweden. PhD Thesis, Univ Stockholm, 138 pp
Lindahl O, Hernroth L (1983) Phyto-zooplankton community in coastal waters of western Sweden – an ecosystem off balance? Mar Ecol Prog Ser 10:119–126
Maestrini SY, Granéli E (1991) Environmental conditions and ecophysiological mechanisms which led to the 1988 *Chrysochromulina polylepis* bloom: an hypothesis. Oceanol Acta 14:397–413
Miller RJ, Daan R (1989) Planktonic predators and copepod abundance near the Dutch coast. J Plankton Res 11:263–282
Möller H (1980) Population dynamics of *Aurelia aurita* medusae in Kiel Bight, Germany (FRG). Mar Biol 60:123–128
Moncheva S, Doncheva V, Kamburska L (2001a) On the long-term response of harmful algal blooms to the evolution of eutrophication off the Bulgarian Black Sea coast: are the recent changes a sign of recovery of the ecosystem – the uncertainties? In: Hallegraeff GM, Blackburn SI, Bolch CJ, Lewis RJ (eds) Harmful algal blooms 2000. IOC-UNESCO, Paris, pp 177–181
Moncheva S, Gotsis-Skretas O, Pagou K, Krastev A (2001b) Phytoplankton blooms in Black Sea and Mediterranean coastal ecosystems subjected to anthropogenic eutrophication: similarities and differences. Est Coast Shelf Sci 53:281–295
Munk P, Nielsen TG (1994) Trophodynamics of the plankton community at Dogger Bank: predatory impact by larval fish. J Plankton Res 16:1225–1245
Myers RA, Worm B (2003) Rapid worldwide depletion of predatory fish communities. Nature 423:280–283
Nejstgaard JC, Gismervik I, Solberg PT (1997) Feeding and reproduction by *Calanus finmarchicus*, and microzooplankton grazing during mesocosm blooms of diatoms and the coccolithophore *Emiliania huxleyi*. Mar Ecol Prog Ser 147:197–217
Oguz T, Ducklow HW, Purcell JE, Malanotte-Rizzoli P (2001) Modeling the response of top-down control exerted by gelatinous carnivores on the Black Sea pelagic food web. J Geophys Res 106:4543–4564
Olsson P, Granéli E, Carlsson P, Abreu P (1992) Structuring of a post spring phytoplankton community by manipulation of trophic interactions. J Exp Mar Biol Ecol 158:249–266
Pauly D, Christensen V, Froese R, Palomares ML (2000) Fishing down aquatic food webs. Am Sci 88:46–51

Pepin P, Penney R (2000) Feeding by a larval fish community: impact on zooplankton. Mar Ecol Prog Ser 204:199–212

Rassoulzadegan F, Laval-Peuto M, Sheldon RW (1988) Partitioning of food ration of marine ciliates between pico- and nanoplankton. Hydrobiologia 159:75–88

Reeve MR, Walter MA (1978) Nutritional ecology of ctenophores – a review of recent research. Adv Mar Biol 15:249–287

Riemann B, Nielsen TG, Horsted SJ, Bjørnsen PK, Pock-Steen J (1988) Regulation of phytoplankton biomass in estuarine enclosures. Mar Ecol Prog Ser 48:205–215

Schneider G, Behrends G (1998) Top-down control in a neritic plankton system by *Aurelia aurita* medusae – a summary. Ophelia 48:71–82

Tomas CR, Deason EE (1981) The influence of grazing by two *Acartia* species on *Olisthodiscus luteus* Carter. PSZNI: Mar Ecol 2:215–223

Turner JT, Granéli E (1992) Zooplankton feeding ecology: grazing during enclosure studies of phytoplankton blooms from the west coast of Sweden. J Exp Mar Biol Ecol 157:19–31

Turner JT, Roff JC (1993) Trophic levels and trophospecies in marine plankton: lessons from the microbial food web. Mar Microb Food Webs 7:225–248

Turner JT, Tester PA (1989) Zooplankton feeding ecology: copepod grazing during an expatriate red tide. In: Cosper EM, Bricelj VM, Carpenter EJ (eds) Novel phytoplankton blooms: causes and impacts of recurrent brown tides and other unusual blooms. Springer, Berlin Heidelberg New York, pp 359–374

Turner JT, Bruno SF, Larson RJ, Staker RD, Sharma GM (1983) Seasonality of plankton assemblages in a temperate estuary. PSZNI: Mar Ecol 4:81–99

Turner JT, Tester PA, Lincoln JA, Carlsson P, Granéli E (1999) Effects of N:P:Si ratios and zooplankton grazing on phytoplankton communities in the northern Adriatic Sea. III. Zooplankton populations and grazing. Aquat Microb Ecol 18:67–75

Yunev OA, Moncheva W, Carstensen J (2005) Long-term variability of vertical chlorophyll-*a* and nitrate profiles in the open Black Sea: eutrophication and climate change. Mar Ecol Prog Ser 294:95–107

Zaitsev YP (1992) Recent changes in the trophic structure of the Black Sea. Fish Oceanogr 1:180–189

28 Climate Change and Harmful Algal Blooms

B. DALE, M. EDWARDS, and P.C. REID

28.1 Introduction

Some authors have suggested that harmful algal blooms (HABs) are increasing throughout the world due to anthropogenic influences (Hallegraeff 1993). Others have stressed that climate variability may be an equally important contributor to the apparent increases (Sellner et al. 2003). Understanding the possible effects of climate change is therefore a critical requirement in the development of the risk assessments needed for the effective management of HABs. This chapter considers the effects of past, present and future climatic variability on HABs.

The one thing we are certain of regarding climate is that it is changing – and it always has been. The geological and historical records show alternations between periods of relative warming and cooling at all timescales: extremes of glaciated to ice-free conditions on scales of millions of years to shorter-term oscillations on scales of a few to tens of years, such as the El Niño-Southern Oscillation (ENSO) and the North Atlantic Oscillation (NAO). The last 1,000 years included periods as warm or warmer than now (at least in some regions) during the Medieval Warm Period (MWP), 550–1300 A.D., and colder than now during the Little Ice Age (LIA) 1300–1900 A.D.

A major scientific effort is currently underway coordinated by the International Panel on Climate Change (IPCC) to assess the extent of human impact on climate, through the burning of fossil fuels, etc. This work has resulted in a much greater appreciation of the complex nature of climate, where temperature is only one of many factors to be considered. This complexity imposes a high degree of uncertainty on the predictive models used. There is a clear and convincing scientific consensus that anthropogenically forced global warming is taking place, but the extent of future warming, especially in the next 100 years, remains uncertain. In considering possible effects on phytoplankton, it is important to realize that at least in some regions, recent global warming has so far involved mainly a warming of winter temperatures. Continued

Ecological Studies, Vol. 189
Edna Granéli and Jefferson T. Turner (Eds.)
Ecology of Harmful Algae

warming could eventually lead to elevated summer temperatures; in the North Sea and northeastern Atlantic, this has already occurred (Edwards and Richardson 2004) with a recognizable impact on plankton.

Water temperature, light, and nutrients, are the main parameters affecting phytoplankton, including HAB species. Each species has a temperature window within which it can survive, and a range within this of low to optimal growth. These are vital factors defining the biogeographic boundaries within which a species can live. Seasonal variations in temperature may be important for determining when optimal growth can occur at any location. The direct effect of global warming – elevated water temperature – therefore may affect both the seasonal composition of the phytoplankton (e.g., increased growth in winter/spring-blooming species as winter temperature increases), including changes in seasonal succession, and the position of biogeographic boundaries (e.g., migration towards higher latitudes for warmer-water species as summer temperatures increase).

However, from physiological experiments it is known that many species show a robust tolerance for moderate shifts in temperature. Most HAB organisms are coastal/estuarine species that may be expected to tolerate at least moderate swings in temperature of several degrees on a daily basis. Their present-day occurrence, in part, is likely to reflect selection pressures from recent climatic oscillations (variations of several degrees average temperatures between the MWP and the LIA within the past 1,000 years). Therefore, temperature as such may not be the dominant factor in phytoplankton response to climate change; other, indirect effects need to be considered. These include:

Stratification increased temperature often causes increased stratification of the upper water column (e.g., favoring flagellates that can swim down to obtain dissolved nutrients from lower levels when surface water nutrients are depleted),

Upwelling climatic variations involve changes in oceanic circulation, including upwelling (e.g., Bakun 1990),

Freshwater run-off from land, and cloud-cover increased temperature often causes increased evaporation and atmospheric precipitation (rain and snow) with a consequent enrichment with nutrients of freshwater run-off from land to aquatic environments; increased cloud cover produced through the same process may reduce available light to the phytoplankton,

Feedback mechanisms blooms of algae take up more heat from the sun leading to a further enhancement of sea temperatures.

28.2 Evidence from the Past

Time-series data are required in order to assess variation in natural phenomena. For HABs, two main sources are important: long-term phytoplankton records, and the microfossil record in bottom sediments. The few available long-term plankton records show important variation in the frequency and intensity of HABs at the decadal time scale (e.g., the recurring blooms of *Karenia brevis* around Florida, Walsh and Steidinger 2001). Since climate also shows decadal variation, at least on a regional scale (e.g., ENSO, and the NAO), records of several to many tens of years are needed to reveal any convincing relationship between climate and HABs.

The ideal basic requirements for phytoplankton records would cover at least 30 consecutive years of consistent monitoring at a site of known HAB problems. The example from Florida cited above has provided the only such data published so far. Several other long-term records cover the required length of time, and presumably will contribute relevant data when published (e.g., from Rhode Island, USA, Sherkin Island, Ireland, the Skagerrak, and Helgoland, Germany). Most data relevant to the topic considered here has come from the Continuous Plankton Recorder (CPR) survey (Warner and Hays 1994; Reid et al. 1998). Although designed mainly to monitor zooplankton, and therefore only providing direct information for a number of larger HAB organisms, the CPR has routinely identified approximately 170 phytoplankton taxa since the 1950s throughout various regions of the North Atlantic.

Long-term plankton records thus provide very few examples linking HABs to climate variation. Most records of HABs cover too short a time period to allow statistically sound comparisons with climate. However, since HAB species are distinguished from other plankton only by their harmful effects to human health and economics, it is worth considering data on the interaction of *any* species of phytoplankton with climate. Also, the list of known causal species is being added to regularly, as more is learned about HABs.

Microfossils produced by some groups of phytoplankton accumulate in bottom sediments and provide an alternative source of integrated time-series data. The main group including HAB species is the dinoflagellates. These produce resting cysts (cysts) protected by fossilizeable cell walls that are strengthened with sporopollenin-like material (comparable to the walls of pollen grains) or calcareous material. The siliceous frustules of diatoms also may fossilize. The main limitation to the cyst record is that only around 10 % of dinoflagellate species are known to form cysts (Dale 1996) although this includes some HAB species. Dinoflagellate cysts, diatoms and other groups of microfossils also provide evidence of other forms of environmental change (e.g., eutrophication and marine pollution), and environmental micropaleontology is now an integral part of environmental sciences (Dale and Dale 2002).

As with plankton records, the greatest challenge is how to differentiate the various environmental signals.

While there certainly is strong evidence that increased phytoplankton biomass and algal blooms are associated with elevated nutrients in some regions of the world (Sellner et al. 2003), assessing larger-scale changes on regional ecosystems is more problematic. For example, while elevated anthropogenic effects in the North Sea have been reported, this has been accompanied by a period of marked climate change over the last few decades. Distinguishing the effects of anthropogenic eutrophication embedded in the climate variability is therefore extremely difficult and needs extensive baseline data. While there have been more studies focusing on the role of eutrophication and HABs, there has been limited work on the role of climate oscillations (Rhodes et al. 1993; Hales et al. 1998; Belgrano et al. 1999; Edwards et al. 2001) and very few on the impact of climate change (Reid et al. 1998; Beaugrand et al. 2002).

28.3 Results from Plankton Records

The relationship between long-term (50 years) phytoplankton changes and the NAO have been discussed in relation to (1) changes in West wind strength and hence water-column stability, (2) changes in sea-surface temperature (SST), and (3) changes in the oceanic inflow into the North Sea (Edwards et al. 2001). According to results from the Continuous Plankton Recorder (CPR) survey, at the regional to oceanic scale, climate variability and regional climatic warming appears to play a dominant role in the long-term changes in phytoplankton assemblages and total biomass (Reid et al. 1998; Edwards et al. 2001).

Regarding HABs and climate, it seems that oceanographic circulation is an important factor, and that changes in circulation patterns often are associated with decadal climate oscillations. The large phytoplankton biomass values and *Dinophysis* blooms recorded off the coast of Sweden in the late 1980s have been associated with inflowing Atlantic water associated with positive NAO values (Belgrano et al. 1999). This period also coincided with one of the most globally known and widely reported bloom events to occur over the last few decades. The *Chrysochromulina* bloom off the Norwegian Skagerrak coast in 1988 not only caused widespread ecological disturbances but was also estimated to have caused an economic loss for the fish farming community of approximately 10 million Euro (Skjodal and Dundas 1991). The bloom in this area was associated with relatively high temperatures and low salinities, resulting in highly stable conditions in the Norwegian Coastal current. This was again associated with the NAO index (Gjøsæter et al. 2000) and coincided with the timing of a regime shift in the planktonic ecosystem of the North Sea (Reid and Edwards 2001).

The NAO has also been implicated as a driver of upwelling-induced blooms along the Spanish coast (Fraga and Bakun 1993). Basin-scale circulation can also act as an effective factor for the transportation of potential HAB taxa. *Alexandrium* populations in the Gulf of Maine, for example, have been associated with transportation via coastal currents (Anderson 1997), and the devastating appearance of the neurotoxic red-tide organism *Karenia brevis* off North Carolina was also associated with current transportation (Tester et al. 1991). Regional climate change therefore, has important consequences for bloom events driven by circulation and meteorology.

While the causative role of changes in SST for long-term phytoplankton dynamics are still rather obscure, there is certainly evidence to suggest that changes in SST can cause a shift from diatoms to flagellates within the phytoplankton community structure. In the context of climate change and HABs this is important, as the majority of HABs are associated with noxious and toxic flagellate species. Recent studies have shown that rising SST has been associated with a shift from diatoms to dinoflagellates in the North Atlantic and North Sea (Edwards et al. 2001) in the southern North Sea (Hickel 1998) and in the Baltic Sea (Wasmund et al. 1998). Therefore, climatic warming may favor flagellate dominance in the phytoplankton community. Rising SST in the North Sea has also been correlated with the earlier appearance of dinoflagellates in the seasonal cycle. Changes in phenology of some dinoflagellate species have resulted in a shift forward of over a month from their typical seasonal peak, whereas the seasonal timing of diatoms has remained relatively static (Edwards and Richardson 2004). It is not yet known, however, why dinoflagellate seasonal timing is correlated with warmer temperatures and whether this is due to a direct physiological response or to an indirect response to SST via enhanced or earlier stratification. It also seems likely that an important environmental impact caused by climate change is an increase in the presence of haline stratification in regions susceptible to such changes, resulting in an increase in dinoflagellate bloom formation (Edwards et al. 2005). For example, in the Grand Banks region there has been an increase in the abundance of dinoflagellate species (notably *Ceratium arcticum*). These changes, since the early 1990s, have been linked to hydro-climatic variations, specifically increased stratification and stability in the region. These indicate a progressive freshening of waters in this region caused by regional climatic warming (Johns et al. 2003).

Recent macroscale research has shown that the increase in regional sea temperatures has triggered a major reorganization in calanoid copepod species composition and biodiversity over the whole North Atlantic Basin (Beaugrand et al. 2002). During the last 40 years there has been a northerly movement of warmer water plankton by 10° latitude in the Northeast Atlantic and a similar retreat of colder water plankton to the north. This geographical movement is much more pronounced than any documented terrestrial study, presumably due to advective processes. Rising SST may also have the potential to shift the biogeographical boundaries of certain phytoplankton species, which may include HAB taxa. However, there have been only limited studies in this area. Warmer-water species of the *Ceratium* genus have certainly been noted further north in the North Atlantic over the last few years (Edwards et al. 2006), and according to Nehring (1998), some thermophilic phytoplankton species have established themselves in the southern North Sea. In experimental studies the physiological response to climatic warming has been studied for a number of HAB taxa (Peperzak 2003). Climate-warming scenarios (expressed as an increase of 4 °C or more), coupled with water-column stratification, led to a doubling of growth rates of potentially harmful dinoflagellates of *Dinophysis* spp. and *Prorocentrum* spp. (Peperzak 2003). Long-term trends in the abundance of *Prorocentrum* spp., as well as the red-tide forming *Noctiluca scintillans*, have been correlated with SST in some regions of the North Sea (Edwards et al. 2006). Biogeographical boundary shifts mediated by climate change in phytoplankton populations, also have the potential in the future to lead to the occurrence of species with southern biogeographical affinities not yet considered to be detrimental in temperate regions.

28.4 Results from the Sedimentary Record of Dinoflagellate Cysts

Dinoflagellates account for many of the different forms of HABs, and their fossil record is a potential source of long-term records of HABs. Unfortunately for science, most toxic species do not produce fossilizeable cysts, and those that do are not always toxic. *Pyrodinium bahamense*, a species responsible for severe PSP in Southeast Asia, produces cysts with a fossil record of at least 20 million years, but seems not to be toxic in parts of its present-day range (e.g., the Caribbean). Reports from the fossil record of this species having caused mass mortality of birds (Emslie et al. 1996) and fish are therefore speculative. *P. bahamense* is a lagoonal species that often heavily dominates cyst assemblages at the present day without causing such dramatic effects. On the other hand, the fossil record is dotted with events such as fish kills, usually thought to have been caused by catastrophic shifts in salinity or oxygen content, for example through isolation of smaller parts of basins. Present day mass mortality of fish and birds is usually associated with species leaving no fossil record (e.g., *Prymnesium parvum* in Lake Koronia in Greece, reported by Moustaka-Gouni et al. 2004).

Long-term records seemingly linking climate change to HAB-producing species of *Alexandrium* have been reported from Atlantic and Pacific Canada (Mudie et al. 2002). When summer SST was up to 5 °C warmer than present during the late glacial–early Holocene, these authors showed a period of sustained high production of "red tide blooms" which strongly implicated global warming in the historical increase in the frequency of HABs. These reports were surprising, since comparable fossil cysts have not been recorded in many studies of sediments from other regions prone to toxic blooms of *Alexandrium*. Follow-up work instead suggests these cysts may be the remains of inner walls of calcareous cysts belonging to a different species with no known connections to HABs (Head et al. 2006).

The distinctive fossilizeable cyst of the HAB species *Gymnodinium catenatum* seems to be a likely candidate for tracing past blooms, but results from the first attempts were surprising. Dale et al. (1993) showed records of very similar cysts from the past few thousands of years in the Kattegat/Skagerrak region of Scandinavia. The species was not recorded from recent plankton there, and the nearest present-day distribution along the coasts of northern Spain and Portugal suggested its need for warmer water. Two periods of unusually high concentrations of these cysts were identified: around 4500 years B.P. during warmer climate, and between 1000 and 500 B.P. during the warmer MWP. This was presumed to pre-date causative human impact, and since the species waned as the cooler LIA set in, and other changes in the cyst assemblage suggested climatic warming, this was identified as the most likely cause of the 'bloom periods' (Dale et al. 1993; Dale and Nordberg 1993; Thorsen and Dale 1998). Subsequent biological studies, however, have shown this to be a non-toxic species separate from, but very similar to *G. catenatum* (Ellegaard and Moestrup 1999). Records of actual *G. catenatum* cysts that are slightly larger, from sediments off the Portuguese coast have been linked to eutrophication rather than climate change (Amorim and Dale 2006).

As yet there are no unequivocal examples of cyst records of HABs, although the literature cited above suggests otherwise. Nevertheless, there are many long-term records of cysts from other species showing assemblage changes corresponding to environmental change, including climatic variation. Such records contain a wealth of information also relevant to considering the HAB species. For example, the Kattegat/Skagerrak work still offers one of the few documented examples of the 'invasion' and subsequent dominance of a warmer-water species due to climatic warming, and the main observations by Mudie et al. (2002), linking climate change to productivity of some dinoflagellates also suggest changes likely to affect HAB species.

In considering the cyst record, it is necessary to distinguish between coastal/shelf records and those from the deep sea. The many long-term deep-sea records attempting to document climate change give poor resolution for the decadal scale changes of interest here, due to low sedimentation rates and uncertainties regarding long-range transport of cysts. The coastal/shelf cyst records covering the past few hundreds of years show a notable amount of variation within the assemblages. There is no reason to suppose that the cyst-forming species alone are affected by environmental change, and the cyst record therefore suggests more change in the phytoplankton than is generally indicated by plankton records. Furthermore, this variation involves distinctive changes previously correlated with known, present-day environmental parameters, such as SST, salinity, nutrients, and water stability (Dale 1996).

While based on few examples so far, first attempts are being made to develop the cysts as indicators, or signals, of different forms of environmental change (Dale and Dale 2002). Again, the main challenge is to separate the various signals within the data. Records from coastal waters so far suggest distinctive cyst signals mainly reflecting influence from the obvious human impact associated with industrialization within the past 150 years, and that any climatic signals are relatively minor by comparison. The timing of these assemblage changes offers support for this interpretation, occurring as they do at clearly different times, sometimes between localities only a few kilometers apart, rather than simultaneously as may be expected from more regional effects of climate.

Records spanning the MWP, roughly 550–1300 A.D., provide some of the most appropriate evidence relevant to predicting possible effects of global warming. The average temperature increase of around 1–2 degrees estimated for the MWP is of the order of changes predicted by climate models for global warming for the next 100 years. Early Holocene temperatures may have been 4–5 degrees warmer than at present, more comparable to the highest estimates within the uncertainties of predicted global warming, but both dating and temperature estimates may well be less reliable for periods of warming prior to the MWP. Although limited to the cyst-forming species, this evidence therefore represents much of what is known regarding phytoplankton response to several degrees of warming.

Records spanning the MWP in Scandinavian waters indicate distinct cyst assemblage changes associated with climatic warming (e.g., Thorsen and Dale 1998; and largely unpublished core data from archeological work in the Oslofjord mentioned by Dale and Dale 2002). The cyst signals involved may be summarized below.

The first signal of climatic warming is a marked increase in the highly cosmopolitan species *Protoceratium reticulatum* (opportunistic development to exploit a changing environment – also seen as a first signal of eutrophication, and with changes to cooler conditions of the LIA). Detailed records often show this as a distinct spike of dominance.

The first stage of climatic warming may be marked by an increase in *Pentapharsodinium dalei*. On first consideration, this seems anomalous, since the species represents the colder element of the assemblage, but is now realized as an indicator of the first phase of warming – affecting largely winter temperatures. Warmer winters give a longer, warmer window of opportunity for the colder water dinoflagellate species blooming in spring (Dale 2001).

Sustained climatic warming leads to increased amounts of the warmer water species blooming in late summer (e.g. some of the *Gonyaulax spinifera* group, and *Lingulodinium polyedrum*). These species may also extend their biogeographic boundaries to higher latitudes, while more-warmer-water species 'invade' by similar expansion from lower latitudes, although this is impossible to prove since even long intervals with no recorded cysts in between the warmer periods cannot be taken as evidence for the complete absence of a species. Thus far, there is a lack of corresponding information from mid-low latitudes.

The cyst records of the past few hundreds of years include the relatively minor oscillations of temperature associated with the NAO. Ironically, some of the cyst records with highest time resolution have proved hardest to relate to climate. Harland et al. (2004) produced one such record from a Swedish fjord on the Skagerrak coast, showing an abrupt major assemblage change from about 1940 to the present, involving a ten-fold increase in the total amounts of cysts, including a substantial increase in *Lingulodinium polyedrum*. This is very similar to the eutrophication signal described elsewhere (Dale 2001), but there was no evidence of eutrophication, despite the obvious influence of Baltic water reported by others to be eutrophic (Elmgren 2001). Instead, Harland et al. (2004) favored a link to the NAO, while admitting that such data may be impossible to interpret fully (the abrupt assemblage change also correlated with a major dredging operation that may have altered the hydrography at the studied site). Comparable records from nearby fjords should suggest if this really is a more regionally expressed climatic signal linked to the NAO.

28.5 Conclusions

The long-term plankton records and the sedimentary records of cysts reported so far can provide only a few examples of direct evidence for the effects of climate change on HABs. This is not surprising, since with only few exceptions long-term monitoring has not been utilized in HAB research, and most of the data cited here were gathered for other scientific purposes. The CPR data is targeted to reveal long-term variability in plankton from the North Atlantic system, and therefore does not include most regions of highest HAB activity, affecting mainly coastal waters, or most of the HAB species which are too small or infrequent in the more oceanic waters covered. The cyst data are mostly from studies of human impact on the more coastal environments, but not specifically aimed at HAB species since many do not fossilize.

Nevertheless, both plankton and cyst records provide complementary information on the effects of climate on other species of phytoplankton, suggesting effects that almost certainly must have influenced at least some past and present HABs. The main effects of interest here are that climatic warming may have increased the occurrence and magnitude of HABs at any particular site through:

(1) Shifts toward relatively more flagellates in some regional ecosystems (many HAB species are flagellates);
(2) Earlier spring blooms of flagellates (increased window of opportunity for these species);
(3) Expansion of the biogeographic ranges of warmer-water species into higher latitudes.

At the same time, it is important to note that such changes may also have the potential to *decrease* the occurrence and magnitude of HABs. None of the effects of climate change discussed here are restricted to just the relatively few HAB species, allowing for the possibility that in at least some cases other species could better exploit the resulting changes in the environment and outcompete the HAB species. There is insufficient evidence to resolve this issue.

Considering the prediction of future HABs, this is one of many examples where scientists are increasingly being asked to predict the effects of global change in environments with few or no records to show the effects of previous change. In this case, the paucity of long-term plankton records has been suggested as one of the main reasons for our limited understanding of HABs and how to predict them despite over 40 years of research (Dale 2005). Without such records, the work discussed here can offer no more than a framework for beginning to assess the possible changes in HABs to be expected from the uncertain predictions of future global warming. Information from the fossil record shows clear links between some bloom species of dinoflagellates and climate, suggesting likely responses to future warming, but as yet no unequivocal records of past HABs. Information from recent and living plankton on HAB – climate links is also still too tenuous to suggest how blooms will respond in the future. It is also difficult to extrapolate from this information, which is largely from the Atlantic, to the rest of the world.

References

Amorim A, Dale B (2006) Historical cyst record as evidence for the recent introduction of the dinoflagellate *Gymnodinium catenatum* in the North Eastern Atlantic. Afr J Mar Sci (in press)

Anderson DM (1997) Bloom dynamics of toxic *Alexandrium* species in the northeastern U.S. Limnol Oceanogr 42:1009–1022

Bakun A (1990) Global climate change and intensification of coastal ocean upwelling. Science 247:198–201

Beaugrand G, Reid PC, Ibanez F, Lindley JA, Edwards M (2002) Reorganisation of North Atlantic marine copepod biodiversity and climate. Science 296:1692–1694

Belgrano A, Lindahl M, Hernroth B (1999) North Atlantic oscillation, primary productivity and toxic phytoplankton in the Gullmar Fjord, Sweden (1985–1996). Proc R Soc Lond, ser B 266:425–430
Dale B (1996) Dinoflagellate cyst ecology: modeling and geological applications. In: Jansonius J, McGregor DC (eds) Palynology: principles and applications. The Am Assoc Stratigr Palynol Found, Publishers Press, Salt Lake City, pp 1249–1276
Dale B (2001) The sedimentary record of dinoflagellate cysts: looking back into the future of phytoplankton blooms. Scientia Marina 65:257–272
Dale B (2005) The sedimentary record shows the need for long-term monitoring of phytoplankton. In: Solbé J (ed) Long-term monitoring. Why, what, where, when & how? Sherkin Island Marine Station, Ireland, pp 106–113
Dale B, Dale A (2002) Dinoflagellate cysts and acritarchs. In: Haslett SK (ed) Quaternary environmental micropalaeontology. Arnold Publishers, London, pp 207–240
Dale B, Nordberg K (1993) Possible environmental factors regulating prehistoric and historic "blooms" of the toxic dinoflagellate *Gymnodinium catenatum* in the Kattegat-Skagerrak region of Scandinavia. In: Smayda TJ, Shimizu Y (eds) Toxic phytoplankton blooms in the sea. Elsevier, Amsterdam, pp 53–57
Dale B, Madson A, Nordberg K, Thorsen TA (1993) Evidence for prehistoric and historic "blooms" of the toxic dinoflagellate *Gymnodinium catenatum* in the Kattegat-Skagerrak region of Scandinavia. In: Smayda TJ, Shimizu Y (eds) Toxic phytoplankton blooms in the sea. Elsevier, Amsterdam, pp 47–52
Edwards M, Richardson A (2004) Impact of climate change on marine pelagic phenology and trophic mismatch. Nature 430:881–884
Edwards M, Reid PC, Planque B (2001) Long-term and regional variability of phytoplankton biomass in the north–east Atlantic (1960–1995). ICES J Mar Sci 58:39–49
Edwards M, Licandro P, John AWG, Johns DG (2005) Ecological status report: results from the CPR survey 2003/2004. SAHFOS Tech Rep ISSN 1744-0750, 2:1–6
Edwards M, Johns DG, Leterme SC, Svendsen E, Richardson AJ (2006) Regional climate change and harmful algal blooms in the north-east Atlantic. Limnol Oceanogr 51(2):820–829
Ellegaard M, Moestrup Ø (1999) Fine structure of the flagellar apparatus and morphological details of *Gymnodinium nolleri* sp. Nov. (Dinophyceae), an unarmoured dinoflagellate producing a microreticulate cyst. Phycologia 38:289–300
Elmgren R (2001) Understanding human impact on the Baltic ecosystem: changing views in recent decades. Ambio 30:222–230
Emslie SD, Allmon WD, Rich FJ, Wrenn JH, de France SD (1996) Integrated taphonomy of an avian death assemblage in marine sediments from the Late Pliocene of Florida. Palaeogeogr Palaeoclimatol Palaeoecol 124:107–136
Fraga S, Bakun A (1993) Global climate change and harmful algal blooms: the example of *Gymnodinium catenatum* on the Galician coast. In: Smayda TJ, Shimizu Y (eds) Toxic phytoplankton blooms in the sea. Elsevier, Amsterdam, pp 59–65
Gjøsæter J, Lekve K, Stenseth NC, Leinaas H P, Christie H, Dahl E, Danielssen DS, Edvardson B, Olsgard F, Oug E, Paasche E (2000) A long–term perspective on the *Chrysochromulina* bloom on the Norwegian Skagerrak coast 1988: a catastrophe or an innocent incident? Mar Ecol Prog Ser 207:201–218
Hales S, Weinstein P, Woodward A (1998) Ciguatera (fish poisoning), El Niño, and Pacific sea surface temperatures. Ecosystem Health 5:20–25
Hallegraeff GM (1993) A review of harmful algal blooms and their apparent global increase. Phycologia 32:79–99
Harland R, Nordberg K, Filipsson H (2004) A higher resolution dinoflagellate cyst record from latest Holocene in Koljö Fjord, Sweden. Rev Palaeobot Palynol 128:119–141

Head MJ, Lewis J, de Vernal A (2006) The cyst of the calcareous dinoflagellate *Scrippsiella trifida*: resolving the fossil record of its organic wall with that of *Alexandrium tamarensis*. J Paleontol 80(1):1–8

Hickel W (1998) Temporal variability of micro- and nannoplankton in the German Bight in relation to hydrographic structure and nutrient changes. ICES J Mar Sci 55:600–609

Johns DG, Edwards M, Richardson A, Spicer JI (2003) Increased blooms of a dinoflagellate in the northwest Atlantic. Mar Ecol Prog Ser 265:283–287

Moustaka-Gouni M, Cook CM, Gkelis S, Michaloudi E, Pantlelidakis K, Pyrovetsi M, Lanaras T (2004) The coincidence of a *Prymesium parvum* bloom and the mass kill of birds and fish in Lake Koronia. Harmful Algae News 26:1–2

Mudie PJ, Rochon A, Levac E (2002) Palynological records of red tide-producing species in Canada: past trends and implications for the future. Palaeogeogr Palaeoclimatol Palaeoecol 180:159–186

Nehring S (1998) Establishment of thermophilic phytoplankton species in the North Sea: biological indicators of climate change? ICES J MarSci 55:818–823

Peperzak L (2003) Climate change and harmful algal blooms in the North Sea. Acta Oecologia 24:139–144

Reid PC, Edwards M, Hunt HG, Warner AJ (1998) Phytoplankton change in the North Atlantic. Nature 391:546

Rhodes LL, Haywood AJ, Ballantine WJ, MacKenzie AL (1993) Algal blooms and climate anomalies in north-east New Zealand, August–December 1992. New Zeal J Mar Fresh 27:419–430

Sellner KG, Doucette GJ, Kirkpatrick GJ (2003) Harmful algal blooms: causes, impacts and detection. J Ind Microbiol Biotechnol 30:383–406

Skjoldal HR, Dundas I (1991) The *Chrysochromulina polylepis* bloom in the Skagerrak and the Kattegat in May–June 1988: environmental conditions, possible causes and effects. ICES Coop Res Rep 175:1–59

Tester PA, Stumpf RP, Vukovich FM, Fowler PK, Turner JT (1991) An expatriate red tide bloom: transportation, distribution and persistence. Limnol Oceanogr 36:1053–1061

Thorsen T, Dale B (1998) Climatically influenced distribution of *Gymnodinium catenatum* during the past 2000 years in coastal sediments of southern Norway. Palaeogeogr Palaeoclimatol Palaeoecol 143:159–177

Walsh JJ, Steidinger KA (2001) Sahara dust and Florida red tides: the cyanophyte connection. J Geophys Res 106 (C6):11597–11612

Warner AJ, Hays GC (1994) Sampling by the Continuous Plankton Recorder survey. Prog Oceanogr 34:237–256

Wasmund N, Nausch G, Matthaus W (1998) Phytoplankton spring blooms in the southern Baltic Sea – spatio-temporal development and long-term trends. J Plankton Res 20:1099–1117

29 Anthropogenic Introductions of Microalgae

G. Hallegraeff and S. Gollasch

29.1 Potential Transport Vectors for Microalgae

The geographic range of microalgal species can expand depending on natural factors (climate change, catastrophic storm events, ocean currents, transport of spores via wind or bird feet) or human-mediated vectors. The present chapter focuses on anthropogenic transport by ships' ballast water, and translocation of aquaculture products such as shellfish. Other potential vectors such as fouling on ships' hulls, and transport via dredging or aquaculture equipment have not been adequately investigated to date with respect to microalgae. The potential impact from escape of aquaculture microalgal feedstocks also remains to be assessed. The global transport of microalgal species via ships' ballast water commenced in the 1870s (Carlton 1985). Annual transport on a worldwide basis is now at an estimated volume of 2–3 billion t, and has received considerable attention in the past two decades (reviewed by Hallegraeff 1998). Not surprisingly, the very first claim, made 100 years ago, of cargo vessel ballast water as a vector in the dispersal of non-indigenous marine organisms refers to a microalga (Ostenfeld 1908). To prove that a particular species of microalga has been introduced, however, is much more complex than, for example, for macroalgae (Wyatt and Carlton 2002). Because of the apparent continuity of the world's oceans, similar hydrological environments in different oceans tend to have morphologically similar phytoplankton assemblages ("latitudinal cosmopolitanism") and some scientists suggest that marine protists have had ample evolutionary time to reach and inhabit all suitable environments (Finlay 2002). While this may be true for oceanic diatoms and dinoflagellates, or for some widespread ecologically tolerant coastal diatoms, other species such as estuarine dinoflagellates can have fastidious nutritional requirements, e.g., with regard to humic substances from land runoff. For cyst-producing estuarine dinoflagellates that cannot cross oceanic boundaries by ocean currents, there is every reason to believe that transport in ballast water has played a role in the apparent global spreading of selected species. If the dogma of widespread cosmopolitanism of microalgae

Ecological Studies, Vol. 189
Edna Granéli and Jefferson T. Turner (Eds.)
Ecology of Harmful Algae

is rejected in favor of largely underestimated microalgal diversity (Sarno et al. 2005), human-mediated translocations are likely to have been seriously underestimated (McCarthy and Crowder 2000). Once successfully established in a new water body, especially when the introduced microalga produces sexual resting stages, it cannot be eradicated. Impacts on shellfish and finfish aquaculture operations may result in cases of toxin-producing microalgae, while broader environmental impacts from microalgal invasions causing altered food webs have not yet been assessed. In this context, we therefore treat the introduction of all non-indigenous microalgae, whether or not toxic, as potentially harmful to the environment.

29.2 Vector Surveys for Microalgae

The first direct examinations of ballast water at the end of a voyage were not made until the 1970s (Medcof 1975, cited in Hallegraeff 1998). Whereas initial ballast tank surveys mostly produced species lists from preserved samples (e.g., http://www.ku.lt/nemo/EuroAquaInvaders.html; including 250 diatoms, 126 dinoflagellates), later studies progressed to culturing and genetic analyses of ballast water algae. In extensive Australian ballast water surveys, 80 % of ships contained up to 30 culturable diatom species (including *Coscinodiscus, Odontella* as well as several potentially toxic *Pseudo-nitzschia* (Forbes and Hallegraeff 2001). Viable cultures of Paralytic Shellfish Poisoning producing dinoflagellates (*Alexandrium catenella, A. tamarense*) were produced from ballast water entering Australia from Japan and Korea (Hallegraeff and Bolch 1992; 5 % of ships), as well as from ships entering British ports (*Alexandrium minutum, A. catenella /tamarense*; Hamer et al. 2001; 17 % of samples). In one case, a single ballast tank was estimated to contain as many as 300 million viable *A. tamarense* cysts (Hallegraeff and Bolch 1992), amply demonstrating the significance of ballast water as a vector in increasing the geographic range of toxic species. Both commercial ships as well as recreational boats have been implicated in the dispersal of the brown tide *Aureococcus anophagefferens* into the North American Lakes (Doblin et al. 2004b). Viable transport of dinoflagellate cells and cysts of *Pfiesteria piscicida, P. shumwayae, Karenia brevis, K. mikimotoi, Alexandrium monilatum, A. tamarense* and *Prorocentrum minimum*, after passage though the digestive tract of shellfish, have also been demonstrated (Scarratt et al. 1993; Shumway et al. 2004). This latter vector can introduce unwanted harmful microalgae directly into sensitive aquaculture areas. The contribution of shellfish translocation to the spread of the dinoflagellate *Heterocapsa circularisquama* in western Japan has been discussed by Imada et al. (2001).

29.3 Evidence for Successful Establishment of Non-Indigenous Microalgae

29.3.1 Absence in Historic Samples

The diatom *Biddulphia* (now *Odontella*) *sinensis*, well known from the tropical and subtropical coasts of the Indo-Pacific, was not reported in European waters until 1903, when it produced dense plankton blooms in the North Sea. Since it was unlikely that this large diatom could have been overlooked previously, and impossible that it could have been carried by currents from distant oceans, Ostenfeld (1908) suggested that this species was introduced by ship. This diatom species was subsequently confirmed by Gollash et al. (2000) from ballast water in a vessel traveling from Singapore to Germany. Whereas the apparent introduction of the diatom *O. sinensis* was without harmful effects, the arrival of the diatom *Coscinodiscus wailesii* (either via ballast or with Japanese oysters) in 1977 in the North Sea for a brief period caused problems to fisheries by clogging fishing nets with extensive mucus (Boalch and Harbour 1977). A range of other microalgae which only recently have bloomed in well-studied Northern European waters (e.g., *Karenia mikimotoi* =*Gyrodinium aureolum* since 1966; *Fibrocapsa japonica* since 1992) do not preserve satisfactorily and their historic absence therefore cannot be confidently verified in this way.

29.3.2 Sediment Cyst Cores

Analyses of dinoflagellate cysts (*Gymnodinium catenatum*) in radionuclide-dated sediment cores in Tasmania, Australia, demonstrated its appearance around 1973 (first plankton blooms in 1980) coinciding with the start-up of a woodchip mill, which initiated the introduction of international ballast water (McMinn et al. 1999). In contrast, sediment cores from the Atlantic coast of Portugal (first PSP outbreak in Spain in 1976) indicated the presence of *G. catenatum* cysts since the beginning of the 20th century (Amorim et al. 2004) and cores from New Zealand (first PSP outbreak in 2000) demonstrated its presence since at least the 1930s (Irwin et al. 2003). Unfortunately, the mucoid cysts of *Alexandrium* dinoflagellates do not routinely preserve in deep sediment cores, with the exception of the study of Mudie et al. (2002) who succeeded in tracking historical *Alexandrium* cyst peaks in eastern Canada back to 13000 B.P.

29.3.3 Increasing Molecular Evidence

Molecular tools increasingly offer the potential to detect non-indigenous microalgal strains and in some cases even track donor source populations (Scholin et al. 1995). A prime example is the detection in the Mediterranean ports of Sete and Barcelona of *Alexandrium catenella* with a temperate Asian ribotype not found anywhere else in Europe (Lilly et al. 2002; Vila et al. 2001; Fig. 29.1). Evidence of introduced ribotypes hiding among indigenous strains is also emerging for Australian *Alexandrium* (De Salas et al. 2001) and North Sea *Fibrocapsa* raphidophytes (Kooistra et al. 2001). The potentially ichthyotoxic dinoflagellate *Pfiesteria piscicida* has been confirmed by molecular probes in ballast water entering US and Australian ports (Doblin et al. 2004a; Parke and Hallegraeff unpubl.), and the brown tide *Aureococcus anophagefferens* was detected in the bilge water of small recreational boats (Doblin et al. 2004b). Genome exchange between resident and invader strains has the potential to generate genetic diversity and possibly even hybrid vigour (Fig. 29.2).

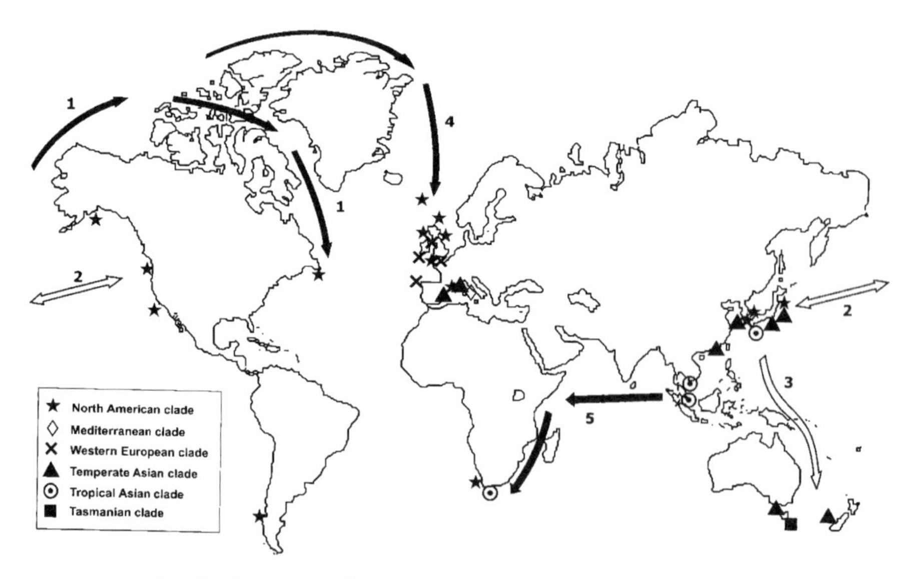

Fig. 29.1. Molecular biogeography of LSU ribotypes in the *Alexandrium tamarense/catenella* dinoflagellate species complex. *Black arrows* indicate natural dispersal, while *clear arrows* suggest human-assisted dispersal. The appearance of temperate Asian ribotype in 1983 in the Mediterranean can only be explained by human-assisted introduction. (After Scholin et al. 1995; Sebastian et al. 2005)

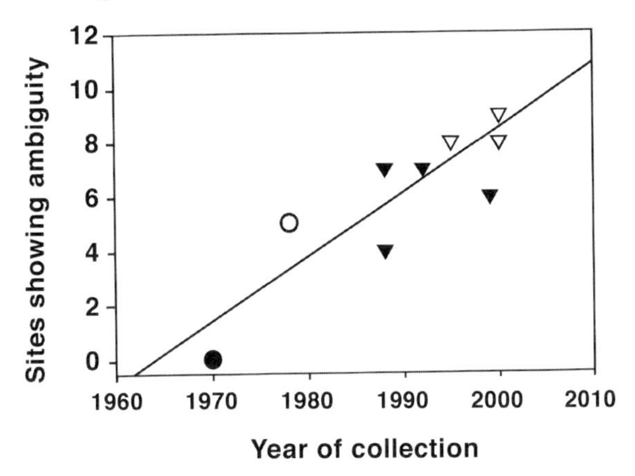

Fig. 29.2. Evidence for a temporal increase of genetic diversity (ambiguities at ten positions in the ITS region) of global cultures of the raphidophyte *Fibrocapsa japonica* (from Kooistra et al. 2001)

29.4 Management Options to Reduce Risk of Introductions

With the growing awareness of the problem of introduced marine pests, in the past two decades a number of national and international regulations have been developed to reduce the risk of transfer of non-indigenous organisms. These include an ICES Code of Practice on the Introductions and Transfer of Marine Organisms (latest version 2003; see www.ices.dk). As an example of an application, during the 2000 *Gymnodinium catenatum* bloom in northern New Zealand, a ban was instigated on the transfer of contaminated shellfish spat (L. MacKenzie, pers. comm.). On a national front, both Canada and Australia introduced mandatory ballast water guidelines in 2001, and the US Coast Guard implemented mandatory ballast water management in 2004. This culminated in the adoption in 2004 of the IMO International Convention for the Control and Management of Ships' Ballast Water and Sediments, which prescribes strict performance standards for ballast water exchange (>95 % volumetric exchange) and treatment processes (discharge shall not exceed ten viable organisms per m^3 >50 μm minimum dimension; 10 viable organisms per mL <50 μm and >10 μm minimum dimension; and a microbial standard). The far-reaching implications of these IMO regulations (see www.imo.org) for global shipping will be explored below.

29.4.1 Warning System for HABs in Ballast-Water-Uptake Zones

Prevention is better than to cure. The most-effective measure to prevent ship uptake of harmful algal blooms would be to avoid taking on ballast water during known harmful bloom events in the world's ports (IMO resolution A. 868(20)). This has been applied e.g., during the 1988 *Chrysochromulina polylepis* bloom in Norwegian coastal waters (K. Tangen, pers. comm.), and the 1993 *Karenia mikimotoi* blooms in New Zealand. Precautionary procedures need also be developed when taking on ballast water in shallow areas with known sediment cyst beds of harmful species.

29.4.2 Ballast Water Exchange Studies on Phytoplankton

The IMO Convention prescribes exchange of ballast water in open oceans more than 200 (or if not possible 50) nautical miles from the shore. Mid-ocean exchange of ballast water is believed to be currently the most reliable method in order to minimize the risk of transfer of unwanted organisms on existing vessels. Compared with coastal waters, deep-ocean waters generally contain fewer organisms, and species occurring in open ocean waters are generally not able to survive in coastal zones and vice versa. Analyses of differences in colored dissolved organic matter (CDOM) between coastal and open-ocean waters are being explored as a method of verification of mid-ocean ballast water exchange (Murphy et al. 2004). The efficiency of removal of organisms as distinct from the original ballast water is a complex issue, which will be affected by the nature and behavior of organisms in the tanks, the design of tanks, mixing within the tanks and the types and behavior of sediments. Another problem is the possible retention of dinoflagellate cysts in ballast-tank sediments. In a study of 32 vessels, which claimed to have exchanged ballast water in mid-ocean, 14 were still found to contain significant amounts of sediment, including dinoflagellate cysts (Hallegraeff and Bolch 1992). There are currently three methods by which a mid-ocean exchange of ballast water may be achieved:

Empty/Refill (Reballasting)

This is usually only possible for smaller ships such as woodchip vessels <40,000 t dead weight, which should be able to achieve a nearly 100 % water exchange. In examining the effectiveness of open ocean reballasting in reducing the number of diatoms and dinoflagellates in ballast water from container ships traveling from California to Hong Kong, Zhang and Dickman (1999) reported that the Master estimated that 95 to 99 % of the original water was removed, and their analyses showed that 87 % of the diatoms and harmful dinoflagellates had been removed. In a similar study (Dickman and Zhang 1999) for container ships traveling from Manzanillo, mid-ocean reballasting resulted in 48 % removal of diatoms and dinoflagellates (again with the Master estimating 95–99 % water replacement). The differences between the two studies were attributed to the different ship ages and varying efficiency of ballast water exchange systems.

Continuous Flow-through of Ballast Water (Ballast Exchange)

A continuous flow-through system allows continuous sea-to-sea circulation of ballast water while the ballast tanks remain filled, i.e., seawater is pumped into the ballast tanks while the tank is simultaneously overflowed from the top. This system does not impose excessive bending moments or shearing forces and minimizes stability problems. Trials have shown that three-times volumetric exchange of ballast water results in approximately 95 % removal of viable algal cells (Rigby and Hallegraeff 1994).

Dilution Method

The dilution method is a variation of the continuous flow-through technique. Continuous ballasting may be carried out from the top of the tanks via one pipe system and at the same time continuous deballasting occurs by a second pipe system at the bottom of the tank. In a trial on an oil carrier a water exchange efficiency of 90 % was achieved with a phytoplankton exchange of 96 %, while chlorophyll *a* exchange was estimated as 86 % (Villac et al. 2001).

Location of Ballast Water Exchange

The precise location of ballast water exchange needs to be carefully chosen. MacDonald and Davidson (1998) reported that during ballast water exchange in European seas the diversity of diatoms and dinoflagellates increased in 69 and 85 % of cases, and abundance increased in 31 and 85 % of cases, respectively. Forbes and Hallegraeff (2001) reported that 80 % of woodchip ships operating between Japan and Tasmania reballast in coastal waters off the Philippines and Papua New Guinea and bring into Tasmania a new viable tropical/cosmopolitan inoculum, mixed with remnants of old Japanese plankton. Monitoring of woodchip vessels, which claimed to have exchanged 100 %, indicated that 80 % of ships still contained up to 30 culturable diatom species (including potentially harmful *Pseudo-nitzschia*).

Is 95 % ballast water exchange of HAB species sufficient?

Even if it is assumed that the efficiency of removal of organisms in ocean exchange is the same as the water replacement efficiency, it is important to realize that large numbers of harmful organisms may still be present in the water discharged into the receiving port. This is especially true when ballasting occurs during an algal bloom. Some *Alexandrium* toxic plankton blooms reach cell densities of 10^5 cells/L, of which approximately 40 % can successfully produce cysts (i.e., resulting in 40,000 cysts/L). Assuming a cargo vessel taking on 60,000 t of ballast water in such bloom conditions, a single ship could theoretically carry up to 2–4x10^{12} cysts. This compares with an actual estimate of 3x10^6 *Alexandrium* cysts contained in a 25,000 t woodchip carrier entering the Australian port of Eden after ballasting during a confirmed *Alexandrium* dinoflagellate bloom in the Japanese port of Muroran (Hallegraeff and Bolch 1992). In a strict sense a *single* viable dinoflagellate cyst would constitute a viable inoculum, but taking into account limited losses from cysts germinating under unfavorable water conditions, 100 to 1,000 cysts would pose an inoculum capable of attempting to colonize their new environment for many years. To prevent such a threshold of dinoflagellate cyst introduction would require a ballast water treatment efficacy better than 99.9 %. This exercise raises serious doubts about the value of ballast water exchange for this particular target species while favoring technologies, which can achieve 100 % efficacy.

29.4.3 Treatment Options

Filtration. Filtration of ballast water is one of the most environmentally sound, albeit expensive, methods available (Table 29.1). Recent investigations with screen sizes of 50 and 100 μm have achieved microalgal removal efficiencies of 90 % at a nominal flow rate of 340 m^3/h (Cangelosi 2002; Parsons and Harkins 2002). It is noted that the flow rates used in these pilot treatments are low compared to the flow requirements on board ship and the efficacy achieved is still lower than that possible with 95 % ballast water exchange. The effectiveness of filter systems can be increased by the use of an additional technique such as heat, biocides or UV as secondary treatment. Cyclonic sep-

Table 29.1. Indicative comparative ballast water treatment costs. Adopted from Taylor et al. (2002)

Treatment option	Costs per m^3 (US$)
Exchange of ocean water (no additional equipment). Costs are reduced by 50 % (empty/refill option) if gravity ballasting can be accomplished	0.01–0.02
Exchange of ocean water including capital costs associated with additional equipment (for safe or effective operation)	up to 0.18
Heating/flushing process	0.03
Using recycled process water	0.04
Heating/flushing process using Hi Tech system involving recirculation, higher temperatures and additional heat exchange equipment	0.05
Hydrocyclones	0.06–0.26
Continuous backflushing filtration (with high capital cost component)	0.07–0.19
UV irradiation	0.10–0.39
Chemical treatment (based on operating cost alone)	0.14–24.10
UV combined with hydrocyclone	0.16–0.58
UV combined with filtration	0.17–0.51
Land based treatment	0.20–8.31
Dedicated treatment ship	0.33
Use of fresh water	0.50–0.72

aration technologies have proved ineffective for removing planktonic organisms from ballast water because most plankton are negatively buoyant with densities too close to water for effective separation.

Heat Treatment. Temperatures of 35–38 °C for a period of 4–5 h effectively killed vegetative cells of most harmful microalgae. With many organisms, the temperature required will generally be lower for longer periods of heating. For example, 30–90 s exposure to temperatures above 40 °C were effective in killing cysts of the dinoflagellates *Gymnodinium catenatum* and *Alexandrium tamarense*, whereas temperatures as low as 35 to 38 °C were sufficient after 4 h of heating. These laboratory findings were confirmed in full-scale shipboard trials, where the ship's pipework was modified to enable waste heat from the main engine-cooling circuit to heat the water in one of the ballast tanks by flushing with the heated water which reached 37–38 °C (Rigby et al. 2004). Heating of ballast water also has the added advantage that organisms contained in sediments would be subjected to these temperatures, but this technology may not be appropriate for short (domestic) voyages or where heat losses to the ocean are high (for example where sea temperatures are low). A successful application of heat treatment was also undertaken to "clean" oyster spat from contaminating dinoflagellate cysts during the 2000 New Zealand *Gymnodinium catenatum* bloom.

Ultraviolet Irradiation. UV irradiation is commonly used for sterilizing large amounts of potable water or wastewater, and for water purification in aquaculture and fisheries. Experiments with phytoplankton showed that a short exposure at high irradiance was found to be more effective than long exposure at low irradiance. Hallegraeff et al. (1997) and Montani et al. (1995) cited in Hallegraeff 1998 demonstrated that germination of cysts of *Alexandrium*, *Gymnodinium*, *Protoperidinium*, *Scrippsiella* and *Gyrodinium* occurred after exposure to UV radiation, most likely because many cyst walls are impermeable to UV. Other problems with UV treatment include the possibility that some smaller organisms could pass the UV unit without any treatment in the shadow of larger organisms or suspended solids, the reduced penetration of UV irradiation in turbid ballast waters, and the recovery of the phytoplankton following exposure to UV irradiation.

Chemical and Biocidal Inactivation of Ballast Water. A large number of chemical disinfectants are commercially available that have been used successfully for many years in land-based potable and wastewater treatment applications. Biocides suggested for use with ballast water include hydrogen peroxide, chlorination (chlorine dioxide, hypochlorite), ozonation, oxygen deprivation using reducing agents (e.g., sulphur dioxide, sodium sulphite), coagulants, antifouling paints as ballast tank coatings, organic biocides (formaldehyde, glutaraldehyde, peracetic acid, vitamin K) and others. Hydrogen peroxide

(100–2,500 ppm) offers considerable potential as an environmentally friendly biocide even for toxic dinoflagellate cysts (Ichikawa et al. 1993; Hallegraeff et al. 1997, cited in Hallegraeff 1998). However, its high cost of application may render it unsuitable as a routine tool, and one would only consider its use in emergency situations with highly contaminated ships. It is expected that the costs of chemical biocides would be significantly reduced when mass production is undertaken to feed the global demand for ballast water treatment. Strict approval processes are currently worked out by IMO to safeguard against any potential side effects of the release of billions of tons of treated water into the environment.

29.5 Conclusions

The potential for transport of non-indigenous marine microalgae via ships' ballast water and by translocation of shellfish has been amply demonstrated. Molecular approaches are increasingly suggesting that global microalgal diversity has been underestimated, and as a result human-mediated translocations are likely to have been seriously underestimated. The broader environmental impacts from microalgal invasions causing altered food webs have not yet been assessed. The dogma of phytoplankton cosmopolitanism has led to false complacency, and more than 100 years after this environmental problem was first raised in the scientific literature, a general consensus has now been reached that not doing anything is no longer an option. Minimizing the risk of ballast water introductions by microalgae and their cysts represents a very significant scientific and technological challenge, which cannot yet be adequately achieved with best currently available technologies and will be high on the research and development agenda in the decade to come.

References

Amorim A, Pereira P, Martins C, Veloso V, Franca S, Dale B (2004) The dinoflagellate *Gymnodinium catenatum* from the coast of Portugal: historical record, toxin profile and molecular characterization. Abstracts 11th Int Conf Harmful Algal Blooms, Cape Town, p 57

Boalch GT, Harbour DS (1977) Unusual diatom off southwest England and its effect on fishing. Nature 269:687–688

Cangelosi A (2002) Filtration as a ballast water treatment measure. In: Leppäkoski E, Gollasch S, Olenin S (eds) Invasive aquatic species of Europe: distribution, impacts and management. Kluwer, Dordrecht, pp 511–519

Carlton JT (1985) Transoceanic and intercontinental dispersal of coastal marine organisms: the biology of ballast water. Oceanogr Mar Biol Ann Rev 23:313–371

De Salas MF, Van Emmerik MJ, Hallegraeff GM, Negri A, Vaillancourt RE, Bolch CJ (2001) Toxic Australian *Alexandrium* dinoflagellates: introduced or indigenous? In: Hallegraeff GM, Blackburn SI, Bolch CJ, Lewis RJ (eds) Harmful algal blooms 2000. IOC-UNESCO, Paris, pp 214–217

Dickman M, Zhang F (1999) Mid-ocean exchange of container vessel ballast water. 1: effects of vessel type in the transport of diatoms and dinoflagellates from Manzanillo, Mexico, to Hong Kong, China. Mar Ecol Prog Ser 176:253–262

Doblin MA, Drake LA, Coyne KJ, Rublee PA, Dobbs FC (2004a) *Pfiesteria* species identified in ships' ballast water and residuals: a possible vector for introductions to coastal areas. In: Steidinger KA, Landsberg JH, Tomas CR, Vargo GA (eds) Harmful algae 2002. Florida Fish and Wildlife Conservation Comm, Fl Inst Oceanogr, IOC-UNESCO, St. Petersburg, USA, pp 317–319

Doblin MA, Popels LC, Coyne KJ, Hutchins DA, Cary SC, Dobbs FC (2004b) Transport of the harmful bloom alga *Aureococcus anophagefferens* by ocean-going ships and coastal boats. Appl Env Microbiol 70:6495–6500

Finlay BJ (2002) Global dispersal of free-living microbial eukaryote species. Science 296:1061–1063

Forbes E, Hallegraeff GM (2001) Transport of potentially toxic *Pseudo-nitzschia* diatom species via ballast water. In: John J (ed) Proc 15th Diatom Symposium, Koeltz Sci Publ, Koenigstein, pp 509–520

Gollash S, Lenz J, Dammer M, Andres H-G (2000) Survival of tropical ballast water organisms during a cruise from the Indian Ocean to the North Sea. J Plankton Res 22:923–937

Hallegraeff GM (1998) Transport of toxic dinoflagellates via ship's ballast water: bioeconomic risk assessment and efficacy of possible ballast water management strategies. Mar Ecol Prog Ser 168:297–309

Hallegraeff GM, Bolch CJ (1992) Transport of dinoflagellate cysts in ship's ballast water: implications for plankton biogeography and aquaculture. J Plankton Res 14:1067–1084

Hamer JP, Lucas IAN, McCollin T (2001) Harmful dinoflagellate resting cysts in ships' ballast tank sediments: potential for introduction into English and Welsh waters. Phycologia 40:246–255

Imada N, Honjo T, Shibata H, Oshima Y, Nagai K, Matsuyama Y, Uchida T (2001) The quantities of *Heterocapsa circularisquama* cells transferred with shellfish consignments and the possibility of its establishment in new areas. In: Hallegraeff GM, Blackburn SI, Bolch CJ, Lewis RJ (eds) Harmful algal blooms 2000. IOC-UNESCO, Paris, pp 474–476

Irwin A, Hallegraeff GM, McMinn A, Harrison J, Heijnis H (2003) Cyst and radionuclide evidence demonstrate historic *Gymnodinium catenatum* populations in Manukau and Hokianga Harbours, New Zealand. Harmful Algae 1:61–74

Kooistra WHCF, De Boer MK, Vrieling EG, Connell LB, Gieskes WWC (2001) Variation along ITS markers across strains of *Fibrocapsa japonica* (Raphidophyceae) suggests hybridisation events and recent range expansion. J Sea Res 46:213–222

Lilly EL, Kulis DM, Gentien P, Anderson DM (2002) Paralytic shellfish poisoning toxins in France linked to a human-introduced strain of *Alexandrium catenella* from the western Pacific: evidence from DNA and toxin analysis. J Plankton Res 24:443–452

MacDonald EM, Davidson RD (1998) The occurrence of harmful algae in ballast discharges to Scottish ports and the effects of mid-water exchange in regional seas. In: Reguera B, Blanco J, Fernandez ML, Wyatt T (eds) Harmful algae. Xunta de Galicia and IOC-UNESCO, Grafisant, Santiago de Compostela, Spain, pp 220–223

McCarthy H, Crowder L (2000) An overlooked scale of global transport: phytoplankton species richness in ships' ballast water. Biol Inv 2:321–322

McMinn A, Hallegraeff GM, Thomson P, Jenkinson AV, Heijnis H (1998) Cyst and radionucleotide evidence for the recent introduction of the toxic dinoflagellate *Gymnodinium catenatum* into Tasmanian waters. Mar Ecol Prog Ser 161:165–172

Mudie PJ, Rochon A, Levac E (2002) Palynological records of red-tide producing species in Canada: past trends and implications for the future. Palaeogeogr Palaeoclimatol Palaeoecol 180:159–186

Murphy K, Boehme J, Coble P, Cullen J, Field P, Moore W, Perry E, Sherrell R, Ruiz G (2004) Verification of mid-ocean ballast water exchange using naturally occurring coastal tracers. Mar Poll Bull 48:711–730

Ostenfeld CJ (1908) On the immigration of *Biddulphia sinensis* Grev. and its occurrence in the North Sea during 1903–1907. Medd f Komm Danmarks Fisk. Hav, Copenhagen, Plankton 1(6) 44 pp

Parsons MG, Harkins RW (2002) Full-scale particle removal performance of three types of mechanical separation devices for the primary treatment of ballast water. Mar Technol 39:211–222

Rigby GR, Hallegraeff GM (1994) The transfer and control of Harmful marine organisms in shipping ballast water: behaviour of marine plankton and ballast water exchange trials on the MV „Iron Whyalla". J Mar Env Eng 1:91–110

Rigby GR, Hallegraeff GM, Taylor AH (2004) Ballast water heating offers a superior treatment option. J Mar Environ Eng 7:217–230

Sarno D, Kooistra WHCF, Medlin L, Percopo I, Zingone A (2005) Diversity in the genus *Skeletonema* (Bacillariophyceae). II. An assessment of the taxonomy of *S. costatum*-like species with the description of four new species. J Phycol 41:151–176

Scarratt AM, Scarratt DJ, Scarratt MG (1993) Survival of live *Alexandrium tamarense* cells in mussel and scallop spat under simulated conditions. J Shellfish Res 12:383–388

Scholin C, Hallegraeff GM, Anderson DM (1995) Molecular evolution of the *Alexandrium tamarense* „species complex" (Dinophyceae) and their dispersal in the North American and West Pacific regions. Phycologia 34:472–485

Sebastian CR, Etheridge SM, Cook PA, O'Ryan, C, Pitcher GC (2005) Phylogenetic analysis of toxic *Alexandrium* (Dinophyceae) isolates from South Africa: implications for the global biogeography of the *Alexandrium tamarense* species complex. Phycologia 44:49–60

Shumway SE, Pate SE, Springer JJ, Burkholder JM (2004) Shellfish as vectors for introduction of harmful algae, with emphasis on the toxic dinoflagellates *Karenia brevis* and *Alexandrium monilatum*. Harmful Algae 3:248–249 (abstract)

Taylor A, Rigby G, Gollasch S, Voigt M, Hallegraeff GM, McCollin T, Jelmert A (2002) Preventive treatment and control techniques for ballast water. In: Leppäkoski E, Gollasch S, Olenin S (eds) Invasive aquatic species of Europe: distribution, impacts and management. Kluwer, Dordrecht, pp 484–507

Vila M, Garcés E, Masó M, Camp J (2001) Is the distribution of the toxic dinoflagellate *Alexandrium catenella* expanding along the NW Mediterranean coast? Mar Ecol Prog Ser 222:73–83

Villac MC, Persich G, Fernandes L, Paranhos R, Bonecker S, Garcia V, Odebrecht C, Tenenbaum D, Tristao ML, de Andrade S, Fadel A (2001) Ballast water exchange: testing the dilution method. In: Hallegraeff GM, Blackburn SI, Bolch CJ, Lewis RJ (eds) Harmful algal blooms 2000. IOC-UNESCO, Paris, pp 470–473

Wyatt T, Carlton JT (2002) Phytoplankton introductions in European coastal waters: why are so few invasions reported? In: CIESM Workshop Monographs, The Mediterranean Science Commission, Monaco, 20:41–46

Zhang F, Dickman M (1999) Mid-ocean exchange of container vessel ballast water. 1. Seasonal factors affecting the transport of harmful diatoms and dinoflagellates. Mar Ecol Prog Ser 176:243–251

30 The Economic Effects of Harmful Algal Blooms

P. HOAGLAND and S. SCATASTA

30.1 Introduction

In this chapter, we focus on how economists approach the problem of measuring the adverse effects of harmful algal blooms (HABs), as one type of natural hazard occurring in the coastal ocean. We start by drawing a distinction between scientific and economic approaches to assessments of the effects of HABs. We note that economists concern themselves with developing measures of changes in value (especially economic losses) as a consequence of HABs. This interest is motivated primarily by society's need to design responses to HABs that could mitigate economic losses at an appropriate scale. In other words, societal responses to HABs should be cost-minimizing, when all the relevant costs are considered. In this chapter, we focus on the costs of the adverse consequences of HABs, not on the costs of societal responses.

We address measures that are used to estimate changes in economic value, noting that some commonly used measures of "economic effects" are not necessarily good measures of changes in economic value. We provide some examples of measures from both published and unpublished studies in the European Union (EU) and in the United States (US). We compare estimates of economic effects from both jurisdictions, finding such estimates roughly comparable for effects in public health, commercial fisheries, and monitoring and management. The economic effects of HABs on tourism apparently are much larger in the EU, where these effects are the consequence of noxious but nontoxic blooms. We conclude with a call for increased attention to the development of estimates of changes in economic value from HABs, so that society's resources can be directed to respond to these hazards more effectively.

Ecological Studies, Vol. 189
Edna Granéli and Jefferson T. Turner (Eds.)
Ecology of Harmful Algae

30.2 Scientific Concerns

Marine scientists measure the effects of natural hazards, such as HABs, in different ways than do economists. Many scientists are concerned primarily with changes in the natural environment or with changes in the behavior of specific organisms or the characteristics of ecosystems. An important scientific concern is to understand the causes and effects of the spatial and temporal distributions of the algal species of which a harmful algal bloom (HAB) is composed.

The concerns of marine scientists are illustrated by a recent widespread bloom of the toxic dinoflagellate *Alexandrium fundyense* in the Gulf of Maine, a large body of water off the coast of New England. News reports have tracked oceanographers as they sample the ocean to determine the number of algal cells per liter of water. The scientists feed their data on algal density and distribution into models of bloom formation, spreading, and eventual decay. The models are designed to describe bloom dynamics by linking environmental factors such as precipitation, riverine nutrient fluxes, wind and current velocities, and insolation with the physical characteristics of blooms. If proven reliable, a model of bloom dynamics might enable future predictions of the timing, scale, and scope of HABs in the Gulf. These predictions could prove useful for shellfishermen, aquaculturists, and other users of the marine environment.

30.3 Economic Concerns

In contrast to scientists, economists focus their attention on changes in the patterns of human activities or resource uses that can result from a natural hazard (this interest in understanding the effects on humans exists regardless of whether the hazard is naturally occurring or is the consequence of anthropogenic activities. Where human activities lead to or exacerbate the effects of HABs, the potential set of responses that could be invoked to mitigate the effects may be different than those used to respond to HABs that occur naturally). These changes, in turn, may affect the value that humans realize from, or assign to coastal and marine activities and uses. For example, if shellfishermen are provided with information about the potential onset of a HAB event, will they respond in a way that mitigates the economic losses that they may incur?

For example, when a HAB in the US or the EU results in sampled saxitoxin concentrations in excess of 80 µg per 100 mg of molluscan tissue, then commercial and recreational fishermen and growers are precluded by law from harvesting and selling shellfish. Cordoned-off thereby from the market, these fishermen and growers will experience economic losses as a consequence of the bloom, while consumers will experience an economic gain from avoiding shellfish poisoning. Economists seek to measure the scale of gains and losses like these (also referred to by economists and others as economic "benefits" and "costs" or "damages") and to express them in monetary terms.

Marine scientists also may share a concern about appropriate societal responses to HAB events. Many of these concerns appear in a recent document outlining the rationale for an expanded national program on HAB research in the US (Ramsdell et al. 2005). Because of their intimate understanding of the physical and biological mechanisms of blooms, scientists may be more likely than economists to come up with ideas for physically controlling either the bloom itself or the factors that lead to bloom formation.

30.4 Why Measure Economic Losses?

Why do economists care about measuring economic losses from HABs? The primary reason is that the scale of the economic losses can tell us something about the appropriate scale of actions taken to prevent or mitigate the losses. At the very least, if one can take some action that removes the threat of a harmful bloom, then more resources (measured in financial terms) should not be spent taking the action than the actual economic losses associated with the bloom itself.

Where the set of feasible actions that can be taken to mitigate a bloom are not of the all-or-none type, economists would argue for more precision in optimizing a societal response to HABs. Applying a body of theory known as "welfare economics," economists would argue for implementing a single action or combination of actions at levels that minimize the total cost of a bloom. The total cost can be decomposed into two main parts: (1) the economic losses caused by a bloom and (2) the costs of mitigating economic losses. Figure 30.1 depicts a stylized economic approach. In the figure, economic losses decline with larger levels of mitigation. At the same time, as more mitigation is undertaken, the unit costs of mitigation increase. When these cost curves intersect, as depicted in Fig. 30.1, a standard result from the economic theory of pollution control is that the cost-minimizing level of mitigation may be such that society might need to bear a non-zero level of loss from a bloom. In other words, part or all of the response to many HAB events is just to put up with them. Depending upon the type of bloom and the affected human uses, a variety of actions may be feasible, from increasing the frequency of environmental monitoring to closing a shellfish bed to relaying shellfish into unaffected areas (or pounds) to spraying clay over the water surface. Scientific research about bloom formation and dynamics and about potential ecological effects is a crucial activity that identifies and complements mitigation actions.

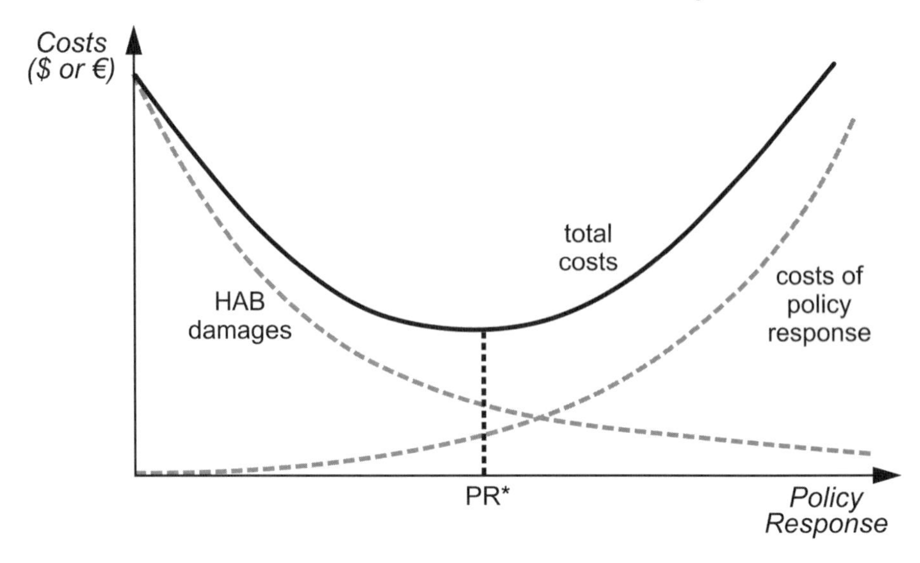

Fig 30.1. Determining the most efficient action or set of actions to respond to a HAB event requires estimates of both the adverse economic effects ("*HAB damages*") and the costs associated with actions taken to mitigate the adverse effects. The optimal response (*PR**) is at that level where the sum of damages and response costs (*total costs*) is minimized

30.5 Economic Losses

The concept of "value" has many definitions in many contexts. In economics, however, value is defined strictly as a surplus. Van den Bergh et al. (2002) present an excellent overview of the economic methodology for measuring surplus changes in the case of HABs introduced through ballast water discharges. There are two kinds of economic surplus. The first kind, known as consumer surplus, is what the purchasers of a good or service are willing to pay, above what they actually have to pay in a market setting. The second kind of surplus, known as a producer surplus, is what firms or other suppliers of a good or service earn in excess of their costs of production.

In a market, the total surplus is the sum of consumer and producer surpluses. In order to estimate the effects of a natural hazard such as a HAB event, economists seek to measure any changes (typically reductions) in total surplus as a consequence of the event. The concepts of consumer and producer surplus are illustrated in Fig. 30.2. In the figure, consumer surplus is the area underneath the demand curve and above the equilibrium price (area P_0bg without HABs). The impact of HABs on consumer surplus is given by the area P_1abP_0. Producer surplus is given by the area above the supply curve and below the equilibrium price (area P_0bd without HABs). The impact of HABs

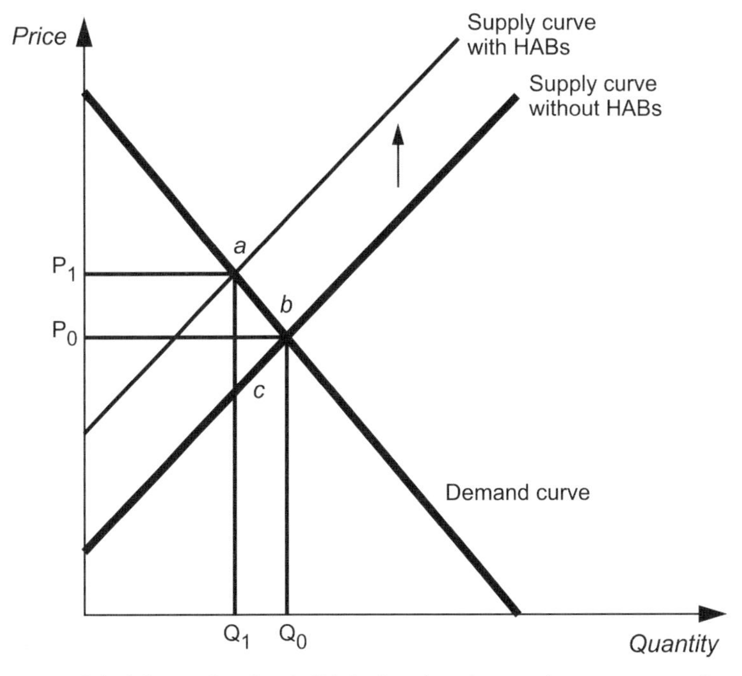

Fig 30.2. A model of the market for shellfish showing changes in consumer and producer surplus before and after a closure due to a HAB event. Closing a portion of the shellfishery results in a contraction of the supply of shellfish to the market (from Q_0 to Q_1), represented by a shift up in the supply curve. The price of shellfish increases as a consequence, rising from P_0 to P_1. All consumers and some producers are affected adversely by the closure

on producer surplus is given by the area P_1ae minus area P_0bd. The change in social welfare is given by area *abc*.

Because surplus changes are apparent only in established markets, economists must first identify the relevant markets that may be affected by HABs. In the case of HABs, these markets typically are linked closely to human activities along the coast or in the ocean. Thus the relevant markets include those for seafood, especially shellfish; labor, in the event of unemployment, morbidities, or mortalities; coastal tourism, including shoreside businesses that benefit from coastal tourism. Economists also recognize that there may be losses associated with the adverse effects of HABs on environmental features, such as marine mammal mortalities, some of which are not traded in established markets. For such losses, economists would still want to measure changes in surplus, but they must do so by looking at changes in real markets that are linked to the environmental features or by modeling hypothetical markets.

One of the earliest examples of a study of surplus losses is one that focused on the consequences of a bloom of brown tide (*Aureococcus anophagefferens*) in 1985 in the Peconic Bay, located on Long Island, New York. The brown tide resulted in the complete loss of an important bay scallop (*Argopecten irradians*) fishery in that estuary. Subsequent efforts to restore bay scallops to the Peconic have met with little success. Kahn and Rockel (1988) estimate annual surplus losses in this fishery on the order of $3 million (2005 dollars). Because this fishery has not been re-established, this loss has been incurred annually for the last two decades.

A more recent study examines economic losses associated with HABs that lead to a wider range of effects, including noxious odors, fishkills, and potential public health concerns. The study was conducted at the Dutch beach resort, Zandvoort, near Haarlem, The Netherlands. Using survey techniques that simulate a market, Nunes and van den Bergh (2004) estimated that tourists might be willing to pay a lump sum between $200 to $300 million (2005 dollars) over 2 years for requiring merchant vessels to treat their ballast waters and for water monitoring programs to protect the beach. Because these programs would mitigate the adverse effects of HABs, they can be interpreted as crude estimates of the scale of economic losses from a red tide at Zandvoort.

A third example concerns an estimate of economic losses associated with the phenomenon known as the "halo effect". A halo effect occurs when economic losses are collateral to (but not directly the consequence of) a HAB event. The origin of the term "halo effect" is somewhat obscure, but it appears to have been first used about 30 years ago by Jensen (1975) in reference to "food scares," such as that associated with a 1972 bloom of *Alexandrium tamarense* in the US Gulf of Maine. Examples of halo effects include reduced consumption of all seafood (not just shellfish), fewer tourist visits to areas known to be experiencing a HAB event, among others. In 1997, a halo effect was measured as the consequence of a bloom of *Pfiesteria piscicida* that resulted in fish kills – mainly of menhaden (*Brevoortia tyrannus*) – along the Maryland coast of Chesapeake Bay. Because of the coincidental appearance of physical and neurological problems in some fishermen, then Maryland Governor Parris Glendening stated publicly that people could be hurt by *Pfiesteria*, and he prohibited recreational and commercial fishing in several Chesapeake tributaries. These actions resulted in lost sales to seafood producers in most seafood categories, including those clearly unrelated to the *Pfiesteria* bloom. A study of the halo effect concluded that the public announcement of a fish kill leads to reductions in demand for seafood. As part of that study, Whitehead et al. (2003) estimated surplus losses to seafood consumers in the mid-Atlantic region of the United States at between $37 and $72 million in the month following the *Pfiesteria* bloom (an important conclusion of the Whitehead et al. (2003) study is that public pronouncements assuring the safety of seafood do little to reduce economic losses. More effective are mandatory programs of seafood inspection).

To date, there have been only a few studies of changes in economic surplus as a consequence of HAB events. There may be many reasons why only a few extant studies exist. These reasons include the lack of data, the low frequency of events in many regions, and the small and localized scales of many events. We conjecture that one of the most important reasons may be the relatively high cost of conducting some types of economic surplus loss studies, particularly those involving the modeling of hypothetical markets. As the scale of the losses resulting from HABs becomes more apparent, and as their frequency increases, we expect that more studies will be forthcoming.

30.6 Economic Impacts

In many instances, crude estimates of economic effects known as "economic impacts" are presented as a measure of the losses associated with HAB events. For example, an estimate of the lost sales of shellfish in Maine and Massachusetts due to closures imposed as a consequence of the 2005 *Alexandrium* bloom is about $ 2.7 million for the month of June. Such an estimate is obtained by examining historical statistics on shellfish harvests and aquaculture production and multiplying by the market price. An economic impact estimate such as this does not measure changes in economic *value* as defined above (viz. Propst and Gavrilis 1987). Nevertheless, estimates of economic impacts can be useful for at least three reasons. First, they are easy to make. Second, they give us a general idea of the scale of the problem in familiar monetary units. Third, they can help us to identify the relevant markets and geographic locations in which economic losses potentially occur.

Estimates of economic impacts can be misleading in various ways, however. Strictly speaking, economic impacts are not measures of surplus changes, and therefore they should not be used to help decide on the most efficient ways in which to respond to HAB events. Under quite normal market conditions, estimates of economic impacts might well exceed estimates of producer surplus losses. Further, estimates of economic impacts do not account for shifts in the distribution of surpluses. For example, a HAB closure might result in the reduction of the supply of softshell clams to the market. As a consequence, we would expect the price of clams to rise. Those firms that are prevented from harvesting clams will experience a loss of producer surplus. Other firms that are harvesting in other open areas, however, may reap more of a surplus because of the price increase. These higher producer surpluses will come at the expense of consumers, who are forced to pay the higher price. These consumer surpluses are not really lost to the economy; they are just transferred from consumers to those firms who remain in the market. Moreover, those producers who are closed out of harvesting do not have to incur harvesting costs. Reports only of economic impacts do not capture all of the subtleties of these shifts in surpluses.

Because of both the widespread use and the perceived usefulness of economic impact estimates, there is a considerable risk that they will be employed to argue for the scale of management responses to HABs. Indeed, such estimates crop up in policy debates over and over again. One classic example concerns an estimate of the surf clam (*Spisula polynyma*) resource in the Bering Sea off the coast of Alaska (Anderson et al. 2000). In 1997, upon introducing the US Harmful Algal Bloom Research and Control Act (S. 1480) (a substitute version of this bill, the "Harmful Algal Bloom and Hypoxia Research and Control Act" [P.L. 105-383] was enacted in 1998) in the US Senate, Senator Olympia Snowe (R-ME) noted that $50 million worth of Alaskan "shellfish resources" remain unexploited because of paralytic shellfish poisoning (PSP) toxicity. This estimate actually emerged in the late 1970s following the results of a US government survey of potentially exploitable quantities of the surf clam in the Bering Sea, and it has reappeared in several policy contexts over the years. Other reasons for the lack of an Alaskan surf clam fishery may be equally or more compelling than the presence of PSP toxins, including the fact that the fishery is unlikely to be commercially viable even in the absence of PSP, a need to protect juvenile spawning grounds for the red king crab (*Paralithodes camtschaticus*) in the same area as the surf clam stock, and concerns expressed by indigenous peoples over the potential impacts of a clam fishery on walrus (*Odobenus rosmarus*) stocks. The act passed without any discussion or apparent concern with the basis for the impact estimate.

30.7 Estimates of National Economic Effects

The EU and the US represent two good examples where estimates of aggregate national level economic effects from HABs have been attempted. Table 30.1 shows roughly similar estimates of broadly defined economic effects of HABs from both the EU and the US. These estimates fall into similar categories, but they have been developed using different methodologies.

Here we present in general terms the nature of the estimates in each of the categories in Table 30.1. Estimates from each of these categories and the totals should be viewed as only very rough estimates of the scale of annual economic effects. We also normalize the estimates by calculating economic effects per kilometer of shoreline, with the caveat that HAB effects are not uniformly distributed either geographically or over time. The US estimates are averaged over the 14-year period from 1987 to 2000; the EU estimates are averaged over the 10-year period from 1989 to 1998. The estimates in Table 30.1 for the US are an unpublished update of earlier estimates reported in Hoagland et al. (2002) and Anderson et al. (2000). This update employs the same methodologies but it averages economic effects over a longer period:

Table 30.1. Estimates of the average annual economic effects of HABs in the European Union and the United States (2005 dollars)[a]

	EU ($\$ \times 10^6$)	US ($\$ \times 10^6$)	EU ($/km)	US ($/km)
Public health	11	37	170	1,856
Commercial fisheries	147	38	2,243	1,912
Recreation and tourism	637	4	9,743	225
Monitoring and management	18	3	273	169
Total	813	82	12,429	4,162

[a] The potential comparability of these estimates is discussed in the text. Values for Europe have been adjusted using an average foreign exchange rate of € 1.13/$1.00 for the year 2003. Values in both columns were converted into 2005 dollars using the US consumer price index. Values in the latter two columns were normalized by dividing by coastline length (65,414 km for the European Union and 19,924 km for the United States)

1987–2000. Estimates from both jurisdictions give us a sense of the scale of average economic effects on an annual basis. We encourage the reader who is interested in more detail on the development of these estimates and their interannual variability to refer to the original publications (Scatasta et al. 2003; Hoagland et al. 2002; Anderson et al. 2000).

The estimates in Table 30.1 have been arranged into four distinct categories. The first category, public health, relates to the costs of morbidities or mortalities that result from eating shellfish contaminated with either PSP, diarrhetic shellfish poisoning (DSP), amnesic shellfish poisoning (ASP), or neurotoxic shellfish poisoning (NSP) or from eating finfish contaminated by ciguatera fish poisoning (CFP). Estimates from each jurisdiction are directly comparable; they employ the same factor to measure the economic impacts of lost productivity (sick days), medical treatments, transportation, and causal investigations. Some minor adjustments are made to distinguish reported from unreported illnesses, and rules-of-thumb are applied to estimate the numbers of the latter. In both jurisdictions, health advisories and fishery closures have reduced greatly the direct economic impacts during the last decade, at some (unknown) cost of halo effects. In the US, the public health effects are dominated by CFP illnesses in tropical jurisdictions, which unlike outbreaks of the other toxins, occur on a continuous basis. In the EU, the public health effects are dominated by DSP illnesses especially in Belgium, France, and Spain. Neither estimate includes the personal (non-market) costs of pain and suffering associated with morbidities.

The second category, commercial fisheries, relates to closures of shellfisheries and aquaculture operations, mortalities of shellfish or fish in aquaculture operations, prohibitions on the commercial sale of recreational fish (due to CFP), untapped fishery resources (fisheries that are arguably viable but that cannot be opened due to HABs), and halo effects, including those that might be precipitated by government policies or pronouncements. Notably, the estimate of annual economic effects in European commercial fisheries are roughly four times greater than those in the US. Estimates from both jurisdictions are not directly comparable, however, as they result from the application of different methodologies. In Europe, the estimate represents the midpoint of a range of combined consumer and producer surpluses. Notably, losses of consumer surplus due to lower perceived quality of shellfish represent 75 % of the total loss in economic surpluses. The relationship between HAB closures and reduced production of shellfish (blue mussels, *Mytilus galloprovincialis*) is apparent only in Galicia, Spain, which produces 18.5 % of the world's supply of aquacultured mussels (data on EU shellfish production can be found only at the national level, while data on harvest closure days are reported at the local level. This mismatch makes the analysis of the relationship between harvest closure days and shellfish production quite difficult. Spain, due to its highly geographically concentrated shellfish production, was the only country for which such a mismatch did not occur).

The European estimate includes also the effects of an EU directive on seafood quality, which benefits producers in France and Ireland and hurts producers in Spain and Italy, as European consumers switch to perceived higher quality mussels. In the US, the estimate is primarily an economic impact estimate (e.g., lost sales). Untapped surf clam resources in Alaska and on Georges Bank may account for as much as one-third of this estimate. The latter estimates do take into account the potential effects of the increased supply of shellfish on the relevant market. Ongoing work on halo effects, as described above, probably would augment the estimate of economic effects in the United States.

The third category represents effects of HABs on recreation and tourism in Europe and the United States. The European estimate is the result of a series of four surveys of coastal tourists in Italy, Ireland, Finland, and France (Scatasta et al. 2003), and is to be interpreted as a measure of potential damages in the EU if conditions found in the surveyed locations were representative of conditions in the EU as a whole (to obtain the European estimate, average economic losses per tourist in surveyed locations were multiplied by the total number of coastal tourists in the EU. Surveyed locations included two areas of high occurrence of HABs (Italy and Finland), one area of medium occurrence of HABs (Ireland) and one area of very low or no occurrence of HABs (France). The surveys attempt to measure the diminished experience of coastal tourism during a HAB event (a "non-market" consumer surplus loss). These surveys reveal that economic losses from HAB events depend upon both the level of tourist income as well as the familiarity of tourists with high-biomass, non-toxic algae blooms (i.e., those causing discoloration of the ocean, high volumes of foam, noxious odors, or bathing beach closings). In particular, these non-market losses were found *not* to be a function of blooms of toxic algae, perhaps because the surveyed tourists were able to substitute foods more easily than destinations. In contrast to Europe, estimates of non-market losses to tourists from HAB events in the United States are markedly absent. The US estimate in Table 30.1 relies mainly upon an incompletely documented economic impact estimate for a 1987 bloom in North Carolina and estimates of reduced expenditures for razor clamming (*Siliqua patula*) in the state of Washington. To date, attempts to account for either surplus losses or the economic effects of blooms of *Karenia brevis* along the Florida coast of the Gulf of Mexico have been mostly inconclusive, with some evidence of reduced beach visitation during a bloom event in certain areas (Adams et al. 2000).

The final category pertains to monitoring and management costs. The values from both jurisdictions in this category are roughly comparable estimates of economic impacts. The estimate for the EU is based upon data from Ireland, Spain, and The Netherlands, which is used to develop an estimate of monitoring and management costs per kilometer of coastline. This unit cost is then used to develop estimates of costs for other EU countries. The estimate for the US, which is an order of magnitude smaller, is based upon data obtained directly from state governments, where no assumptions are made about expenditures in states where no data exists. Government expenditures on scientific research on HABs would fall into this category, but they are included in neither estimate. An interesting interpretation of monitoring and management costs is that, unlike the estimates in other categories, they represent crude estimates of the costs of policy responses to HABs.

30.8 Conclusions

Much work remains to develop reliable estimates of the economic effects of HABs. As this work proceeds, attention should be directed at the rationale for developing these estimates. While government officials and others might solicit economic estimates of any kind in order to justify idiosyncratic public health or scientific agendas, attention should be directed at developing estimates of true economic losses, i.e., surplus changes. Based upon our experience with the field, although the number of studies of economic losses or impacts is limited, they outnumber studies of the economic costs of societal responses to HAB events. In other words, societal responses to HABs have been debated, formulated, and implemented with an inadequate understanding of the net benefits of such responses. Further efforts on the economics of HABs should focus on identifying the array of societal responses and characterizing the cost minimizing combination of management actions.

Acknowledgements. Woods Hole Oceanog. Inst. Contribution No. 11,414. ECOHAB Contribution No. 192. This paper is a result of research fundet in part by theee National Oceanic and Atmospheric Administration Coastal Ocean Program under award #NA04NOS4780270 to teh Woods Hole Oceanographic Institution.

References

Adams CM, Mulkey D, Hodges A, Milon JW (2000) Development of an economic impact assessment methodology for occurrence of red tide. SP 00-12. Inst Food Agric Sci, Food Resource Econ Dept, Univ Florida, Gainesville, Florida, 58 pp

Anderson DM, Hoagland P, Kaoru Y, White AW (2000) Estimated annual economic impacts from harmful algal blooms (HABs) in the United States. WHOI-2000-11Woods Hole Oceanogr Inst, Woods Hole, Massachusetts. http://www.whoi.edu/redtide/pertinentinfo/Economics_report.pdf

Hoagland P, Anderson DM, Kaoru Y, White AW (2002) The economic effects of harmful algal blooms in the United States: estimates, assessment issues, and information needs. Estuaries 25:677–695

Jensen AC (1975) The economic halo of a red tide. In: LoCicero VR (ed) Proc 1st Intl Conf on Toxic Dinoflagellate Blooms. The Massachusetts Sci Technol Found, Wakefield, Massachusetts, pp 507–516

Kahn JR, Rockel M (1988) Measuring the economic effects of brown tides. J Shellfish Res 7:677–682

Nunes PA, van den Bergh JC (2004) Can people value protection against invasive marine species? Evidence from a joint TC-CV survey in the Netherlands. Environ Resour Econ 28(4):517–532

Propst DB, Gavrilis DG (1987) Role of economic impact assessment procedures in recreational fisheries management. Trans Am Fish Soc 116:450–460

Ramsdell JS, Anderson DM, Glibert PM (eds) (2005) Harmful algal research and response (HARNESS): a national environmental science strategy. Ecol Soc Am, Washington, DC, 82 pp

Scatasta S, Stolte W, Granéli E, Weikard HP, van Ierland E (2003) The socio-economic impact of harmful algal blooms in European marine waters and description of future risks. ECOHARM project, 6 FP, EU. Environ Nat Resour Group, Wageningen Univ, Wageningen, NL http://www.bom.hik.se/ecoharm/index.html

Van den Bergh JC, Nunes PA, Dotinga HM, Kooistra WH, Vrieling E, Peperzak L (2002) Exotic harmful algae in marine ecosystems: an integrated biological-economic-legal analysis of impacts and policies. Mar Policy 26:59–74

Whitehead JC, Haab TC, Parsons GR (2003) Economic effects of *Pfiesteria*. Ocean Coast Manage 46:845–858

Subject Index

A

accumulation 43, 117, 128, 130, 131, 135, 142, 196, 217, 229, 233, 234, 238, 261, 263, 266, 283, 285, 286, 290
across-shelf currents 133, 134
acute dermal lesions 97
advection 43, 44, 128, 130, 131, 133, 263, 355
AFLP (amplified fragment length polymorphisms) 222, 223, 305
Akashiwo sanguinea 166, 177, 276–278
Alexandrium 6, 11, 12, 24, 26, 39, 42, 43, 45, 167, 189, 190, 193, 197, 222, 231, 237, 245, 248, 261, 262, 276, 278, 312, 329, 370, 372, 381, 382, 385, 387, 397
– *ostenfeldii* 25, 26, 131, 177, 179, 190, 193, 217, 219–222, 278, 318, 319
– *tamarense* 11, 12, 17, 24, 26, 45, 145, 177, 179, 190, 193, 197, 217, 220, 222–224, 231, 233, 235, 237, 319, 346, 380, 382, 387, 396
– *tamiyavanichii* 25
algal
– bloom dynamics 127, 246
– exudates 248
– growth 127, 198, 203, 205–208, 237, 243, 247
– prey 60, 181
algicidal activity 245, 246, 273, 274, 284
algicidal bacteria 249, 251, 273–275, 279, 330
algicides 105
alkaloids 104
allelochemicals 59, 157–159, 185, 189, 191–193, 196–198, 225
allelopathic 59, 70, 76, 86, 129, 157, 184, 189, 190–198, 222, 261
allelopathy 4, 71, 157, 182, 189, 191–198
along-shore currents 133
alpha Proteobacteria 244, 246, 273
Alteromonas 85, 237, 245, 273–275
amino acids 74, 104, 114, 163, 164, 166, 169, 237, 250
Amoebophrya 276–278, 329, 330
Amphidinium 14, 155, 190, 192, 219, 221
Amphora 17
Anabaena 14, 15, 31, 96, 99, 104, 190, 194, 230, 235, 276
– *oscillarioides* 104
Anabaenopsis 99
anatoxin 105, 191, 216, 236
anthropogenic 4, 5, 31, 99, 168, 171, 247, 332, 342, 348, 367, 369, 370, 379, 392
antibiotics 218, 250
antibody detection 304
antibody probes 302, 304
Aphanizomenon 14, 15, 18, 96, 131, 169, 190, 194, 217, 235
aplysiatoxins 97
aponin 190, 330
aquaculture 61, 76, 95, 157, 159, 272, 275, 327, 335, 343, 379, 380, 387, 397, 399
asexual stage 42
Aureococcus anophagefferens 111, 156, 166, 169, 264, 272, 305, 347, 380, 382, 395
Aureoumbra lagunensis 111, 348
autecology 69, 74
autotrophic growth 181
auxotrophy 56
axenic culture 166, 182, 248, 249

B
Bacillariophyceae 144
bacteria 4, 6, 9, 41, 43, 45, 59, 85, 88, 102, 104, 106, 166, 171, 179–181, 204, 218, 229, 238, 243, 244
bacterial-algal 246, 249, 252
bacterial community 243, 249, 251, 252
– composition 249
bacterial degradation 196, 249
bacterial diversity 244
bacterial metacommunity 244
bacterial physiology 244
bacterial prey 181
bacterial toxin metabolism 250
Bacteroides 245, 273, 275
Bacteroidetes 246
bait protein 303, 304
ballast water 25, 27, 60, 159, 279, 335, 379–388, 394, 396
Baltic Sea 14, 15, 18, 68, 95, 99, 100, 102, 103, 131, 169, 170, 278, 357, 358, 371
benthic 3, 12, 14, 15, 37, 40, 42, 44, 45, 53, 57, 59, 60, 62, 70, 76, 83, 95–97, 102, 103, 106, 113, 114, 118, 119, 129, 134, 157, 178, 259, 262, 283, 286–288, 290, 331, 333, 358
– dinoflagellates 178
– filter feeders 106, 118, 333
– flux 114
biogeochemical 71, 81, 208, 250
biogeographical distribution 25
biogeography 23–25, 29, 328, 382
biological manipulation 106
biological species concept 23
biomanipulation 106, 107, 333
bio-remediation 328, 333
bivalves 4, 113, 118, 119, 272, 284, 286–288, 290, 330
Black Sea 345, 357, 359, 360
bloom 3–5, 15, 27, 29, 38–41, 43–46, 53, 56, 59–62, 67–72, 74–76, 83, 85, 86, 88, 95, 99, 101–106, 111–119, 127, 129, 131–135, 143, 149, 156–159, 163, 168–170, 177, 180, 182, 184, 185, 198, 203, 205, 206, 208–211, 215, 225, 231, 232, 244–246, 248, 260–262, 264–266, 271–273, 276, 279, 288, 299, 302, 311, 321, 327–334, 341, 344–349, 356–360, 362, 363, 367–370, 372, 373, 376, 381, 383–385, 391–393, 400, 401
– decline 4, 245
– development 39, 43, 45, 114, 118, 208, 246, 260, 263, 266, 350
– dynamics 62, 76, 86, 102, 105, 117, 127, 184, 246, 299, 302, 329, 334, 392
– formation 55, 72, 177, 215, 349, 371, 392, 393
brevetoxin 59, 193, 217, 236, 237, 249
brown tide 4, 111, 113–119, 169, 180, 262, 264, 265, 348, 380, 382, 395, 396
buoyancy 98, 103, 130, 133

C
C limited 102
C, N and P budgets 181
Calanus finmarchicus 285
Calothrix 96, 102
carbohydrates 104
cell death 75, 153, 154, 155, 156, 158, 283
Ceratium furca 145, 177, 181, 359, 363
Chattonella 6, 30, 67, 72–75, 193, 274
– *antiqua* 29, 30, 72–76, 327, 330
– *globosa* 29
– *marina* 29, 30, 72–76, 190
– *ovata* 29, 72, 73, 75
– *verruculosa* 29, 72, 73
chemical biocides 388
Chrysochromulina 3, 6, 29, 67, 146, 164, 178, 181–183, 234, 362, 370
– *brevifilum* 178, 272
– *ericina* 156, 183, 272
– *kappa* 178
– *leadbeateri* 70, 178, 180, 234, 236
– *polylepis* 29, 67–71, 157, 178, 180–182, 191, 192, 194, 196–198, 206, 217, 218, 220, 221, 230, 231, 233, 234, 236, 356, 362, 363, 383
– *strobilus* 178
chytrids 275, 279
ciguatera fish poisoning 288, 399
ciguatoxins 59, 288, 289
ciliates 10, 129, 182, 183, 196, 197, 207, 260, 261, 264, 265, 329, 330, 360–363
Citharichthys sordidus 288
cladocerans 105, 106, 357, 358
clay 328, 330–335, 393
climate change 159, 360, 367, 368, 370–373, 375, 376, 379
climate variation 369
coastal embayments 97
cobalt 100, 206

Cochlodinium polykrikoides 14, 177, 179, 181, 182, 327, 329–332
commercial fisheries 327, 334, 391, 396, 399, 400
community ecology 184
competition 59, 60, 71, 101, 103, 140, 149, 157, 185, 191, 192, 198, 215
competitors 4, 141, 177, 182, 184, 189, 191, 225
complementary DNA (cDNA) 87, 215, 221, 224, 302, 303, 304, 306, 307
conserved 10, 13, 244, 302, 304, 312
consumer 4, 106, 259, 260, 266, 286, 289, 362, 393, 395–397, 400
– surplus 394, 397, 400
contingency plan 328, 334
Continuous Plankton Recorder 369, 370
convergent fronts 128
copepods 83, 105, 106, 160, 207, 259–265, 285, 329, 330, 355–363, 371
copper 100, 106, 205–207, 209, 210, 329, 330
– sulfate 106, 330
cost-minimizing 391, 393, 401
crayfish 106
cryptophyte plastids 180
ctenophores 355–360
cyanobacteria 4, 6, 9, 14–17, 31, 32, 37, 42, 95, 97, 99, 100, 102–106, 127, 131, 143, 144, 169, 180, 189, 194, 197, 203–206, 216, 222, 225, 231, 232, 246, 276, 283, 315, 348, 355
CyanoHABs 95–97, 100, 102–106
cyanotoxins 105, 106, 283
cyclins 220, 304
cyclonic eddies 131
Cylindrospermopsis raciborskii 24, 31, 99, 101
cyst 25, 27, 30, 38, 40, 42–45, 54, 58, 72–76, 130, 134, 149, 158, 191, 197, 245, 248, 300, 301, 305, 306, 331, 369, 372–375, 379–381, 384–388
– formation 40, 44, 75, 130, 149, 245, 248
cysteine 330
Cytophaga 245, 273, 274, 275, 319
Cytophaga-Flavobacteria-Bacteriodes (CFB) 245
cytostome 183

D
dark 72, 170, 181, 194, 219
darkness 57, 75, 181
debromoaplysiatoxin 97
degenerate nucleic acid primers 304
degenerate PCR primers 304
deposit feeders 287
diarrhetic shellfish poisoning 3, 146, 147, 170, 234, 286, 399
diatom 4, 9, 17, 27, 38, 40, 41, 45, 59, 60, 62, 76, 81, 82, 85–88, 103, 116, 127, 129, 130, 132, 134, 135, 143, 160, 166, 180, 181, 189, 193, 197, 203, 206–208, 210, 215, 216, 218, 231, 232, 237, 244, 246, 249, 263, 274–278, 319, 328, 333, 344, 346, 348, 349, 355–358, 361, 362, 369, 371, 379–381, 384, 385
DIC 102, 103, 191, 278
diel vertical migration 76, 278
diffusion 128, 131, 142, 165, 300, 314, 332
dinoflagellate 3, 4, 9, 10, 12, 17, 18, 25, 26, 32, 38, 40–45, 53–63, 127–135, 143, 147–149, 155, 164, 166, 168–170, 177–184, 189, 193, 194, 196, 197, 206, 210, 211, 215–219, 220–222, 224, 225, 230, 232, 233, 244–249, 260–262, 264, 265, 272, 274, 276, 277, 278, 313, 329–333, 346–348, 350, 357, 359–363, 369, 371–374, 376, 379–382, 384–388, 392
– cyst 45, 369, 372, 381, 384, 386–388
– sexuality 248
Dinophyceae 145
Dinophysis 6, 11, 26, 39, 41, 132, 147, 149, 170, 182, 190, 234, 236, 370, 372
– *acuminata* 11, 17, 145, 177, 179, 234, 236, 238, 261
– *norvegica* 11, 17, 145, 177, 179, 278
diplontic life cycle 73
direct control 328, 335
dissolved inorganic carbon (DIC) 102, 103, 191, 278
dissolved organic matter/material (DOM) 70, 101, 102, 114, 117, 164, 167, 184, 237, 245, 247, 274, 384
dissolved toxin 106, 249, 283, 284, 286
distribution 16, 23–30, 32, 43, 44, 46, 55, 57, 68, 69, 73, 74, 127, 128, 134, 231, 243, 249, 288, 289, 301, 311, 320, 345, 346, 373, 392, 397
Ditylum brightwellii 24, 154, 274

DNA microarrays 87, 220, 224, 305, 306, 316–318
dolphins 289
DOM (dissolved organic matter/material) 70, 101, 102, 114, 117, 164, 167, 184, 237, 245, 247, 274, 384
domoic acid 17, 27, 82, 83, 193, 208–210, 217, 218, 230, 236, 248, 249, 263, 283–285, 289
downwelling 133, 134, 135
drinking water 106

E
early warning 311, 321, 335
ecological niche 61, 216, 243
ecological strategy 182
ecology 5, 6, 37, 53, 54, 57, 61, 67, 81, 82, 84, 87, 95, 127, 139, 140, 184, 246, 252, 266, 274, 275, 299, 328
economic effects 391, 394, 397–401
economic impact 4, 30, 61, 113, 159, 397–401
economic loss 67, 327, 335, 370, 391–394, 396, 397, 400, 401
economic surplus 394, 396, 397, 400
ecophysiology 69, 74, 76, 239
Electra pilosa 178
emergent measures 334
Emiliania huxleyi 206, 272
encystment 43–45, 73, 75, 76
endotoxins 105
Engraulis mordax 288
estuarine 40, 53, 55–59, 95–97, 99, 102–104, 119, 204, 207, 262, 345, 349, 359, 368, 379
eubacteria 104
eutrophication 6, 43, 46, 95, 97, 132, 149, 159, 163, 168, 247, 332, 333, 335, 341, 343–345, 347, 348, 360, 369, 370, 373–375
evolution 11, 17, 24, 25, 95, 170, 183
- of toxicity 183
excystment 38, 40, 43–45, 76, 247

F
faecal pellets 286
feeding 3, 56, 57, 102, 105, 106, 113, 170, 178–181, 183, 184, 231, 261, 263, 264, 285, 286, 288, 289, 349, 357, 362
- frequency 183
- rates 56
Fibrocapsa 67, 72, 73, 382
- *japonica* 29, 30, 73, 193, 230, 236, 275, 381, 383
filter feeders 106, 118, 288, 333
Firmicutes 273
fish 3–5, 37, 58, 59, 70–73, 75, 83, 95, 97, 104–106, 112, 115, 179, 180, 183, 192, 194, 234, 259, 266, 287–289, 299–301, 303–305, 311–317, 327–329, 331–335, 355–359, 362, 363, 370, 372, 381, 387, 391, 399
- kill 3, 6, 29, 39, 67, 68, 72–75, 95, 105, 146, 147, 149, 192, 193, 327, 331, 332, 334, 372, 396
fishing 76, 106, 355, 356, 363, 381, 396
Flavobacteriaceae 351
flocculants 329–331
flow cytometry 154, 155, 158, 167, 300, 314, 315
fluorescent in situ hybridization 299, 311, 312
fluorescent-labelled bacteria 181
flushing 97, 106, 131, 334, 387
food capture 183
food selection 105, 118
food vacuoles 56, 170, 179, 180, 182
food web 3, 57, 59, 60, 81, 83, 84, 95, 106, 159, 171, 177, 259, 266, 279, 283, 285–288, 290, 347, 356, 359, 360, 363, 380, 388
Fragilariopsis 25
free fatty acid 75, 191
free-living bacteria 248
freshwater clams 106
fungi 86, 104, 218, 225, 271, 275, 276, 279

G
Gambierdiscus toxicus 24, 178, 179, 221, 247, 288
gamma-Proteobacteria 245, 250, 273–275
gastropods 106, 288
Gelidium 275
genetic diversity 24, 28, 32, 37, 87, 103, 111, 222, 223, 382, 383
genetic recombination 72, 305
genomics 215, 223, 244
germination 42, 75, 76, 333, 387
global spreading 379
glutathione 250
glycocalyx 74

Gonyaulax polygramma 177, 181
gonyautoxin 236, 250
grazer 4, 6, 54, 71, 72, 76, 81, 106, 112, 113, 116, 118, 119, 127, 141, 142, 177, 182, 183, 189, 192, 197, 207, 208, 229, 239, 243, 259, 260, 263–266, 279, 285, 287, 349, 355, 356, 358, 360, 361, 363
grazing 4, 43, 44, 60, 71, 97, 104–106, 113, 116–119, 140–143, 147, 149, 153, 159, 177, 179, 191, 208, 260, 263–266, 279, 287, 329, 330, 333, 349, 355–363
– impact 182, 260, 262, 264–266
– interactions 183
gross growth rate 140, 143
growth 5, 6, 12, 25, 29, 39, 41–45, 53–55, 57, 61, 67–71, 74, 76, 81, 84–88, 95, 97, 99–106, 112–117, 119, 128, 139–141, 143, 147, 149, 153, 156, 157, 159, 160, 163, 166–170, 179–181, 184, 189, 191–194, 196–198, 203, 205–210, 215–218, 220, 222–224, 229–233, 235–237, 239, 245, 247, 248, 250, 251, 261, 262, 266, 274, 284, 301, 302, 305, 306, 333, 344–349, 356, 361, 362, 368
– rate 41, 42, 54, 55, 57, 60, 69, 70, 85, 86, 98, 103, 104, 115–117, 119, 127, 129, 130, 139–144, 147–149, 153, 168, 177, 180, 181, 184, 192, 198, 207, 215, 217–219, 223, 230, 235, 243, 264, 265, 277, 301, 302, 358, 372
– stimulation 247
Gymnodinium catenatum 27, 40, 128–131, 177, 179, 221, 230, 232, 236, 247, 274, 346, 373, 381, 383, 387
Gymnodinium impudicum 177
Gymnodinium instriatum 245
Gymnodinium microreticulatum 27
Gymnodinium nolleri 27, 40
Gyrodinium resplendens 181

H

halo effect 396, 399, 400
haptonema 67, 183
haptophyte 4, 29, 43, 69, 71, 72, 177, 178, 180–184, 238, 262
harmful algae 3–5, 9, 10, 23, 32, 38, 53, 61, 88, 118, 131, 147, 148, 153, 157, 162, 163, 177, 180, 181, 184, 189, 195, 215, 216, 222–224, 239, 243, 244, 259, 260, 263, 264, 311, 312, 318, 320, 321, 330, 333–335, 356
harmful algal bloom (HAB) species 4, 6, 86, 116, 130, 134, 139, 140, 141, 147, 149, 156, 157, 158, 159, 160, 166, 168, 171, 177, 180, 189, 194, 203, 208, 210, 222, 234, 243, 244, 284, 299, 300, 301, 302, 304, 305, 307, 334, 335, 341, 344, 345, 347, 348, 349, 350, 355, 356, 360, 362, 363, 368, 369, 373, 375, 376, 385
harmful algal blooms (HABs) 3–5, 113, 127, 139, 163, 184, 203, 215, 260, 262–266, 271, 273, 279, 283, 299, 311, 327, 330, 341, 355, 356, 359, 360, 362, 363, 367, 383, 391, 392, 398
harmful cyanobacterial blooms 95
harmful dinoflagellate 53–57, 59–62, 177, 179, 274, 277, 333, 372, 384
harmful microalgae 37, 38, 271, 274, 279, 331, 380, 387
HaV 272, 273
HcRNAV 272, 273
heat treatment 387
Heterocapsa circularisquama 61, 189, 190, 196, 197, 271–273, 327, 330, 380
Heterocapsa triquetra 131, 146, 177, 178, 194, 276
heterocysts 15, 16, 42, 104
Heterokontophyta 72
Heterosigma 67, 72, 73
– *akashiwo* 6, 29, 30, 73, 75, 156, 157, 180, 197, 236, 264, 271–275, 327, 358
heterotrophic 14, 53, 54, 57, 102, 104, 114, 167, 179, 180, 182, 183, 204, 262, 350
– dinoflagellates 57, 60, 164, 180, 183, 189, 193, 260, 261, 264, 265, 350
– protists 53, 182, 363
heterotrophs 42, 57, 59, 164, 167, 177, 347
heterotrophy 57, 102, 163
high biomass blooms 59, 134, 157, 159, 177, 185
high light conditions 72, 181
human influences 60, 61, 344
humic substances 56, 167, 168, 181, 238, 379
hydrogen peroxide 205, 314, 330, 387
hydroxamate chelators 99
hypnozygote 38, 40, 41, 45
hypoxia 59, 95, 100, 149, 398

I

immobilization 183, 193
IMO 383, 384, 388
in situ growth rate 42, 139, 140–144, 147, 148, 277
in situ hybridization 299, 300, 306, 311
infection 58, 72, 119, 153, 154, 156, 157, 159, 271–273, 275, 277–279
ingestion 55, 70, 163, 177, 179–182, 237, 238, 261–265, 283, 284, 286–288
inorganic nutrients 114, 116, 117, 163, 164, 166, 168, 181, 184, 231, 347, 348, 350
intergenic transcribed spacers (ITS) 10, 11, 13–17, 28, 30, 383
internal nutrient ratios 181
International Ballast Water 381
interspecific interaction 192, 271
intracellular bacteria 229, 245
introductions 27, 29, 379, 383, 388
iron (Fe) 56, 70, 74, 84, 85, 88, 97, 99, 100, 102, 184, 203–211, 235, 236, 247, 331, 348
irradiance 38, 43, 55, 63, 98, 101, 115, 156, 170, 181, 387

K

Karenia 13, 288
– *brevis* 13, 39, 56, 135, 168, 190, 193, 210, 211, 217, 220, 221, 236, 246, 249, 251, 304, 327, 330, 347, 348, 369, 370, 380, 401
– *mikimotoi* 13, 41, 56, 129, 190, 192, 197, 236, 251, 274–276, 327, 329, 348, 357, 359, 361–363, 380, 381, 384
Karlodinium micrum 13, 56, 57, 166, 177–179, 181, 184, 190, 193, 278
K-selected 61, 140, 141, 143, 147–149

L

Laguna Madre 4, 111–113, 115–117, 119
Lake Michigan 95
life cycle 15, 25, 29, 37–39, 41, 43–45, 57, 67, 68, 73, 75, 76, 81, 111, 130, 262, 299–303, 305, 306
life cycle stage 46, 68, 72, 73, 159, 299–304, 306, 307,
life stage 37, 39, 40, 42, 44, 46, 301, 307
light 4, 40–45, 54–56, 61, 67, 68, 70–72, 74, 75, 81, 103, 112, 113, 115, 116, 130, 140, 153, 154, 168–170, 179, 181, 182, 184, 191, 194, 196, 205, 206, 217, 219, 230, 235, 236, 238, 243, 299, 301, 311, 313, 314, 316, 368
– limitation 170, 181
limitation 75, 84, 88, 99, 101, 103, 153, 166, 181, 184, 191, 196, 198, 203, 206, 208–210, 218, 231–238, 279, 300, 301, 318, 343, 369
limiting elements 184
Lingulodinium polyedra 177, 179–181
lipopolysaccharide 104, 236
live/dead 154, 155, 157, 158
Long Island 4, 5, 112, 118, 119, 169, 264, 333, 396
long-term phytoplankton 368, 370, 371
Lyngbya 96, 97, 102, 104
lyngbyatoxin 97

M

macroaggregates 274, 275
magnesium hydroxide 331
maintenance cost 167, 180
management 6, 106, 107, 163, 333, 335, 367, 383, 391, 398, 399, 401
manatees 289
manganese 70, 100, 206
marine mammals 3–5, 83, 84, 208, 259, 266, 289, 395
Marinobacter 245, 247
market 392, 394–397, 399–401
maximum growth rate 57, 69, 115, 184, 218
mechanical contact 183
medusae 355–360
mesocosm 114, 115, 118, 159, 168, 208, 355, 357, 359–363
mesozooplankton 71, 113, 119, 259, 260, 262, 264–266, 355–358, 360
metagenomics 344
microalgae 37, 38, 40–42, 44, 45, 69, 70, 114, 193, 215, 217–222, 224, 225, 263, 271–279, 283, 286, 315, 327, 331, 379–381, 387, 388
microbes 101, 243, 317, 318
microbial food web 279
microcystin 16, 105, 190, 191, 216, 225, 232, 236, 287, 289
Microcystis 15, 16, 96, 102, 103, 105, 190, 225, 235, 276
– *aeruginosa* 15, 16, 105, 232, 289

Micromonas pusilla 272
micronutrient 56, 74, 98, 99, 203, 205, 252
microzooplankton 113, 116–119, 142, 259, 260, 264–266, 349
mitigation 6, 149, 279, 321, 327, 328, 332, 334, 335, 393
mixotrophic 56, 70, 71, 115, 163, 182, 189, 260
– dinoflagellate 55, 181, 182, 184
– flagellate 180
mixotrophy 56, 57, 60, 114, 163, 167, 170, 171, 181, 192, 237, 350
model 38, 46, 60–62, 130, 140, 159, 163, 169, 171, 208, 221, 225, 230, 237, 252, 290, 302, 327–329, 343, 345, 360, 367, 374, 392, 395
molecular and imaging techniques 299
molecular taxonomy 9, 17, 18
molluscan shellfish 82–84
molybdenum 100, 101, 206
monitoring 45, 46, 88, 154, 167, 244, 311, 314, 318, 320, 321, 328, 335, 369, 375, 385, 393, 396
– and management 391, 399, 401
monocultures 184
mono-specific blooms 177
montmorillonite 331
morphology 10–15, 23, 31, 42, 46, 61, 67, 68, 72, 73, 87, 111, 133, 301, 312
mortality 3, 58, 73, 84, 105, 112, 117–119, 140, 153–156, 158, 159, 260–262, 264, 272, 273, 276, 277, 288, 289, 334, 359, 372
motility 54, 71, 113, 128, 130
muciferous bodies 183
mucus 74, 75, 183, 344, 381
Mya arenaria 284
Mytilus edulis 113, 193, 284, 286, 287
Mytilus galloprovincialis 287, 400

N
N or P limitation 181, 236
N:P ratio 71, 98, 99, 234, 237, 346–348
N_2 fixing 95–99, 101–103
– cyanobacteria 97, 99, 102, 210
– gene 95, 96
natural hazard 391, 392, 394
net growth rate 55, 115, 129, 143
neurotoxic shellfish poisoning 3, 288, 399
neurotoxin 27, 75, 82, 208
Neuse River Estuary 99, 100, 104
nitrogen (N) 4, 16, 41, 42, 45, 56, 70–72, 74, 75, 81, 86, 97–99, 101–104, 114, 115, 117, 131, 154, 164, 166, 168, 169, 181, 184, 196, 198, 203, 205, 206, 208, 210, 211, 230–238, 248, 259, 287, 333, 342, 343, 345–349
nitrogenase 100, 101, 207
Nitzschia 17, 82, 83, 144, 145, 231
Nodularia 14, 15, 95, 96, 99, 102, 104, 235
– *spumigena* 15, 102, 131, 169, 190, 194, 232, 238, 287
nodularin 216, 232, 235, 284, 287, 289
non-cultivable bacteria 244
non-N_2 fixing cyanobacteria 97
North Atlantic 26, 68, 69, 82, 84, 88, 285, 367, 369, 371, 375
North Sea 28, 74, 76, 159, 223, 317, 344, 357, 359, 368, 370–372, 381, 382
Nostoc 96, 102
nucleic acid probes 299, 300
nutrient 4, 6, 14, 29, 40–43, 45, 54, 56, 57, 60–63, 71, 74, 97, 103, 104, 106, 113–117, 128–140, 149, 154, 156, 159, 163–169, 171, 181, 182, 184, 191, 192, 198, 203–206, 208, 219, 231–233, 237, 332, 333, 341, 343, 347, 348, 350, 368, 369, 373
– acquisition 54, 184, 216
– deficiency 194, 196, 218, 238, 349
– limitation 75, 103, 153, 181, 184, 196, 198, 218, 279, 343
– limited 184, 349
– ratio 181, 231, 333, 347, 348, 360
nutrition 53, 56, 102, 105, 113, 163, 166, 167, 237, 350
nutritional strategy 41, 164, 185

O
obligate phagotrophy 182
oligonucleotide 224, 299, 302, 306, 312, 314, 315
Olisthodiscus luteus 72, 358
organic growth factors 181
organic matter 97, 100, 167, 168, 171, 235–237, 246, 247, 274, 279
Oscillatoria 16, 96, 97, 102–104, 190, 231, 232, 235
osmotrophy 57, 163–165, 167, 168, 170, 171

Ostreopsis lenticularis 178, 179, 190
Ostreopsis ovata 24, 178, 179, 236
Ostreopsis siamensis 178, 179
over-summering 73
over-wintering 73
oxygen depletion 81, 105, 157
oxygenation 328
Oxyrrhis marina 183, 193
ozone 330

P

P or N limited 181
paralytic shellfish poisoning 3, 146, 148, 193, 248, 284, 345, 346, 380, 398
parasite 6, 43, 57, 140, 271, 275–279, 329, 330
parasitic fungi 86, 275, 276
parasitic protests 276, 278
parasitism 53, 56, 271, 278, 279
particulate organic material 250
Parvilucifera 277, 278, 329, 330
pathogen 72, 154, 156, 243, 246, 271, 276, 279, 283
peduncle 57, 58, 180
Pelagophyceae 111
pelagophyte 4, 45, 111, 115, 180
peptide 104, 166, 169, 191, 216, 222, 225, 300
perimeter skirt 330, 334
Pfiesteria 42, 57, 59, 164, 180, 183, 193, 247, 347
- *piscicida* 6, 42, 56, 169, 177, 179, 182, 183, 262, 380, 382, 396
- *shumwayae* 58, 177, 179, 380
pH 71, 85, 191, 192, 194, 196, 231, 235, 236
Phaeocystis 4, 6, 38, 43, 67, 69–72, 159, 259, 263, 266, 344
- *globosa* 43, 68–71, 146, 156, 157, 159, 272, 344
- *pouchetii* 68–71, 156, 157, 272
phagotrophy 6, 56, 57, 70, 149, 163, 164, 167, 170, 177–185, 238, 350
phosphate deficiency 181
phosphorus (P) 45, 56, 60, 70–72, 74, 81, 84, 97–99, 101–103, 114, 115, 117, 166, 169, 181, 184, 196, 198, 203, 205, 206, 208, 209, 218, 230–238, 248, 333, 342, 343, 346–348
photoautotrophs 177
phycotoxin 81, 82, 215–217, 221, 223, 249, 259, 283–286, 288–290
phylogenetic species concept 23
phylogeny 9, 16, 17
physiological ecology 5, 84, 184
phytoflagellates 104, 179, 182, 215, 216, 218, 356, 361–363
phytoplankton 4, 5, 18, 24, 25, 39, 43, 54, 61, 62, 71, 76, 81, 83, 86, 88, 97, 99, 100–103, 105, 113, 116, 127–132, 134, 139–141, 143, 147–149, 153–157, 159, 160, 165, 167, 170, 180, 183–185, 189, 192, 193, 196, 198, 203–206, 208, 215, 229, 231, 234, 239, 246, 247, 259–266, 276, 283, 287, 299, 312–314, 317, 345, 349, 355–363, 367–375, 379, 384, 385, 387, 388
Phytoplankton mandala 129
pinocytosis 165–167
planktivorous fish 83, 105, 106, 259, 266, 287, 356, 359
plastid 9, 10, 13, 14, 170, 177, 180
ploidy levels 301
polymerase chain reaction 220, 248, 300
population dynamics 45, 46, 153, 159, 246, 250, 301, 302, 307, 328
population ecology 139, 184
population models 302
post dormancy 76
potential growth rates 142, 184
precautionary impact prevention 328
predation 43, 44, 53, 105, 129, 177, 183, 185, 192, 260, 261, 265, 266, 349, 355–363
predator 41, 43, 54, 55, 57, 59, 61, 182–184, 225, 239, 349, 355–357, 361, 362
prediction 61, 86, 163, 207, 222, 328, 329, 335, 376, 392,
pre-encystment small cell 73–76
prevention of HABs 275
prey capture 57, 183, 184
prey ingestion 182
prey protein 303
probe array 311, 316, 319, 320
producer surplus 394, 395, 397, 400
Prorocentrum arenarium 12, 178, 179
Prorocentrum belizeanum 178, 179
Prorocentrum donghaiense 12, 13, 177
Prorocentrum micans 12, 131, 166, 177
Prorocentrum minimum 12, 13, 54, 146, 166, 168, 169, 177, 179, 230, 236, 346–348, 380

Prorocentrum triestinum 131, 132, 146, 177, 331
protease 155, 166, 274
protein 10, 15, 17, 56, 114, 142, 166, 171, 193, 206, 220–222, 224, 225, 250, 302–305, 314, 342
– kinase C activators 97
Proteobacteria 56, 244–246, 250, 273–275
proteomic 222–224, 252, 304
protist 37, 53, 56, 57, 169, 170, 179, 180, 182, 183, 189, 194, 196, 215, 220, 221, 259, 260, 264, 271, 278, 279, 363, 379
protistan grazers 182, 183, 189, 197
Protoceratium reticulatum 177, 179, 374
Protoperidinium crassipes 12, 59, 177, 179, 180, 182
protozoan 71, 104, 113
Prymnesiophyceae 67, 146
Prymnesium 3, 6, 29, 67–70, 72, 182, 183, 197, 261
– *parvum* 29, 67–70, 158, 178, 180–183, 191, 192, 194–198, 233, 234, 236, 238, 262, 372
– *parvum* f. *patelliferum* 29, 68, 178, 180, 236
Pseudoalteromonas 245, 246, 273, 274
Pseudomonas 284
– *aeruginosa* 104
Pseudo-nitzschia 6, 17, 24, 27, 28, 41, 81–87, 129, 193, 206, 208–210, 218, 231, 232, 248, 263, 344, 380, 385
– *australis* 17, 27–29, 82–87, 209, 210, 236, 288, 289
– *delicatissima* 27–29, 82, 85, 236
– *multiseries* 17, 27–29, 41, 82, 84–87, 209, 210, 224, 230, 232, 236, 237, 248, 249
– *pungens* 17, 27, 28, 82, 190, 217, 223, 236
– *seriata* 17, 27–29, 82, 84, 85, 236
public health 61, 97, 250, 327, 335, 391, 396, 399, 401
pycnoclines 148

Q
qRT-PCR (quantitative reverse-transcriptase PCR) 305, 306
quorum sensing 225, 274

R
radio-labeled bacteria 179, 180
RAPD (random amplification of polymorphic DNA) 305
Raphidophyceae 29, 67, 72
raphidophyte 4, 30, 43, 45, 67, 72, 74–76, 127, 180, 189, 193, 217, 230, 274, 382, 383
reactive oxygen species 75, 205, 236
recreation 399, 400
red-tide 3, 5, 61, 62, 73–75, 129, 130, 132, 177, 182, 193, 272–276, 327, 330–333, 345, 370, 372, 396
reef 97, 288
regulatory binding domain 303
regurgitation 286
removal rate 331, 332
reservoir 68, 72, 97, 103, 106, 131, 246, 273
residence time 97, 98, 115, 117, 133
resting cell 37, 40–42, 44, 45
resting cyst 25, 40, 300, 369
retention time 106
Rhizoselenia-Richelia 103
Rhodomonas baltica 157, 158, 181, 195
ribosomal DNA (rDNA) 9–17, 31, 111, 248
ribosomal RNA (rRNA) 10, 17, 111, 311, 312, 314, 316, 317, 319, 320
– genes 9–11, 15, 17, 111, 300, 312
river 68, 76, 97, 127, 133, 168, 204, 208, 343
Roseobacter 244, 319
rotifer 105, 106, 119, 260–262, 265
r-selected 61, 141, 143, 149

S
salinity 42, 44, 55, 68, 71, 72, 74, 85, 95, 98, 100, 102, 113, 115, 116, 140, 217, 230, 231, 235, 236, 243, 261, 349, 372, 373
sandwich hybridization 300, 305, 318, 320
Saprospira 273
saxitoxin 59, 216–219, 222, 236, 250, 283, 285, 287, 289, 392
Scandinavian coastal waters 68, 206
Scrippsiella trochoidea 177, 197
Scytonema 96, 102
seabirds 3–5, 208, 259, 289
seaweed 273, 275

secondary dormancy 76
sedimentary record 344, 372, 375
seed population 76
Seto Inland Sea 74–76, 273, 274, 330, 333
sexual (meiotic) stage 42, 85, 299, 301, 304
sexual reproduction 25, 37, 40, 85, 149, 219, 302, 305
shear 61, 103, 104, 131
shellfish 3, 4, 39, 59, 61, 71, 82–84, 112, 113, 118, 146–148, 159, 170, 193, 194, 234, 248–250, 259, 284–288, 327, 328, 331, 334, 335, 345, 346, 379, 380, 383, 388, 392, 393, 395–400
siderophore 99, 204, 209, 247
silicon 231
sloppy feeding 102, 286
societal response 391, 393, 401
sodium hypochlorite 330
solid-phase cytometry 315
species concept 23, 25, 31
spine scales 183
spontaneous dormancy 76
springs 97
ssu rDNA 9
stage-specific proteins 303–305
suffocation 4, 75, 334
sulfate 101, 106, 250, 330
superoxide 75, 205, 206
survival strategy 76
swarming 274, 275
swimming 54, 58, 128, 131, 135, 149, 215, 261, 275, 277

T
Takayama 13
temperature 29, 30, 39, 40, 42, 44, 45, 55, 67–70, 73–76, 88, 97, 101, 113, 115, 116, 117, 128, 140, 159, 217, 230, 231, 235, 236, 238, 243, 266, 313, 318, 349, 350, 367, 368, 370, 371, 374, 375, 386, 387
Tilapia 106
tintinnids 260, 261, 285
top-down predation 355, 357, 359, 362, 363
tourism 159, 391, 395, 399, 400
toxic bloom 3, 27, 95, 210, 232, 286, 332, 362, 372, 391
toxic secondary metabolites 97
toxic zoospores 183
toxicity 3, 26, 29, 31, 45, 53, 67, 69–71, 75, 83, 85, 105, 117, 119, 177, 182–184, 194, 196, 197, 203, 207, 210, 216–218, 221, 229–331, 244, 259, 261, 285–288, 290, 313, 331, 347–349, 358, 398
toxin 3, 4, 6, 31, 32, 38, 45, 53, 54, 59, 70, 71, 83, 104–106, 146, 149, 156, 164, 177, 182–185, 192–194, 196, 197, 217–221, 225, 229–233, 235–237, 247–250, 259–266, 283–290, 316, 321, 328, 345, 398, 399
- biotransformation 287
- composition 285
- content 166, 218, 229, 330–334, 337, 338, 347
- production 45, 59, 60, 67, 72, 83, 87, 184, 210, 215, 217–221, 223, 225, 229–235, 237–239, 243, 248, 249, 279, 283
- profile 12, 31, 222, 223, 229, 285, 287
- trophic transfer 249, 283, 284, 290
trace elements 6, 70, 163, 247
trace metal 60, 81, 84, 85, 99, 100, 102, 103, 203, 205, 207–211
transcription factor 303, 304
trichocysts 72, 183
Trichodesmium 16, 99, 168, 210, 211
trophic cascade 18, 355, 358, 360, 362, 363
trophic transfer 105, 249, 266, 283, 284, 290, 362
trophodynamics 260
tumor 97
turbulence 6, 38, 42–44, 55, 58, 60–63, 81, 102–104, 127–130, 132, 216, 217
two-hybrid system 303

U
Ulva 275
upwelling 43, 44, 54, 63, 83, 84, 133–135, 159, 208, 210, 368, 370
- system 38, 131, 133, 134, 203, 208, 209

V
vertical mixing 98, 106, 128
Vibrio 284
- *cholerae* 246
virus 43, 86, 106, 119, 140, 156, 157, 159, 271, 273, 279, 329, 330, 347

virus-like particles 119, 271
vitamin 56, 70, 74, 191, 206, 247, 387

W
W. Atlantic Ocean 99, 101
water discoloration 3, 4
water quality 60, 95, 99, 107
whale 84, 285, 289

Z
zebra mussel 105
zinc 100, 203, 205, 206
zooplankton 4, 44, 71, 81, 102, 104–106, 113, 117–119, 157, 159–261, 263–266, 285–288, 290, 355–359, 361, 363, 369
zoospores 58, 183, 275, 277–279

Ecological Studies

Volumes published since 2001

Volume 151
Ecological Comparisons of Sedimentary Shores (2001)
K. Reise (Ed.)

Volume 152
Global Biodiversity in a Changing Environment: Scenarios for the 21st Century (2001)
F.S. Chapin, O. Sala, and E. Huber-Sannwald (Eds.)

Volume 153
UV Radiation and Arctic Ecosystems (2002)
D.O. Hessen (Ed.)

Volume 154
Geoecology of Antarctic Ice-Free Coastal Landscapes (2002)
L. Beyer and M. Bölter (Eds.)

Volume 155
Conserving Biological Diversity in East African Forests: A Study of the Eastern Arc Mountains (2002)
W.D. Newmark

Volume 156
Urban Air Pollution and Forests: Resources at Risk in the Mexico City Air Basin (2002)
M.E. Fenn, L. I. de Bauer, and T. Hernández-Tejeda (Eds.)

Volume 157
Mycorrhizal Ecology (2002, 2003)
M.G.A. van der Heijden and I.R. Sanders (Eds.)

Volume 158
Diversity and Interaction in a Temperate Forest Community: Ogawa Forest Reserve of Japan (2002)
T. Nakashizuka and Y. Matsumoto (Eds.)

Volume 159
Big-Leaf Mahogany: Genetic Resources, Ecology and Management (2003)
A. E. Lugo, J. C. Figueroa Colón, and M. Alayón (Eds.)

Volume 160
Fire and Climatic Change in Temperate Ecosystems of the Western Americas (2003)
T. T. Veblen et al. (Eds.)

Volume 161
Competition and Coexistence (2002)
U. Sommer and B. Worm (Eds.)

Volume 162
How Landscapes Change: Human Disturbance and Ecosystem Fragmentation in the Americas (2003)
G.A. Bradshaw and P.A. Marquet (Eds.)

Volume 163
Fluxes of Carbon, Water and Energy of European Forests (2003)
R. Valentini (Ed.)

Volume 164
Herbivory of Leaf-Cutting Ants: A Case Study on *Atta colombica* in the Tropical Rainforest of Panama (2003)
R. Wirth, H. Herz, R.J. Ryel, W. Beyschlag, B. Hölldobler

Volume 165
Population Viability in Plants: Conservation, Management, and Modeling of Rare Plants (2003)
C.A Brigham, M.W. Schwartz (Eds.)

Volume 166
North American Temperate Deciduous Forest Responses to Changing Precipitation Regimes (2003)
P. Hanson and S.D. Wullschleger (Eds.)

Volume 167
Alpine Biodiversity in Europe (2003)
L. Nagy, G. Grabherr, Ch. Körner, D. Thompson (Eds.)

Volume 168
Root Ecology (2003)
H. de Kroon and E.J.W. Visser (Eds.)

Volume 169
Fire in Tropical Savannas: The Kapalga Experiment (2003)
A.N. Andersen, G.D. Cook, and R.J. Williams (Eds.)

Volume 170
Molecular Ecotoxicology of Plants (2004)
H. Sandermann (Ed.)

Volume 171
Coastal Dunes: Ecology and Conservation (2004)
M.L. Martínez and N. Psuty (Eds.)

Volume 172
Biogeochemistry of Forested Catchments in a Changing Environment: A German Case Study (2004)
E. Matzner (Ed.)

Volume 173
Insects and Ecosystem Function (2004)
W.W. Weisser and E. Siemann (Eds.)

Volume 174
Pollination Ecology and the Rain Forest: Sarawak Studies (2005)
D. Roubik, S. Sakai, and A.A. Hamid (Eds.)

Volume 175
Antarctic Ecosystems: Environmental Contamination, Climate Change, and Human Impact (2005)
R. Bargagli

Volume 176
Forest Diversity and Function: Temperate and Boreal Systems (2005)
M. Scherer-Lorenzen, Ch. Körner, and E.-D. Schulze (Eds.)

Volume 177
A History of Atmospheric CO_2 and its Effects on Plants, Animals, and Ecosystems (2005)
J.R. Ehleringer, T.E. Cerling, and M.D. Dearing (Eds.)

Volume 178
Photosynthetic Adaptation: Chloroplast to Landscape (2005)
W.K. Smith, T.C. Vogelmann, and C. Chritchley (Eds.)

Volume 179
Lamto: Structure, Functioning, and Dynamics of a Savanna Ecosystem (2006)
L. Abbadie et al. (Eds.)

Volume 180
Plant Ecology, Herbivory, and Human Impact in Nordic Mountain Birch Forests (2005)
F.E. Wielgolaski (Ed.) and P.S. Karlsson, S. Neuvonen, D. Thannheiser (Ed. Board)

Volume 181
Nutrient Acquisition by Plants: An Ecological Perspective (2005)
H. BassiriRad (Ed.)

Volume 182
Human Ecology: Biocultural Adaptations in Human Cummunities (2006)
H. Schutkowski

Volume 183
Growth Dynamics of Conifer Tree Rings: Images of Past and Future Environments (2006)
E.A. Vaganov, M.K. Hughes, and A.V. Shashkin

Volume 184
Reindeer Management in Northernmost Europe: Linking Practical and Scientific Knowledge in Social Ecological Systems (2006)
B.C. Forbes, M. Bölter, L. Müller-Wille, J. Hukkinen, F. Müller, N. Gunslay, and Y. Konstantinov (Eds.)

Volume 185
Ecology and Conservation of Neotropical Montane Oak Forests (2006)
M. Kappelle (Ed.)

Volume 186
Biological Invasions in New Zealand (2006)
R.B. Allen and W.G. Lee (Eds.)

Volume 187
Managed Ecosystems and CO_2: Case Studies, Processes, and Perspectives (2006)
J. Nösberger, S.P. Long, R.J. Norby, M. Stitt, G.R. Hendrey, and H. Blum (Eds.)

Volume 188
Boreal Peatland Ecosystem (2006)
R.K. Wieder and D.H. Vitt (Eds.)

Volume 189
Ecology of Harmful Algae (2006)
E. Granéli and J.T. Turner (Eds.)